JOHNSON/EVINRUDE

Outboards
1971-89 REPAIR MANUAL
1¼- 60 HORSEPOWER, 1 AND 2 CYLINDER

SELOC

Managing Partners	Dean F. Morgantini, S.A.E. Barry L. Beck
Executive Editor	Kevin M. G. Maher, A.S.E.
Manager-Marine/Recreation	James R. Marotta, A.S.E.
Production Specialist	Melinda Possinger
Authors	Joan and Clarence Coles

Manufactured in USA
© 1989 Seloc Publications
1025 Andrew Drive, Suite 102
West Chester, PA 19380
ISBN 0-89330-008-X
567890123 87654321

Other titles
Brought to you by

Title	Part #
Chrysler Outboards, All Engines, 1962-84	018-7
Force Outboards, All Engines, 1984-96	024-1
Honda Outboards, All Engines 1988-98	1200
Johnson/Evinrude Outboards, 1-2 Cyl, 1956-70	007-1
Johnson/Evinrude Outboards, 1-2 Cyl, 1971-89	008-X
Johnson/Evinrude Outboards, 1-2 Cyl, 1990-95	026-8
Johnson/Evinrude Outboards, 3, 4 & 6 Cyl, 1973-91	010-1
Johnson/Evinrude Outboards, 3-4 Cyl, 1958-72	009-8
Johnson/Evinrude Outboards, 4, 6 & 8 Cyl, 1992-96	040-3
Kawasaki Personal Watercraft, 1973-91	032-2
Kawasaki Personal Watercraft, 1992-97	042-X
Marine Jet Drive, 1961-96	029-2
Mariner Outboards, 1-2 Cyl, 1977-89	015-2
Mariner Outboards, 3, 4 & 6 Cyl, 1977-89	016-0
Mercruiser Stern Drive, Type I, Alpha/MR, Bravo I & II, 1964-92	005-5
Mercruiser Stern Drive, Alpha I (Generation II), 1992-96	039-X
Mercruiser Stern Drive, Bravo I, II & III, 1992-96	046-2
Mercury Outboards, 1-2 Cyl, 1965-91	012-8
Mercury Outboards, 3-4 Cyl, 1965-92	013-6
Mercury Outboards, 6 Cyl, 1965-91	014-4
Mercury/Mariner Outboards, 1-2 Cyl, 1990-94	035-7
Mercury/Mariner Outboards, 3-4 Cyl, 1990-94	036-5
Mercury/Mariner Outboards, 6 Cyl, 1990-94	037-3
Mercury/Mariner Outboards, All Engines, 1995-99	1416
OMC Cobra Stern Drive, Cobra, King Cobra, Cobra SX, 1985-95	025-X
OMC Stern Drive, 1964-86	004-7
Polaris Personal Watercraft, 1992-97	045-4
Sea Doo/Bombardier Personal Watercraft, 1988-91	033-0
Sea Doo/Bombardier Personal Watercraft, 1992-97	043-8
Suzuki Outboards, All Engines, 1985-99	1600
Volvo/Penta Stern Drives 1968-91	011-X
Volvo/Penta Stern Drives, Volvo Engines, 1992-93	038-1
Volvo/Penta Stern Drives, GM and Ford Engines, 1992-95	041-1
Yamaha Outboards, 1-2 Cyl, 1984-91	021-7
Yamaha Outboards, 3 Cyl, 1984-91	022-5
Yamaha Outboards, 4 & 6 Cyl, 1984-91	023-3
Yamaha Outboards, All Engines, 1992-98	1706
Yamaha Personal Watercraft, 1987-91	034-9
Yamaha Personal Watercraft, 1992-97	044-6
Yanmar Inboard Diesels, 1975-98	7400

SAFETY NOTICE

Proper service and repair procedures are vital to the safe, reliable operation of all marine engines, as well as the personal safety of those performing repairs. This manual outlines procedures for servicing and repairing stern drives using safe, effective methods. The procedures contain many NOTES, CAUTIONS and WARNINGS which should be followed, along with standard procedures, to eliminate the possibility of personal injury or improper service which could damage the vessel or compromise its safety.

It is important to note that repair procedures and techniques, tools and parts for servicing marine engines, as well as the skill and experience of the individual performing the work, vary widely. It is not possible to anticipate all of the conceivable ways or conditions under which these engines may be serviced, or to provide cautions as to all possible hazards that may result. Standard and accepted safety precautions and equipment should be used during cutting, grinding, chiseling, prying, or any other process that can cause material removal or projectiles.

Some procedures require the use of tools specially designed for a specific purpose. Before substituting another tool or procedure, you must be completely satisfied that neither your personal safety, nor the performance of the marine engine, will be compromised.

Although information in this manual is based on industry sources and is complete as possible at the time of publication, the possibility exists that some vehicle manufacturers made later changes which could not be included here. While striving for total accuracy, Nichols Publishing cannot assume responsibility for any errors, changes or omissions that may occur in the compilation of this data.

PART NUMBERS

Part numbers listed in this reference are not recommendations by Nichols Publishing for any product brand name. They are references that can be used with interchange manuals and aftermarket supplier catalogs to locate each brand sup-plier's discrete part number.

SPECIAL TOOLS

Special tools are recommended by the marine manufacturer to perform a specific task. Use has been kept to a minimum, but, where absolutely necessary, they are referred to in the text by the part number of the tool manufacturer. These tools can be purchased, under the appropriate part number, from your local dealer or regional distributor, or an equivalent tool can be purchased locally from a tool supplier or parts outlet. Before substituting any tool for the one recommended, read the SAFETY NOTICE at the top of this page.

ALL RIGHTS RESERVED

ACKNOWLEDGMENTS

Nichols Publishing expresses sincere appreciation to Outboard Marine Corporation for their assistance in the production of this manual.

Nichols Publishing would like to express thanks to all of the fine companies who participate in the production of our books:
- Hand tools supplied by Craftsman are used during all phases of our vehicle teardown and photography.
- Many of the fine specialty tools used in our procedures were provided courtesy of Lisle Corporation.
- Lincoln Automotive Products (1 Lincoln Way, St. Louis, MO 63120) has provided their industrial shop equipment, including jacks (engine, transmission and floor), engine stands, fluid and lubrication tools, as well as shop presses.
- Rotary Lifts (1-800-640-5438 or www.Rotary-Lift.com), the largest automobile lift manufacturer in the world, offering the biggest variety of surface and in-ground lifts available, has fulfilled our shop's lift needs.
- Much of our shop's electronic testing equipment was supplied by Universal Enterprises Inc. (UEI).
- Safety-Kleen Systems Inc. has provided parts cleaning stations and assistance with environmentally sound disposal of residual wastes.
- United Gilsonite Laboratories (UGL), manufacturer of Drylock® concrete floor paint, has provided materials and expertise for the coating and protection of our shop floor.

TABLE OF CONTENTS

11 MAINTENANCE

APPENDIX

1
SAFETY

1-1 INTRODUCTION

Today, a boat and power unit represents a sizeable investment for the owner. In order to protect this investment and to receive the maximum amount of enjoyment from the boat, it must be cared for properly while being used and when it is out of the water. Always store the boat with the bow higher than the stern and be sure to remove the transom drain plug and the inner hull drain plugs. If any type cover is used to protect the boat, plastic, canvas, whatever, be sure to allow for some movement of air through the hull. Proper ventilation will assure evaporation of any condensation due to changes in temperature and humidity.

1-2 CLEANING, WAXING, AND POLISHING

An outboard boat should be washed with clear water after each use to remove surface dirt and any salt deposits from use in salt water. Regular rinsing will extend the time between waxing and polishing. It will also give you "pride of ownership", by having a sharp looking piece of equipment. Elbow grease, a mild detergent, and a brush

will be required to remove stubborn dirt, oil, and other unsightly deposits.

Stay away from harsh abrasives or strong chemical cleaners. A white buffing compound can be used to restore the original gloss to a scratched, dull, or faded area. The finish of your boat should be thoroughly cleaned, buffed, and polished at least once each season. Take care when buffing or polishing with a marine cleaner not to overheat the surface you are working, because you will burn it.

A small outboard engine mounted on an aluminum boat should be removed from the boat and stored separately. Under all circumstances, any outboard engine must **ALWAYS** be stored with the powerhead higher than the lower unit and exhaust system. This position will prevent water trapped in the lower unit from draining back through the exhaust ports into the powerhead.

DRAIN PLUG

Whenever the boat is stored, for long or short periods, the bow should be slightly higher than the stern and the drain plug in the transom removed to ensure proper drainage of rain water.

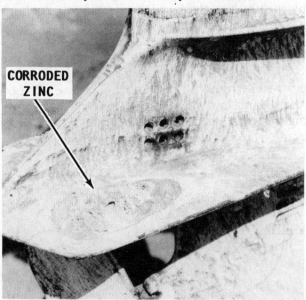

CORRODED ZINC

Lower unit badly corroded because the zinc was not replaced. Once the zinc is destroyed, more costly parts will be damaged. Attention to the zinc condition is extremely important during boat operation in salt water.

A new zinc prior to installation. This inexpensive item will save corrosion on more valuable parts.

Most outboard engines have a flat area on the back side of the powerhead. When the engine is placed with the flat area on the powerhead and the lower unit resting on the floor, the engine will be in the proper altitude with the powerhead higher than the lower unit.

1-3 CONTROLLING CORROSION

Since man first started out on the water, corrosion on his craft has been his enemy. The first form was merely rot in the wood and then it was rust, followed by other forms of destructive corrosion in the more modern materials. One defense against corrosion is to use similar metals throughout the boat. Even though this is difficult to do in designing a new boat, particularily the undersides, similar metals should be used whenever and wherever possible.

A second defense against corrosion is to insulate dissimilar metals. This can be done by using an exterior coating of Sea Skin or by insulating them with plastic or rubber gaskets.

Using Zinc

The proper amount of zinc attached to a boat is extremely important. The use of too much zinc can cause wood burning by placing the metals close together and they become "hot". On the other hand, using too

small a zinc plate will cause more rapid deterioration of the the metal you are trying to protect. If in doubt, consider the fact that it is far better to replace the zincs than to replace planking or other expensive metal parts from having an excess of zinc.

When installing zinc plates, there are two routes available. One is to install many different zincs on all metal parts and thus run the risk of wood burning. Another route, is to use one large zinc on the transom of the boat and then connect this zinc to every underwater metal part through internal bonding. Of the two choices, the one zinc on the transom is the better way to go.

Small outboard engines have a zinc plate attached to the cavitation plate. Therefore, the zinc remains with the engine at all times.

1-4 PROPELLERS

As you know, the propeller is actually what moves the boat through the water. This is how it is done. The propeller operates in water in much the manner as a wood screw does in wood. The propeller "bites" into the water as it rotates. Water passes between the blades and out to the rear in the shape of a cone. The propeller "biting" through the water in much the same manner as a wood auger is what propels the boat.

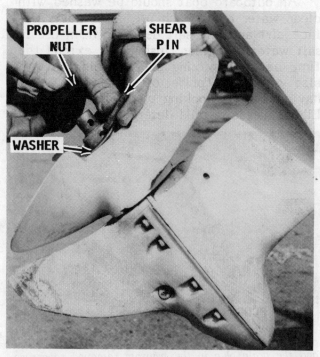

Propeller and associated parts in order, washer, shear pin, and nut, ready for installation.

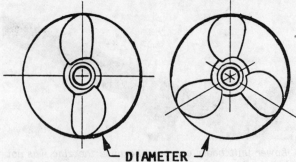

Diameter and pitch are the two basic dimensions of a propeller. The diameter is measured across the circumference of a circle scribed by the propeller blades, as shown.

Arrangement of propeller and associated parts, in order, for a small horsepower engine.

Diameter and Pitch

Only two dimensions of the propeller are of real interest to the boat owner: the diameter and the pitch. These two dimensions are stamped on the propeller hub and always appear in the same order: the diameter first and then the pitch. For instance, the number 15-19 stamped on the hub, would mean the propeller had a diameter of 15 inches with a pitch of 19.

The diameter is the measured distance from the tip of one blade to the tip of the other as shown in the accompanying illustration.

The pitch of a propeller is the angle at which the blades are attached to the hub. This figure is expressed in inches of water travel for each revolution of the propeller. In our example of a 15-19 propeller, the propeller should travel 19 inches through the water each time it revolves. If the propeller action was perfect and there was no slippage, then the pitch multiplied by the propeller rpms would be the boat speed.

Most outboard manufacturers equip their units with a standard propeller with a diameter and pitch they consider to be best suited to the engine and the boat. Such a propeller allows the engine to run as near to the rated rpm and horsepower (at full throttle) as possible for the boat design.

The blade area of the propeller determines its load-carrying capacity. A two-blade propeller is used for high-speed running under very light loads.

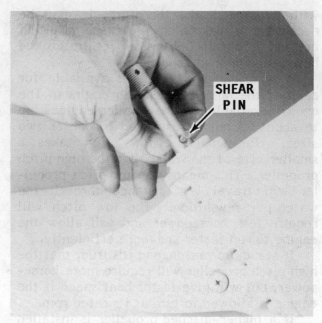

Shear pin installed behind the propeller instead of in front of the propeller.

A four-blade propeller is installed in boats intended to operate at low speeds under very heavy loads such as tugs, barges, or large houseboats. The three-blade propeller is the happy medium covering the wide range between the high performance units and the load carrying workhorses.

Propeller Selection

There is no standard propeller that will do the proper job in very many cases. The list of sizes and weights of boats is almost endless. This fact coupled with the many boat-engine combinations makes the propeller selection for a specific purpose a difficult job. In fact, in many cases the propeller is changed after a few test runs. Proper selection is aided through the use of charts set up for various engines and boats. These charts should be studied and understood when buying a propeller. However, bear in mind, the charts are based on average boats

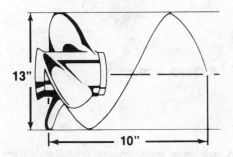
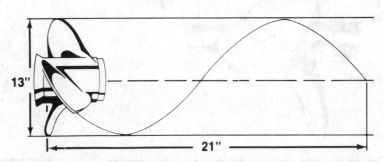

Diagram to explain the pitch dimension of a propeller. The pitch is the theoretical distance a propeller would travel through the water if there was no slippage.

with average loads, therefore, it may be necessary to make a change in size or pitch, in order to obtain the desired results for the hull design or load condition.

A wide range of pitch is available for each of the larger horsepower engines. The choice available for the smaller engines, up to about 25 hp, is restricted to one or two sizes. Remember, a low pitch takes a smaller bite of the water than the high pitch propeller. This means the low pitch propeller will travel less distance through the water per revolution. The low pitch will require less horsepower and will allow the engine to run faster and more efficiently.

It stands to reason, and it's true, that the high pitch propeller will require more horsepower, but will give faster boat speed if the engine is allowed to turn at its rated rpm.

If a higher-pitched propeller is installed on a boat, in an effort to get more speed, extra horsepower will be required. If the extra power is not available, the rpms will be reduced to a less efficient level and the actual boat speed will be less than if the lower-pitched propeller had been left installed.

All engine manufacturers design their units to operate with full throttle at, or slightly above, the rated rpm. If you run your engine at the rated rpm, you will increase spark plug life, receive better fuel economy, and obtain the best performance

from your boat and engine. Therefore, take time to make the proper propeller selection for the rated rpm of your engine at full throttle with what you consider to be an average load. Your boat will then be correctly balanced between engine and propeller throughout the entire speed range.

A reliable tachometer must be used to measure engine speed at full throttle to ensure the engine will achieve full horsepower and operate efficiently and safely. To test for the correct propeller, make your run in a body of smooth water with the lower unit in forward gear at full throttle. Observe the tachometer at full throttle.

NEVER run the engine at a high rpm when a flush attachment is installed. If the reading is above the manufacturer's recommended operating range, you must try propellers of greater pitch, until you find the one that allows the engine to operate continually within the recommended full throttle range.

If the engine is unable to deliver top performance and you feel it is properly tuned, then the propeller may not be to blame. Operating conditions have a marked effect on performance. For instance, an engine will lose rpm when run in very cold water. It will also lose rpm when run in salt water as compared with fresh water. A hot, low-barometer day will also cause your engine to lose power.

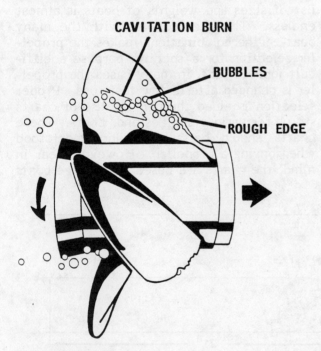

Cavitation (air bubbles) formed at the propeller. Manufacturers are constantly fighting this problem, as explained in the text.

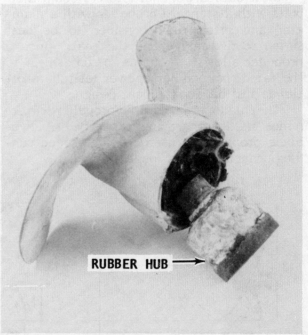

A corroded hub on a small engine propeller. Replacement of this propeller will be less expensive than the cost of a rebuild.

Ventilation

Ventilation is the forming of voids in the water just ahead of the propeller blades. Marine propulsion designers are constantly fighting the battle against the formation of these voids due to excessive blade tip speed and engine wear. The voids may be filled with air or water vapor, or they may actually be a partial vacuum. Ventilation may be caused by installing a piece of equipment too close to the lower unit, such as the knot indicator pickup, depth sounder, or bait tank pickup.

Vibration

Your propeller should be checked regularly to be sure all blades are in good condition. If any of the blades become bent or nicked, this condition will set up vibrations in the drive unit and the motor. If the vibration becomes very serious it will cause a loss of power, efficiency, and boat performance. If the vibration is allowed to continue over a period of time it can have a damaging effect on many of the operating parts.

Vibration in boats can never be completely eliminated, but it can be reduced by keeping all parts in good working condition and through proper maintenance and lubrication. Vibration can also be reduced in some cases by increasing the number of blades. For this reason, many racers use two-blade props and luxury cruisers have four- and five-blade props installed.

Shock Absorbers

The shock absorber in the propeller plays a very important role in protecting the shafting, gears, and engine against the shock of a blow, should the propeller strike an underwater object. The shock absorber allows the propeller to stop rotating at the instant of impact while the power train continues turning.

How much impact the propeller is able to withstand before causing the clutch hub to slip is calculated to be more than the force needed to propel the boat, but less than the amount that could damage any part of the power train. Under normal propulsion loads of moving the boat through the water, the hub will not slip. However, it will slip if the propeller strikes an object with a force that would be great enough to stop any part of the power train.

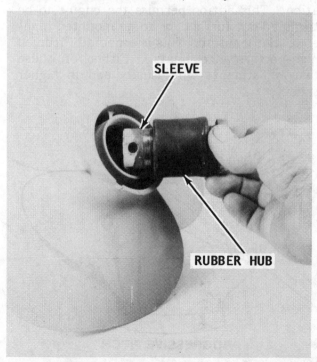

Rubber hub removed from a propeller. This hub was removed because the hub was slipping in the propeller.

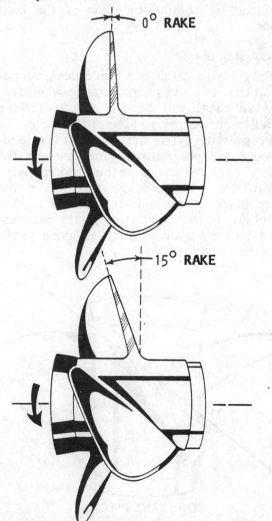

Illustration depicting the rake of a propeller, as explained in the text.

If the power train was to absorb an impact great enough to stop rotation, even for an instant, something would have to give and be damaged. If a propeller is subjected to repeated striking of underwater objects, it would eventually slip on its clutch hub under normal loads. If the propeller would start to slip, a new hub and shock absorber would have to be installed.

Propeller Rake

If a propeller blade is examined on a cut extending directly through the center of the hub, and if the blade is set vertical to the propeller hub, as shown in the accompanying illustration, the propeller is said to have a zero degree (0°) rake. As the blade slants back, the rake increases. Standard propellers have a rake angle from 0° to 15°.

A higher rake angle generally improves propeller performance in a cavitating or ventilating situation. On lighter, faster boats, higher rake often will increase performance by holding the bow of the boat higher.

Progressive Pitch

Progressive pitch is a blade design innovation that improves performance when forward and rotational speed is high and/or the propeller breaks the surface of the water.

Progressive pitch starts low at the leading edge and progressively increases to the trailing edge, as shown in the accompanying illustration. The average pitch over the entire blade is the number assigned to that propeller. In the illustration of the progressive pitch, the average pitch assigned to the propeller would be 21.

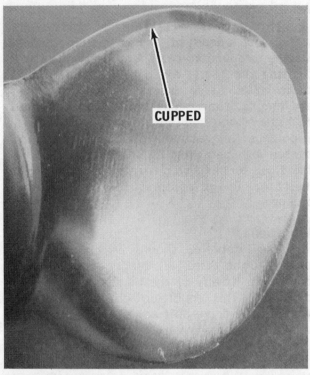

Propeller with a "cupped" leading edge. "Cupping" gives the propeller a better "hold" in the water.

Cupping

If the propeller is cast with a edge curl inward on the trailing edge, the blade is said to have a cup. In most cases, cupped blades improve performance. The cup helps the blades to **"HOLD"** and not break loose, when operating in a cavitating or ventilating situation. This action permits the engine to be trimmed out further, or to be mounted higher on the transom. This is especially true on high-performance boats. Either of these two adjustments will usually add to higher speed.

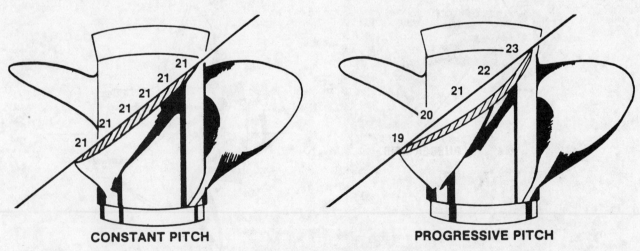

Comparison of a constant and progressive pitch propeller. Notice how the pitch of the progressive pitch propeller, right, changes to give the blade more thrust and therefore, the boat more speed.

The cup has the effect of adding to the propeller pitch. Cupping usually will reduce full-throttle engine speed about 150 to 300 rpm below the same pitch propeller without a cup to the blade. A propeller repair shop is able to increase or decrease the cup on the blades. This change, as explained, will alter engine rpm to meet specific operating demands. Cups are rapidly becoming standard on propellers.

In order for a cup to be the most effective, the cup should be completely concave (hollowed) and finished with a sharp corner. If the cup has any convex rounding, the effectiveness of the cup will be reduced.

Rotation

Propellers are manufactured as right-hand rotation (RH), and as left-hand rotation (LH). The standard propeller for outboards is RH rotation.

A right-hand propeller can easily be identified by observing it as shown in the accompanying illustration. Observe how the blade slants from the lower left toward the upper right. The left-hand propeller slants in the opposite direction, from upper left to lower right, as shown.

When the propeller is observed rotating from astern the boat, it will be rotating clockwise when the engine is in forward gear. The left-hand propeller will rotate counterclockwise.

Propeller Modification

If poor acceleration is experienced on hard-to-plane boats, OMC suggests a slight modification be performed. The modification involves drilling three 6 mm (7/32") holes through the outer shell in a precise pattern. The holes allow exhaust gasses to bleed onto the propeller blades causing controlled ventilation during the acceleration period. This action will allow the motor to turn at a higher rpm under acceleration, thus providing more power to plane the boat.

Layout the exact position of each hole 16 $\pm$2mm (5/8" $\pm$1/16") back from the inner lip and the same amount in a **CLOCKWISE** direction from the base of each blade, as shown in the accompanying illustration.

If an inner rib is located under the position of any one of the holes, another propeller must be used. If the holes are properly positioned, and the correct size is drilled, there will be no affect on top speed, maximum rpm, or ventilation in turns. Improper location or size holes will have no affect on performance, particularly in turns.

1-5 FUEL SYSTEM

With Built-in Fuel Tank

All parts of the fuel system should be selected and installed to provide maximum service and protection against leakage. Reinforced flexible sections should be installed in fuel lines where there is a lot of motion, such as at the engine connection. The flaring of copper tubing should be annealed after it is formed as a protection against hardening. **CAUTION:** Compression fittings should **NOT** be used because they are so easily overtightened, which places them under a strain and subjects them to fatigue.

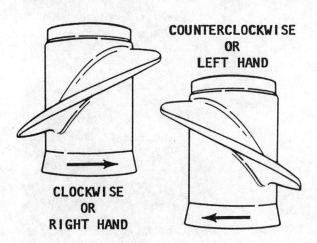

Right- and left-hand propellers showing how the angle of the blades is reversed. Right-hand propellers are by far the most popular.

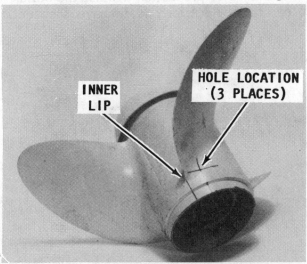

Layout for drilling three holes through the propeller shell for increased performance to provide more power to plane the boat, as explained in the text.

Such conditions will cause the fitting to leak after it is connected a second time.

The capacity of the fuel filter must be large enough to handle the demands of the engine as specified by the engine manufacturer.

A manually-operated valve should be installed if anti-siphon protection is not provided. This valve should be installed in the fuel line as close to the gas tank as possible. Such a valve will maintain anti-siphon protection between the tank and the engine.

The supporting surfaces and hold-downs must fasten the tank firmly and they should be insulated from the tank surfaces. This insulation material should be non-abrasive and nonabsorbent material. Fuel tanks installed in the forward portion of the boat should be especially well secured and protected because shock loads in this area can be as high as 20 to 25 g's ("g" equals force of gravity).

Static Electricity

In very simple terms, static electricity is called frictional electricity. It is generated by two dissimilar materials moving over each other. One form is gasoline flowing through a pipe or into the air. Another form is when you brush your hair or walk across a synthetic carpet and then touch a metal object. All of these actions cause an electrical charge. In most cases, static electricity is generated during very dry weather conditions, but when you are filling the fuel tank on a boat it can happen at any time.

Fuel Tank Grounding

One area of protection against the build-up of static electricity is to have the fuel

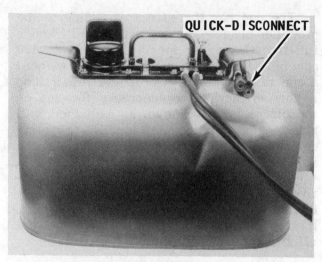

Old style pressure-type tank showing the fuel line to the engine and quick-disconnect fitting.

tank properly grounded (also known as bonding). A direct metal-to-metal contact from the fuel hose nozzle to the water in which the boat is floating. If the fill pipe is made of metal, and the fuel nozzle makes a good contact with the deck plate, then a good ground is made.

As an economy measure, some boats use rubber or plastic filler pipes because of compound bends in the pipe. Such a fill line does not give any kind of ground and if your boat has this type of installation and you do

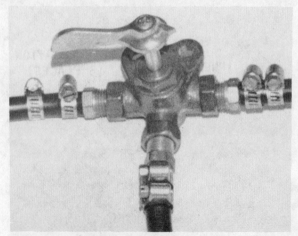

A three-position valve permits fuel to be drawn from either tank or to be shut off completely. Such an arrangement prevents accidental siphoning of fuel from the tank.

Adding fuel to a six-gallon OMC fuel tank. Some fuel must be in the tank before oil is added to prevent the oil from accumulating on the tank bottom.

not want to replace the filler pipe with a metal one, then it is possible to connect the deck fitting to the tank with a copper wire. The wire should be 8 gauge or larger.

The fuel line from the tank to the engine should provide a continuous metal-to-metal contact for proper grounding. If any part of this line is plastic or other non-metallic material, then a copper wire must be connected to bridge the non-metal material. The power train provides a ground through the engine and drive shaft, to the propeller in the water.

Fiberglass fuel tanks pose problems of their own. One method of grounding is to run a copper wire around the tank from the fill pipe to the fuel line. However, such a wire does not ground the fuel in the tank. Manufacturers should imbed a wire in the fiberglass and it should be connected to the intake and the outlet fittings. This wire would avoid corrosion which could occur if a wire passed through the fuel. **CAUTION: It is not advisable to use a fiberglass fuel tank if a grounding wire was not installed..**

Anything you can feel as a "shock" is enough to set off an explosion. Did you know that under certain atmospheric conditions you can cause a static explosion yourself, particularly if you are wearing synthetic clothing. It is almost a certainty you could cause a static spark if you are **NOT** wearing insulated rubber-soled shoes.

As soon as the deck fitting is opened, fumes are released to the air. Therefore, to be safe you should ground yourself before opening the fill pipe deck fitting. One way to ground yourself is to dip your hand in the water overside to discharge the electricity in your body before opening the filler cap. Another method is to touch the engine block or any metal fitting on the dock which goes down into the water.

1-6 LOADING

In order to receive maximum enjoyment, with safety and performance, from your boat, take care not to exceed the load capacity given by the manufacturer. A plate attached to the hull indicates the U.S. Coast Guard capacity information in pounds for persons and gear. If the plate states the maximum person capacity to be 750 pounds and you assume each person to weigh an average of 150 lbs., then the boat could carry five persons safely. If you add another 250 lbs. for motor and gear, and the maximum weight capacity for persons and gear is 1,000 lbs. or more, then the five persons and gear would be within the limit.

Try to load the boat evenly port and starboard. If you place more weight on one side than on the other, the boat will list to the heavy side and make steering difficult. You will also get better performance by placing heavy supplies aft of the center to keep the bow light for more efficient planing.

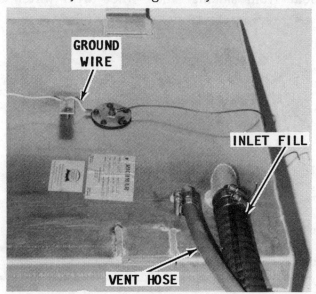

A fuel tank properly grounded to prevent static electricity. Static electricity could be extremely dangerous when taking on fuel.

U.S. Coast Guard plate affixed to all new boats. When the blanks are filled in, the plate will indicate the Coast Guard's recommendations for persons, gear, and horsepower to ensure safe operation of the boat. These recommendations should not be exceeded, as explained in the text.

Clarification

Much confusion arises from the terms, certification, requirements, approval, regulations, etc. Perhaps the following may clarify a couple of these points.

1- The Coast Guard does not approve boats in the same manner as they "Approve" life jackets. The Coast Guard applies a formula to inform the public of what is safe for a particular craft.

2- If a boat has to meet a particular regulation, it must have a Coast Guard certification plate. The public has been led to believe this indicates approval of the Coast Guard. Not so.

3- The certification plate means a willingness of the manufacturer to meet the Coast Guard regulations for that particular craft. The manufacturer may recall a boat if it fails to meet the Coast Guard requirements.

4- The Coast Guard certification plate, see accompanying illustration, may or may not be metal. The plate is a regulation for the manufacturer. It is only a warning plate and the public does not have to adhere to the restrictions set forth on it. Again, the plate sets forth information as to the Coast Guard's opinion for safety on that particular boat.

5- Coast Guard Approved equipment is equipment which has been approved by the Commandant of the U.S. Coast Guard and has been determined to be in compliance with Coast Guard specifications and regulations relating to the materials, construction, and performance of such equipment.

1-7 HORSEPOWER

The maximum horsepower engine for each individual boat should not be increased by any great amount without checking requirements from the Coast Guard in your area. The Coast Guard determines horsepower requirements based on the length, beam, and depth of the hull. **TAKE CARE NOT** to exceed the maximum horsepower listed on the plate or the warranty and possibly the insurance on the boat may become void.

1-8 FLOTATION

If your boat is less than 20 ft. overall, a Coast Guard or BIA (Boating Industry of America) now changed to NMMA (National Marine Manufacturers Association) requirement is that the boat must have buoyant material built into the hull (usually foam) to keep it from sinking if it should become swamped. Coast Guard requirements are mandatory but the NMMA is voluntary.

"Kept from sinking" is defined as the ability of the flotation material to keep the boat from sinking when filled with water

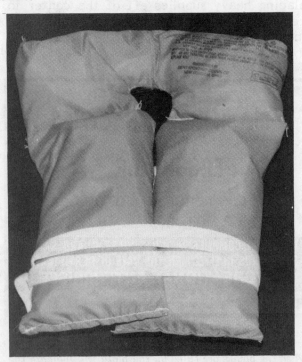

Type I PFD Coast Guard Approved life jacket. This type flotation device provides the greatest amount of buoyancy. **NEVER** *use them for cushions or other purposes.*

A Type IV PFD cushion device intended to be thrown to a person in the water. If air can be squeezed out of the cushion it is no longer fit for service as a PFD.

and with passengers clinging to the hull. One restriction is that the total weight of the motor, passengers, and equipment aboard does not exceed the maximum load capacity listed on the plate.

Life Preservers —Personal Flotation Devices (PFDs)

The Coast Guard requires at least one Coast Guard approved life-saving device be carried on board all motorboats for each person on board. Devices approved are identified by a tag indicating Coast Guard approval. Such devices may be life preservers, buoyant vests, ring buoys, or buoyant cushions. Cushions used for seating are serviceable if air cannot be squeezed out of it. Once air is released when the cushion is squeezed, it is no longer fit as a flotation device. New foam cushions dipped in a rubberized material are almost indestructible.

Life preservers have been classified by the Coast Guard into five type categories. All PFDs presently acceptable on recreational boats fall into one of these five designations. All PFDs **MUST** be U.S. Coast Guard approved, in good and serviceable condition, and of an appropriate size for the persons who intend to wear them. Wearable PFDs **MUST** be readily accessible and throwable devices **MUST** be immediately available for use.

Type I PFD has the greatest required buoyancy and is designed to turn most **UNCONSCIOUS** persons in the water from a face down position to a vertical or slightly backward position. The adult size device provides a minimum buoyancy of 22 pounds and the child size provides a minimum buoyancy of 11 pounds. The Type I PFD provides the greatest protection to its wearer and is most effective for all waters and conditions.

Type II PFD is designed to turn its wearer in a vertical or slightly backward position in the water. The turning action is not as pronounced as with a Type I. The device will not turn as many different type persons under the same conditions as the Type I. An adult size device provides a minimum buoyancy of 15½ pounds, the medium child size provides a minimum of 11 pounds, and the infant and small child sizes provide a minimum buoyancy of 7 pounds.

Type III PFD is designed to permit the wearer to place himself (herself) in a vertical or slightly backward position. The Type III device has the same buoyancy as the Type II PFD but it has little or no turning ability. Many of the Type III PFD are designed to be particularly useful when water skiing, sailing, hunting, fishing, or engaging in other water sports. Several of this type will also provide increased hypothermia protection.

Type IV PFD is designed to be thrown to a person in the water and grasped and held by the user until rescued. It is **NOT** designed to be worn. The most common Type IV PFD is a ring buoy or a buoyant cushion.

Type V PFD is any PFD approved for restricted use.

Coast Guard regulations state, in general terms, that on all boats less than 16 ft. overall, one Type I, II, III, or IV device shall be carried on board for each person in the boat. On boats over 26 ft., one Type I, II, or III device shall be carried on board for each person in the boat **plus** one Type IV device.

It is an accepted fact that most boating people own life preservers, but too few actually wear them. There is little or no excuse for not wearing one because the modern comfortable designs available today do not subtract from an individual's boating pleasure. Make a life jacket available to

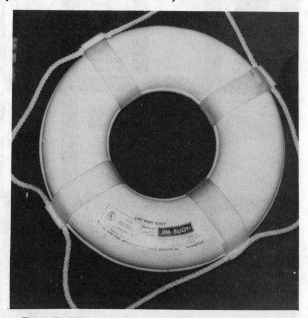

Type IV PFD ring buoy designed to be thrown. On ocean cruisers, this type device usually has a weighted pole with flag, attached to the buoy.

your crew and advise each member to wear it. If you are a crew member ask your skipper to issue you one, especially when boating in rough weather, cold water, or when running at high speed. Naturally, a life jacket should be a must for non-swimmers any time they are out on the water in a boat.

1-9 EMERGENCY EQUIPMENT

Visual Distress Signals
The Regulation

Since January 1, 1981, Coast Guard Regulations require all recreation boats when used on coastal waters, which includes the Great Lakes, the territorial seas and those waters directly connected to the Great Lakes and the territorial seas, up to a point where the waters are less than two miles wide, and boats owned in the United States when operating on the high seas to be equipped with visual distress signals.

The only exceptions are during daytime (sunrise to sunset) for:

Recreational boats less than 16 ft. (5 meters) in length.

Boats participating in organized events such as races, regattas or marine parades.

Open sailboats not equipped with propulsion machinery and less than 26 ft. (8 meters) in length.

Manually propelled boats.

The above listed boats need to carry night signals when used on these waters at night.

Pyrotechnic visual distress signaling devices **MUST** be Coast Guard Approved, in serviceable condition and stowed to be readily accessible. If they are marked with a date showing the serviceable life, this date must not have passed. Launchers, produced before Jan. 1, 1981, intended for use with approved signals are not required to be Coast Guard Approved.

USCG Approved pyrotechnic visual distress signals and associated devices include:

Pyrotechnic red flares, hand held or aerial.

Pyrotechnic orange smoke, hand held or floating.

Launchers for aerial red meteors or parachute flares.

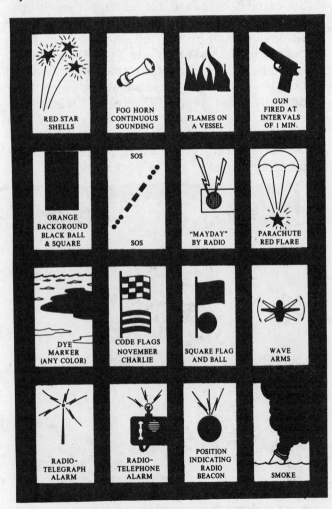

Internationally accepted distress signals.

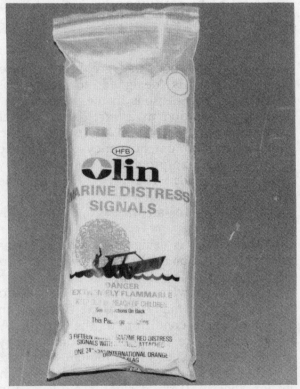

Moisture-protected flares should be carried on board for use as a distress signal.

Non-pyrotechnic visual distress signaling devices must carry the manufacturer's certification that they meet Coast Guard requirements. They must be in serviceable condition and stowed so as to be readily accessible.

This group includes:

Orange distress flag at least 3 x 3 feet with a black square and ball on an orange background.

Electric distress light -- not a flashlight but an approved electric distress light which **MUST** automatically flash the international **SOS** distress signal (. . . - - - . . .) four to six times each minute.

Types and Quantities

The following variety and combination of devices may be carried in order to meet the requirements.

1- Three hand-held red flares (day and night).

2- One electric distress light (night only).

3- One hand-held red flare and two parachute flares (day and night).

4- One hand-held orange smoke signal, two floating orange smoke signals (day) and one electric distress light (day and night).

If young children are frequently aboard your boat, careful selection and proper stowage of visual distress signals becomes especially important. If you elect to carry pyrotechnic devices, you should select those in tough packaging and not easy to ignite should the devices fall into the hands of children.

Coast Guard Approved pyrotechnic devices carry an expiration date. This date can **NOT** exceed 42 months from the date of manufacture and at such time the device can no longer be counted toward the minimum requirements.

SPECIAL WORDS

In some states the launchers for meteors and parachute flares may be considered a firearm. Therefore, check with your state authorities before acquiring such a launcher.

First Aid Kits

The first-aid kit is similar to an insurance policy or life jacket. You hope you don't have to use it but if needed, you want it there. It is only natural to overlook this essential item because, let's face it, who likes to think of unpleasantness when planning to have only a good time. However, the prudent skipper is prepared ahead of time, and is thus able to handle the emergency without a lot of fuss.

Good commercial first-aid kits are available such as the Johnson and Johnson "Marine First-Aid Kit". With a very modest expenditure, a well-stocked and adequate kit can be prepared at home.

Any kit should include instruments, supplies, and a set of instructions for their use. Instruments should be protected in a watertight case and should include: scissors, tweezers, tourniquet, thermometer, safety

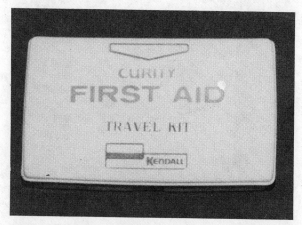

An adequately stocked first-aid kit should be on board for the safety of crew and guests.

A sounding device should be mounted close to the helmsperson for use in sounding an emergency alarm.

pins, eye-washing cup, and a hot water bottle. The supplies in the kit should include: assorted bandages in addition to the various sizes of "band-aids", adhesive tape, absorbent cotton, applicators, petroleum jelly, antiseptic (liquid and ointment), local ointment, aspirin, eye ointment, antihistamine, ammonia inhalent, sea-sickness pills, antacid pills, and a laxative. You may want to consult your family physician about including antibiotics. Be sure your kit contains a first-aid manual because even though you have taken the Red Cross course, you may be the patient and have to rely on an untrained crew for care.

Fire Extinguishers

All fire extinguishers must bear Underwriters Laboratory (UL) "Marine Type" approved labels. With the UL certification, the extinguisher does not have to have a Coast Guard approval number. The Coast Guard classifies fire extinguishers according to their size and type.

Type B-I or B-II Designed for extinguishing flammable liquids. Required on all motorboats.

The Coast Guard considers a boat having one or more of the following conditions as a "boat of closed construction" subject to fire extinguisher regulations.

1- Inboard engine or engines.
2- Closed compartments under thwarts and seats wherein portable fuel tanks may be stored.
3- Double bottoms not sealed to the hull or which are not completely filled with flotation materials.
4- Closed living spaces.
5- Closed stowage compartments in which combustible or flammable material is stored.
6- Permanently installed fuel tanks.

Detailed classification of fire extinguishers is by agent and size:

B-I contains 1-1/4 gallons foam, 4 pounds carbon dioxide, 2 pounds dry chemical, and 2-1/2 pounds freon.

B-II contains 2-1/2 gallons foam, 15 pounds carbon dioxide, and 10 pounds dry chemical.

The class of motorboat dictates how many fire extinguishers are required on board. One B-II unit can be substituted for two B-I extinguishers. When the engine compartment of a motorboat is equipped with a fixed (built-in) extinguishing system, one less portable B-I unit is required.

Dry chemical fire extinguishers without

A suitable fire extinguisher should be mounted close to the helmsman for emergency use.

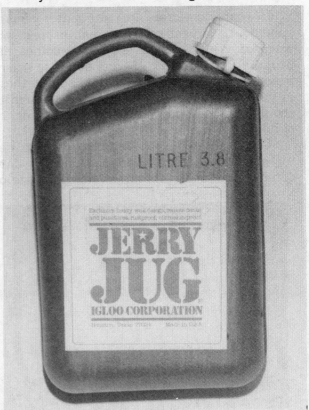

At least one gallon of emergency fuel should be kept on board in an approved container.

gauges or indicating devices must be weighed and tagged every 6 months. If the gross weight of a carbon dioxide (CO_2) fire extinguisher is reduced by more than 10% of the net weight, the extinguisher is not acceptable and must be recharged.

READ labels on fire extinguishers. If the extinguisher is U.L. listed, it is approved for marine use.

DOUBLE the number of fire extinguishers recommended by the Coast Guard, because their requirements are a bare **MINIMUM** for safe operation. Your boat, family, and crew, must certainly be worth much more than "bare minimum".

1-10 COMPASS

Selection

The safety of the boat and her crew may depend on her compass. In many areas weather conditions can change so rapidly that within minutes a skipper may find himself "socked-in" by a fog bank, a rain squall, or just poor visibility. Under these conditions, he may have no other means of keeping to his desired course except with the compass. When crossing an open body of water, his compass may be the only means of making an accurate landfall.

During thick weather when you can neither see nor hear the expected aids to navigation, attempting to run out the time on a given course can disrupt the pleasure of the cruise. The skipper gains little comfort in a chain of soundings that does not match those given on the chart for the expected area. Any stranding, even for a short time, can be an unnerving experience.

A pilot will not knowingly accept a cheap parachute. A good boater should not accept a bargain in lifejackets, fire extinguishers, or compass. Take the time and spend the few extra dollars to purchase a compass to fit your expected needs. Regardless of what the salesman may tell you, postpone buying until you have had the chance to check more than one make and model.

Lift each compass, tilt and turn it, simulating expected motions of the boat. The compass card should have a smooth and stable reaction.

The card of a good quality compass will come to rest without oscillations about the lubber's line. Reasonable movement in your hand, comparable to the rolling and pitching

The compass is a delicate instrument and deserves respect. It should be mounted securely and in position where it can be easily observed by the helmsman.

of the boat, should not materially affect the reading.

Installation

Proper installation of the compass does not happen by accident. Make a critical check of the proposed location to be sure compass placement will permit the helmsman to use it with comfort and accuracy. First, the compass should be placed directly in front of the helmsman and in such a position that it can be viewed without body stress as he sits or stands in a posture of relaxed alertness. The compass should be in the helmsman's zone of comfort. If the compass is too far away, he may have to bend forward to watch it; too close and he must rear backward for relief.

Do not hesitate to spend a few extra dollars for a good reliable compass. If in doubt, seek advice from fellow boaters.

Second, give some thought to comfort in heavy weather and poor visibilty conditions during the day and night. In some cases, the compass position may be partially determined by the location of the wheel, shift lever, and throttle handle.

Third, inspect the compass site to be sure the instrument will be at least two feet from any engine indicators, bilge vapor detectors, magnetic instruments, or any steel

"Innocent" objects close to the compass, such as diet coke in an aluminum can, may cause serious problems and lead to disaster, as these three photos and the accompanying text illustrate.

or iron objects. If the compass cannot be placed at least two feet (six feet would be better) from one of these influences, then either the compass or the other object must be moved, if first order accuracy is to be expected.

Once the compass location appears to be satisfactory, give the compass a test before installation. Hidden influences may be concealed under the cabin top, forward of the cabin aft bulkhead, within the cockpit ceiling, or in a wood-covered stanchion.

Move the compass around in the area of the proposed location. Keep an eye on the card. A magnetic influence is the only thing that will make the card turn. You can quickly find any such influence with the compass. If the influence can not be moved away or replaced by one of non-magnetic material, test to determine whether it is merely magnetic, a small piece of iron or steel, or some magnetized steel. Bring the north pole of the compass near the object, then shift and bring the south pole near it. Both the north and south poles will be attracted if the compass is demagnetized. If the object attracts one pole and repels the other, then the compass is magnetized. If your compass needs to be demagnetized, take it to a shop equipped to do the job **PROPERLY**.

After you have moved the compass around in the proposed mounting area, hold it down or tape it in position. Test everything you feel might affect the compass and cause a deviation from a true reading. Rotate the wheel from hard over to hard over. Switch on and off all the lights, radios, radio direction finder, radio telephone, depth finder and the shipboard intercom, if one is installed. Sound the electric whistle, turn on the windshield wipers, start the engine (with water circulating through the engine), work the throttle, and move the gear shift lever. If the boat has an auxiliary generator, start it.

If the card moves during any one of these tests, the compass should be relocated. Naturally, if something like the windshield wipers cause a slight deviation, it may be necessary for you to make a different deviation table to use only when certain pieces of equipment is operating. Bear in mind, following a course that is only off a degree or two for several hours can make considerable difference at the end, putting you on a reef, rock, or shoal.

Check to be sure the intended compass site is solid. Vibration will increase pivot wear.

Now, you are ready to mount the compass. To prevent an error on all courses, the line through the lubber line and the compass card pivot must be exactly parallel to the keel of the boat. You can establish the fore-and-aft line of the boat with a stout cord or string. Use care to transfer this line to the compass site. If necessary, shim the base of the compass until the stile-type lubber line (the one affixed to the case and not gimbaled) is vertical when the boat is on an even keel. Drill the holes and mount the compass.

Magnetic Items After Installation

Many times an owner will install an expensive stereo system in the cabin of his boat. It is not uncommon for the speakers to be mounted on the aft bulkhead up against the overhead (ceiling). In almost every case, this position places one of the speakers in very close proximity to the compass, mounted above the ceiling.

As we all know, a magnet is used in the operation of the speaker. Therefore, it is very likely that the speaker, mounted almost under the compass in the cabin will have a very pronounced affect on the compass accuracy.

Consider the following test and the accompanying photographs as prove of the statements made.

First, the compass was read as 190 degrees while the boat was secure in her slip.

Next a full can of diet coke in an **aluminum** can was placed on one side and the compass read as 204 degrees, a good 14 degrees off.

Next, the full can was moved to the opposite side of the compass and again a reading was observed. This time as 189 degrees, 11 degrees off from the original reading.

Finally the contents of the can were consumed, the can placed on both sides of the compass with **NO** affect on the compass reading.

Two very important conclusions can be drawn from these tests.

1- Something must have been in the contents of the can to affect the compass so drastically.

2- Keep even "innocent" things clear of the compass to avoid any possible error in the boat's heading.

REMEMBER, a boat moving through the water at 10 knots on a compass error of just 5 degrees will be almost 1.5 miles off course in only **ONE** hour. At night, or in thick weather, this could very possibly put the boat on a reef, rock, or shoal, with disastrous results.

1-11 STEERING

USCG or BIA certification of a steering system means that all materials, equipment, and installation of the steering parts meet or exceed specific standards for strength, type, and maneuverability. Avoid sharp bends when routing the cable. Check to be sure the pulleys turn freely and all fittings are secure.

1-12 ANCHORS

One of the most important pieces of equipment in the boat next to the power plant is the ground tackle carried. The engine makes the boat go and the anchor and its line are what hold it in place when the boat is not secured to a dock or on the beach.

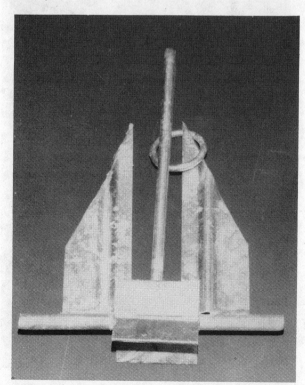

The weight of the anchor MUST be adequate to secure the boat without dragging.

The anchor must be of suitable size, type, and weight to give the skipper peace of mind when his boat is at anchor. Under certain conditions, a second, smaller, lighter anchor may help to keep the boat in a favorable position during a non-emergency daytime situation.

In order for the anchor to hold properly, a piece of chain must be attached to the anchor and then the nylon anchor line attached to the chain. The amount of chain should equal or exceed the length of the boat. Such a piece of chain will ensure that the anchor stock will lay in an approximate horizontal position and permit the flutes to dig into the bottom and hold.

1-13 MISCELLANEOUS EQUIPMENT

In addition to the equipment you are legally required to carry in the boat and those previously mentioned, some extra items will add to your boating pleasure and safety. Practical suggestions would include: a bailing device (bucket, pump, etc.), boat

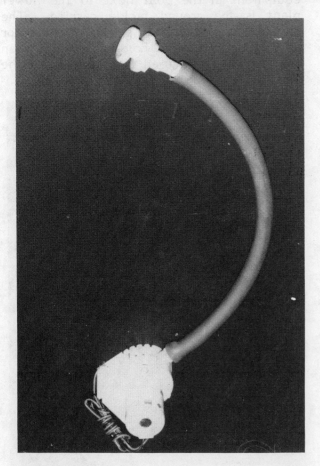

The bilge pump line must be cleaned frequently to ensure the entire bilge pump system will function properly in an emergency.

hook, fenders, spare propeller, spare engine parts, tools, an auxiliary means of propulsion (paddle or oars), spare can of gasoline, flashlight, and extra warm clothing. The area of your boating activity, weather conditions, length of stay aboard your boat, and the specific purpose will all contribute to the kind and amount of stores you put aboard. When it comes to personal gear, heed the advice of veteran boaters who say, "Decide on how little you think you can get by with, then cut it in half".

Bilge Pumps

Automatic bilge pumps should be equipped with an overriding manual switch. They should also have an indicator in the operator's position to advise the helmsman when the pump is operating. Select a pump that will stabilize its temperature within the manufacturer's specified limits when it is operated continuously. The pump motor should be a sealed or arcless type, suitable for a marine atmosphere. Place the bilge pump inlets so excess bilge water can be removed at all normal boat trims. The intakes should be properly screened to prevent the pump from sucking up debris from the bilge. Intake tubing should be of a high quality and stiff enough to resist kinking and not collapse under maximum pump suction condition if the intake becomes blocked.

To test operation of the bilge pump, operate the pump switch. If the motor does not run, disconnect the leads to the motor. Connect a voltmeter to the leads and see if voltage is indicated. If voltage is not indicated, then the problem must be in a blown fuse, defective switch, or some other area of the electrical system.

If the meter indicates voltage is present at the leads, then remove, disassemble, and inspect the bilge pump. Clean it, reassemble, connect the leads, and operate the switch again. If the motor still fails to run, the pump must be replaced.

To test the bilge pump switch, first disconnect the leads from the pump and connect them to a test light or ohmmeter. Next, hold the switch firmly against the mounting location in order to make a good ground. Now, tilt the opposite end of the switch upward until it is activated as indicated by the test light coming on or the ohmmeter showing continuity. Finally, lower the switch slowly toward the mounting position until it is deactivated. Measure the distance between the point the switch was activated and the point it was deactivated. For proper service, the switch should deactivate between 1/2-inch and 1/4-inch from the planned mounting position. **CAUTION: The switch must never be mounted lower than the bilge pump pickup.**

1-14 BOATING ACCIDENT REPORTS

New federal and state regulations require an accident report to be filed with the nearest State boating authority within 48 hours if a person is lost, disappears, or is injured to the degree of needing medical treatment beyond first aid.

Accidents involving only property or equipment damage **MUST** be reported within 10 days if the damage is in excess of $200. Some States require reporting of accidents with property damage less than $200 or total boat loss.

A **$500 PENALTY** may be asessed for failure to submit the report.

WORD OF ADVICE

Take time to make a copy of the report to keep for your records or for the insurance company. Once the report is filed, the Coast Guard will not give out a copy, even to the person who filed the report.

The report must give details of the accident and include:

1- The date, time, and exact location of the occurrence.

2- The name of each person who died, was lost, or injured.

3- The number and name of the vessel.

4- The names and addresses of the owner and operator.

If the operator cannot file the report for any reason, each person on board **MUST** notify the authorities, or determine that the report has been filed.

1-15 NAVIGATION

Buoys

In the United States, a buoyage system is used as an assist to all boaters of all size craft to navigate our coastal waters and our navigable rivers in safety. When properly read and understood, these buoys and markers will permit the boater to cruise with comparative confidence that he will be able to avoid reefs, rocks, shoals, and other hazards.

In the spring of 1983, the Coast Guard began making modifications to U.S. aids to navigation in support of an agreement sponsored by the International Associaiton of Lighthouse Authorities (IALA) and signed by representatives from most of the maritime nations of the world. The primary purpose of the modifications is to improve safety by making buoyage systems around the world more alike and less confusing.

The modifications shown in the accompanying illustrations should be completed by the end of 1989.

Lights

The following information regarding lights required on boats between sunset and sunrise or during restricted visibility is taken directly from a U.S. Coast Guard publication dated 1984.

The terms **"PORT"** and **"STARBOARD"** are used to refer to the left and right side of the boat, when looking forward. One easy way to remember this basic fundamental is to consider the words "port" and "left" both have four letters and go together.

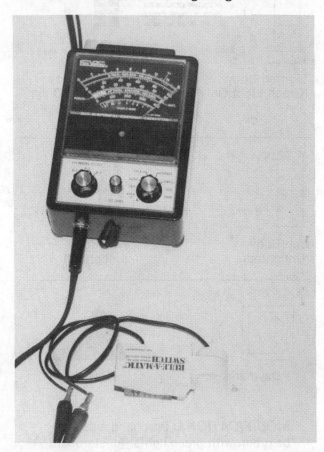

Hookup for testing an automatic bilge pump switch.

Waterway Rules

On the water, certain basic safe-operating practices must be followed. You should learn and practice them, for to **know**, is to be able to handle your boat with confidence and safety. Knowledge of what to do, and not do, will add a great deal to the enjoyment you will receive from your boating investment.

Rules of the Road

The best advice possible and a Coast Guard requirement for boats over 39' 4" (12 meters) since 1981, is to obtain an official copy of the "Rules of the Road", which includes Inland Waterways, Western Rivers, and the Great Lakes for study and ready reference.

The following two paragraphs give a **VERY** brief condensed and abbreviated --

almost a synopsis of the rules and should not be considered in any way as covering the entire subject.

Powered boats must yield the right-of-way to all boats without motors, except when being overtaken. When meeting another boat head-on, keep to starboard, unless you are too far to port to make this practical. When overtaking another boat, the right-of-way belongs to the boat being overtaken. If your boat is being passed, you must maintain course and speed.

When two boats approach at an angle and there is danger of collision, the boat to port must give way to the boat to starboard. Always keep to starboard in a narrow channel or canal. Boats underway must stay clear of vessels fishing with nets, lines, or trawls. (Fishing boats are not allowed to fish in channels or to obstruct navigation.)

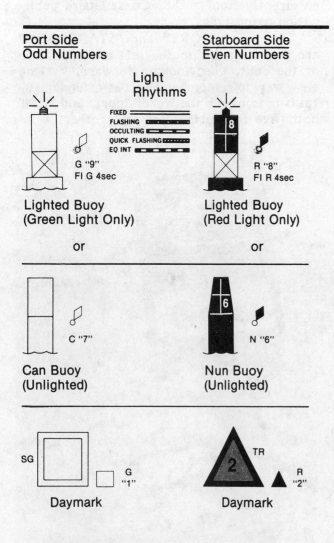

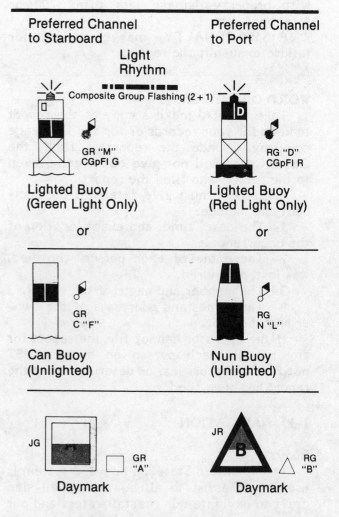

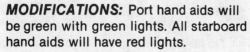

MODIFICATIONS: Port hand aids will be green with green lights. All starboard hand aids will have red lights.

MODIFICATIONS: Green will replace black. Light rhythm will be changed to Composite Gp Fl (2 + 1).

2
TUNING

2-1 INTRODUCTION

The efficiency, reliability, fuel economy and enjoyment available from engine performance are all directly dependent on having it tuned properly. The importance of performing service work in the sequence detailed in this chapter cannot be over emphasized. Before making any adjustments, check the Specifications in the Appendix. **NEVER** rely on memory when making critical adjustments.

Before beginning to tune any engine, check to be sure the engine has satisfactory compression. An engine with worn or broken piston rings, burned pistons, or badly scored cylinder walls, cannot be made to perform properly no matter how much time and expense is spent on the tune-up. Poor compression must be corrected or the tune-up will not give the desired results.

The opposite of poor compression would be to consider good compression as evidence of a satisfactory cylinder. However, this is not necessarily the case, when working on an outboard engine. As the professional mechanic has discovered, many times the compression check will indicate a satisfactory cylinder, but after the head is pulled

DAMAGED PISTON

Damaged piston, probably caused by inaccurate fuel mixture, or improper point setting.

A clean boat and engine appearance reflects this owner's pride in his unit. Keeping the interior well lubricated and properly adjusted will give him the enjoyment deserved for his investment.

and an inspection made, the cylinder will require service.

A practical maintenance program that is followed throughout the year, is one of the best methods of ensuring the engine will give satisfactory performance at any time.

The extent of the engine tune-up is usually dependent on the time lapse since the last service. A complete tune-up of the entire engine would entail almost all of the work outlined in this manual. A logical sequence of steps will be presented in general terms. If additional information or detailed service work is required, the chapter containing the instructions will be referenced.

Each year higher compression ratios are built into modern outboard engines and the electrical systems become more complex, especially with electronic (capacitor discharge) units. Therefore, the need for reliable, authoratative, and detailed instructions becomes more critical. The information in this chapter and the referenced chapters fulfill that requirement.

2-2 TUNE-UP SEQUENCE

If twenty different mechanics were asked the question, "What constitutes a major and minor tune-up?", it is entirely possible twenty different answers would be given. As the terms are used in this manual and other Seloc outboard books, the following work is normally performed for a minor and major tune-up.

Minor Tune-up

Lubricate engine.
Drain and replace gear oil.
Adjust points.
Adjust carburetor.
Clean exterior surface of engine.
Tank test engine for fine adjustments.

Major Tune-up

Remove head.
Clean carbon from pistons and cylinders.
Clean and overhaul carburetor.
Clean and overhaul fuel pump.
Rebuild and adjust ignition system.
Lubricate engine.
Drain and replace gear oil.
Clean exterior surface of engine.
Tank test engine for fine adjustments.

During a major tune-up, a definite sequence of service work should be followed to return the engine to the maximum performance desired. This type of work should not be confused with attempting to locate problem areas of "why" the engine is not performing satisfactorily. This work is classified as "troubleshooting". In many cases, these two areas will overlap, because many times a minor or major tune-up will correct the malfunction and return the system to normal operation.

The time, effort, and expense of a tune-up will not restore an engine to satisfactory performance, if the pistons are damaged.

A boat and lower unit covered with marine growth. Such a condition is a serious hinderance to satisfactory performance.

The following list is a suggested sequence of tasks to perform during the tune-up service work. The tasks are merely listed here. Generally procedures are given in subsequent sections of this chapter. For more detailed instructions, see the referenced chapter.

1- Perform a compression check of each cylinder, see next section.

2- Inspect the spark plugs to determine their condition. Test for adequate spark at the plug, see Section 2-4.

3- Start the engine in a body of water and check the water flow through the engine. See Chapter 8.

4- Check the gear oil in the lower unit. See Chapter 8.

5- Check the carburetor adjustments and the need for an overhaul. See Chapter 4.

6- Check the fuel pump for adequate performance and delivery. See Chapter 4.

7- Make a general inspection of the ignition system. See Chapter 5.

8- Test the starter motor and the solenoid. See Chapter 7.

9- Check the internal wiring.

10- Check the timing and synchronization. See Chapter 5.

2-3 COMPRESSION CHECK

A compression check is extremely important, because an engine with low or uneven compression between cylinders **CANNOT** be tuned to operate satisfactorily. Therefore, it is essential that any compression problem be corrected before proceeding with the tune-up procedure. See Chapter 3.

If the powerhead shows any indication of overheating, such as discolored or scorched paint, especially in the area of the top (No. 1) cylinder, inspect the cylinders visually thru the transfer ports for possible scoring. A more thorough inspection can be made if the head is removed. It is possible for a cylinder with satisfactory compression to be scored slightly. Also, check the water pump. The overheating condition may be caused by a faulty water pump.

An overheating condition may also be caused by running the engine out of the water. For unknown reasons, many operators have formed a bad habit of running a small engine without the lower unit being submerged. Such a practice will result in an overheated condition in a matter of seconds. It is interesting to note, the same operator would never operate or allow anyone else to run a large horsepower engine without water circulating through the lower unit for cooling. Bear-in-mind, the laws governing operation and damage to a large unit **ALL** apply equally as well to the small engine.

Removing the spark plugs for inspection. Worn plugs are one of the major contributing factors to poor engine performance.

COMPRESSION GAUGE

A compression check should be taken in each cylinder before spending time and money on tune-up work. Without adequate compression, efforts in other areas to regain engine performance will be wasted.

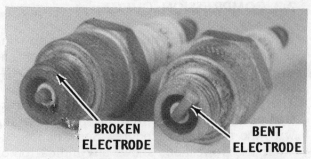

Damaged spark plugs. Notice the broken electrode on the left plug. The broken part must be found and removed before returning the engine to service.

Checking Compression

Remove the spark plug wires. **ALWAYS** grasp the molded cap and pull it loose with a twisting motion to prevent damage to the connection. Remove the spark plugs and keep them in **ORDER** by cylinder for evaluation later. Ground the spark plug leads to the engine to render the ignition system inoperative while performing the compression check.

Insert a compression gauge into the No. 1, top, spark plug opening. Crank the engine with the starter, or pull on the starter· cord, thru at least 4 complete strokes with the throttle at the wide-open position, or until the highest possible reading is observed on the gauge. Record the reading. Repeat the test and record the compression for each cylinder. A variation between cylinders is far more important than the actual readings. A variation of more than 5 psi between cylinders indicates the lower compression cylinder may be defective. The problem may be worn, broken, or sticking piston rings, scored pistons or worn cylinders. These problems may only be

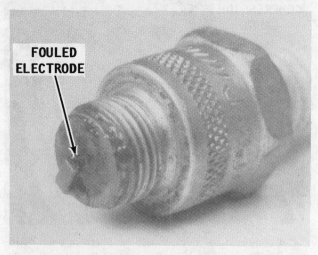

A fouled spark plug. The condition of this plug indicates problems in the air/fuel mixture or the amount of oil added to the mixture.

determined after the head has been removed. Removing the head on an outboard engine is not that big a deal and may save many hours of frustration and the cost of purchasing unnecessary parts to correct a faulty condition.

2-4 SPARK PLUG INSPECTION

Inspect each spark plug for badly worn electrodes, glazed, broken, blistered, or lead fouled insulators. Replace all of the plugs, if one shows signs of excessive wear.

Make an evaluation of the cylinder performance by comparing the spark condition with those shown in Chapter 5. Check each spark plug to be sure they are all of the same manufacturer and have the same heat range rating.

Inspect the threads in the spark plug opening of the head and clean the threads before installing the plug. If the threads are damaged, the head should be removed and and a Heli-coil insert installed. If an attempt is made to drill out the opening with the head in place, some of the filings may fall into the cylinder and cause damage to the cylinder wall during operation. Because the head is made of aluminum, the filings cannot be removed with a magnet.

When purchasing new spark plugs, **ALWAYS** ask the marine dealer if there has been a spark plug change for the engine being serviced.

Crank the engine through several revolutions to blow out any material which might have become dislodged during cleaning.

Install the spark plugs and tighten them to the proper torque value. **ALWAYS** use a new gasket and wipe the seats in the block clean. The gasket must be fully compressed

*Today, numerous type spark plugs are available for service. **ALWAYS** check with your local marine dealer to be sure you are purchasing the proper plugs for the engine being serviced.*

Worn ignition points are a common problem area contributing to poor engine performance.

on clean seats to complete the heat transfer process and to provide a gas tight seal in the cylinder. If the torque value is too high, the heat will dissipate too rapidly. Conversely, if the torque value is too low, heat will not dissipate fast enough.

2-5 IGNITION SYSTEM

Four different ignition systems are used on outboard engines covered in this manual: A flywheel magneto; a low-tension magneto; a capacitor discharge (CD) system with timer base; and a CD with a sensor coil.

If engine performance is less than expected, and the ignition is diagnosed as the problem area, refer to Chatper 5 for detailed service procedures. To properly synchronize the ignition system with the fuel system, see appropriate Section in Chapter 5.

*The fuel and ignition systems on any engine **MUST** be properly synchronized before maximum performance can be obtained from the unit.*

Breaker Points

Engines equipped with either the flywheel magneto or low-tension magneto systems utilize breaker points. Breaker points are **NOT** used in the magneto capacitor discharge (CD) ignition system.

Rough or discolored contact surfaces are sufficient reason for replacement. The cam follower will usually have worn away by the time the points have become unsatisfactory for efficient service.

Check the resistance across the contacts. If the test indicates **ZERO** resistance, the points are serviceable. A slight resistance across the points will affect idle operation. A high resistance may cause the ignition system to malfunction and loss of spark. Therefore, if any resistance across the points is indicated, the point set should be replaced.

2-6 SYNCHRONIZING

The timing on small OMC (Johnson and Evinrude) outboard engines is controlled through adjustment of the points. On the 40hp, 50hp, 55hp, and 60hp engines, and some newer smaller models, the timing is adjustable through the synchronization, see Chapter 5. If the points are adjusted too closely, the spark plugs will fire early; if adjusted with excessive gap, the plugs will fire too late, for efficient operation.

Therefore, correct point adjustment and synchronization are essential for proper engine operation. An engine may be in apparent excellent mechanical condition, but perform poorly, unless the points and synchronization have been adjusted precisely, according to the Specifications in the Appendix. To synchronize the engine, see Chapter 5.

*The battery **MUST** be located near the engine in a well-ventilated area. It must be secured in such a manner that absolutely no movement is possible in any direction under the most violent action of the boat.*

2-7 BATTERY SERVICE

Many owner/operators are not fully aware of the role a battery performs with a magneto ignition system outboard engine. To clarify: With a magneto ignition system, a battery is only used to crank the engine for starting purposes. Once the engine is running properly, the battery could very well be removed without affecting engine operation. Therefore, if the battery is completely dead, the engine may be hand started with a pull cord and operate efficiently.

If a battery is used for starting, inspect and service the battery, cables and connections. Check for signs of corrosion. Inspect the battery case for cracks or bulges, dirt, acid, and electrolyte leakage. Check the electrolyte level in each cell.

Fill each cell to the proper level with distilled water or water passed thru a demineralizer.

Clean the top of the battery. The top of a 12-volt battery should be kept especially clean of acid film and dirt, because of the high voltage between the battery terminals. For best results, first wash the battery with a diluted ammonia or baking soda solution to

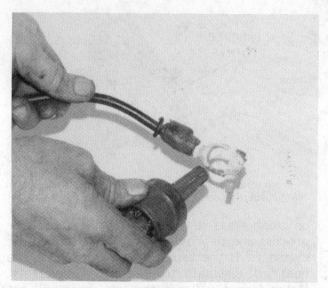

An inexpensive brush should be purchased and used to clean the battery terminals. Clean terminals will ensure a proper connection.

neutralize any acid present. Flush the solution off the battery with clean water. Keep the vent plugs tight to prevent the neutralizing solution or water from entering the cells.

Check to be sure the battery is fastened securely in position. The hold-down device should be tight enough to prevent any movement of the battery in the holder, but not so tight as to place a strain on the battery case.

A check of the electrolyte in the battery should be a regular task on the maintenance schedule on any boat.

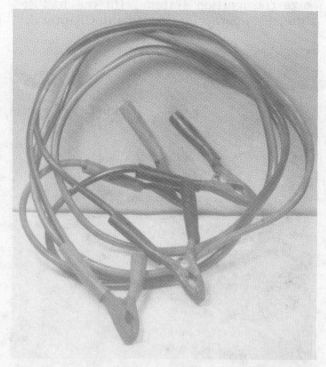

Common set of jumper cables for using a second battery to crank and start the engine. EXTREME care should be used when using a second battery, as explained in the text.

If the battery posts or cable terminals are corroded, the cables should be cleaned separately with a baking soda solution and a wire brush. Apply a thin coating of Multi-purpose Lubricant to the posts and cable clamps before making the connections. The lubricant will help to prevent corrosion.

If the battery has remained under-charged, check for high resistance in the charging circuit. If the battery appears to be using too much water, the battery may be defective, or it may be too small for the job.

Jumper Cables

If booster batteries are used for starting an engine the jumper cables must be connected correctly and in the proper sequence to prevent damage to either battery, or the alternator diodes.

ALWAYS connect a cable from the positive terminals of the dead battery to the positive terminal of the good battery **FIRST**. **NEXT,** connect one end of the other cable to the negative terminals of the good battery and the other end on the **ENGINE** for a good ground. By making the ground connection on the engine, if there is an arc when you make the connection it will not be near the battery. An arc near the battery could cause an explosion, destroying the battery and causing serious personal injury.

DISCONNECT the battery ground cable before replacing an alternator or before connecting any type of meter to the alternator.

If it is necessary to use a fast-charger on a dead battery, **ALWAYS** disconnect one of the boat cables from the battery first, to prevent burning out the diodes in the alternator.

NEVER use a fast charger as a booster to start the engine because the diodes in the alternator will be **DAMAGED.**

Generator Charging

Normally a generating system is not standard equipment on the smaller horsepower engines, up to the 40 hp model. However, a generator kit may be purchased and installed on the 40 hp engines for battery charging while the engine is operating. A generator system is standard equipment of the 40 hp electric shift model.

When the battery is partially discharged, the ammeter should change from discharge to charge between 1500 to 1800 rpm for all models. If the battery is fully-charged, the rpm will be a little higher.

With the engine running, in gear, in the water, increase the throttle until the rpm is approximately 5200 rpm. The ammeter reading should meet the Alternator Specifications in the Appendix. With a fully-charged battery the ammeter reading will be a bit lower because of the self-regulating characteristics of the generating systems. Before disconnecting the ammeter, remove the red harness lead connected to the positive battery terminal.

Alternator Charging

When the battery is partially discharged, the ammeter should change from discharge to charge between 800 to 1000rpm for all models. If the battery is fully-charged, the rpm will be a little higher.

A 40 hp engine with a starter installed on the starboard side and a generator on the port side. The generator, in kit form, is available from the local OMC dealer.

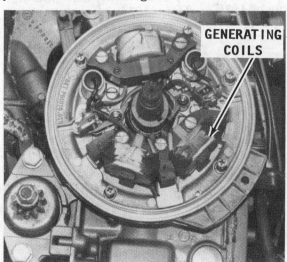

View of the armature plate with the flywheel removed to show the generating coils mounted on the plate.

With the engine running, increase the throttle to approximately 5200 rpm. The ammeter reading should be approximately equal to the amperage rating of the alternator installed. With a fully-charged battery, the ammeter reading will be a bit lower because of the self-regulating characteristics of the generating systems. Before disconnecting the ammeter, reconnect the red harness lead to the positive battery terminal and install the wing nut.

2-8 CARBURETOR ADJUSTMENTS

Fuel and Fuel Tanks

Take time to check the fuel tank and all of the fuel lines, fittings, couplings, valves, flexible tank fill and vent. Turn on the fuel supply valve at the tank, if the engine is equipped with a self-contained fuel tank. If the gas was not drained at the end of the previous season, make a careful inspection for gum formation. When gasoline is allowed to stand for long periods of time, particularly in the presence of copper, gummy deposits form. This gum can clog the filters, lines, and passageway in the carburetor.

If the condition of the fuel is in doubt, drain, clean, and fill the tank with fresh fuel.

Fuel pressure at the carburetor should be checked whenever a lack of fuel volume at the carburetor is suspected.

High-speed Adjustment

The high-speed needle valve is adjustable on some models covered in this manual through 1974. After 1975, the high-speed orifice is fixed at the factory and is **NOT**

An OMC six-gallon fuel tank with the fuel line connected through a quick-disconnect fitting. Such a fitting is handy when the tank is removed from the boat for filling.

adjustable. However, larger or smaller orifices may be installed for different elevations. On all Johnson/Evinrude engines, the high-speed needle valve, or orifice, is the lower valve on the carburetor. The upper needle valve is always the idle adjustment.

A beginning "rough" adjustment for the high-speed needle valve is 3/4 turn out (counterclockwise) from the lightly seated (closed) position. **TAKE CARE** not to seat the valve firmly to prevent damage to the valve or the carburetor.

To make the high-speed adjustment:

a- Mount the engine in a test tank or body of water, preferably with a test wheel. Engines up to 40 hp may be operated in the high rpm range in a test tank without sustaining damage.

NEVER, AGAIN NEVER, operate the engine at high speed with a flush device attached. The engine, operating at high speed with such a device attached, would **RUN-A-WAY** from lack of a load on the propeller, causing extensive damage.

b- Connect a tachometer to the engine.

CAUTION: Water must circulate through the lower unit to the engine any time the engine is run to prevent damage to the water pump in the lower unit. Just five seconds without water will damage the water pump.

c- Start the engine and allow it to warm to operating temperature.

d- Shift the engine into forward gear.

e- With the engine running in forward

Small horsepower engine mounted in a test tank with the low- and high-speed adjustments indicated.

gear, advance the throttle to the wide open position, and then very **SLOWLY** turn the high-speed needle valve inward (**CLOCKWISE**) until the engine begins to loose rpm. Now, **SLOWLY** rotate the needle valve outward (**COUNTERCLOCKWISE**) until the engine peaks out at the highest rpm.

If the high-speed needle valve adjustment is too lean, the low-speed adjustment will be affected. Under certain conditions it may be necessary to adjust the high-speed needle valve just a bit richer in order to obtain a satisfactory idle adjustment.

After the high-speed needle adjustment has been obtained, proceed with the idle adjustment as outlined in the next paragraphs.

Idle Adjustment

Due to local conditions, it may be necessary to adjust the carburetor while the engine is running in a test tank or with the boat in a body of water. For maximum performance, the idle mixture and the idle rpm should be adjusted under actual operating conditions.

Set the idle mixture screw at the specified number of turns open from a lightly seated position. In most cases this is from 1 to 1½ turns open from close.

Start the engine and allow it to warm to operating temperature.

CAUTION: Water must circulate through the lower unit to the engine any time the engine is run to prevent damage to the water pump in the lower unit. Just five seconds without water will damage the water pump.

A 40hp powerhead with the idle adjustment screw and the high speed orifice plug identified.

NEVER, AGAIN NEVER, operate the engine at high speed with a flush device attached. The engine, operating at high speed with such a device attached, would **RUN-A-WAY** from lack of a load on the propeller, causing extensive damage.

With the engine running in forward gear, slowly turn the idle mixture screw **COUNTERCLOCKWISE** until the affected cylinders start to load up or fire unevenly, due to an over-rich mixture. Slowly turn the idle mixture screw **CLOCKWISE** until the cylinders fire evenly and engine rpm increases. Continue to slowly turn the screw **CLOCKWISE** until too lean a mixture is obtained and the rpms fall off and the engine begins to misfire. Now, set the idle mixture screw one-quarter (1/4) turn out (counterclockwise) from the lean-out position. This adjustment will result in an approximate true setting. A too-lean setting is a major cause of hard starting a cold engine. It is better to have the adjustment on the rich side rather than on the lean side. Stating it another way, do not make the adjustment any leaner than necessary to obtain a smooth idle.

If the engine hesitates during acceleration after adjusting the idle mixture, the mixture is too lean. Enrich the mixture slightly, by turning the adjustment screw inward until the engine accelerates correctly.

With the engine running in forward gear, rotate the nylon idle adjustment screw, located on the portside of the engine, until the engine idles at the recommended rpm, as given in the Specifications in the Appendix. This idle adjustment screw is always exposed on the outside of the shroud.

Repairs and Adjustments

For detailed procedures to disassemble, clean, assemble, and adjust the carburetor, see the appropriate section in Chapter 4 for the carburetor type on the engine being serviced.

2-9 FUEL PUMPS

Many times, a defective fuel pump diaphragm is mistakenly diagnosed as a problem in the ignition system. The most common problem is a tiny pin-hole in the diaphragm. Such a small hole will permit gas to enter the crankcase and wet foul the spark plug at idle-speed. During high-speed

Two of the many types of fuel pumps installed on OMC outboard engines. These fuel pumps cannot be rebuilt, as explained in the text.

operation, gas quantity is limited, the plug is not foul and will therefore fire in a satisfactory manner.

If the fuel pump fails to perform properly, an insufficient fuel supply will be delivered to the carburetor. This lack of fuel will cause the engine to run lean, lose rpm or cause piston scoring.

When a fuel pressure gauge is added to the system, it should be installed at the end of the fuel line leading to the upper carburetor. To ensure maximum performance, the fuel pressure must be 2 psi or more at full throttle.

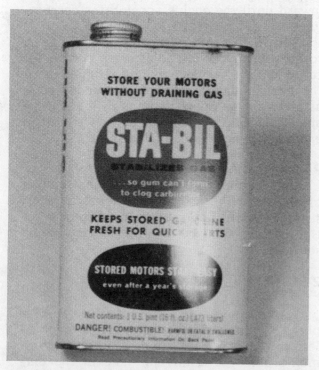

Commercial additives, such as Sta-bil, may be used to keep the gasoline in the fuel tank fresh. Under favorable conditions, such additives will prevent the fuel from "souring" for up to twelve months.

Tune-up Task

Most fuel pumps are equipped with a fuel filter. The filter may be cleaned by first removing the cap, then the filter element, cleaning the parts and drying them with compressed air, and finally installing them in their original position.

A fuel pump pressure test should be made any time the engine fails to perform satisfactorily at high speed.

NEVER use liquid Neoprene on fuel line fittings. Always use Permatex when making fuel line connections. Permatex is available at almost all marine and hardware stores.

Only one Johnson/Evinrude fuel pump may be rebuilt, see accompanying illustration. All others pumps must be replace as a unit. For fuel pump service, see Chapter 4.

2-10 STARTER AND SOLENOID

Starter Motor Test

Check to be sure the battery has a 70-ampere rating and is fully charged. Would you believe, many starter motors are needlessly disassembled, when the battery is actually the culprit.

Lubricate the pinion gear and screw shaft with No. 10 oil.

Starter Bendix drive mechanism. Good maintenance practices should include just a drop of 10-weight oil periodically to the area shown.

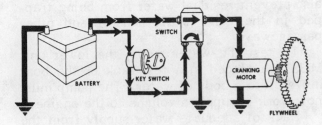

Functional diagram of a typical cranking circuit.

Connect one lead of a voltmeter to the positive terminal of the starter motor. Connect the other meter lead to a good ground on the engine. Check the battery voltage under load by turning the ignition switch to the **START** position and observing the voltmeter reading.

If the reading is 9-1/2 volts or greater, and the starter motor fails to operate, repair or replace the starter motor. See Chapter 7.

Solenoid Test

An ohmmeter is the only instrument required to effectively test a solenoid. Test the ohmmeter by connecting the red and black leads together. Adjust the pointer to the right side of the scale.

On all Johnson/Evinrude engines the case of the solenoid does **NOT** provide a suitable ground to the engine. Hundreds of solenoids

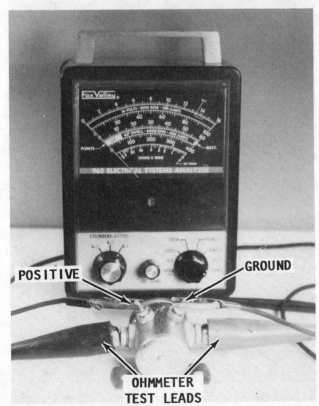

Proper hook-up of an ohmmeter in preparation to testing a starter solenoid.

have been discarded because of the erroneous belief the case is providing a ground and the unit should function when 12-volts is applied. Not so! One terminal of the solenoid is connected to a 12-volt source. The other terminal is connected via a white wire to a cutout switch on top of the engine. This cutout switch provides a safety to break the ground to the solenoid in the event the engine starts at a high rpm. Therefore, the solenoid ground is made and broken by the cutout switch.

NEVER connect the battery leads to the large terminals of the solenoid, or the test meter will be damaged. Connect each lead of the test meter to each of the large terminals on the solenoid.

Using battery jumper leads, connect the positive lead from the positive terminal of the battery to the the small **"S"** terminal of the solenoid. Connect the negative lead to the the **"I"** terminal of the solenoid. Connect the other end of the jumper lead to the negative battery terminal. If the meter indicates continuity, the solenoid is serviceable. If the meter fails to indicate continuity, the solenoid must be replaced.

2-11 INTERNAL WIRING HARNESS

An internal wiring harness is only used on the larger horsepower engines covered in this manual. If the engine is equipped with a wiring harness, the following checks and test will apply.

Check the internal wiring harness if problems have been encountered with any of the electrical components. Check for frayed or chafed insulation and/or loose connections between wires and terminal connections.

Check the harness connector for signs of corrosion. If the harness shows any evidence of damage or corrosion, the problem must be corrected before proceeding with any harness testing.

Convince yourself a good electrical connection is being made between the harness connector and the remote control harness.

2-12 WATER PUMP CHECK

FIRST A GOOD WORD: The water pump **MUST** be in very good condition for the engine to deliver satisfactory service. The pump performs an extremely important function by supplying enough water to properly cool the engine. Therefore, in most

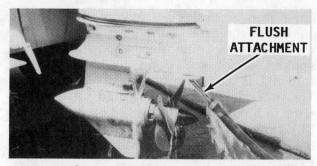

Using a flush attachment with a garden hose hook-up to clean the engine water circulation system with fresh water. This arrangement may also be used while operating the engine at idle speeds to make adjustments.

cases, it is advisable to replace the complete water pump assembly at least once a year, or anytime the lower unit is disassembled for service.

Sometimes during adjustment procedures, it is necessary to run the engine with a flush device attached to the lower unit. **NEVER** operate the engine over 1000 rpm with a flush device attached, because the engine may **"RUNAWAY"** due to the no-load condition on the propeller. A "runaway" engine could be severely damaged. As the name implies, the flush device is primarily used to flush the engine after use in salt water or contaminated fresh water. Regular use of the flush device will prevent salt or silt deposits from accumulating in the water passageway. During and immediately after flushing, keep the motor in an upright position until all of the water has drained from the drive shaft housing. This will prevent water from entering the power head by way of the drive shaft housing and the exhaust ports, during the flush. It will

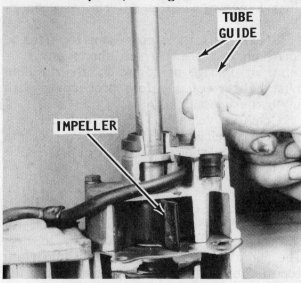

Water pump installed on a 50 hp lower unit.

also prevent residual water from being trapped in the drive shaft housing and other passageways.

To test the water pump, the lower unit **MUST** be placed in a test tank or the boat moved into a body of water. The pump must now work to supply a volume to the engine.

Lack of adequate water supply from the water pump thru the engine will cause any number of powerhead failures, such as stuck rings, scored cylinder walls, burned pistons, etc.

2-13 PROPELLER

Check the propeller blades for nicks, cracks, or bent condition. If the propeller is damaged, the local marine dealer can make repairs or send it out to a shop specializing in such work.

Remove the cotter key, propeller nut, shear pin, and the propeller from the shaft. Check the propeller shaft seal to be sure it is not leaking. Check the area just forward of the seal to be sure a fish line is not wrapped around the shaft.

When installing the propeller, **ALWAYS** use an OMC or approved seal compound on the propeller shaft splines to prevent the propeller from seizing onto the shaft.

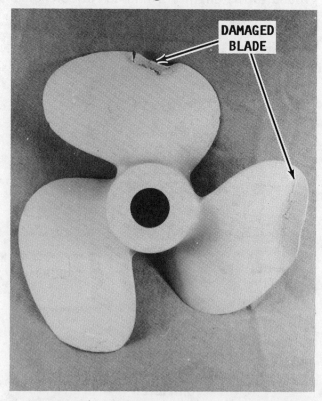

Example of a damaged propeller. This unit should have been replaced long before this amount of damage was sustained.

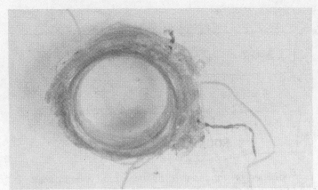

Considerable amount of fish line entangled around the propeller shaft. Some of the fish line actually melted, giving it the appearance of a washer.

Operation At Recommended RPM

Check with the local OMC dealer, or a propeller shop for the recommended size and pitch for a particular size engine, boat, and intended operation. The correct propeller should be installed on the engine to enable operation at recommended rpm.

Two rpm ranges are usually given. The lower rpm is recommended for large, heavy slow boats, or for commercial applications. The higher rpm is recommended for light, fast boats. The wide rpm range will result in greater satisfaction because of maximum performance and greater fuel economy. If the engine speed is above the recommended rpm, try a higher pitch propeller or the same pitch cupped. See Chapter 1 for explanation of propeller terms, pitch, diameter, cupped, etc.

For a dual engine installation, the next higher pitch propeller may prove the most satisfactory condition for water skiing.

2-14 LOWER UNIT

NEVER remove the vent or filler plugs when the lower unit is hot. Expanded lubricant would be released through the plug

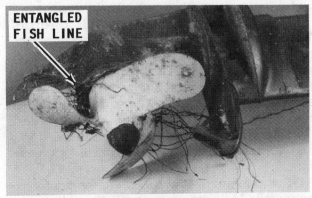

Considerable debris entangled behind the propeller. This type of maintenance neglect will seriously affect powerhead efficiency and boat performance.

New propeller ready for installation. Rebuilding a small propeller is not economical when balanced against the minor difference of purchasing a new unit.

hole. Check the lubricant level after the unit has been allowed to cool. Add only OMC approved gear lubricant. **NEVER** use regular automotive-type grease in the lower unit, because it expands and foams too much. Outboard lower units do not have provisions to accommodate such expansion.

If the lubricant appears milky brown, indicating the presence of water, a check should be made to determine how the water entered. If large amounts of lubricant must be added to bring the lubricant up to the full mark, a thorough inspection should be made to find the cause of the lubricant loss.

Draining Lower Unit

The fill/drain plug on Johnson/Evinrude lower units may be located towards the bottom of the unit on the port side, starboard side, or on the leading edge of the

The gear oil in the lower unit should be checked on a daily basis during the season of operation. The oil should be drained and replenished with new oil every 100 hours of operation.

lower unit. On many models a Phillips screw will be found very close to the fill/ drain plug. **NEVER** remove this Phillips screw because the lower unit would then have to be disassembled in order to return the cradle for the shift dog back in place.

Remove the drain plug and then remove the vent plug located just above the anti-cavitation plate.

Filling Lower Unit

Position the drive unit approximately vertical and without a list to either port or starboard. Insert the lubricant tube into the **FILL/DRAIN** hole at the bottom plug hole, and inject lubricant until the excess begins to come out the **VENT** hole. Install the **VENT** plug first then replace the **FILL** plug with **NEW** gaskets. Check to be sure the gaskets are properly positioned to prevent water from entering the housing. Many times some of the gear lubricant is lost during installation of the plugs. Therefore, if the vent plug is removed again, and more lubricant added very **SLOWLY** using a small-spout oil can to allow air to pass out the opening, the unit will be filled to capacity.

For detailed lower unit service procedures, see Chapter 8. For lower unit lubrication capacities, see the Appendix.

Filling the lower unit with new gear oil. Notice the unit is filled through the lower drain plug. The vent plug MUST be removed to allow trapped air to escape.

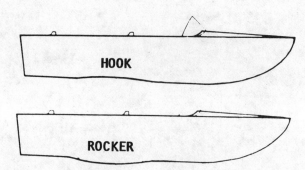

Boat performance will be drastically hampered, if the bottom is damaged.

Repairs and Adjustments

For detailed procedures to disassemble, clean, assemble, and adjust the carburetor, see the appropriate section in Chapter 4 for the carburetor type on the engine being serviced.

2-15 BOAT TESTING

Hook and Rocker

Before testing the boat, check the boat bottom carefully for marine growth or evidence of a "hook" or a "rocker" in the bottom. Either one of these conditions will greatly reduce performance.

Performance

Mount the motor on the boat. Install the remote control cables and check for proper adjustment.

Make an effort to test the boat with what might be considered an average gross load. The boat should ride on an even keel, without a list to port or starboard. Adjust the motor tilt angle, if necessary, to permit the boat to ride slightly higher than the stern. If heavy supplies are stowed aft of the center, the bow will be light and the boat will "plane" more efficiently. For this test the boat must be operated in a body of water.

Check the engine rpm at full throttle. The rpm should be within the Specifications in the Appendix. All OMC engine model serial number identification plates indicate the horsepower rating and rpm range for the engine. If the rpm is not within specified range, a propeller change may be in order. A higher pitch propeller will decrease rpm, and a lower pitch propeller will increase rpm.

For maximum low speed engine performance, the idle mixture and the idle rpm should be readjusted under actual operating conditions.

3
POWERHEAD

3-1 INTRODUCTION

The carburetion and ignition principles of two-cycle engine operation **MUST** be understood in order to perform a proper tune-up on an outboard motor. Therefore, it would be well worth the time, to study the principles of two-cycle engines, as outlined in this section.

A Polaroid, or equivalent instant-type camera is an extremely useful item providing the means of accurately recording the arrangement of parts and wire connections **BEFORE** the disassembly work begins. Such a record is most valuable during the assembly work.

Tags are handy to identify wires after they are disconnected to ensure they will be connected to the same terminal from which they were removed. These tags may also be used for parts where marks or other means of identification is not possible.

THEORY OF OPERATION

The two-cycle engine differs in several ways from a conventional four-cycle (automobile) engine.

1- The method by which the fuel-air mixture is delivered to the combustion chamber.
2- The complete lubrication system.
3- In most cases, the ignition system.
4- The frequency of the power stroke.

These differences will be discussed briefly and compared with four-cycle engine operation.

Intake/Exhaust

Two-cycle engines utilize an arrangement of port openings to admit fuel to the combustion chamber and to purge the exhaust gases after burning has been completed. The ports are located in a precise

pattern in order for them to be open and closed off at an exact moment by the piston as it moves up and down in the cylinder. The exhaust port is located slightly higher than the fuel intake port. This arrangement opens the exhaust port first as the piston starts downward and therefore, the exhaust phase begins a fraction of a second before the intake phase.

Actually, the intake and exhaust ports are spaced so closely together that both open almost simultaneously. For this reason, the pistons of most two-cycle engines have a deflector-type top. This design of the piston top serves two purposes very effectively.

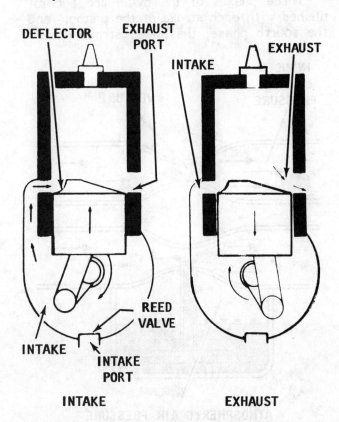

Drawing to depict the intake and exhaust cycles of a two-cycle engine.

First, it creates turbulence when the incoming charge of fuel enters the combustion chamber. This turbulence results in more complete burning of the fuel than if the piston top were flat. The second effect of the deflector-type piston crown is to force the exhaust gases from the cylinder more rapidly.

This system of intake and exhaust is in marked contrast to individual valve arrangement employed on four-cycle engines.

Lubrication

A two-cycle engine is lubricated by mixing oil with the fuel. Therefore, various parts are lubricated as the fuel mixture passes through the crankcase and the cylinder. Four-cycle engines have a crankcase containing oil. This oil is pumped through a circulating system and returned to the crankcase to begin the routing again.

Power Stroke

The combustion cycle of a two-cycle engine has four distinct phases.
1- Intake
2- Compression
3- Power
4- Exhaust

Three phases of the cycle are accomplished with each stroke of the piston, and the fourth phase, the power stroke occurs with each revolution of the crankshaft. Compare this system with a four-cycle engine. A stroke of the piston is required to accomplish each phase of the cycle and the power stroke occurs on every other revolution of the crankshaft. Stated another way, two revolutions of the four-cycle engine crankshaft are required to complete one full cycle, the four phases.

Physical Laws

The two-cycle engine is able to function because of two very simple physical laws.

One: Gases will flow from an area of high pressure to an area of lower pressure. A tire blowout is an example of this principle. The high-pressure air escapes rapidly if the tube is punctured.

Two: If a gas is compressed into a smaller area, the pressure increases, and if a gas expands into a larger area, the pressure is decreased.

If these two laws are kept in mind, the operation of the two-cycle engine will be easier understood.

Actual Operation

Beginning with the piston approaching top dead center on the compression stroke:

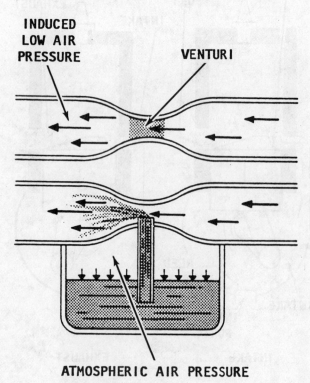

Air flow principle for a modern carburetor.

Adding OMC approved oil into the fuel tank.

The intake and exhaust ports are closed by the piston; the reed valve is open; the spark plug fires; the compressed fuel-air mixture is ignited; and the power stroke begins. The reed valve was open because as the piston moved upward, the crankcase volume increased, which reduced the crankcase pressure to less than the outside atmosphere.

As the piston moves downward on the power stroke, the combustion chamber is filled with burning gases. As the exhaust port is uncovered, the gases, which are under great pressure, escape rapidly through the exhaust ports. The piston continues its downward movement. Pressure within the crankcase increases, closing the reed valves against their seats. The crankcase then becomes a sealed chamber. The air-fuel mixture is compressed ready for delivery to the combustion chamber. As the piston continues to move downward, the intake port is uncovered. Fresh fuel rushes through the intake port into the combustion chamber striking the top of the piston where it is deflected along the cylinder wall. The reed valve remains closed until the piston moves upward again.

When the piston begins to move upward on the compression stroke, the reed valve opens because the crankcase volume has been increased, reducing crankcase pressure to less than the outside atmosphere. The

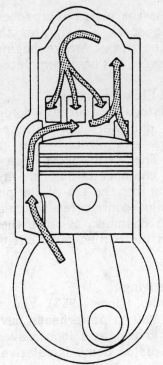

Drawing to depict fuel flow of the "loop charge" while the piston is on the down stroke.

intake and exhaust ports are closed and the fresh fuel charge is compressed inside the combustion chamber.

Pressure in the crankcase decreases as the piston moves upward and a fresh charge of air flows through the carburetor picking

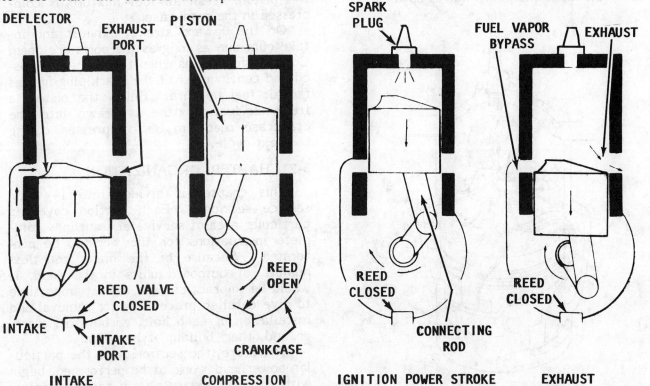

Complete piston cycle of a two-cycle engine, depicting intake, power, and exhaust.

up fuel. As the piston approaches top dead center, the spark plug ignites the air-fuel mixture, the power stroke begins and one complete cycle has been completed.

Cross Fuel Flow Principle

OMC pistons are a defector dome type. The design is necessary to deflect the fuel charge up and around the combustion chamber. The fresh fuel mixture enters the combusion chamber through the intake ports and flows across the top of the piston. The piston design contributes to clearing the combustion chamber, because the incoming fuel pushes the burned gases out the exhaust ports.

Loop Scavenging

The 40 hp (since 1983), and the 50 hp, 55 hp, and 60 hp powerheads have what is commonly known as a loop scavenging system. The piston dome is relatively flat on top with just a small amount of crown. Pressurized fuel in the crankcase is forced up through the skirt of the piston and out through irregular shaped openings cut in the skirt. After the fuel is forced out the piston skirt openings it is transfered upward through long deep grooves molded into the cylinder wall. The fuel then enters the combustion portion of the cylinder and is compressed, as the piston moves upward.

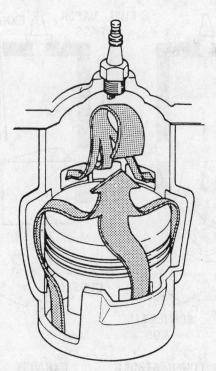

Drawing to depict the exhaust leaving the cylinder and fuel entering through the three ports in the piston.

This particular powerhead does not have intake cover plates, because the intake passage is molded into the cylinder wall as described in the previous paragraph. Therefore, if these engines are being serviced, disregard the sections covering intake cover plates.

Timing

The exact time of spark plug firing depends on engine speed. At low speed the spark is retarded -- fires later than when the piston is at or beyond top dead center. Therefore, the timing is advanced as the magneto armature plate advances.

At high speed, the spark is advanced -- fires earlier than when the piston is at top dead center.

The 40 hp (since 1983), 50 hp, 55 hp, and 60 hp powerheads have a timing adjustment for low and high speeds. Procedures for making the timing adjustment will be found in Chapter 5.

Summary

More than one phase of the cycle occurs simultaneously during operation of a two-cycle engine. On the downward stroke, power occurs above the piston while the ports are closed. When the ports open, exhaust begins and intake follows. Below the piston, fresh air-fuel mixture is compressed in the crankcase.

On the upward stroke, exhaust and intake continue as long as the ports are open. Compression begins when the ports are closed and continues until the spark plug ignites the air-fuel mixture. Below the piston, a fresh air-fuel mixture is drawn into the crankcase ready to be compressed during the next cycle.

3-2 CHAPTER ORGANIZATION

This chapter is divided into 14 main service sections. Each section covers a particular area of service and outlines complete instructions for the work to be performed. Because of the many countless number of outboard units in the field, it would be impractical and almost impossible to give detailed procedures for removal and installation of each bolt, carburetor, starter, and other "buildup" type units.

Therefore, the sections, for the particular powerhead work to be performed, begin with the preliminary access tasks completed. As an example, disassembly of the powerhead begins with the necessary hood, cowling, and accessories removed.

The information is presented in a logical sequence for complete powerhead overhaul. The instructions can be followed generally for almost any size horsepower engine. In rare cases, where the procedures differ depending on the model being serviced, separate steps are included. One example is the three different type of crankshaft installations.

The illustrations accompanying the text are from different size units and the captions clearly identify which model is covered.

Exploded drawings, showing principle parts, for the various size powerheads are included at the end of the chapter.

Special tools may be called out in certain instances. These tools may be purchased from the local Johnson/Evinrude dealer or directly from Customer Services Department, Outboard Marine Corporation (OMC), Waukegan, Illinois, 60085.

The chapter ends with Break-in Procedures, Section 3-15, to be performed after the powerhead has been assembled, all accessories installed, and the powerhead mounted on the exhaust housing.

Torque Values

All torque values must be met when they are specified. Many of the outboard castings and other parts are made of aluminum. The torque values are given to prevent stretching the bolts, but more importantly to protect the threads in the aluminum. It is extremely important to tighten the connecting rods to the proper torque value to ensure proper service. The head bolts are probably the next most important torque value.

Powerhead Components

Service procedures for the carburetors, fuel pumps, starter, and other powerhead components are given in their respective chapters of this manual. See the Table of Contents.

Reed Installation

All reeds on Johnson/Evinrude engines covered in this manual are installed just behind the carburetor behind the intake manifold.

Cleanliness

Make a determined effort to keep parts and the work area as clean as possible.

Parts **MUST** be cleaned and thoroughly inspected before they are assembled, installed, or adjusted. Use proper lubricants, or their equivalent, whenever they are recommended.

Keep rods and rod caps together as a set to ensure they will be installed as a pair and in the proper sequence.

Needle bearings **MUST** remain as a complete set. **NEVER** mix needles from one set with another. If only one needle is damaged, the complete set **MUST** be replaced.

3-3 POWERHEAD DISASSEMBLING

Preliminary Work

Before the powerhead can be disassembled, the battery must be disconnected; fuel lines disconnected; and the carburetor, generator, starter, flywheel, and magneto, all removed. If in doubt as to how these items are to be removed, refer to the appropriate chapter.

After the accessories have been removed, remove the bolts in the front and rear of the powerhead securing the powerhead to the exhaust housing. Lift the powerhead free.

BAD NEWS

If the unit is several years old, or if it has been operated in salt water, or has not had proper maintenance, or shelter, or any number of other factors, then separating the

Cleaning the pistons while they remain in the powerhead. The pistons should be carefully inspected for burned areas and the cylinder walls for scoring.

powerhead from the exhaust housing may not be a simple task. An air hammer may be required on the studs to shake the corrosion loose; heat may have to be applied to the casting to expand it slightly; or other devices employed in order to remove the powerhead. One very serious condition would be the driveshaft "frozen" with the crankshaft. In this case, a circular plug-type hole must be drilled and a torch used to cut the driveshaft. Let's assume the powerhead will come free on the first attempt.

The following procedures pickup the work after these preliminary tasks have been completed.

3-4 HEAD SERVICE

Usually the head is removed and an examination of the cylinders made to determine the extent of overhaul required. However, if the head has not been removed, back out all of the head bolts and lift the head free of the powerhead.

Many, but not all, heads have a thermostat installed. In addition to the thermostat, the engine may have a thermostat bypass valve. These two items are easily removed, inspected and cleaned.

Normally, if a thermostat is not functioning properly, it is almost always stuck in the open position. An engine operating at too low a temperature is almost as much a problem as an engine running too hot.

Therefore, during a major overhaul, good shop practice dictates to replace the thermostat and eliminate this area as a possible problem at a later date.

Lay a piece of fine sandpaper or emery paper on a flat surface (such as a piece of glass) with the abrasive side facing up. With

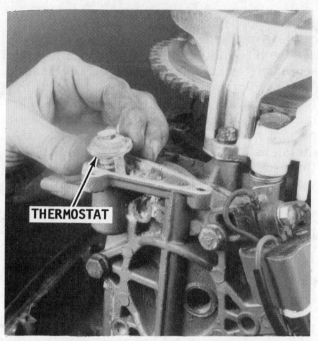

Removing the thermostat from the head.

the machined face of the head on the sandpaper, move the head in a circular motion to dress the surface. This procedure will also indicate any "high" or "low" spots.

Check the spark plug opening/s to be sure the threads are not damaged. Most marine dealers can insert a heli-coil into a spark plug opening if the threads have been damaged.

On many engines, a sending unit is installed in the head to warn the operator if the engine begins to run too hot. The light

Cylinder block water passages corroded preventing proper circulation of coolant water.

Operating an engine without the thermostat and thermostat cover installed to check the coolant water flow.

on the dash can be checked by turning the ignition switch to the **ON** position, and then ground the wire to the sending unit. The light should come on. If it does not, replace the bulb and repeat the test.

3-5 REED SERVICE

DESCRIPTION

All two-cycle engines have individual ignition and fuel delivery for each cylinder. This means the cylinder is operating independently of the others. The cylinder consists of a top seal, the cylinder, the center seal, and the lower seal. This means each cylinder is completely sealed off from the others.

Therefore, with a two-cylinder powerhead, two sets of reeds are installed, one for each cylinder. These reeds may be installed on a reed plate or with a reed box, depending on the model engine. One carburetor provides fuel to both sets of reeds.

The reed arrangement operates in much the same manner as the reed in a saxophone or other wind instrument. At rest, the reed is closed and seals the opening to which it is attached. In the case of an outboard engine, this opening is between the crankcase and the carburetor. The reeds are mounted in the intake manifold, just behind the carburetor.

Actual Operation

The piston creates vacuum and pressure as it moves up and down in the cylinder. As

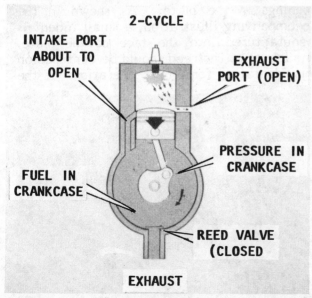

Diagram to illustrate operation of a two-cycle engine.

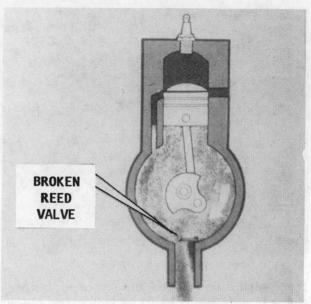

Cross-section of a cylinder to illustrate a broken reed in the crankcase.

the piston moves upward, a vacuum is created in the crankcase pulling the reed open. On the compression stroke, when the piston moves downward, the reed is forced closed.

Reed Designs

A wide range of reeds, reed plates, and reed box installations may be found on an outboard unit, due to the varying designs of the engines. All installations employ the same principle and there is no difference in their operation.

Broken Reed

A broken reed is usually caused by metal fatigue over a long period of time. The failure may also be due to the reed flexing too far because the reed stop has not been adjusted properly or the stop has become distorted. If the reed is broken, the loose

Reed box with a broken reed.

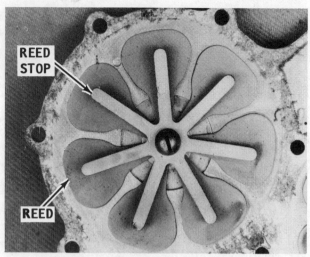

Reed stops centered over the reeds.

piece **MUST** be located and removed, before the engine is returned to service. The piece of reed may have found its way into the crankcase, behind the bypass cover. If the broken piece cannot be located, the powerhead must be completely disassembled until it is located and removed.

The accompanying illustration depicts how a broken reed will cause a backflow through the carburetor.

An excellent check for a broken reed on an operating engine is to hold an ordinary business card in front of the carburetor. Under normal operating conditions, a very small amount of fine mist will be noticeable, but if fuel begins to appear rapidly on the card from the carburetor, one of the reeds is broken and causing the backflow through the carburetor and onto the card.

A broken reed will cause the engine to operate roughly and with a "pop" back through the carburetor.

Reed Stops

If the reed stops have become distorted, the most effective corrective action is to replace the stop instead of making an attempt at adjustment.

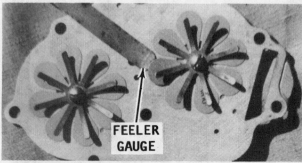

Using a feeler gauge to measure the clearance between the reed tip and the reed plate.

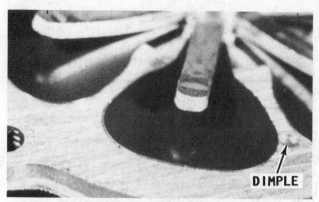

Close view showing the dimple on the reed plate. The reed leaves must straddle the dimple and be centered over the openings for proper operation.

Reed to Base Plate Check

The specified clearance of the reed from the base plate, when the reed is at rest, is 0.010" (0.254 mm) at the tip of the reed.

An alternate method of the checking the reed clearance is to hold the reed up to the sunlight and look through the back side. Some air space should be visible, but not a great amount. If in doubt, check the reed at the tip with a feeler gauge. The maximum clearance should not exceed 0.010" (0.254 mm).

The reeds must **NEVER** be turned over in an attempt to correct a problem. Such action would cause the reed to flex in the opposite direction and the reed would break in a very short time.

REED VALVE ADJUSTMENT

In many instances, the reed is placed on the reed plate in such a manner to cover the openings in the plate. As shown in the accompanying illustration, a small indent is manufactured into the face of the plate. The leaves of the reed should be centered on this identation for proper operation. If the

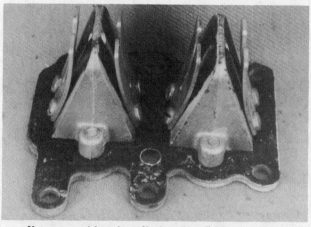

V-type reed box installed on the 9.5 hp engines.

reed is being replaced, both reeds **AND** the reed stops should be replaced as a set.

V-Type Reed Boxes

As the name implies, these reed boxes are shaped in a "V" with a set of reeds and stops on both arms of the "V". If a problem develops with this type reed box, it is strongly recommended that the complete assembly be replaced -- reeds, box, and stops. The assembly may be purchased as a complete unit and the cost will usually not exceed the time, effort, and problems encountered in an attempt to replace only one part.

CLEANING AND SERVICE

Always handle the reeds with the utmost care. Rough treatment will result in the reeds becoming distorted and will affect their performance.

Wash the reeds in solvent, and blow them dry with compressed air from the **BACK SIDE ONLY**. Do not blow air through the reed from the front side. Such action would cause the reed to open and fly up against the reed stop. Wipe the front of the reed dry with a lint free cloth.

Clean the base plate thoroughly by removing any old gasket material.

Secure the reed blocks together with screws and nuts tightened to the torque value given in the Appendix.

Check for chipped or broken reeds. Observe that the reeds are not preloaded or standing open. Satisfactory reeds will not adhere to the reed block surface, but still

Front view of a reed plate with the two sets of reeds and reed stops in place.

there is not more than 0.010" (0.254 mm) clearance between the reed and the block surface.

DO NOT remove the reeds, unless they are to be replaced. **ALWAYS** replace reeds in sets. **NEVER** turn used reeds over to be used a second time.

Check the reed location over the reed block, or plate openings to be sure the reed is centered.

The reed assemblies are then ready for installation.

Small Engines

Disassemble the reed block by first removing the screws securing the reed stops and reeds to the reed block, and then lifting the reed stops and reeds from the block.

Clean the gasket surfaces of the reed block or plate. Check the surfaces for deep grooves, cracks, or any distortion that could

Front view of the reed box showing the Phillips screws that must be removed before the box can be removed from the plate.

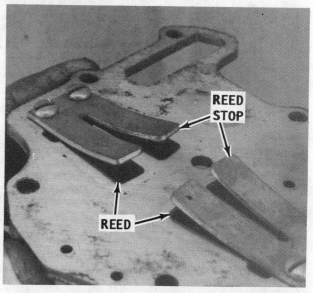

Reed stop installation with flat bars acting as the reed stop. The bars are the same width as the reed.

cause leakage. Replace the reed block or plate if it is damaged.

After new reeds have been installed, and the reed stop and attaching screws have been tightened to the required torque value, check the new reeds as outlined in the following paragraphs.

Check to be sure the reeds are not preloaded. They should not adhere to the block or plate, and still the clearance between the reed and the block surface, should not be more than 0.010" (0.254 mm). **DO NOT** remove the reeds, unless they are to be replaced. **ALWAYS** replace reeds in sets. **NEVER** turn used reeds over to be used a second time.

Lay the reeds on a flat surface and measure all the reed stops. If there is a great difference between the stops, the entire reed stop assembly should be replaced. Any attempt to bend and get all the stops equal and level would be almost impossible.

INSTALLATION

Procedures to install the reeds to the powerhead will be found in Section 3-14, Cylinder Block Service, under Reed Box Installation.

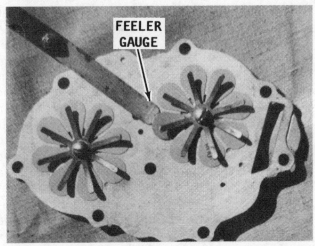

Using a feeler gauge to measure the clearance between the reed tip and the reed plate.

3-6 BYPASS COVERS

On some small horsepower units the powerhead does not contain bypass covers. The bypass cover actually covers the passageway the fuel travels from the crankcase up the side of the powerhead and into the cylinder.

Seldom does a bypass cover cause any problem. On some models, a fuel pump may be attached to one of the bypass covers.

During a normal overhaul, the bypass covers should be removed, cleaned, and new gaskets installed. Identify the covers to ensure installation in the same location from which they are removed.

Close view showing the dimple on the reed plate. The reed leaves must straddle the dimple and be centered over the openings for proper operation.

Removing the bypass cover from a typical powerhead.

INSTALLATION

Procedures to install the bypass covers to the powerhead will be found in Section 3-14, Cylinder Block Service, under Bypass Cover and Exhaust Cover Installation.

3-7 EXHAUST COVER

The exhaust covers are one of the most neglected items on any outboard engine. Seldom are they checked and serviced. Many times an engine may be overhauled and returned to service without the exhaust covers ever having been removed.

One reason the exhaust covers are not removed is because the attaching bolts usually become corroded in place. This means they are very difficult to remove, but the work should be done. Heat applied to the bolt head and around the exhaust cover will help in removal. However, some bolts may still be broken. If the bolt is broken it must be drilled out and the hole tapped with new threads.

The exhaust covers are installed over the exhaust ports to allow the exhaust to leave the powerhead and be transferred to the exhaust housing. If the cover was the only item over the exhaust ports, they would become so hot from the exhaust gases they might cause a fire or a person would be severely burned if they came in contact with the cover.

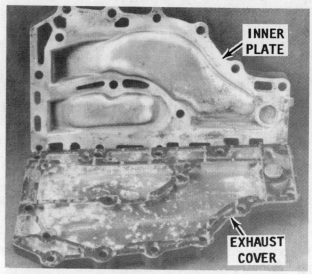

The inner exhaust plate and exhaust cover from a 40hp, 50hp, 55hp, or 60hp powerhead.

Therefore, an inner plate is installed to help dissipate the exhaust heat. Two gaskets are installed -- one on either side of the inner plate. Water is channeled to circulate between the exhaust cover and the inner plate. This circulating water cools the exhaust cover and prevents it from becoming a hazard.

On some early model outboards, the inner plate was constructed of aluminum. Unfortunately, the aluminum would corrode through, especially in a salt water enviroment, and then water could enter the lower cylinder and cause a powerhead failure. The accompanying illustration clearly shows an inner plate corroded

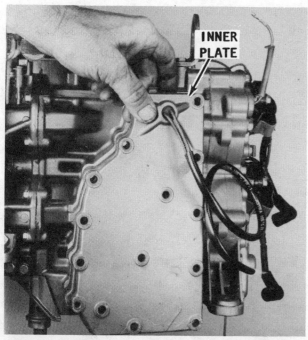

Removing the inner plate from a 40hp, 50hp, 55hp, or 60hp powerhead.

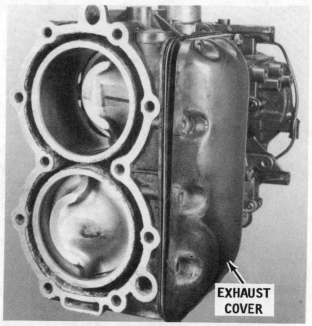

Removing the exhaust cover from an early powerhead.

through, allowing water to enter the lower cylinder. To correct this corrosion problem, the inner plate is now made of stainless steel material.

A thorough cleaning of the inner plate behind the exhaust covers should be performed during a major engine overhaul. If the integrity of the exhaust cover assembly is in doubt, replace the complete cover including the inner plate.

On powerheads equipped with the heat/-electric choke, a baffle is installed on the inside surface of the inner plate. This baffle is heated from the engine exhaust gases. Air passing through the baffle heats the choke and allows the choke to open as engine temperature rises.

CLEANING

Clean any gasket material from the cover and inner plate surfaces. Check to be sure the water passages in the cover and plate are clean to permit adequate passage of cooling water.

Inspect the inlet and outlet hole in the powerhead to be sure they are clean and free of corrosion. The openings in the powerhead may be cleaned with a small size screwdriver.

Clean the area around the exhaust ports and in the webs running up to the exhaust ports. Carbon has a habit of forming in this area.

The exhaust area of the powerhead, open for inspection and cleaning.

INSTALLATION

Procedures to install the exhaust covers will be found in Section 3-14, Cylinder Block Service, under Bypass Cover and Exhaust Cover Installation.

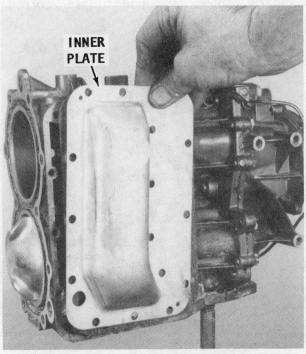

Removing the inner plate from an early powerhead.

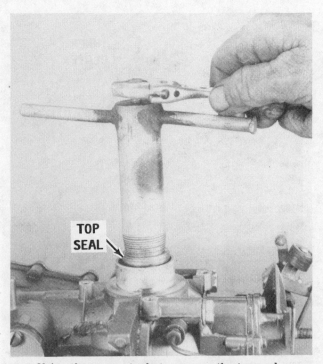

Using the proper tools to remove the top seal.

3-8 SEAL — TOP AND BOTTOM

The top seal maintains vacuum and pressure in the crankcase at the top cylinder.

REMOVAL

This seal can only be removed using one of two methods.

The first, is by using a special puller while the powerhead is still assembled. If the puller is used, thread the end of the puller into the seal. After the puller is secured to the seal, remove the seal from the powerhead by tightening the center screw on the puller. **DO NOT** attempt to use any other type of tool to remove this seal or the powerhead flanges will be damaged. If the flanges are damaged, the block must be replaced.

The second method is to remove the seal during powerhead disassembling. After the crankcase cover has been removed, the seal will be loose and can be easily removed by holding onto the bearing and prying the seal out.

INSTALLATION

To install the seal with the powerhead assembled, coat the outside diameter of the

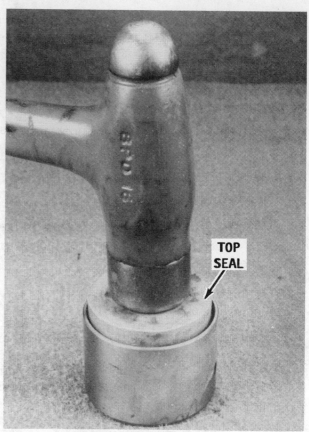

Pressing a new seal into a bearing.

seal with OMC Seal Compound. Use the special tool to tap the seal **EVENLY** into place around the crankshaft.

If the powerhead has been disassembled, a socket or other similar type tool of equal diameter as the seal may be used to tap it into the bearing. If the powerhead being

Using the proper tools to install the top seal.

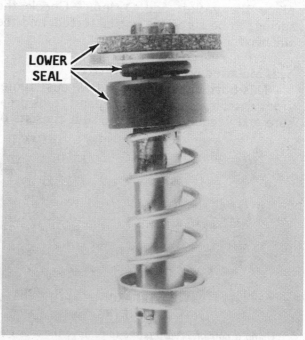

Powerhead lower seal installed on the driveshaft.

serviced does not have the seal in the bearing, lay the seal in the recess of the block. When the crankcase cover is installed, the seal will be in place.

BOTTOM SEAL

The bottom seal has equal importance as the top seal. This seal is installed to maintain vacuum and pressure in the lower half of the crankcase for the lower cylinder.

The bottom seal will vary, depending on the model engine being serviced. The following procedures and accompanying illustrations cover the most common Johnson/Evinrude bottom seal installed.

Seal Mounted on the Driveshaft

When the powerhead is removed, observe around the driveshaft at the lower end, and the seal will be visible. The seal consists of a gasket, plate, an O-ring, lower seal bearing, spring, washer, and a pin. The pin is installed through the driveshaft and holds the seal upward and in place.

As the powerhead is lowered down over the driveshaft during installation, the seal is held in place and will hold the vacuum and pressure created when the engine is operating.

Removal

With the powerhead assembled, it is a simple matter to reach into the exhaust housing and remove the seal and then replace the gasket and O-ring. Check the spring, to be sure it is not distorted, and the washer for damage.

Seal Mounted on the Crankshaft

This bottom seal prevents exhaust fumes from entering the crankcase, and holds pressure and vacuum inside. The seal consists of

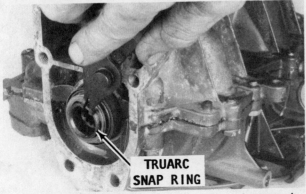

Using a pair of Truarc pliers to remove the snap ring from the crankshaft.

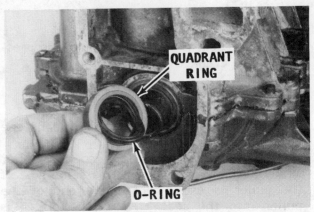

Removing the O-ring from the quadrant ring. These two items form the bottom crankshaft seal.

a quadrant ring, O-ring, retainer washer, spring, another washer, and a snap ring.

Removal

To remove this seal from the lower end of the crankshaft, use a pair of Tru-arc pliers and **CAREFULLY** remove the snap ring. **TAKE CARE** not to lose any of the parts due to the spring pressure against the snap ring. Notice how the quadrant O-ring fits inside the seal. This ring is also removable. Observe how the seal has a raised edge on one side. This raised edge **MUST** face upward when the seal is installed.

INSPECTION

Check to be sure the spring has good tension. Check to be sure the washers are not distorted. The quadrant ring should be **DISCARDED** and a new one installed.

Good shop practice dictates the quadrant seal be replaced each time the lower seal is serviced.

Check the groove in the lower end of the crankshaft where the truarc ring fits. If the groove is not clean, the ring will snap out and the lower sealing qualities will be lost. If the groove is badly corroded, the crankshaft must be replaced.

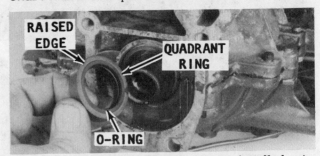

A new O-ring and quadrant ring are installed onto the crankshaft with the raised edge of the quadrant ring facing UPWARD when installed.

Remove the seal and O-ring from the cap of a 40hp, 50hp, 55hp, or 60hp powerhead.

Seal in a Cap

Many of the larger horsepower powerheads have the seal installed in a cap. The cap is bolted to the bottom of the powerhead. The four bolts securing the cap must be removed before the crankcase cover can be removed. The cap is then removed after the cover has been removed from the cylinder block. The seal can be punched out and new ones installed without difficulty.

3-9 CENTERING PINS

All Johnson/Evinrude outboard engines have at least one, and in most cases two, centering pins installed through the crankcase cover. These pins index into matching holes in the powerhead block when the crankcase cover is installed. These pins center the crankcase cover on the powerhead block.

Cylinder block with the two centering pins installed.

Removing a centering pin from the cylinder block.

The centering pins are tapered. The pins must be carefully checked to determine how they are to be removed from the cover. In most cases the pin is removed by using a center punch and tapping the pin towards the carburetor or intake manifold side of the crankcase.

When removing a centering pin, hold the punch securely onto the pin head, then strike the punch a good hard forceful blow. **DO NOT** keep beating on the end of the pin, because such action would round the pin head until it would not be possible to drive it out of the cover.

Centering pins are the first item to be installed in the cover when replacing the crankcase cover.

3-10 MAIN BEARING BOLTS AND CRANKCASE SIDE BOLTS

The main bearing bolts are installed through the crankcase cover into the powerhead block. Most engines have two bolts installed for the top main bearing, two for the center main bearing, and two for the lower main bearing.

Removing the main bearing bolts from the powerhead.

In many cases the upper and lower main bearing bolts are **DIFFERENT** lengths. Therefore, take time to tag and identify the bolts to ensure they will be installed in the same location from which they were removed.

The crankcase side bolts are installed along the edge of the crankcase cover to secure the cover to the cylinder block. These bolts usually have a 7/16" head and all must be removed before the crankcase cover can be removed. Remove the crankcase side bolts.

Remove the main bearing bolts. Two bolts installed in the center are behind the reeds. Normally these two are not actually bolts, but Allen head screws. All six main bearing bolts must be removed before the crankcase cover can be removed.

INSTALLATION

Main bearing bolt and the crankcase side bolt installation is given in Section 3-14, Cylinder Block Assembling, under Main Bearing and Crankcase Side Bolt Installation.

3-11 CRANKCASE COVER

REMOVAL

After all side bolts and main bearing bolts have been removed, use a soft-headed mallet and tap on the bottom side of the crankshaft. A soft, hollow sound should be heard indicating the cover has broken loose

Removing the crankcase cover from a 40hp, 50hp, 55hp, or 60hp powerhead.

from the crankcase. If this sound is not heard, check to be sure all the side bolts and main bearing bolts have been removed. **NEVER** pry between the cover and the crankcase or the cover will surely be distorted. If the cover is distorted, it will fail to make a proper seal when it is installed.

Once the crankshaft has been tapped, as described, and the proper sound heard, the cover will be jarred loose and may be removed.

CLEANING AND INSPECTING

Wash the cover with solvent, and then dry it thoroughly. Check the mating surface to the cylinder block for damage that may affect the seal.

Cylinder block after the crankcase cover has been removed.

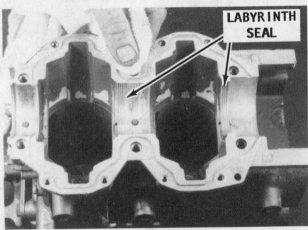
Cleaning the powerhead surface of a 40hp, 50hp, 55hp, or 60hp powerhead. Notice the labyrinth seal at the center and bottom main bearings.

Crankcase cover with the labyrinth seal area clearly visible.

Inspect the labyrinth seal grooves at the center main bearing area to be sure they are clean and not damaged in any manner.

INSTALLATION

Installation procedures for the crankcase cover are given in Section 3-14, Cylinder Block Service, under Crankcase Cover Installation.

3-12 CONNECTING RODS AND PISTONS

The connecting rods and their rod caps are a **MATCHED set.** They absolutely **MUST** be identified, kept, and installed as a set. Under no circumstances should the connecting rod and caps be interchanged. Therefore, on a multiple piston engine, **TAKE TIME AND CARE** to tag each rod and rod cap; to keep them together as a set while they are on the bench; and to install them into the same cylinder from which they were removed as a set.

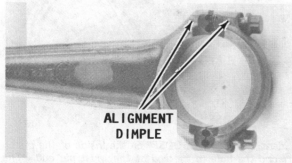

Rod and rod cap with the two alignment dimples shown.

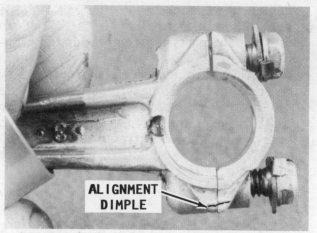

Rod and rod cap with the alignment line marks shown.

The connecting rod and its cap on 15 hp to early 40 hp engines are manufactured as a set —as a single unit. After the complete rod and cap have been made, two holes are drilled through the side of the cap and rod, and the cap is then fractured from the rod. Therefore, the cap must always be installed with its original rod. The cap half of the break can **ONLY** be matched with the other half of the break on the **ORIGINAL** rod.

The rods and caps on the smaller horsepower engines are made of aluminum with babbitt inserts. These rods and caps are manufactured as two separate items.

Inspect the rod and the rod cap before removing the cap from the crankshaft. Under normal conditions, a line or a dot is

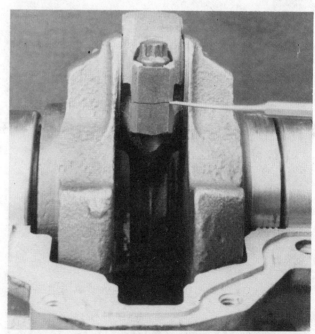

A punch points to the fractured break of a rod and its cap. The rod and cap must be matched during installation.

visible on the top side of the rod and the cap. This identification is an assist to assemble the parts together and in the proper location.

Observe into the block and notice how the rods have a "trough". Also notice the hole in the rod near where the wrist pin passes through the piston. On many rods there is also a hole in the rod at the crank end. These two holes **MUST ALWAYS** face upward during installation.

REMOVAL

To remove the rod bolts from the cap, it is recommended to loosen each bolt just a little at-a-time and alternately. This procedure will prevent one bolt from being completely removed while the other is still tightened to its recommended torque value. Such action may very likely warp the cap.

Remove the bolts as described in the previous paragraph, and then **CAREFULLY** remove the rod cap to prevent loosing the needle bearings installed under the cap, if used.

Remove the needle bearings and cages, if used, from around the crankshaft. Count the needle bearings and insert them into a separate container -- one container for each rod, with the container clearly identified to ensure they will be installed with the proper rod at the crankshaft journal from which they were removed.

Removing the bolts from the rod cap.

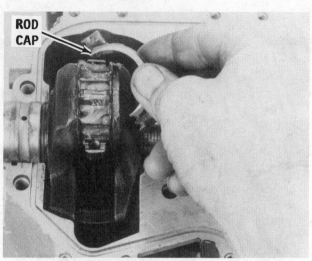

Removing the rod cap from the rod.

Tap the piston out of the cylinder from the crankshaft side. Immediately attach the proper rod cap to the rod and hold it in place with the rod bolts. The few minutes involved in securing the cap with the rod will ensure the matched cap remains with its mating rod during the cleaning and assembling work.

Identify the rod to ensure it will be installed into the cylinder from which it was removed.

Remove and identify the other rod caps, needle bearings and cages, and rods with pistons, in the same manner.

DISASSEMBLING

Before separating the piston from the rod, notice the location of the piston in relation to the rod. Observe the hole in the rod trough on one side of the rod near the wrist pin opening and another at the lower end. These holes must face toward the **TOP** of the engine during installation.

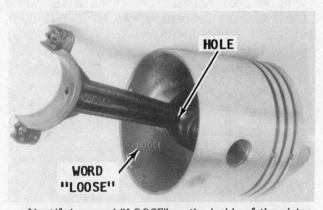

Identifying word "LOOSE" on the inside of the piston skirt and the hole in the rod at the wrist pin end. The wrist pin must be driven from the loose side of the piston out the tight side, as described in the text.

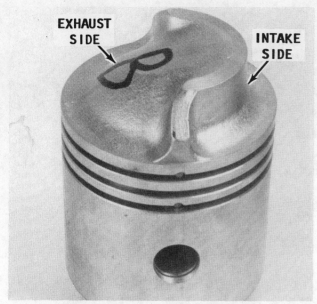

Close view of a piston with the slanted edge and sharp edges identified. The piston can only be installed one way for proper operation.

Observe the slanted edge and the sharp edge of the dome-type piston. The slanted edge **MUST** face toward the exhaust side of the cylinder and the sharp edge toward the intake side during installation.

Pistons installed in the block of a 40hp, 50hp, 55hp or 60hp powerhead. The word "UP" embossed on the piston must be at the top of the cylinder, as shown.

Pistons installed in the powerhead of a 40hp, 50hp, 55hp, or 60hp unit. The word "UP" embossed on the piston must be at the top of the cylinder, as shown.

If servicing a 40 hp, 50 hp, 55 hp, or 60 hp powerhead, carefully observe the hole in the rod near the wrist pin and the relationship of the irregular cutouts in the piston skirt. Only in this position will this relationship exist. The rod and piston **MUST** be assembled in this manner or the engine will run **VERY** poorly.

When the rod is installed to the piston, the relationship of the rod can only be one way. The rod holes must face upward and the piston must face as described in the previous paragraph.

Observe into the piston skirt. On most model pistons, notice the **"L"** stamped on the boss through which the wrist pin passes. The letter mark identifies the "loose" side

Piston with the word "LOOSE" embossed on the inside surface of the skirt.

*A broken rod, possibly caused by inadequate oil delivery or the powerhead operated in a **RUNAWAY** condition (excessive rpm under a "No Load" condition).*

of the piston and indicates side of the piston from which the wrist pin must be driven out without damaging the piston. Some pistons may have the full word "LOOSE" stamped on the inside of the piston skirt.

If the piston does not have the "L" or the word "LOOSE" stamped, the wrist pin may be driven out in either direction.

It may be necessary to heat the piston in a container of boiling water in order to press the wrist pin free.

Remove the retaining clips from each end of the wrist pin. Some clips are spring wire type and may be worked free of the piston using a screwdriver. Other model pistons have a truarc snap ring. This type of ring can only be successfully removed using a pair of truarc pliers.

Place the piston in an arbor press using the **PROPER** size cradle for the piston being serviced, and with the **LOOSE** side of the piston facing **UPWARD**.

The wrist pin must be driven out **FROM** the loose side. This may not seem reasonable, but there is a very simple explanation.

Heating a piston in hot water to expand the metal slightly.

By placing the piston in the arbor press cradle with the tight side down, and the arbor ram pushing from the loose side, the piston has good suport and will not be distorted. If the piston is placed in the arbor press with the loose side down, the piston would be distorted and unfit for further service.

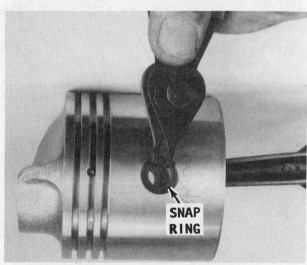

Removing the wrist pin Truarc snap ring from the piston.

SNAP RING

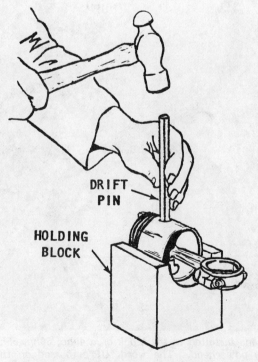

DRIFT PIN

HOLDING BLOCK

Removing the wrist pin using a holding block.

Many rods have a wrist pin bearing. Some are caged bearings and other are not. **TAKE CARE** not to lose any of the bearings when the wrist pin is driven free of the piston.

Alternate Removal Method

If the piston does not have the **"L"** or the word **"LOOSE"** stamped, the wrist pin may be driven out in either direction.

If an arbor press or cradle is not available, proceed as follows: Heat the piston in a container of very hot water for about ten minutes. Heating the piston will cause the metal to expand ever so slightly, but ease the task of driving the pin out. Assume a sitting position in a chair, on a box, whatever. Next, lay a couple towels over your legs. Hold your legs tightly together to form a cradle for the piston above your knees. Set the piston between your legs with the **"LOOSE"** side of the piston facing upward. Now, drive the wrist pin free using a drift pin with a shoulder. The drift pin will fit into the hole through the wrist pin and the shoulder will ride on the edge of the wrist pin. Use sharp hard blows with a

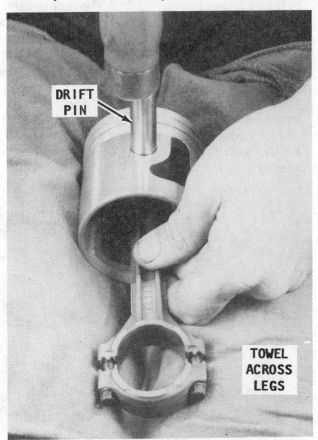

Removing the wrist pin using a drift pin. The piston can be supported between your legs as described in the text.

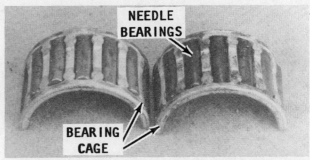

Needle bearings and cages unfit for further service.

hammer. Your legs will absorb the shock without damaging the piston. If this method is used on a regular basis during the busy season, your legs will develop black-and-blue areas, but no problem, the marks will disappear in a few days.

ROD INSPECTION AND SERVICE

If the rod has needle bearings, the needles should be replaced anytime a major overhaul is performed. It is not necessary to replace the cages, but a complete **NEW** set of needles should be purchased and installed.

Place each connecting rod on a surface plate and check the alignment. If light can be seen under any portion of the machined surfaces, or if the rod has a slight wobble on the plate, or if a 0.002" feeler gauge can be inserted between the machined surface and the surface plate, the rod is bent and unfit for further service.

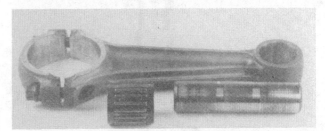

Rod, rod cap, wrist pin, and wrist pin bearing, after removal.

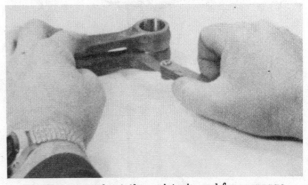

Testing two rods at the wrist pin end for warpage.

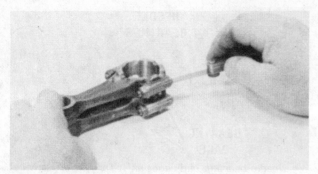

Testing two rods at the rod cap end for warpage.

Badly rusted and corroded crankshaft from a submerged engine. This crankshaft is no longer fit for service.

Inspect the connecting rod bearings for rust or signs of bearing failure. **NEVER** intermix new and used bearings. If even one bearing in a set needs to be replaced, all bearings at that location **MUST** be replaced.

Inspect the bearing surface of the rod and the rod cap for rust and pitting.

Inspect the bearing surface of the rod and the rod cap for water marks. Water marks are caused by the bearing surface being subjected to water contamination, which causes "etching". The "etching" will worsen **VERY** rapidly.

Inspect the bearing surface of the rod and rod cap for signs of spalling. Spalling is the loss of bearing surface, and resembles

flaking or chipping. The spalling condition will be most evident on the thrust portion of the connecting rod in line with the I-beam. Bearing surface damage is usually caused by improper lubrication.

Check the bearing surface of the rod and rod cap for signs of chatter marks. This condition is identified by a rough bearing surface resembling a tiny washboard. The condition is caused by a combination of low-speed low-load operation in cold water, and is aggravated by inadequate lubrication and improper fuel. Under these conditions, the crankshaft journal is hammered by the connecting rod. As ignition occurs in the cylinder, the piston pushes the connecting rod with tremendous force, and this force is transferred to the connecting rod journal.

Since there is little or no load on the crankshaft, it bounces away from the connecting rod. The crankshaft then remains immobile for a split second, until the piston travel causes the connecting rod to catch up to the waiting crankshaft journal, then hammers it.

A crankshaft cleaned and ready for installation.

*Installing the rod cap onto the rod in preparation for cleaning the inside surface. The cap **MUST** always be kept with its matching rod.*

In some instances, the connecting rod crankpin bore becomes highly polished.

While the engine is running, a "whirr" and/or "chirp" sound may be heard when the engine is accelerated rapidly from idle speed to about 1500 rpm, then quickly returned to idle. If chatter marks are discovered, the crankshaft and the connecting rods should be replaced.

Inspect the bearing surface of the rod and rod cap for signs of uneven wear and possible overheating. Uneven wear is usually caused by a bent connecting rod. Overheating is identified as a bluish bearing surface color and is caused by inadequate lubrication or operating the engine at excessive high rpm.

Inspect the needle bearings, if installed. A bluish color indicates the bearing became very hot and the complete set for the rod **MUST** be replaced, no question.

Service the connecting rod bearing surfaces according to the following procedures and precautions:

a- Align the etched marks on the knob side of the connecting rod with the etched marks on the connecting rod cap.

b- Tighten the connecting rod cap attaching bolts securely.

c- Use **ONLY** crocus cloth to clean bearing surface at the crankshaft end of the connecting rod. **NEVER** use any other type of abrasive cloth.

d- Insert the crocus cloth in a slotted 3/8" diameter shaft. Chuck the shaft in a drill press and operate the press at high speed and at the same time, keep the connecting rod at a 90° angle to the slotted shaft.

e- Clean the connecting rod **ONLY** enough to remove marks. **DO NOT** continue once the marks have disappeared.

f- Clean the piston pin end of the connecting rod using the method described in Steps d and e, but using 320 grit Carborundum cloth instead of crocus cloth.

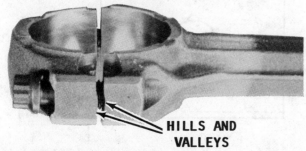

HILLS AND VALLEYS

Rod cap separated slightly from its matching rod. Notice the matching hills and valleys.

WRIST PIN

Testing the wrist pin end of the rod prior to installation.

g- Thoroughly wash the connecting rods to remove abrasive grit. After washing, check the bearing surfaces a second time.

h- If the connecting rod cannot be cleaned properly, it should be replaced.

i- Lubricate the bearing surfaces of the connecting rods with light-weight oil to prevent corrosion.

PISTON AND RING INSPECTION AND SERVICE

Inspect each piston for evidence of scoring, cracks, metal damage, cracked piston pin boss, or worn pin boss. Be especially critical during inspection if the engine has been submerged.

Carefully check each wrist pin to be sure it is not the least bit bent. If a wrist pin is bent, the pin and piston **MUST** be replaced as a set, because the pin will have damaged the boss when it was removed.

SCORED AREA

Piston badly scored and no longer fit for service.

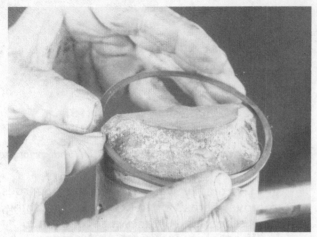

Removing the rings from the piston.

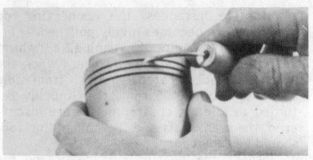

Cleaning the piston ring grooves. An automotive type ring groove cleaner should NEVER be used.

Check the wrist pin bearings. If the bearing is the pressed-in type, use your finger and determine the bearing is in good condition with no indication of binding or "rough" spots. If the wrist pin bearing is the removable type, the needle should be replaced.

Grasp each end of the ring with either a ring expander or your thumbnails, open the ring and remove it from the piston. Many times, the ring may be difficult to remove because it is "frozen" in the piston ring groove. In such a case, use a screwdriver and pry the ring free. The ring may break, but if it is difficult to remove, it **MUST** be replaced.

OBSERVE the pin in each ring groove of the piston. The ends of the ring **MUST** straddle this pin. The pin prevents the ring from rotating while the engine is operating. This fact is the direct opposite of a four-cycle engine where the ring must rotate. In a two-cycle engine, if the ring is permitted to rotate, at one point, the opening between the ring ends would align with either the intake or exhaust port in the cylinder. At that time, the ring would expand very slightly, catch on the edge of the port, and **BREAK**.

Therefore, when checking the condition of the piston, **ALWAYS** check the pin in each groove to be sure it is tight. If one pin is the least bit loose, the piston **MUST** be replaced, without question. Never attempt to replace the pin, it is **NEVER** successful.

Check the piston ring grooves for wear, burns, distortion or loose locating pins. During an overhaul, the rings should be replaced to ensure lasting repair and proper engine performance after the work has been completed.

Clean the piston dome, ring grooves and the piston skirt. Clean the piston skirt with a crocus cloth.

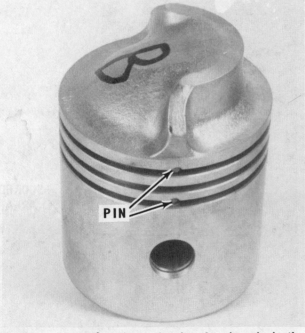

PIN

Close view of a piston showing the ring pin in the groove.

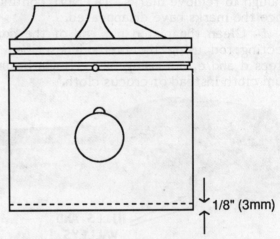

1/8" (3mm)

Measure the diameter of the piston at 1/8" (3mm) above the bottom edge.

Clean carbon deposits from the top of the piston using a soft wire brush, carbon removal solution, or by sand blasting. If a wire brush is used, **TAKE CARE** not to burr or round machined edges.

Wear a pair of good gloves for protection against sharp edges, and clean the piston ring grooves using the recessed end of the proper broken ring as a tool. **NEVER** use a rectangular ring to clean the groove for a tapered ring, or use a tapered ring to clean the groove for a rectangular ring.

NEVER use an automotive-type ring groove cleaner to clean piston ring grooves, because this type of tool could loosen the piston ring locating pins. **TAKE CARE** not to burr or round the machined edges. Inspect the piston ring locating pins to be sure they are tight. There is one locating pin in each ring groove. If one locating pin is loose, the piston must be replaced. Never attempt to replace the pin, it is **NEVER** successful.

Oversize Pistons and Rings

Scored cylinder blocks can be saved for further service by reboring and installing oversize pistons and piston rings. **ONE MORE WORD:** Oversize pistons and rings are not available for all engines. At the time of this printing, the sizes listed in the Appendix were available. Check with the parts department at your local dealer for the model engine you are servicing, and to be sure the factory has not deleted a size from their stock.

ASSEMBLING

CRITICAL WORDS

Two conditions absolutely **MUST** exist when the piston and rod assembly are installed into the cylinder block.

The slanted side of the piston must face toward the exhaust side of the cylinder.

*An automotive ring compressor should **NEVER** be used to install the rings for a two-cycle engine.*

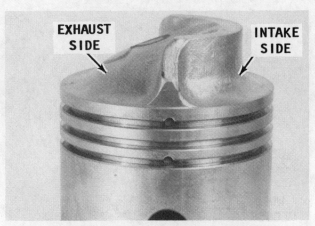

*The slanted side of the piston **MUST** face the exhaust port and the sharp edge face the intake port.*

The hole in the rod near the wrist pin opening and at the lower end of the rod must face **UPWARD**.

Therefore, the rod and piston **MUST** be assembled correctly in order for the assembly to be properly installed into the cylinder. Soak the piston in a container of very hot water for about ten minutes. Heating the piston will cause it to expand ever so slightly, but enough to allow the wrist pin to be pressed through without difficulty.

Before pressing the wrist pin into place, hold the piston and rod near the cylinder block and check to be sure both will be facing in the right direction when they are installed.

Pack the wrist pin needle bearing cage with needle bearing grease, or a good grade

Piston with the wrist pin ready for installation. Notice the ring groove pin. For proper operation and lubrication, the rod must be installed to the piston and the piston into the cylinder as described in the text.

Wrist pin entering the piston from the "LOOSE" side.

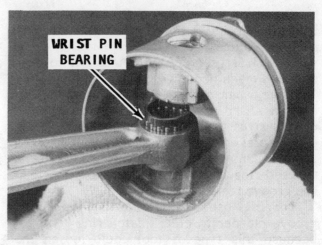

Installing the rod and wrist pin bearing into the piston.

of petroleum jelly. Load the bearing cage with needles and insert it into the end of the rod.

Slide the rod into the piston boss and check a second time to be sure the slanted side of the piston is facing toward the exhaust side of the cylinder and the hole in the rod is facing upward.

Place the piston and rod in the arbor press with the "LOOSE" or stamped "L" side of the piston facing UPWARD. Press the wrist pin through the piston and rod. Continue to press the wrist pin through until the groove in the wrist pin for the lock ring is visible on

both ends of the pin. Remove the assembly from the arbor press. Install the retaining ring onto each end of the wrist pin. Some models have a wire ring, and others have a truarc ring. Use a pair of truarc pliers to install the truarc ring.

Fill the piston skirt with a rag, towel, shop cloths, or other suitable material. The rag will prevent the rod from coming in contact with the piston skirt while it is laying on the bench. If the rod is allowed to strike the piston skirt, the skirt may become distorted.

Assemble the other pistons, rods, and wrist pins in the same manner. Fill the skirt with rags as protection until the assembly is installed.

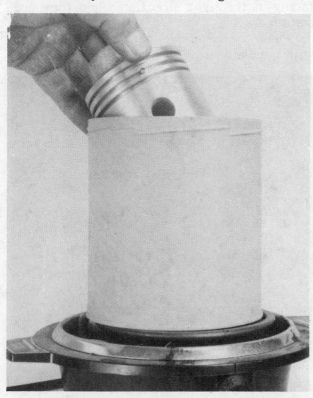

Heating the piston in hot water to expand the metal slightly as an assist to installing the wrist pin.

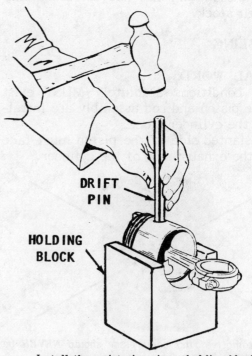

Install the wrist pin using a holding block.

Alternate Assembling Method

If an arbor press is not available, the piston may be assembled to the rod in much the same manner as described for disassembling.

First, soak the piston in a container of very hot water for about ten minutes.

Before pressing the wrist pin into place, hold the piston and rod near the cylinder block and check to be sure both will be facing in the right direction when they are installed.

Pack the wrist pin needle bearing cage with needle bearing grease, or a good grade of petroleum jelly. Load the bearing cage with needles and insert it into the end of the rod.

Slide the rod into the piston boss and check a second time to be sure the slanted side of the piston is facing toward the exhaust side of the cylinder and the hole in the rod is facing upward.

Now, assume a sitting position and lay a couple towels over your lap. Hold your legs tightly together to form a cradle for the piston above your knees. Set the piston between your legs with the **LOOSE** side of the piston facing upward. Now, drive the wrist pin through the piston using a drift pin with a shoulder. The drift pin will fit into the hole through the wrist pin and the shoulder will ride on the end of the wrist pin. Use sharp hard blows with a hammer. Your legs will absorb the shock without damaging the piston. If this method is used on a regular basis during the busy season, your legs will develop black-and-blue areas, but no problem, the marks will disappear in a few days.

Continue to drive the wrist pin through the piston until the groove in the wrist pin for the lockring is visible at both ends. Install the retaining spring wire or truarc ring onto each end of the wrist pin.

Fill the piston skirt with a rag, towel, shop cloths, or other suitable material. The rag will prevent the rod from coming in contact with the piston skirt while it is laying on the bench. If the rod is allowed to strike the piston skirt, the skirt may become distorted.

Assemble the other pistons, rods, and wrist pins in the same manner. Fill the skirt with rags as protection until the assembly is installed.

INSTALLATION

Piston and rod assembly installation procedures will be found in Section 3-14, Cylinder Block Service under Piston Installation.

3-13 CRANKSHAFT

REMOVAL

Lift the crankshaft assembly from the block. On some models, especially the

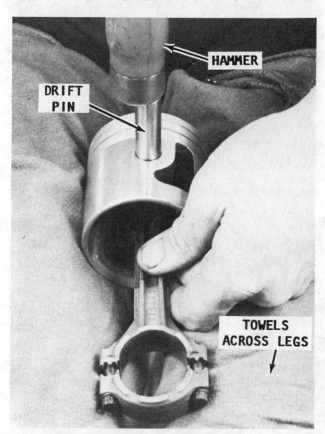

Installing the wrist pin without an arbor press. The piston can be held in your lap as described in the text.

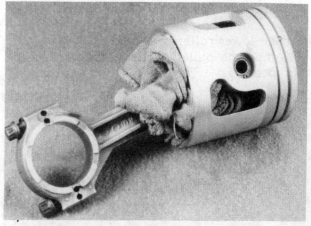

The inside of the piston should be filled with rags or shop cloths to protect the skirt from being struck by the rod while the piston is out of the cylinder.

Crankshaft with the upper, center, and lower main bearings ready to be removed.

larger horsepower engines, it may be necessary to use a soft-headed mallet and tap on the bottom side of the crankshaft to jar it loose. As the crankshaft is lifted, **TAKE CARE** to work the center main bearing loose. This center bearing is a split bearing held together with a snap wire ring. On some models, the bottom half of the bearing may be stuck in the cylinder block. Therefore, the crankshaft and the center main bearing must be worked free of the block together.

If servicing a 15 hp to 40 hp powerhead (40 hp thru 1983), observe how the center main bearing, and the top and bottom main bearings all have a hole in the outside circumference. Notice the locating pins in the cylinder block. The purpose of this arrangement is to prevent the bearing shell from rotating. During assembling, the holes in the bearings **MUST** index with the pins in the block. Also notice the grooves in the block on one side of the center main bearing. Observe the grooves in the crankcase cover. This arrangement of grooves forms what is commonly known as a "labyrinth" seal. The grooves fill with oil and/or fuel creating a seal between the cylinders.

The 40 hp since 1983, and the 50 hp, 55 hp, and 60 hp powerheads have a pressed in place lower roller bearing. A clamp-type puller is required to removed this bearing. The bearing need not be

Removing the crankshaft from the block of a 40hp, 50hp, 55hp or 60hp powerhead.

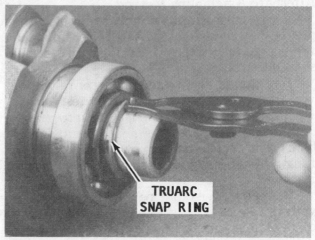

Removing the Truarc snap ring from the lower bearing of a 40hp, 50hp, 55hp or 60hp powerhead.

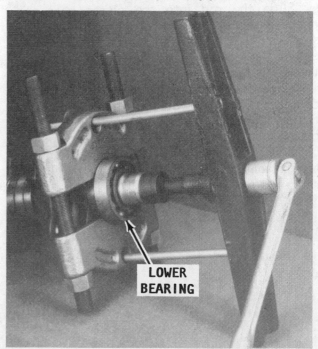

Using a puller to remove the lower bearing of a 40hp, 50hp, 55hp or 60hp unit.

Using a sleeve-type tool over the end of the drive-shaft to install the bearing.

removed for cleaning and inspection. Remove this bearing **ONLY** if the determination has been made that it is unfit for further service.

To install a new bearing, place the bearing onto the shaft and press it into place using an arbor press. If an arbor press is not available a socket large enough to fit over the crankshaft could be used to drive the bearing into place.

On the smaller horsepower engines, babbitt bearings are used for the center main with needle bearings installed for the upper and lower main bearings.

CLEANING AND INSPECTION

Inspect the splines for signs of abnormal wear. Check the crankshaft for straightness. Inspect the crankshaft oil seal surfaces to be sure they are not grooved, pitted or scratched. Replace the crankshaft if it is severely damaged or worn. Check all crankshaft bearing surfaces for rust, water marks, chatter marks, uneven wear or overheating. Clean the crankshaft surfaces with crocus cloth.

Clean the crankshaft and crankshaft bearing with solvent. Dry the parts, but **NOT** the bearing, with compressed air. Check the crankshaft surfaces a second time. Replace the crankshaft if the surfaces cannot be cleaned properly for satisfactory service. If the crankshaft is to be installed for service, lubricate the surfaces with light oil.

The top and lower bearing may be easily removed from the crankshaft. The center

Badly rusted and corroded crankshaft from a submerged engine. This crankshaft is no longer fit for service.

main bearing has a spring steel wire securing the two halves together. Remove the wire, and then the outer sleeve, then the needle bearings. **TAKE CARE** not to lose any of the needles. The outer shell is a fractured break type unit. Therefore, the two halves of the shell **MUST** absolutely be kept as a set.

Check the crankshaft bearing surfaces to be sure they are not pitted or show any signs of rust or corrosion. If the bearing surfaces are pitted or rusted, the crankshaft and bearings must be replaced.

During an engine overhaul to this degree, it is a good practice to remove the seal from the top main bearing. If the same type of seal is used in the bottom main bearing, remove that seal also.

Crankcase cover with the labyrinth seal area clearly visible.

A crankshaft cleaned and ready for installation.

A crankshaft with a badly corroded "throw". This crankshaft is unfit for further service.

ASSEMBLING

Insert the proper number of needle bearings into the center main bearing cage. Install the outer sleeve over the bearing cage. Check to be sure the two halves of the outer sleeve are matched. Again, these two halves are manufactured as a single unit and then broken. Therefore, the hills and valleys of the break absolutely **MUST** match during installation.

Snap the retaining ring into place around the bearing. Slide the upper bearing onto the crankshaft journal at the upper end and the lower bearing onto the lower end. Rotate the installed bearings to be sure there is no evidence of binding or rough spots. The crankshaft is now ready for installation.

INSTALLATION

Installation procedures are given in Section 3-14, Cylinder Block Service, under Crankshaft Installation.

3-14 CYLINDER BLOCK SERVICE

Inspect the cylinder block and cylinder bores for cracks or other damage. Remove

The needles and cage of the center main bearing ready for the outside shell to be installed.

Crankshaft with the upper, center, and lower main bearings installed. Check to be sure the snap ring on the center main bearing is installed.

carbon with a fine wire brush on a shaft attached to an electric drill or use a carbon remover solution.

Use an inside micrometer or telescopic gauge and micrometer to check the cylinders for wear. Check the bore for out-of-round and/or oversize bore. If the bore is tapered, out-of-round or worn more than 0.003" - 0.004" (0.076 mm - 0.102 mm) the cylinders should be rebored and oversize pistons and rings installed.

GOOD WORDS:

Oversize piston weight is approximately the same as a standard size piston. Therefore, it is **NOT** necessary to rebore all cylinders in a block just because one cylinder requires reboring. The APBA (American Power Boat Association) accepts and permits the use of 0.015" (0.381 mm) oversize pistons.

Hone the cylinder walls lightly to seat the new piston rings, as outlined in the Honing Procedures Section in this chapter. If the cylinders have been scored, but are not out-of-round or the sleeve is rough, clean the surface of the cylinder with a cylinder hone as described in Honing Procedures, next section.

SPECIAL WORD

Cylinder sleeves may be installed on some models, but the cost is very high.

HONING PROCEDURES

To ensure satisfactory engine performance and long life following the overhaul work, the honing work should be performed with patience, skill, and in the following sequence:

a- Follow the hone manufacturer's recommendations for use of the hone and for cleaning and lubricating during the honing operation.

b- Pump a continuous flow of honing oil into the work area. If pumping is not practical, use an oil can. Apply the oil generously and frequently on both the stones and work surface.

c- Begin the stroking at the smallest diameter. Maintain a firm stone pressure against the cylinder wall to assure fast stock removal and accurate results.

d- Expand the stones as necessary to compensate for stock removal and stone wear. The best cross-hatch pattern is obtained using a stroke rate of 30 complete cycles per minute. Again, use the honing oil generously.

e- Hone the cylinder walls **ONLY** enough to de-glaze the walls.

f- After the honing operation has been completed, clean the cylinder bores with hot water and detergent. Scrub the walls with a stiff bristle brush and rinse thoroughly with hot water. The cylinders **MUST** be cleaned well as a prevention against any abrasive material remaining in the cylinder bore. Such material will cause rapid wear of new piston rings, the cylinder bore, and the bearings.

g- After cleaning, swab the bores several times with engine oil and a clean cloth, and then wipe them dry with a clean cloth. **NEVER** use kerosene or gasoline to clean the cylinders.

Checking the ring gap clearance by inserting the ring in the cylinder, as described in the text.

h- Clean the remainder of the cylinder block to remove any excess material spread during the honing operation.

WORDS OF ADVICE

If new rings are to be installed, each ring from the package **MUST** be checked in the cylinder. Errors happen. Men and machines can make mistakes. The wrong size ring can be included in a package with the proper part number.

Therefore, check **EACH** ring, one at-a-time as follows: Turn the ring sideways and lower it a couple inches into the cylinder bore. Now, turn the ring horizontal in the cylinder. It is now in its normal operating position, but without the piston. Next, use a feeler gauge and measure the distance (the gap) between the ends of the ring. The maximum and minimum allowable ring gap is listed in the Specifications in the Appendix.

Turn the piston upside down and slide it in and out of the cylinder. The piston should slide without any evidence of binding.

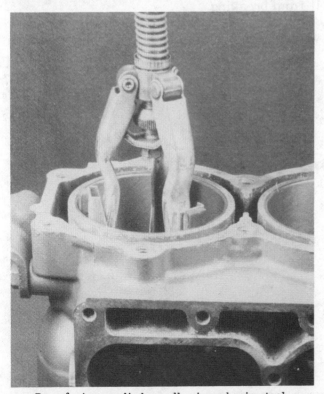

Resurfacing a cylinder wall using a honing tool.

Using a feeler gauge to check the ring end gap with the ring in the cylinder.

ASSEMBLING

SPECIAL WORD

The cylinder block assembling work should proceed quickly and without interruptions. If the work is partially completed and then left for any period of time, sealant may become hard, parts may be moved and their identity for a particular cylinder lost, or an important step may be bypassed, overlooked, or forgotten.

The following procedures pickup the work of assembling the cylinder block AFTER the various parts have been serviced and assembled. Procedures for each area are found in this chapter under separate headings.

PISTON AND ROD ASSEMBLY INSTALLATION

Several different methods are possible to install the piston and rod assembly into the cylinder. The following procedures are outlined for the do-it-yourselfer, working at home without the advantage of special tools.

First, purchase a special hose clamp with a strip of metal inside the clamp, as shown in the accompanying illustration. This piece of metal on the inside allows the outside portion of the clamp to slide on the inner strip without causing the ring to rotate.

Actually, to our knowledge a Mercruiser dealer is the only place such a clamp may be purchased. At the Mercruiser marine dealer, ask for an exhaust bellows hose clamp. The design of this hose clamp prevents the clamp and the piston ring from turning as the clamp is tightened. DO NOT attempt to use an ordinary hose clamp from an automotive parts house because such a clamp will cause the piston ring to rotate as the clamp is tightened. The ring MUST NOT rotate, because the ring ends must remain on either side of the dowel pin in the ring groove.

Next, coat the inside surface of the cylinder with a film of light-weight oil. Coat the exterior surface of the piston with the oil.

TAKE TIME

Take just a minute to notice how the piston rings are manufactured. Each end of the ring has a small cutout on the inside circumference. Now, visualize the ring installed in the piston groove. The ring ends must straddle the pin installed in each piston groove. As the ring is tightened around the piston, the ends will begin to come together. When the piston is installed into the cylinder bore, the two ends of the ring will come together and the cutout edge will be up against the pin. For this reason, CARE must be exercised when installing the rings onto the piston and when the piston is installed into the cylinder.

Install only the bottom ring into the bottom piston groove. Do not expand the ring any further than necessary, to prevent it from breaking.

Install the ring into the piston groove with the ends of the ring straddling the pin

Proper hose clamp to install the rings if a ring compressor is not available.

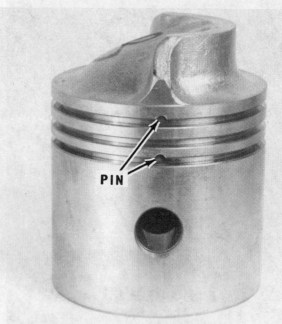

Piston ring groove pins. The ends of the ring must straddle the pin.

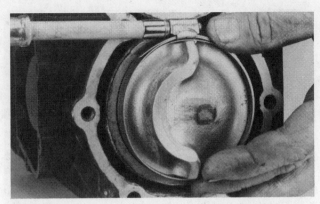

Installing the hose clamp over the ring prior to moving the piston further into the cylinder.

Tapping a piston into the cylinder of a 40hp, 50hp, 55hp, or 60hp powerhead. Notice the proper type hose clamp used to compress the rings, as explained in the text.

in the groove. The ring ends **MUST** straddle the pin to prevent the ring from rotating during engine operation. In a two-cycle engine, if the ring is permitted to rotate, at one point the opening between the ring ends would align with either the intake or exhaust port in the cylinder, the ring would expand very slightly, catch on the edge of the port, and **BREAK**.

CAREFULLY insert the rod and the piston skirt down into the cylinder.

GOOD WORDS

The following four areas must be checked at this point in the assembling work.

a- The piston and rod are being installed into the same cylinder from which they were removed.

b- The hole in the rod is facing **UPWARD**.

c- The slanted side of the piston is **TOWARD** the exhaust side of the cylinder.

d- The ends of each piston ring **MUST** straddle the pin in the piston groove.

Push the piston into the cylinder until the bottom ring, just installed, is about an inch from the surface of the cylinder block.

Install the hose clamp over the piston and bottom ring. Tighten the hose clamp with one hand and at the same time rotate the clamp back-and-forth slightly with the other hand. This "rocking" motion of the clamp as it is tightened will convince you the ring ends are properly positioned on either side of the pin. Continue to tighten the clamp, and "rocking" the clamp until the clamp is against the piston skirt. At this point, the ring ends will be together and the cutout on each ring end will be against the pin.

Tap the piston with the end of a wooden tool handle until the ring enters the cylinder. Remove the hose clamp.

Install the remaining rings in the same manner, one at-a-time, making sure the ends of each ring straddle the pin in the piston groove.

Notice how the ring pins are staggered from one groove to the next, by 180°.

After the last ring has been installed and the clamp removed, tap the piston into the bore until the crown is about even with the cylinder block surface.

Tapping the piston into the cylinder with a soft-headed mallet.

Using a ring installer to expand the ring during installation into the piston ring groove.

When installing the piston into a small horsepower powerhead, it is possible to compress the ring with the fingers of each hand, and then to push the piston into the cylinder with your thumbs.

Install the other pistons in exactly the same manner.

Turn the cylinder block upside down with the top of the block to your **LEFT.** Remove the bolts and rod caps from each rod. Set each rod cap in a definite position to ensure each will be installed onto the rod from which it was removed.

Checking the flexibility of the rings through intake port.

Checking the flexibility of the rings through the exhaust port.

Both pistons installed in the powerhead. The slanted edge of each piston is facing toward the exhaust port.

Both pistons installed into a 40hp, 50hp, 55hp or 60hp powerhead. The embossed word "UP" on the piston must be at the top of the cylinder, as shown.

Checking the ring tension with a small screwdriver through the exhaust port of a 40hp, 50hp, 55hp, or 60hp powerhead.

CRANKSHAFT INSTALLATION NEEDLE MAIN AND ROD BEARINGS

The following procedures outline steps to install a crankshaft with needle upper, center, and lower main bearings. The upper and lower mains are complete bearings and cannot be disassembled. The center bearing is caged. Installation procedures for small horsepower crankshafts with babbitt upper, lower, and center main bearings and with babbitt rod bearings are given in the following sections.

Observe the pin installed in each main bearing recess. Notice the hole in each main bearing outer shell. During installation, the hole in each bearing shell **MUST** index over the pin in the cylinder block.

Crankshaft with the upper, lower, and center main bearings installed. Notice the hole in each bearing. Matching pins in the cylinder block must index into these holes during crankshaft installation.

Hold the crankshaft over the cylinder block with the upper end to your **LEFT**. Now, lower the crankshaft into the block, and at the same time, align the hole in each bearing to enable the pin in the block to index with the hole. Rotate each bearing slightly until all pins are properly indexed with the matching bearing hole. Once all pins are indexed, the crankshaft will be properly seated.

Apply needle bearing grease to each bearing cage. Coat the rod half of the bearing area with needle bearing grease. Needle bearing grease **MUST** be used because other types of grease will not thin out and dissipate. The grease must disipate to allow the gasoline and oil mixture to enter and lubricate the bearing. If needle bearing grease is not available, use a good grade of petroleum jelly (Vasoline).

Insert the proper number of needle bearings into each cage. Set the bearing cage into the bottom half of the rod. With your

Bearing locating pins in the cylinder block. Each pin must index into a hole in the bearing shown in the illustration at the top of this column.

Installing the crankshaft into the block of a 40hp, 50hp, 55hp, or 60hp powerhead. The holes in the top and center main bearings must index with the pin in the block.

Cage and needle bearings installed into the lower portion of the rod.

fingers on each side of the rod, pull up on the rod and bring the rod up to the bottom side of the crankshaft. Put one needle bearing on each side of the crankshaft. Using needle bearing grease load the other cage and install the needle bearings into the cage. Lower the cage onto the crankshaft journal.

Install the proper rod cap to the rod with the identifying mark or dimple properly aligned to ensure the cap is being installed in the same position from which it was removed. Tighten the rod bolts fingertight, and then just a bit more.

Use a "scratchall", pick, or similar tool and move it back-and-forth on the outside surface of the rod and cap. Make the

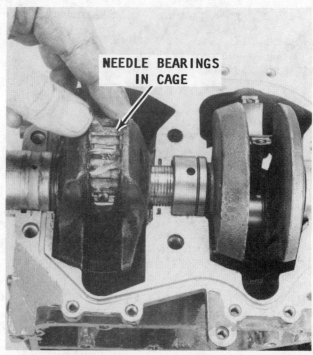

Lowering the cage and needle bearings over the top of the crankshaft.

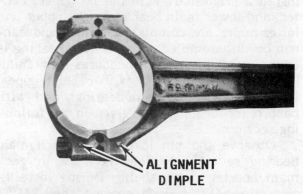

Rod and cap showing the alignment dimples.

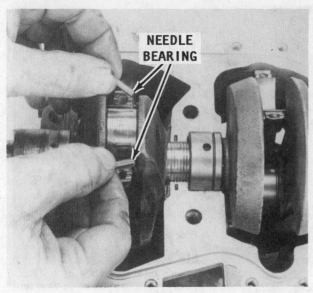

Installing a needle bearing on each side of the crankshaft.

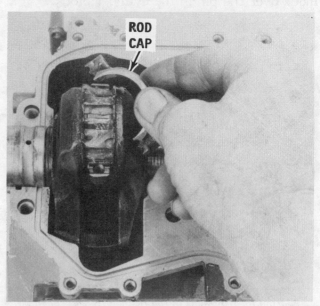

Installing the rod cap over the needle bearings and cage.

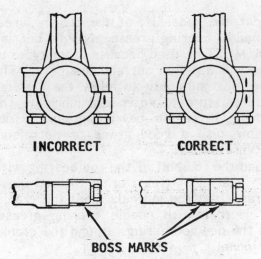

INCORRECT   CORRECT

BOSS MARKS

Correct and incorrect rod cap alignment.

Checking the flexibility of the rings through the exhaust port on a small horsepower powerhead.

movement across the mating line of the rod and cap. The tool should not catch on the rod or on the cap. The rod cap must seat squarely with the rod. If not, tap the cap until the "scratchall" will move back-and-forth on the rod and cap across the mating line without any feeling of catching. Any step on the outside will mean a step on the inside of the rod and cap. Just a whisker of a lip, will cause one of the needle bearings to catch and fail to rotate. The needle will quickly flatten, and the rod will begin to "knock". Needle bearings **MUST** rotate or the function of the bearing is lost.

Tighten the rod cap bolts alternately and evenly in three rounds to the torque value given in the Torque Table in the Appendix.

Tighten the bolts to 1/2 the torque value on the first round, to 3/4 the torque value on the second round, and to the full torque value on the third and final round. On each round, check with the pick to be sure the cap remains seated squarely.

Install the other rod cap/s in the same manner.

After the rods have been connected to the crankshaft, rotate the crankshaft until the rings on one cylinder are visible through the exhaust port. Use a screwdriver and push on each ring to be sure it has spring tension. It will be necessary to move the piston slightly, because all of the rings will not be visible at one time. If there is no spring tension, the ring was broken during installation. The piston must be removed and a new ring installed. Repeat the tension test at the intake port. Check the other cylinder/s in the same manner.

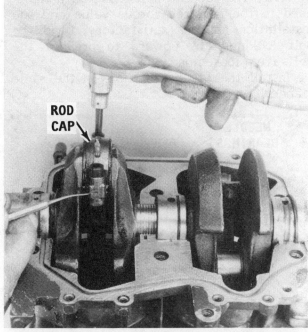

ROD CAP

Installing the rod cap bolts, and at the same time checking the cap alignment with the rod using a pick.

Checking the flexibility of the rings through the exhaust port on a 40hp, 50hp, 55hp, or 60hp powerhead, as described in the text.

CRANKSHAFT INSTALLATION
UNITS W/TOP NEEDLE MAIN BRG AND BABBITT CTR & BOTTOM
UNITS W/TOP & BOTTOM NEEDLE BRG AND CENTER BABBITT BRG
UNITS W/ALL BABBITT MAIN BRGS

This section provides detailed instructions to install a small horsepower crankshaft with any of the bearing combinations listed in the heading. Some of the powerheads covered in these paragraphs have rod liners, others do not.

The procedures pickup the work after the piston/s have been installed, as described earlier in this section.

ADVICE
Before installing the crankshaft, check to be sure each "throw" is clean and shiny. There should be no evidence of corrosion that might damage the "throw" during engine operation.

Lower the crankshaft into place in the cylinder block with the long threaded shank end at the top of the cylinder block. (It is a known fact, in more than just a few shops around the country, because of haste, the crankshaft installation work has proceeded with the short end at the top.)

The hole in the upper and lower main bearing **MUST** index into the pin in the cylinder block.

Some engines may have a lining arrangement as listed in the heading of this section. The lining is made in two parts. Install the liner half into the rod, then install the bearings as described in the next paragraph. The matching liner is to be installed into the rod cap.

Coat the rod half, of the bearing area, with needle bearing grease. Needle bearing grease **MUST** be used because other types of grease will not thin out and dissipate. The grease must dissipate to allow the gasoline and oil mixture to enter and lubricate the bearing. If needle bearing grease is not available, use a good grade of petroleum jelly (Vasoline).

Load the rod half of the rod bearing with needle bearings. Next, bring the rod up to the crankshaft rod journal. Coat the crankshaft journal with needle bearing grease. Place the needle bearings around the crankshaft jounal.

Position the rod cap, with the liners (if used) over the needle bearings. Install the rod cap bolts and lockwashers. Bring the bolts up fingertight, and then just a bit more.

If the liners are used, the cap and rod automatically align properly. If liners are not used, a dowel pin is installed in the rod cap. This pin will index into a hole in the rod for proper alignment.

Tighten the rod cap bolts alternately and evenly in three rounds to the torque value given in the Specifications in the Appendix. Tighten the bolts to 1/2 the torque value on the first round, to 3/4 the torque value on the second round, and to the full torque value on the third and final round. On each round, check with the pick to be sure the cap remains seated squarely.

After the rod cap bolts have been tightened to the required torque value and the installation appears satisfactory, bend the bolt locking tabs upward to prevent the bolts from loosening.

Install the other rod cap/s in the same manner.

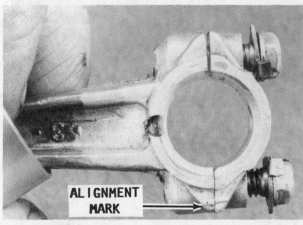

Rod and cap with the alignment marks visible.

Needle bearings installed in the rod cap liner and around the crankshaft.

Rod cap with liner ready for installation around the crankshaft.

After the rods have been connected to the crankshaft, rotate the crankshaft until the rings on one cylinder are visible through the exhaust port. Use a screwdriver and push on each ring to be sure it has spring tension. It will be necessary to move the piston slightly, because all of the rings will not be visible at one time. If there is no spring tension, the ring was broken during installation. The piston must be removed and a new ring installed. Repeat the tension test at the intake port. Check the other cylinder/s in the same manner.

The locking tabs must be bent upward after the rod cap bolts have been tightened to the proper torque value.

Rod and cap with the alignment marks visible.

CRANKSHAFT INSTALLATION BABBITT MAIN AND ROD BEARINGS

This section provides detailed instructions to install a small horsepower crankshaft with babbitt upper, lower, and center main bearings, and with babbitt rod bearings.

The procedures pickup the work after the piston/s have been installed, as described earlier in this section.

Lower the crankshaft into place in the cylinder block with the long threaded shank end at the top of the cylinder block. (It is a known fact, in more than just a few shops around the country, because of haste, the crankshaft installation work has proceeded with the short end at the top.)

Pull the rod up to the crankshaft journal. Position the rod cap over the crankshaft

Tightening the rod cap bolts to the proper torque value.

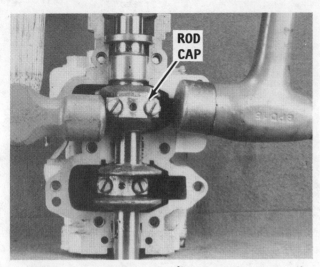

Using two hammers to fit the rod cap to the crankshaft.

journal. Install the rod cap bolts and lockwashers. Bring the bolts up fingertight, and then just a bit more.

Tighten the rod cap bolts alternately and evenly in three rounds to the torque value given in the Torque Table in the Appendix. Tighten the bolts to 1/2 the torque value on the first round, to 3/4 the torque value on the second round, and to the full torque value on the third and final round. On each round, check with the pick to be sure the cap remains seated squarely.

Repeat the procedure for the other rod and cap.

After the other rod cap has been installed and the bolts tightened to the proper torque value, hold one hammer on one side

Checking movement of the pistons and crankshaft with the flywheel temporarily installed.

of the rod and cap, and at the same time tap the other side of the rod and cap with the other hammer. Tap lightly on the top of the cap. Reverse the hammer positions and tap the opposite sides of the rod and cap. This procedure will "fit" the rod and cap to the crankshaft journal.

Repeat the "fitting" procedure for the other rod and cap.

Once the installation procedure appears satisfactory and all work has been completed, bend the bolt locking tabs upward to prevent the bolts from loosening.

CRANKCASE COVER INSTALLATION

First, check to be sure the mating surfaces of the crankcase cover and the cylinder block are clean. Pay particular attention to the labyrinth seal grooves in the center main bearing area. The mating surfaces and the seal grooves **MUST** be free of any old sealing compound or other foreign material.

CRITICAL WORDS

The remainder of the cylinder block installation work should be performed **WITHOUT** interruption. Do not begin the work if a break in the sequence is expected -- coffee, lunch, tea, whatever. Both types of

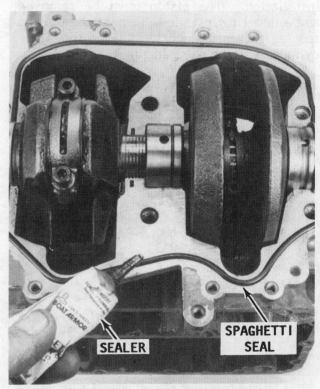

Installing sealer to the "spaghetti" seal in the crankcase cover.

Installing sealer . to the cylinder block when the "spaghetti" seal is not used.

Two tapered pins installed in the cylinder block.

sealer will begin to set almost immediately, therefore, the crankcase cover installation, main bearing bolt installation and tightening, and the side bolt installation and tightening **MUST** move along **RAPIDLY.**

Apply just a small amount of 1000 Sealer into the groove in the cylinder block to hold the "spaghetti" seal in place, if used. Install a new seal into the groove. After the seal on both sides of the cylinder block has been installed, apply a light coating of 1000 Sealer to the outside edge of the "spaghetti" seal.

SPECIAL NOTE

Since 1980, the cylinder and crankcase assemblies do not have the grooves on the mating surfaces for the neoprene "spaghetti" seals. The new sealing method is to use OMC Gel-Seal (P/N 322702). This new seal is to be used on all 2-cylinder through 6-cylinder models. **DO NOT** use similar appearing jel-type sealants, since some of them contain fillers that have a shimming affect. Such shimming could cause improper bearing location, bearing misalignment, or tight armature plate bearings.

Lay down a small bead of the Gel-Seal along one flange of the crankcase, as shown. Take care to apply sealant inside all bolt holes. Keep the sealant at least 1/4" from the labyrinth seals. If the motor is to be operated the same day, the surface opposite the one with the Gel-Seal should be sprayed with OMC LocQuic Primer. Wait several hours before starting the engine. If LocQuic Primer is not used, the assembly should sit overnight before the engine is started.

The remaining installation instructions apply to all powerheads.

Next, lower the crankcase cover into place on the cylinder block. Install the two

guide centering pins through the cover and into the block. The centering pins are tapered. Therefore, check the crankcase and notice which side has the large hole and which has the small hole. The pin must be inserted into the large hole first. If the pin is installed into the small hole first, the crankcase cover or the cylinder block will break.

MAIN BEARING BOLT AND CRANKCASE SIDE BOLT INSTALLATION

Apply a coating of 1000 Sealer to the threads of the main bearing bolts. Install and tighten the main bearing bolts finger-tight and then just a bit more.

Tighten the main bearing bolts alternately and evenly in three rounds to the torque value given in the Torque Table in

Installing the main bearing bolts through the crankcase cover into the cylinder block.

the Appendix. Be sure to check the Specifications in the Appendix for the engine being
serviced.

Tighten the bolts to 1/2 the total torque
value on the first round, to 3/4 the total
torque value on the second round, and to the
full torque value on the third and final
round.

As an example: If the total torque value
specified is 200 ft-lbs, the bolts should be
tightened to 100 ft-lbs on the first go-
around; to 150 ft-lbs on the second round;
and to the full 200 ft-lbs on the third round.

Install and tighten the crankcase side
bolts to the torque value given in the Appendix.

Install the Woodruff key in the crankshaft. Slide the flywheel onto the crankshaft. Rotate the flywheel through several
revolutions and check to be sure all moving
parts indicate smooth operation without evidence of binding or "rough" spots.

Remove the flywheel and the Woodruff
key.

LOWER SEAL INSTALLATION
TYPE ATTACHED TO LOWER
END OF CRANKSHAFT

This type of seal is attached to the lower
end of the crankshaft. On the smaller
engines, a seal is used on the crankshaft
with a spring, O-ring, and gasket. These
items push up against the bottom of the
powerhead to affect the seal.

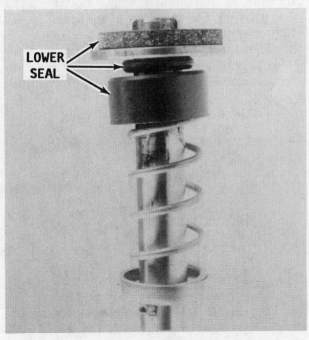

Powerhead lower seal installed on the driveshaft.

Installing the lower main seal assembly to the crankshaft.

Install the quadrant O-ring into the quadrant retainer.

Apply a small amount of light-weight oil
onto the quadrant retainer and O-ring, and
then slide them onto the crankshaft with the
lip or raised edge of the retainer facing
UPWARD upon installation.

Slide the large washer, spring, and small
washer, onto the crankshaft, and secure
them in place with the truarc snap ring.

LOWER SEAL INSTALLATION
40 HP SINCE 1983
ALL 50 HP, 55 HP, and 60 HP

After the crankshaft and the crankcase
cover have been installed, install an O-ring
onto the cap. Cover the O-ring and the
crankshaft with some light-weight oil. Slip
the cap over the crankshaft as far as possible. Coat the threads of the attaching
bolts with Loctite. Secure the cap in place
with the four bolts. Tighten the bolts
alternately and evenly.

EXHAUST COVER AND
BYPASS COVER INSTALLATION

Coat both sides of a NEW gasket with
1000 Sealer, and then place the gasket in

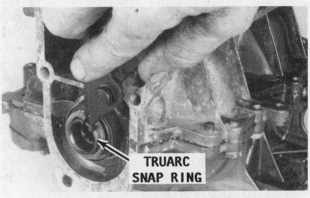

Using a pair of Truarc pliers to install the Truarc snap ring onto the crankshaft to secure the bottom seal in place.

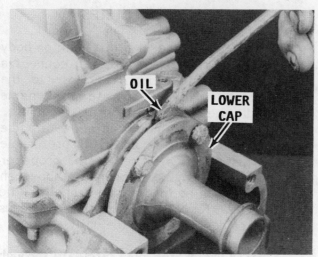

Installing the lower main bearing cap onto the crankshaft of a 40hp, 50hp, 55hp, or 60hp powerhead.

position on the exhaust side of the cylinder block. Install the inner plate. Coat both sides of another **NEW** gasket with sealer,and then install the gasket and exhaust cover.

Secure the exhaust cover in place with the attaching hardware.

Coat both sides of a **NEW** gasket with sealer, and then place it in position on the cylinder block. Install the bypass covers and secure them in place with the attaching hardware. If a fuel pump is used, be sure the same bypass cover is installed in the position from which it was removed.

REED BOX INSTALLATION

Install the reed box and intake manifold onto the cylinder block. A gasket is usually installed on both sides of the reed box. The reeds and reed stops face inward toward the cylinder.

On 15 hp to 40 hp powerheads (40 hp thru 1983), a screw is installed in the center of the reed box into the cylinder block. This center screw is installed first, then the intake manifold is installed and secured in place.

Installing the exhaust cover.

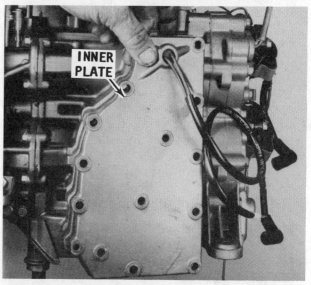

Installing the inner exhaust plate on a 40hp, 50hp, 55hp, or 60hp powerhead.

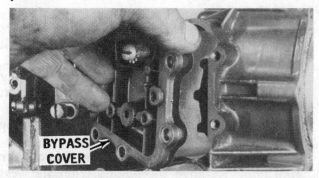

Installing the intake bypass covers. One is already in place.

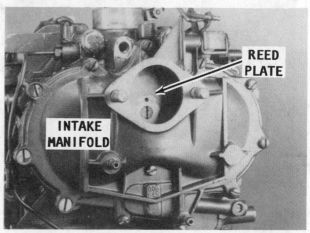

Installing the intake manifold over the reed plate.

HEAD AND POWERHEAD INSTALLATION

Place a **NEW** head gasket in place on the cylinder block. **NEVER** use automotive type head gasket sealer. The chemicals in the sealer will cause electrolytic action and eat the aluminum faster than you can get to the bank for money to buy a new cylinder block.

Install the head bolts and tighten them fingertight, then just a bit more. Tighten the bolts alternately and evenly in three rounds to the torque value specified in the Appendix. On the first round tighten the bolts to 1/2 the total torque value, on the second round to 3/4 the total torque value, and to the full torque value on the third and final round.

Install the assembled powerhead to the exhaust housing and tighten the attaching bolts alternately and evenly in three rounds to the torque value specified in the Appendix. Tighten the bolts to 1/2 the torque value on the first round, to 3/4 the total torque value on the second round, and to the full torque value on the third and final round.

Install all powerhead accessories including the flywheel, carburetor, magneto, starter, etc.. If any doubts or difficulties are encountered, follow the procedures outlined in the chapter covering the particular component. Connect the fuel lines, wiring, and battery cables.

The complete outboard unit is now ready to be started and "broke-in" according to the procedures outlined in the next section.

Tightening the head bolts to the proper torque value.

3-15 BREAK-IN PROCEDURES

Mount the engine in a test tank or body of water. If this is not possible, connect a flush attachment and garden hose to the lower unit. **NEVER** operate the engine above idle speed using the flush attachment.

If the engine is operated above an idle speed, the unit must be **IN GEAR,** preferable with a test wheel attached to the propeller shaft. If the engine is operated above an idle speed with no load on the propeller, the engine could **RUNAWAY** resulting in serious damage or destruction of the unit.

CAUTION: Water must circulate through the lower unit to the engine any time the engine is run to prevent damage to the water pump in the lower unit. Just five seconds without water will damage the water pump.

a- **ALWAYS** use OMC or BIA certified TC-W oil lubricant.

b- For all 1971-84 engines in this manual: use 50:1 oil mixture.

c- Since 1985 -- the Colt, Junior, Ultra, and 2 hp thru 35 hp: first 5 hours of break-in use 50:1 oil mixture. Recommendation is 1/6 pint of oil per gallon of gasoline (50:1). After "break-in", use 100:1 oil mixture.

d- Since 1985 -- without VRO: during break-in and after, use 50:1 oil mixture.

e- **NEVER** use: automotive oils, premixed fuel of unknown oil quantity, or premixed fuel richer than 50:1

f- When engine starts, **CHECK** water pump operation. Prior to 1977, a water mist should discharge from the exhaust relief holes at rear of drive shaft housing. Since 1977, a stream of water should discharge from the starboard side.

During the first 5 hours of operation, **DO NOT** operate the engine at full throttle (except for **VERY** short periods). Perform the break-in as follows:

1- Operate at 1/2 throttle for 2 hours.

2- Operate at any speed after 2 hours **BUT NOT** at sustained full throttle until another 3 hours of operation.

3- While the engine is operating during the initial period, check the fuel, exhaust, and water systems for leaks.

4- Refer to Chapter 5 for synchronizing procedures.

After the test period, disconnect the fuel line. Remove the engine from the test tank. Install the engine hood.

OMC approved oil for Johnson and Evinrude outboard engines. Only quality grade oil should be used for
any engine. The added expense is ridiculously small
compared to the cost of the outboard.

Filling a six-gallon fuel tank with small quantity of
fuel prior to adding the recommended amount of oil.

OMC rust preventative to be injected into the
engine through the carburetor prior to storage. The
small cost of this product is well worth the preservation
afforded the engine through its use.

Adding oil to the fuel in a built-in tank.

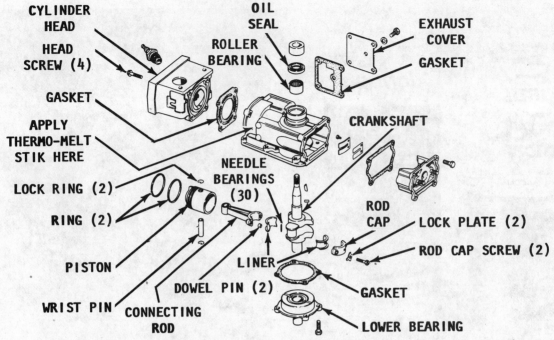

Exploded drawing of a typcial single cylinder powerhead with major parts identified.

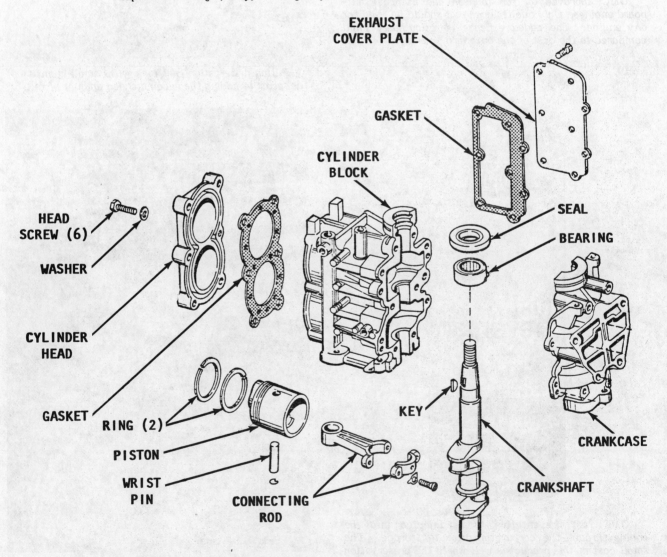

Exploded drawing of a 4 hp -- 1971-77 powerhead with principle parts identified.

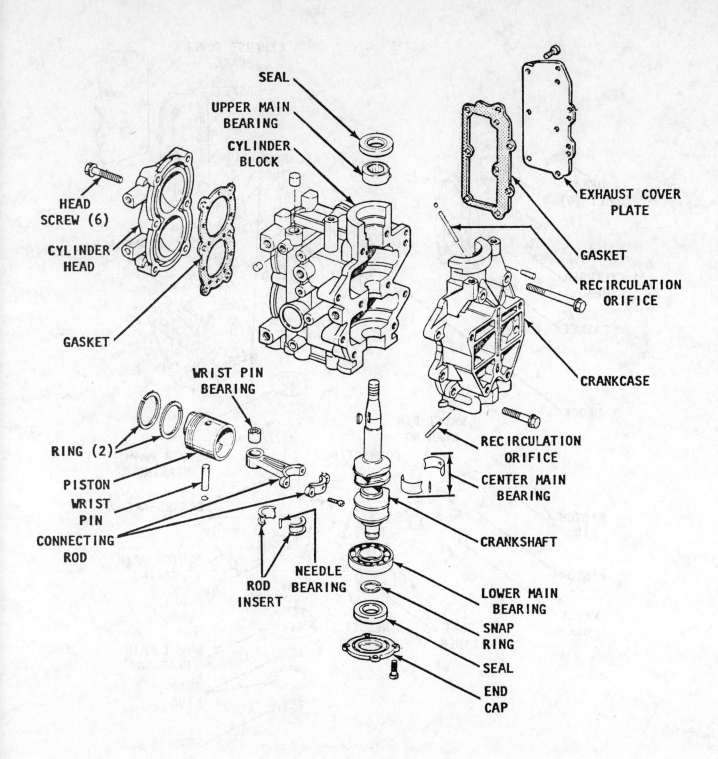

SEAL

UPPER MAIN
BEARING

CYLINDER
BLOCK

HEAD
SCREW (6)

CYLINDER
HEAD

GASKET

EXHAUST COVER
PLATE

GASKET

RECIRCULATION
ORIFICE

CRANKCASE

WRIST PIN
BEARING

RING (2)

PISTON

WRIST
PIN

CONNECTING
ROD

ROD
INSERT

NEEDLE
BEARING

RECIRCULATION
ORIFICE

CENTER MAIN
BEARING

CRANKSHAFT

LOWER MAIN
BEARING

SNAP
RING

SEAL

END
CAP

Exploded drawing of a 2.5hp and 4hp powerhead -- 1978 and on. Major parts are identified.

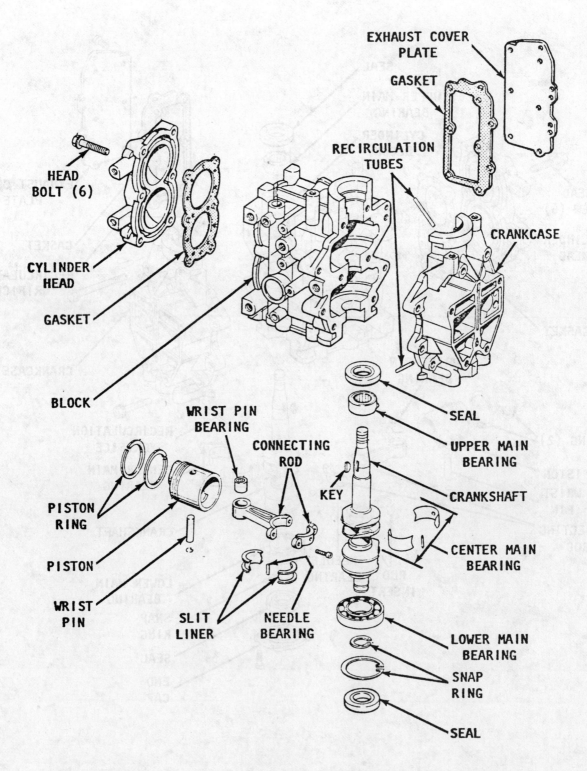

HEAD BOLT (6)

CYLINDER HEAD

GASKET

BLOCK

PISTON RING

PISTON

WRIST PIN

WRIST PIN BEARING

CONNECTING ROD

KEY

SLIT LINER

NEEDLE BEARING

EXHAUST COVER PLATE

GASKET

RECIRCULATION TUBES

CRANKCASE

SEAL

UPPER MAIN BEARING

CRANKSHAFT

CENTER MAIN BEARING

LOWER MAIN BEARING

SNAP RING

SEAL

Exploded drawing of a 4.5 hp -- 1980-85 and 7.5 hp -- 1980-83 powerhead with principle parts identified.

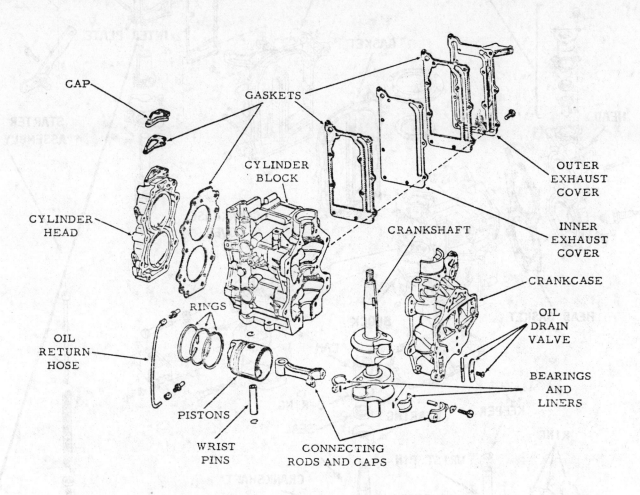

CAP

GASKETS

CYLINDER
BLOCK

OUTER
EXHAUST
COVER

CYLINDER
HEAD

INNER
EXHAUST
COVER

CRANKSHAFT

CRANKCASE

OIL
DRAIN
VALVE

RINGS

OIL
RETURN
HOSE

BEARINGS
AND
LINERS

PISTONS

WRIST
PINS

CONNECTING
RODS AND CAPS

Exploded powerhead drawing of a 6hp -- 1971 and on: 8hp -- 1984 and on. Major parts are identified.

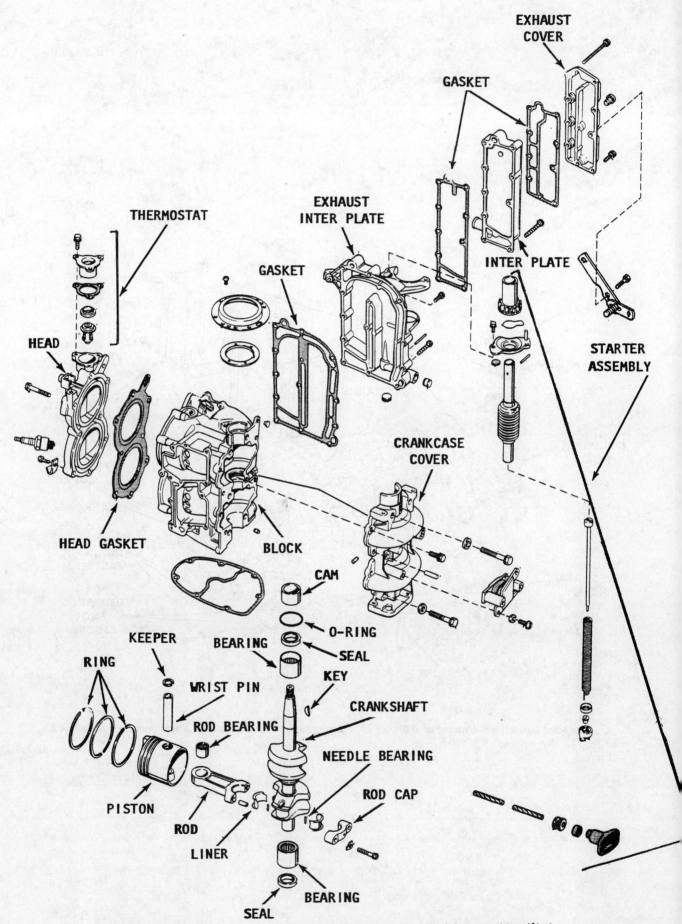

Exploded drawing of a 9.5 hp -- 1971-73 powerhead with principle parts identified.

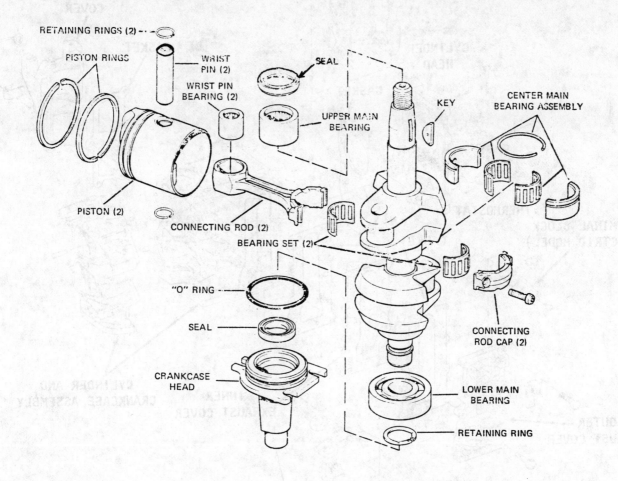

RETAINING RINGS (2)

PISTON RINGS

WRIST PIN (2)

WRIST PIN BEARING (2)

SEAL

UPPER MAIN BEARING

KEY

CENTER MAIN BEARING ASSEMBLY

PISTON (2)

CONNECTING ROD (2)

BEARING SET (2)

"O" RING

SEAL

CRANKCASE HEAD

CONNECTING ROD CAP (2)

LOWER MAIN BEARING

RETAINING RING

Exploded drawing of a 9.9hp and 15hp crankshaft assembly -- 1974 and on. Major parts are identified. The block is shown on the next page.

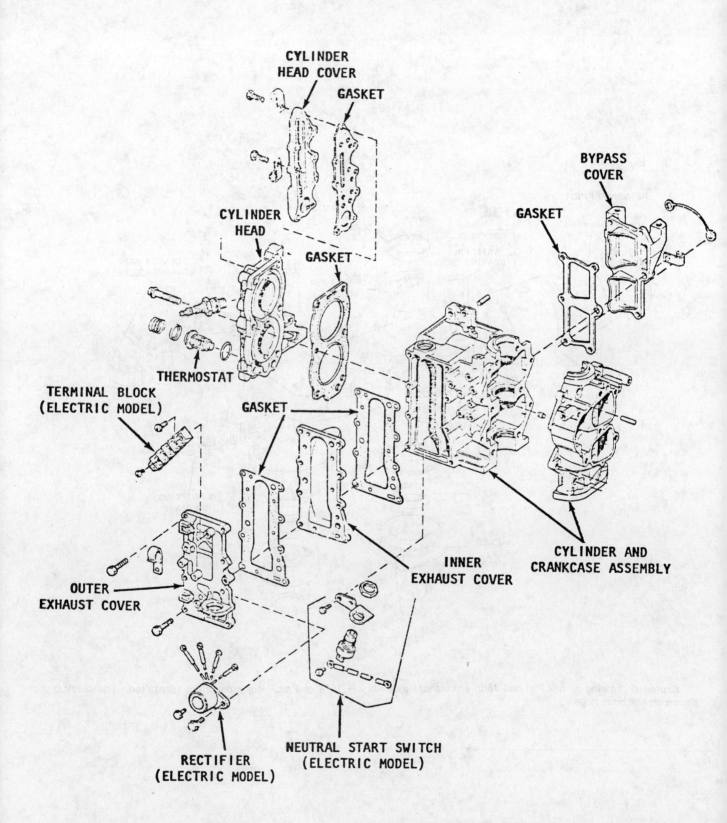

CYLINDER
HEAD COVER

GASKET

BYPASS
COVER

CYLINDER
HEAD

GASKET

GASKET

THERMOSTAT

TERMINAL BLOCK
(ELECTRIC MODEL)

GASKET

OUTER
EXHAUST COVER

INNER
EXHAUST COVER

CYLINDER AND
CRANKCASE ASSEMBLY

RECTIFIER
(ELECTRIC MODEL)

NEUTRAL START SWITCH
(ELECTRIC MODEL)

Exploded drawing of a 9.9hp and 15hp cylinder block -- 1974 and on. Major parts are identified. The crankshaft assembly is shown on the previous page.

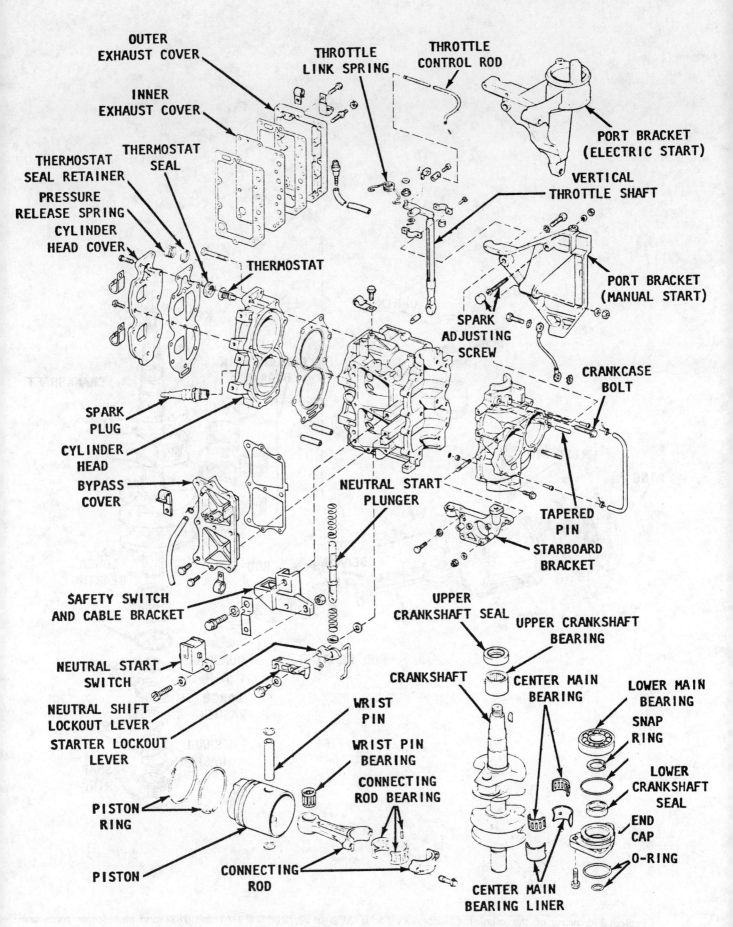

Exploded powerhead drawing -- 25hp -- 1971 and on and 35hp -- 1976-84. Major parts are identified.

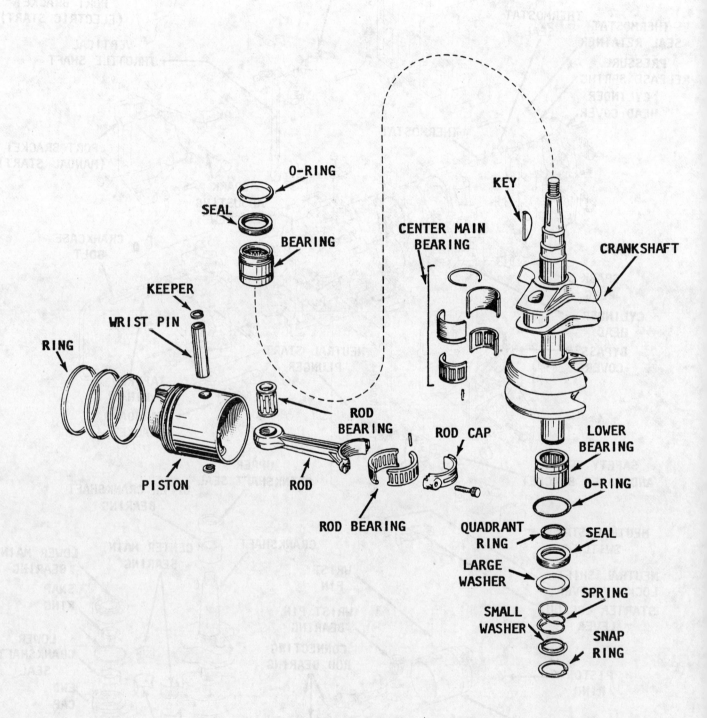

Exploded drawing of the crankshaft assembly for a 20hp powerhead 1971-73 and 1981 and on. Major parts are identified.

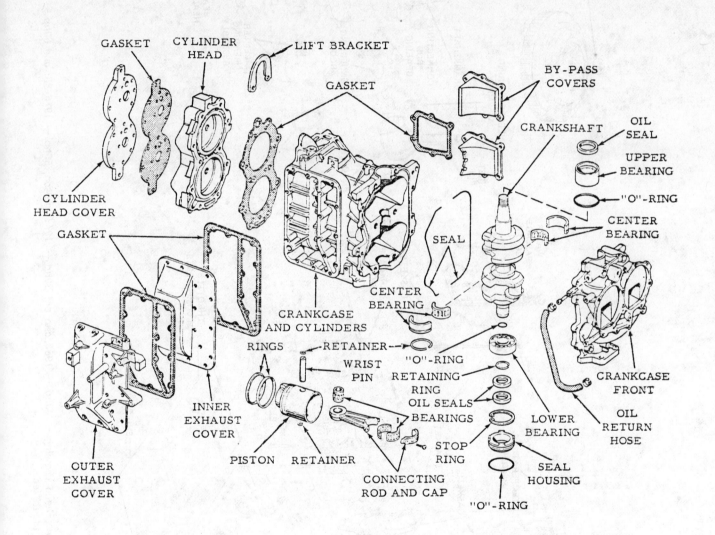

GASKET   CYLINDER HEAD   LIFT BRACKET   GASKET   BY-PASS COVERS   CRANKSHAFT   OIL SEAL   UPPER BEARING   "O"-RING   CENTER BEARING   CYLINDER HEAD COVER   GASKET   SEAL   CENTER BEARING   CRANKCASE AND CYLINDERS   CRANKCASE FRONT   OIL RETURN HOSE   RINGS   RETAINER   "O"-RING   WRIST PIN   RETAINING RING   OIL SEALS   BEARINGS   LOWER BEARING   INNER EXHAUST COVER   PISTON   RETAINER   STOP RING   SEAL HOUSING   OUTER EXHAUST COVER   CONNECTING ROD AND CAP   "O"-RING

Exploded powerhead drawing -- 40hp -- 1971-76 and 1981-1982. Major parts are identified.

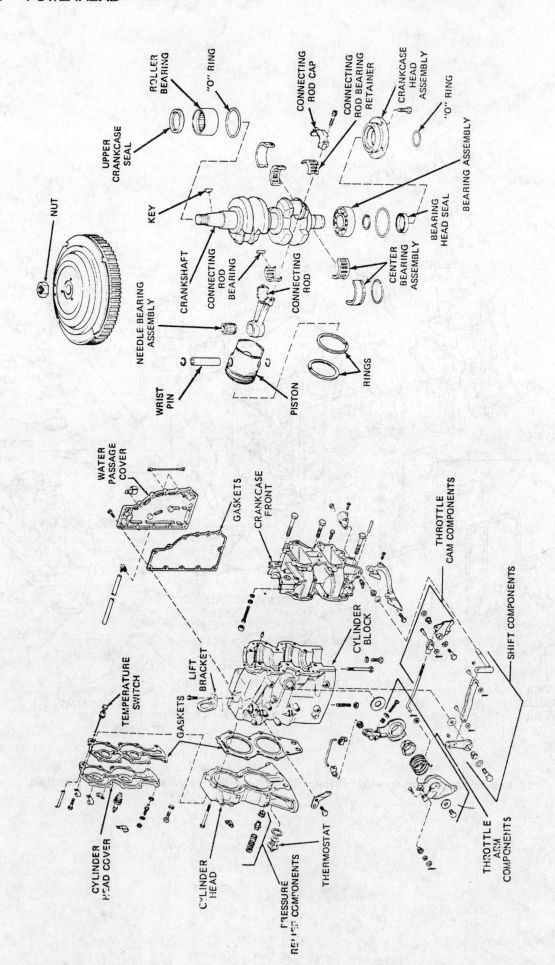

Exploded powerhead drawing — 40hp — 1983 and on: 50hp — 1971 and on; 55hp — 1976-83; and 60hp — 1980-85. Major parts are identified.

4
FUEL

4-1 INTRODUCTION

The carburetion and ignition principles of two-cycle engine operation **MUST** be understood in order to perform a proper tune-up on an outboard motor.

If you have any doubts concerning your understanding of two-cycle engine operation, it would be best to study the operation theory section in the first portion of Chapter 3, before tackling any work on the fuel system.

Newer units (since about 1985), have been equipped with either an electric or manual primer system. The electric primer system is covered in Section 4-13. The manual primer system is covered in Section 4-14.

Since 1985, all 40hp, and some 25hp and 30hp units have been equipped with a Variable Ratio Oil System, commonly referred to as simply VRO.

Since 1986 all non-electric start 25hp and 30hp units and all 9.9hp and 15hp units have been equipped with an oil injection system known as AutoBlend. In 1987, the name was changed to AccuMix. The Accu-Mix system is available as an optional accessory on all smaller models down to the 4hp.

Both oil injection systems are covered in Section 4-15, the last section of this chapter.

4-2 GENERAL CARBURETION INFORMATION

The carburetor is merely a metering device for mixing fuel and air in the proper proportions for efficient engine operation. At idle speed, an outboard engine requires a mixture of about 8 parts air to 1 part fuel.

At high speed or under heavy duty service, the mixture may change to as much as 12 parts air to 1 part fuel.

Float Systems

A small chamber in the carburetor serves as a fuel reservoir. A float valve admits fuel into the reservoir to replace the fuel consumed by the engine.

Fuel level in each chamber is extremely critical and must be maintained accurately. Accuracy is obtained through proper adjustment of the float. This adjustment will provide a balanced metering of fuel to each cylinder at all speeds.

Following the fuel through its course, from the fuel tank to the combustion chamber of the cylinder, will provide an appreciation of exactly what is taking place. In order to start the engine, the fuel must be moved from the tank to the carburetor by a squeeze bulb installed in the fuel line.

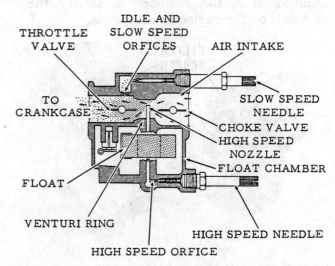

Fuel flow through the venturi, showing principle and related parts controlling intake and outflow.

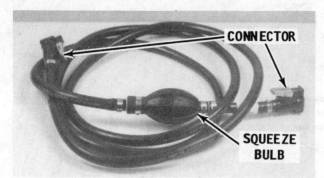

Typical fuel line with squeeze bulb and quick-disconnect fitting at each end. These items may be purchased as an assembled unit.

All powerheads covered in this manual are equipped with a manually-operated squeeze bulb in the line to transfer fuel from the tank to the engine until the engine starts.

After the engine starts, the fuel passes through the fuel pump to the carburetor. All systems have some type of filter installed somewhere in the line between the tank and the carburetor.

At the carburetor, the fuel passes through the inlet passage to the needle and seat, and then into the float chamber (reservoir). A float in the chamber rides up and down on the surface of the fuel. After fuel enters the chamber and the level rises to a predetermined point, a tang on the float closes the inlet needle and the flow of entering the chamber is cutoff. When fuel leaves the chamber as the engine operates, the fuel level drops and the float tang allows the inlet needle to move off its seat and fuel once again enters the chamber. In this manner a constant reservoir of fuel is maintained in the chamber to satisfy the demands of the engine at all speeds.

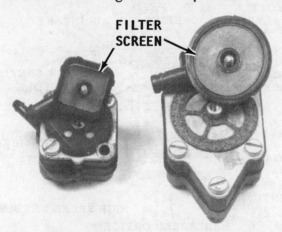

Two different type fuel pumps with the covers removed to show the filter screen. These pumps cannot be rebuilt. The only service possible is to clean the screens with solvent and then blow them dry.

A fuel chamber vent hole is located near the top of the carburetor body to permit atmospheric pressure to act against the fuel in each chamber. This pressure assures an adequate fuel supply to the various operating systems of the engine.

Air/Fuel Mixture

A suction effect is created each time the piston moves upward in the cylinder. This suction draws air through the throat of the carburetor. A restriction in the throat, called a venturi, controls air velocity and has the effect of reducing air pressure at this point.

The difference in air pressures at the throat and in the fuel chamber, causes the fuel to be pushed out metering jets extending down into the fuel chamber. When the fuel leaves the jets, it mixes with the air passing through the venturi. This air/fuel mixture should then be in the proper proportion for burning in the cylinder/s for maximum engine performance.

In order to obtain the proper air/fuel mixture for all engine speeds, high- and low-speed needle valves are installed. On late-model engines, the high-speed needle valve was replaced with a high-speed orifice. There is no adjustment with the orifice type. These needle valves are used to

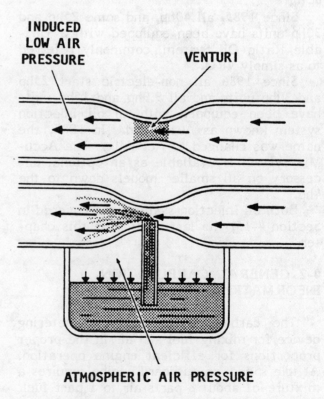

Air flow principle of a modern carburetor.

compensate for changing atmospheric conditions. Only 15% to 20% of the engines covered in this manual have an adjustable high- and low-speed needle valve.

Engine operation at sea level compared with performance at high altitudes is quite noticeable. A throttle valve controls the volume of air/fuel mixture drawn into the engine. A cold engine requires a richer fuel mixture to start and during the brief period it is warming to normal operating temperature. A choke valve is placed ahead of the metering jets and venturi to provide the extra amount of air required for start and while the engine is cold.

When this choke valve is closed, a very rich fuel mixture is drawn into the engine.

The throat of the carburetor is usually referred to as the "barrel." Carburetors installed on engines covered in this manual all have a single metering jet with a single throttle and choke plate. Single barrel carburetors are fed by one float and chamber.

4-3 FUEL SYSTEM

The fuel system includes the fuel tank, fuel pump, fuel filters, carburetor, connecting lines, with a squeeze bulb, and the associated parts to connect it all together. Regular maintenance of the fuel system to obtain maximum performance, is limited to changing the fuel filter at regular intervals and using **FRESH** fuel. Even with the high price of fuel, removing gasoline that has

Damaged piston, possibly caused by insufficient oil mixed with the fuel; using too-low an octane fuel; or using fuel that had "soured" (stood too long without a preservative additive).

been standing unused over a long period of time is still the easiest and least expensive preventive maintenance possible.

In most cases this old gas, even with some oil mixed with it, can be used without harmful effects in an automobile using regular gasoline.

If a sudden increase in gas consumption is noticed, or if the engine does not perform properly, a carburetor overhaul, including boil-out, or replacement of the fuel pump may be required.

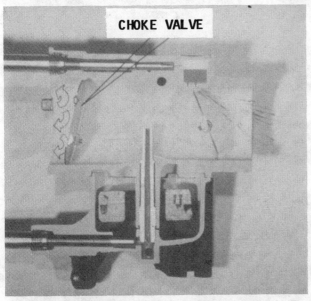

CHOKE VALVE

Choke valve location in the carburetor venturi. The choke valve on most Johnson/Evinrude carburetors is located in front of the venturi.

STORE YOUR MOTORS
WITHOUT DRAINING GAS

STA-BIL

...so gum can't form
to clog carburetor

KEEPS STORED GASOLINE
FRESH FOR QUICK STARTS

STORED MOTORS START EASY
even after a year's storage

Net contents 1 U.S. pint (16 fl. oz) (473 liters)
DANGER! COMBUSTIBLE!

Commercial additives, such as Sta-bil, may be used to keep the fuel in the tank fresh. Under favorable conditions, such additives will prevent the fuel from "souring" for up to twelve months.

4-4 TROUBLESHOOTING

The following paragraphs provide an orderly sequence of tests to pinpoint problems in the system. If an engine has not been used for some time and fuel has remained in the carburetor, it is possible that varnish may have formed. Such a condition could be the cause of hard starting, or complete failure of the engine to operate.

Fuel Problems

Many times fuel system troubles are caused by a plugged fuel filter, a defective fuel pump, or by a leak in the line from the fuel tank to the fuel pump. Aged fuel left in the carburetor and the formation of varnish could cause the needle to stick in its seat and prevent fuel flow into the bowl. A defective choke may also cause problems. **WOULD YOU BELIEVE,** a majority of starting troubles, which are traced to the fuel system, are the result of an empty fuel tank or aged fuel.

Fuel will begin to sour in three to four months and will cause engine starting problems. Therefore, leaving the motor setting idle with fuel in the carburetor, lines, or tank during the off-season, usually results in very serious problems. A fuel additive such as Sta-Bil may be used to prevent gum from forming during storage or prolonged idle periods.

For many years there has been the widespread belief that simply shutting off the fuel at the tank and then running the engine

Fouled spark plug, possibly caused by operator's habit of over-choking or a malfunction holding the choke closed. Either of these conditions delivered a too-rich fuel mixture to the cylinder.

until it stops is the proper procedure before storing the engine for any length of time. Right? **WRONG.**

First, it is **NOT** possible to remove all fuel in the carburetor by operating the engine until it stops. Considerable fuel is trapped in the float chamber and other passages and in the line leading to the carburetor. The **ONLY** guaranteed method of removing **ALL** fuel is to take the time to remove the carburetor, and drain the fuel.

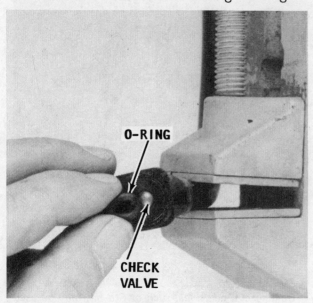

Fuel connector with the O-ring visible. These O-rings have a relative short life and must be replaced at regular intervals, as detailed in the text.

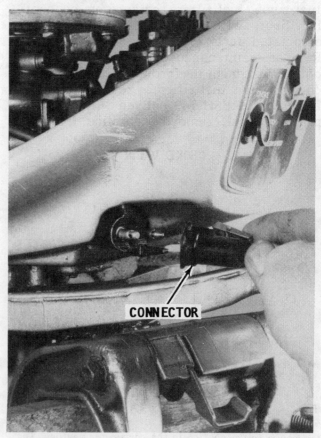

Female fuel line connector ready to be mated with the male portion of the connector.

Secondly, if the engine is operated with the fuel supply shut off until it stops the fuel and oil mixture inside the engine is removed, leaving bearings, pistons, rings, and other parts without any protective lubricant.

Proper procedure involves: Shutting off the fuel supply at the tank; disconnecting the fuel line at the tank; operating the engine until it begins to run **ROUGH;** then stopping the engine, which will leave some fuel/oil mixture inside; and finally removing and draining the carburetor. By disconnecting the fuel supply, all **SMALL** passages are cleared of fuel even though some fuel is left in the carburetor. A light oil should be put in combustion chamber as instructed in the Owners Manual. On some model carburetors, the high-speed orifice plug can be removed to drain the fuel from the carburetor.

Choke Problems

When the engine is hot, the fuel system can cause starting problems. After a hot engine is shut down, the temperature inside the fuel bowl may rise to 200°F and cause the fuel to actually boil. All carburetors are vented to allow this pressure to escape to the atmosphere. However, some of the fuel may percolate over the high-speed nozzle.

If the choke should stick in the open position, the engine will be hard to start. If the choke should stick in the closed position,

the engine will flood making it **VERY** difficult to start.

In order for this raw fuel to vaporize enough to burn, considerable air must be added to lean out the mixture. Therefore, the only remedy is to remove the spark plug/s; ground the leads; crank the engine through about 10 revolutions; clean the plugs; install the plugs again; and start the engine.

If the needle valve and seat assembly is leaking, an excessive amount of fuel may enter the intake manifold in the following manner: After the engine is shut down, the pressure left in the fuel line will force fuel past the leaking needle valve. This extra fuel will raise the level in the fuel bowl and cause fuel to overflow into the intake manifold.

A continuous overflow of fuel into the intake manifold may be due to a sticking inlet needle or to a defective float which would cause an extra high level of fuel in the bowl and overflow into the intake manifold.

FUEL PUMP TESTS

CAUTION: Gasoline will be flowing in the engine area during this test. Therefore, guard against fire by grounding the high-tension wire to prevent it from sparking.

Testing System with Squeeze Bulb
An adequate safety method, is to ground

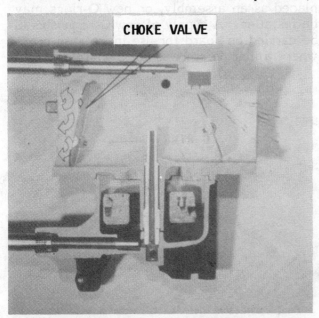

The choke plays a most important role during engine start and in controlling the amount of air entering the carburetor, under various load conditions.

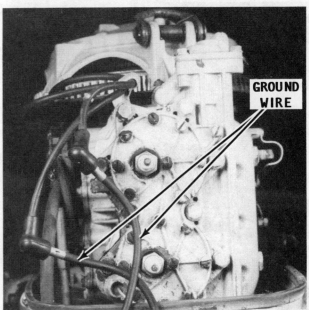

Grounding the spark plug wires during fuel tests to prevent an accidental spark from igniting fuel or fuel vapors. Exercise care when working on the fuel system.

each spark plug lead. Disconnect the fuel line at the carburetor. Place a suitable container over the end of the fuel line to catch the fuel discharged. Insert a small screwdriver into the end of the line to open the check valve, and then squeeze the primer bulb and observe if there is satisfactory fuel flow from the line.

If there is no fuel discharged from the line, the check valve in the squeeze bulb may be defective, or there may be a break or obstruction in the fuel line.

If there is a good fuel flow, then crank the engine. If the fuel pump is operating properly, a healthy stream of fuel should pulse out of the line.

Continue cranking the engine and catching the fuel for about 15 pulses to determine if the amount of fuel decreases with each pulse or maintains a constant amount. A decrease in the discharge indicates a restriction in the line. If the fuel line is plugged, the fuel stream may stop. If there is fuel in the fuel tank but no fuel flows out the fuel line while the engine is being cranked, the problem may be in one of several areas:

1- Plugged fuel line from the fuel pump to the carburetor.

2- Defective O-ring in fuel line connector into the fuel tank.

3- Defective O-ring in fuel line connector into the engine.

4- Defective fuel pump.

5- The line from the fuel tank to the

Fuel line connection at the carburetor on a 6.0 horsepower engine.

fuel pump may be plugged; the line may be leaking air; or the squeeze bulb may be defective.

6- Defective fuel tank.

7- If the engine does not start even though there is adequate fuel flow from the fuel line, the fuel inlet needle valve and the seat may be gummed together and prevent adequate fuel flow.

FUEL LINE TEST

On most installations, the fuel line is provided with quick-disconnect fittings at the tank and at the engine. If there is reason to believe the problem is at the quick-disconnects, the hose ends can be replaced as an assembly, or new O-rings may be installed. A supply of new O-rings should be carried on board for use in isolated areas where a marine store is not available. For a

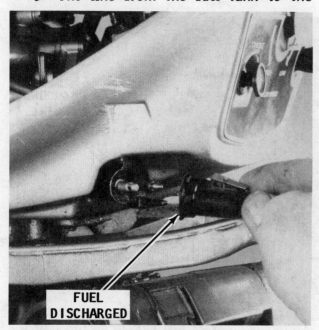

The fuel line quick-disconnect fitting at the engine separated in preparation to making a fuel flow test.

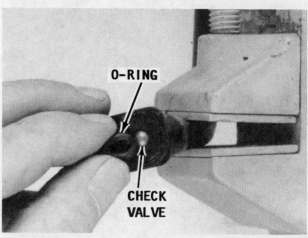

Fuel connector with the O-ring visible. These O-rings have a relative short life and may be the source of fuel problems. The O-rings should be replaced on a regular basis.

small additional expense, the entire fuel line can be replaced and eliminate this entire area as a problem source for many future seasons.

The primer squeeze bulb can be replaced in a short time. A squeeze bulb assembly kit, complete with the check valves installed, may be obtained from the local Johnson/Evinrude dealer. The replacement kit will also include two tie straps to secure the bulb properly in the line.

An arrow is clearly visible on the squeeze bulb to indicate the direction of fuel flow. The squeeze bulb **MUST** be installed correctly in the line because the check valves in each end of the bulb will allow fuel to flow in **ONLY** one direction. Therefore, if the squeeze bulb should be installed backwards, in a moment of haste to get the job done, fuel will not reach the carburetor.

To replace the bulb, first unsnap the clamps on the hose at each end of the bulb. Next, pull the hose out of the check valves at each end of the bulb. New clamps are included with a new squeeze bulb.

A typical squeeze bulb with the directional arrow clearly visible. The squeeze bulb must be installed with the arrow pointing in the direction of fuel flow, toward the powerhead.

If the fuel line has been exposed to considerable sunlight, it may have become hardened, causing difficulty in working it over the check valve. To remedy this situation, simply immerse the ends of the hose in boiling water for a few minutes to soften the rubber. The hose will then slip onto the check valve without further problems. After the lines on both sides have been installed, snap the clamps in place to secure the line. Check a second time to be sure the arrow is pointing in the fuel flow direction, **TOWARDS** the engine.

Integral Fuel Tank System

On small engines equipped with an integral fuel tank above the flywheel, or elsewhere on the engine, the filter in the tank may be plugged. To check the fuel flow, disconnect the fuel line at the carburetor and fuel should flow from the line if the shut-off valve is open. If fuel is not present, the usual cause is a plugged filter in the fuel tank. It is very difficult to determine if a porcelain-type filter is plugged. The

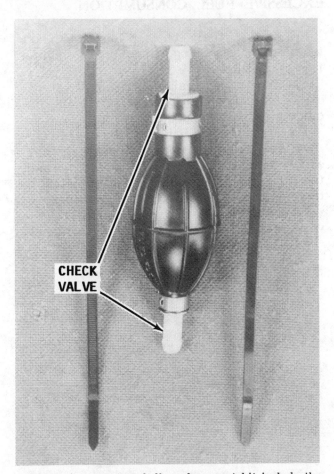

Parts in a squeeze bulb replacement kit include the squeeze bulb, two check valves, and two tie straps to secure the bulb in the line.

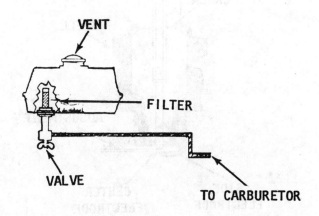

Sectional view of a engine mounted fuel tank showing the filter inside the tank and the external shut-off valve.

general appearance that the filter is satisfactory may be a false indication. Therefore, if the filter is suspected, the best remedy is replacement. The cost is very modest and this one area is thus eliminated as a problem source.

Would you believe, many times lack of fuel at the carburetor is caused because the vent on the fuel tank was not opened.

To remove the filter, first remove the fuel tank filler cap, turn the engine upside down and drain the tank. Next, disconnect the fuel line to the carburetor. Remove the fuel shut-off valve. In most cases, the filter will remain with the valve and can be separated from the valve by removing the fitting. If the fitting and filter do not remain with the valve when it is removed, then remove the fitting and filter from the tank.

ROUGH ENGINE IDLE

If an engine does not idle smoothly, the most reasonable approach to the problem is to perform a tune-up to eliminate such areas as: defective points; faulty spark plugs; and synchronization out of adjustment.

Other problems that can prevent an engine from running smoothly include: An air leak in the intake manifold; uneven compression between the cylinders; and sticky or broken reeds.

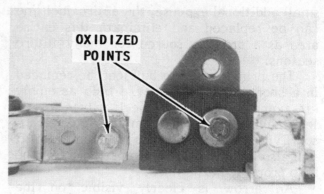

A set of points unfit for service due to oxidation.

Of course any problem in the carburetor affecting the air/fuel mixture will also prevent the engine from operating smoothly at idle speed. These problems usually include: Too high a fuel level in the bowl; a heavy float; leaking needle valve and seat; defective automatic choke; and improper idle and high-speed needle valve adjustments.

"Sour" fuel (fuel left in a tank without a preservative additive) will cause an engine to run rough and idle with great difficulty.

EXCESSIVE FUEL CONSUMPTION

Excessive fuel consumption can result from one of three conditions, or a combination of all three.

1- Inefficient engine operation.

2- Damaged condition of the hull, including excessive marine growth.

3- Poor boating habits of the operator.

If the fuel consumption suddenly increases over what could be considered normal, then the cause can probably be attributed to the engine or boat and not the operator.

Marine growth on the hull can have a very marked effect on boat performance.

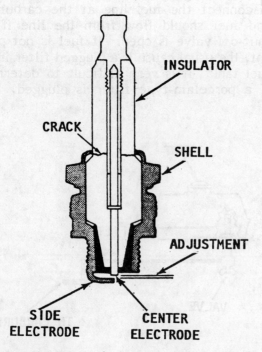

Cross-section view of a spark plug showing the principle parts with important comments for satisfactory service.

Inspection of the spark plugs should become a "first" whenever the engine fails to perform properly.

Carburetor installed on most 6 hp engines.

This is why sail boats always try to have a haul-out as close to race time as possible. While you are checking the bottom take note of the propeller condition. A bent blade or other damage will definitely cause poor boat performance.

If the hull and propeller are in good shape, then check the fuel system for possible leaks. Check the line between the fuel pump and the carburetor while the engine is running and the line between the fuel tank and the pump when the engine is not running. A leak between the tank and the pump many times will not appear when the engine is operating, because the suction created by the pump drawing fuel will not allow the fuel to leak. Once the engine is turned off and the suction no longer exists, fuel may begin to leak.

If a minor tune-up has been performed and the spark plugs, points, and synchronization are properly adjusted, then the problem most likely is in the carburetor, indicating an overhaul is in order. Check for leaks at the needle valve and seat. Use extra care when making any adjustments affecting the fuel consumption, such as the float level or automatic choke.

ENGINE SURGE

If the engine operates as if the load on the boat is being constantly increased and decreased, even though an attempt is being made to hold a constant engine speed, the problem can most likely be attributed to the fuel pump.

Operational description and service procedures for the fuel pump are given in Section 4-11.

4-5 JOHNSON/EVINRUDE CARBURETORS

This section provides complete detailed procedures for removal, disassembly, cleaning and inspecting, assembling including bench adjustments, installation, and operating adjustments for the four OMC carburetors installed on engines covered in this manual. The Type I carburetor is installed on the 1.25 hp (Colt, Junior and Ultra) to 15 hp engines; the Type II on the 18 to 40 hp models; and the Type III carburetor is used on the 9.5 hp model; Type IV on the 50 hp, 55 hp and 60 hp engines. These carburetors are equipped with either a manual choke, or

Marine growth allowed to accumulate on the lower unit will create "drag" and seriously hamper boat performance, and corrode the metal if it is not removed.

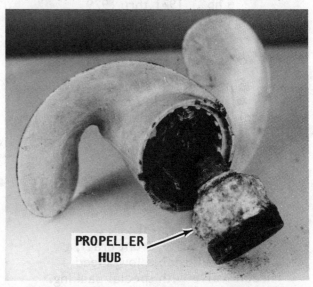

PROPELLER HUB

A corroded hub on a small engine propeller. Replacement of this propeller will be less expensive than the cost of a rebuild.

an electric choke with a manual backup. The manual backup permits the operator to operate the choke in the event the battery is dead. The following table lists the types of carburetors, the horsepower and model engines equipped with each.

CARBURETOR INSTALLATIONS

Type I
Single barrel, front draft, with adjustable low-speed and high-speed needle valves.

1.25 hp	1986 and on
(Colt, Junior and Ultra)	
2 hp	1971 thru 1985
2.5 hp	1987 and on
4 hp	1971 thru 1977

Type I
Single barrel, front draft, with adjustable low-speed and high-speed fixed orifice in a removable nozzle.

4 hp	1978 and on
4.5 hp	1980 thru 1984

Type I
Single barrel, front draft, with adjustable low-speed needle valve. High-speed is fixed orifice in float bowl.

5 hp	1984 thru 1985
6 hp	1971 thru 1979
6 hp	1982 thru 1985
8 hp	1984 and on

Type I
Single barrel, front draft, with adjustable low-speed needle valve with special packing. High-speed is fixed orifice in a nozzle.

9.9 & 15 hp	1974 and on
7.5 hp	1980 thru 1983

Type II
Single barrel, front draft, with adjustable low-speed needle valve with special packing. High-speed is fixed orifice in the float bowl

18 & 20 hp	1971 thru 1973
20 hp	1981 and on
25 hp	1971 and on
35 hp	1976 thru 1984

Type II
Single barrel, front draft, with adjustable low-speed and high-speed needle valves.

40 hp	1971

Type II
Single barrel, front draft with adjustable low-speed needle valve. High-speed is fixed orifice in the float bowl.

40 hp	1972 thru 1976

Type III
Single barrel, down draft, with adjustable low-speed needle valve. High-speed is fixed orifice in float bowl.

9.5 hp	1971 thru 1973

Type IV
Single barrel, front draft, with adjustable low-speed needle valve and special packing. High-speed is fixed orifice in the float bowl

50 hp	1971 thru 1975
50 hp	1978
55 hp	1976

Type IV
Single barrel, front draft, with fixed low-speed orifice, fixed intermediate orifice, and fixed high-speed orifice.

50 hp	1979
55 hp	1979 thru 1983
60 hp	1980 thru 1985

Type IV
Single barrel, front draft, with fixed low-speed orifice and fixed high-speed orifice

40 hp	1984 and on
50 hp	1980 and on
55 hp	1977 thru 1978

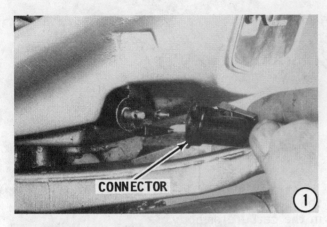

CONNECTOR ①

4-6 TYPE I CARBURETOR

REMOVAL

1- On engines using an electric cranking motor, remove the battery leads from the battery terminals. Disconnect the fuel line from the engine or from the fuel tank at the quick-disconnect fitting.

2- Remove the hood assembly from the engine. Remove the choke and throttle linkage to the carburetor.

3- Remove the fuel line from the carburetor. This may be accomplished by either one of two methods: On engines equipped with a self-contained fuel tank, close the fuel shut-off valve, located just below the tank. Disconnect both ends, and then remove the copper fuel line between the shut-off valve and the carburetor.

4- On engines utilizing a separate fuel tank, remove the tie-strap or clamp securing the rubber hose connecting the fuel connector to the carburetor. Remove the rubber hose from the carburetor. Remove the nuts securing the carburetor to the crankcase. The carburetor may have to be

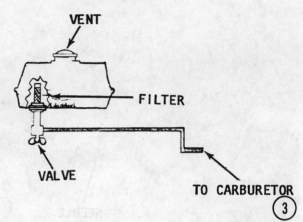

VENT

FILTER

VALVE

TO CARBURETOR ③

moved slightly forward as the nuts are loosened in order to obtain clearance for the nuts to clear the studs. After the nuts are clear, lift the carburetor from the intake manifold.

5- Remove and **DISCARD** the gasket from the intake manifold or the carburetor, if the gasket adhered to the carburetor when it was removed. A new gasket is included in a carburetor repair kit.

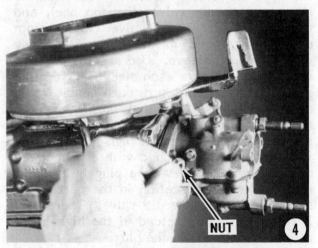

NUT ④

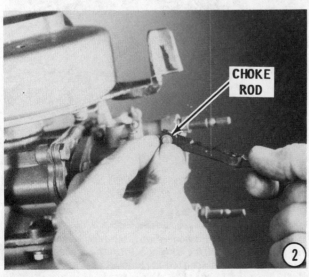

CHOKE ROD ②

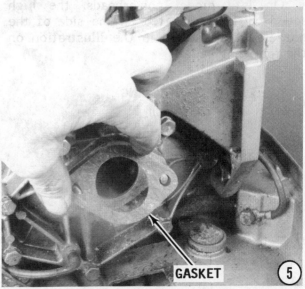

GASKET ⑤

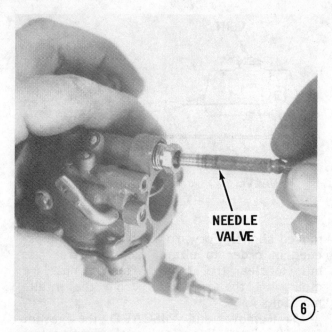

7- Use a small screwdriver, and remove the packing from the needle valve cavities in the carburetor body.

8- If installed, remove the high-speed orifice, using the proper size screwdriver.

DISASSEMBLING

6- On early powerheads equiped with this type carburetor: Observe how both the low-speed needle valve (the top one), and the high-speed needle valve (the bottom one), are secured in the carburetor body with a packing sleeve. Use a 7/16" box-end wrench and remove each sleeve. After the sleeves are removed, turn each needle valve counterclockwise until they are free of the carburetor body. If the carburetor being serviced does not have a high-speed needle valve, then it is equipped with a high-speed orifice and covered with a plug in approximately the same location in the carburetor as the high-speed needle valve. Therefore, if there is a plug, instead of the high-speed needle valve, remove the plug.

On late model powerheads, the high speed orifice is located on the side of the nozzel well as shown in the illustration on the following page.

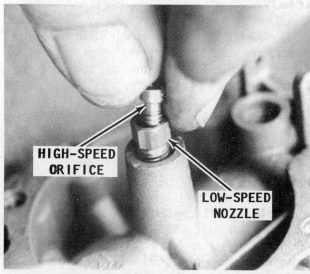

Late model carburetor with removable high-speed nozzle. The high-speed orifice threads into the center of the nozzle. On this model, both the nozzle and orifice are removable.

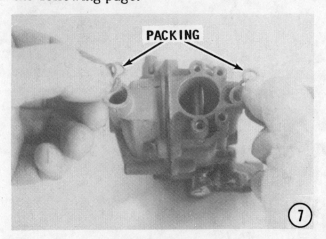

Carburetor with a non-removable high-speed nozzle. The high-speed orifice threads into the center of the nozzle, but only the orifice is removable.

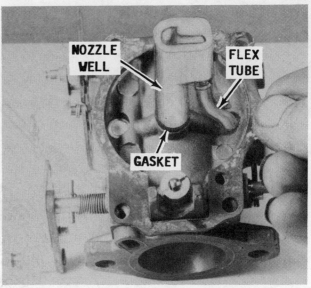

*Nozzle design used on Type I carburetors installed on **LATE** model 5hp, 6hp, and 8hp powerheads.*

9- Turn the carburetor upside down. Remove the five attaching screws, and then lift the bowl free of the carburetor.

10- Remove the hinge pin and lift the float from the carburetor bowl cavity. As the float is lifted from the carburetor, observe the small spring attached to the needle. The needle will come out with the float assembly. Remove the needle seat from the carburetor.

A GOOD WORD

Further disassembly of the carburetor is not necessary. The nozzle in the center of the carburetor does **NOT** have to be removed in order to properly clean the carburetor. However, the nozzle can be removed with a screwdriver if it is damaged and needs to be replaced.

CLEANING AND INSPECTING

NEVER dip rubber parts, plastic parts, diaphragms, or pump plungers in carburetor cleaner. These parts should be cleaned **ONLY** in solvent, and then blown dry with compressed air.

Place all metal parts in a screen-type tray and dip them in carburetor cleaner until they appear completely clean, then blow them dry with compressed air.

GOOD WORDS

Since 1983 OMC states that carburetor parts should **NOT** be submerged in carburetor cleaner, as has been the practice since carburetors were invented. Their approved procedure is to place the parts in a shallow tray and then spray them with an aerosol carburetor cleaner.

A syringe, short section of clear plastic hose, and Isopropyl Alcohol should be used to clear passages and jets.

Blow out all passages in the castings with compressed air. Check all parts and passages to be sure they are not clogged or contain any deposits. **NEVER** use a piece of

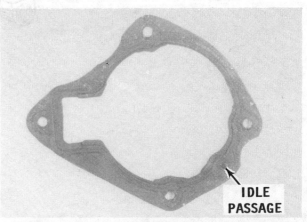

Gasket used between the bowl and the carburetor showing the opening to allow fuel to leave the bowl and enter the idle passage of the carburetor.

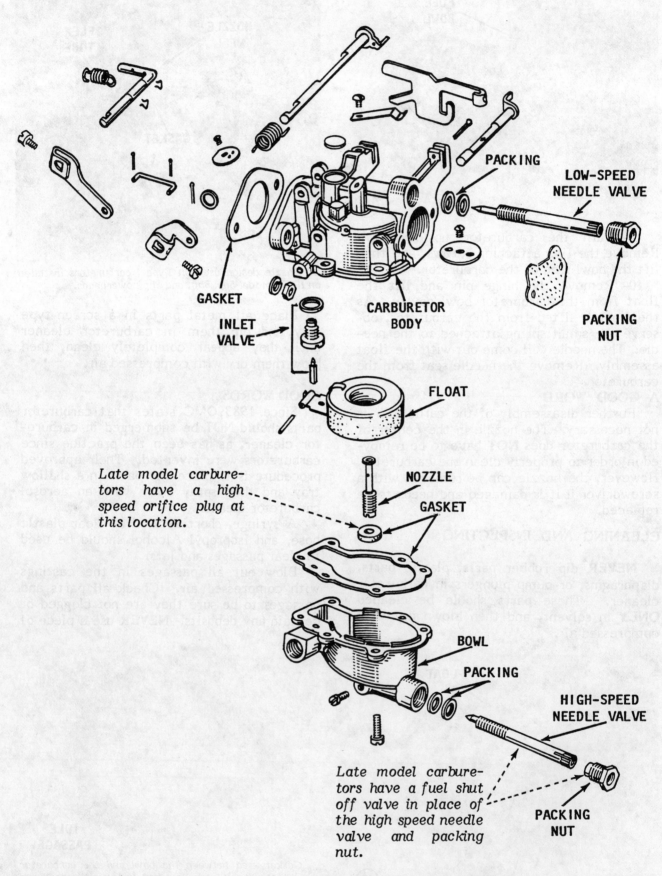

PACKING

LOW-SPEED
NEEDLE VALVE

PACKING
NUT

GASKET

CARBURETOR
BODY

INLET
VALVE

*Late model carbure-
tors have a high
speed orifice plug at
this location.*

FLOAT

NOZZLE

GASKET

BOWL

PACKING

HIGH-SPEED
NEEDLE VALVE

*Late model carbure-
tors have a fuel shut
off valve in place of
the high speed needle
valve and packing
nut.*

PACKING
NUT

*Exploded view of a Type I carburetor installed on small Johnson/Evinrude engines equipped with the fuel tank
attached to the engine. The filter on these units may be in the tank or in the fuel pump.*

wire or any type of pointed instrument to clean drilled passages or calibrated holes in a carburetor.

Move the throttle shaft back-and-forth to check for wear. If the shaft appears to be too loose, replace the complete throttle body because individual replacement parts are **NOT** available.

If any part of the float is damaged, the unit must be replaced. Check the float arm needle contacting surface and replace the float if this surface has a groove worn in it.

Inspect the tapered section of the idle adjusting needles and replace any that have developed a groove.

If a high-speed orifice is installed on the carburetor being serviced, check the orifice for cleanliness. The orifice has a stamped number. This number represents a drill size. Check the orifice with the shank of the proper size drill to verify the proper orifice is used. The correct size orifice for the engine and carburetor being serviced may be obtained from the local OMC dealer.

Most of the parts that should be replaced during a carburetor overhaul are included in overhaul kits available from your local marine dealer. One of these kits will contain a matched fuel inlet needle and seat. This combination should be replaced each time the carburetor is disassembled as a precaution against leakage.

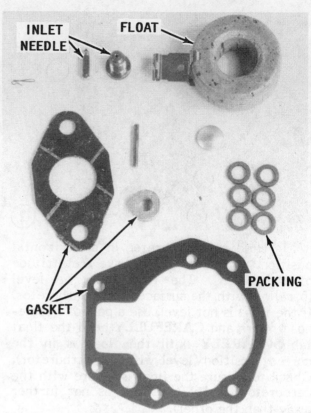

Parts necessary to properly rebuild a carburetor are included in a Johnson/Evinrude repair kit.

ASSEMBLING TYPE I CARBURETOR

1- Install a **NEW** inlet nozzle seat with a **NEW** gasket into the carburetor. Slide a **NEW** gasket down over the inlet nozzle onto the surface of the carburetor.

2- Attach a **NEW** inlet needle onto the spring included in the carburetor repair kit, and slip the spring over the edge of the float, as shown. Apply just a drop of oil to the inlet needle and then lower the needle into the seat. Install a **NEW** hinge pin included in the kit.

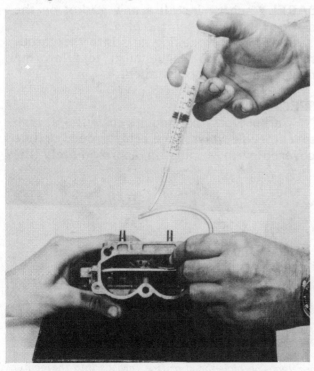

Using a Syringe, piece of clear plastic hose, and Isopropyl Alcohol to clean the idle air bleed system.

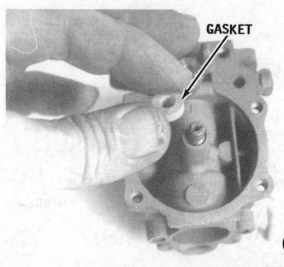

①

3- Hold the carburetor in a horizontal position, as shown, and observe the attitude of the float. The float must be level (parallel) with the surface of the carburetor. If the float is not level, use a pair of needle-nose pliers and **CAREFULLY** bend the float tab **SQUARELY** until the float is in the correct position (level with the carburetor). Check to be sure the float is square with the carburetor cavity (one side is not further away than the other).

Measuring Float Drop

4- Allow the float to drop under its own weight, and then measure the distance between the base of the carburetor body and the lowest edge of the float. Dimension should be as listed.

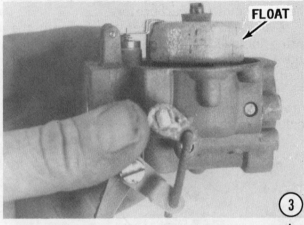

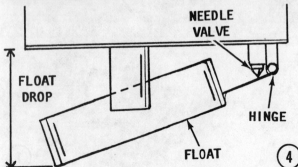

Carburetor with a non-removeable high-speed nozzle. The high-speed orifice threads into the center of the nozzle, but only the orifice is removeable.

1.25 thru 4.5hp: 1-1/8" to 1-1/2" (28-33mm)
5 thru 8hp: 1" to 1-3/8" (25 to 35mm)
9.9 thru 15hp: 1-1/8" to 1-1/2" (28-33mm)

5- Place the bowl gasket in position on the carburetor, and then position the bowl on top of the gasket. Secure the bowl in place with the five screws.

6- Insert three new packing washers into the high-speed and low-speed cavities. Place a white plastic washer into the high- and low-speed cavities. If a high-speed orifice is used, install the orifice with the proper size screwdirver until the orifice just **BARELY** seats. Install a new gasket on the orifice plug, and then the plug.

7- Start the packing nuts into the carburetor, but **ONLY** far enough to allow the needle valves to be installed.

SPECIAL WORDS

Carburetors not equipped with a high speed needle valve have a high speed orifice covered with a plug in approximately the

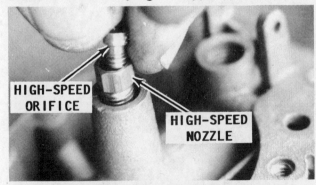

Late model carburetor with removeable high-speed nozzle. The high-speed orifice threads into the center of the nozzle. On this model, both the nozzle and orifice are removeable.

BOWL

same location as the high speed needle valve.

If the carburetor has a high speed needle vavle perform Steps 8 and 9, then jump to Step 12.

If the carburetor has a high speed orifice, skip Step 8, but perform all the other steps.

8- Install the high-speed needle valve into the bottom cavity by rotating it **CLOCKWISE.** Allow the needle valve to seat **LIGHTLY,** then back it out **(COUNTER-CLOCKWISE)** 3/4 turn. Now, tighten the sleeve nut only until it is difficult to turn the needle valve by hand.

9- Install the low-speed needle valve into the upper cavity in the same manner as the high-speed needle valve in Step 8. After the valve is seated **LIGHTLY,** back it out **(COUNTERCLOCKWISE)** 1-1/2 turns. Now, tighten the packing nut until there is drag on the needle valve, but the valve may still be rotated by hand, but with just a little difficulty.

10- Install the high-speed orifice, if one is used, into the float bowl. **ALWAYS** take time to use the proper size screwdriver to prevent damaging the orifice. Tighten the orifice only until it just seats.

11- Install the high-speed orifice plug using a **NEW** gasket. Tighten the plug securely.

NYLON WASHER

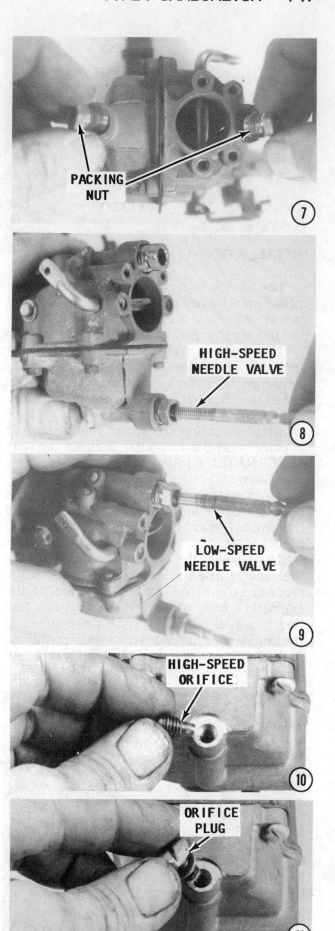

PACKING NUT

HIGH-SPEED NEEDLE VALVE

LOW-SPEED NEEDLE VALVE

HIGH-SPEED ORIFICE

ORIFICE PLUG

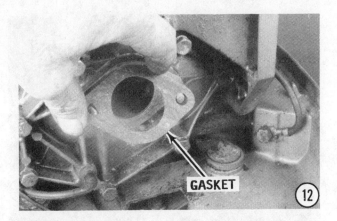

GASKET

12

INSTALLATION

12- Check to be sure the surface of the intake manifold is clean and free of any old gasket material. Place a **NEW** gasket in position on the intake manifold.

13- Connect the fuel line to the carburetor, or on engines equipped with a self-contained fuel tank, replace the line between the shut-off valve and the carburetor.

14- Slide the carburetor over the intake manifold mounting studs, and then secure it in place with the two nuts. Tighten the nuts **ALTERNATELY** to avoid warping the carburetor body.

15- Connect the manual choke rod and the front cover onto the carburetor. Thread the knobs onto the low- and high-speed needle valves.

16- Check the synchronization of the fuel and ignition systems according to the procedures outlined in Chapter 5. Mount the engine in a body of water. Connect the fuel line to a fuel source. Prime the engine. If the engine is equipped with a self-contained fuel tank, open the fuel shut-off valve and allow the carburetor to fill with fuel.

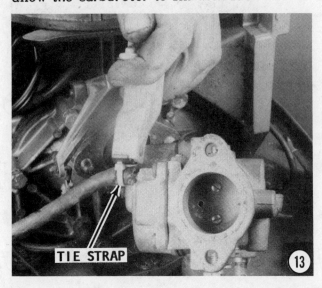

TIE STRAP

13

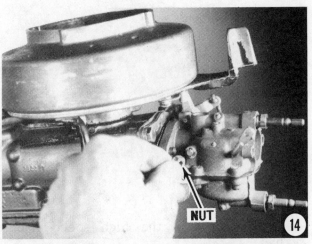

NUT

14

CAUTION: Water must circulate through the lower unit to the engine any time the engine is run to prevent damage to the water pump in the lower unit. Just five seconds without water will damage the water pump.

17- Start the engine and allow it to warm to operating temperature. Shift the engine into **FORWARD** gear. Adjust the

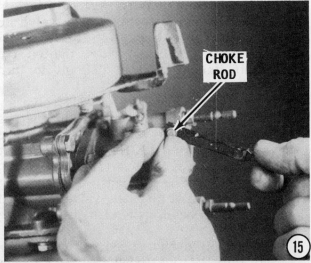

CHOKE ROD

15

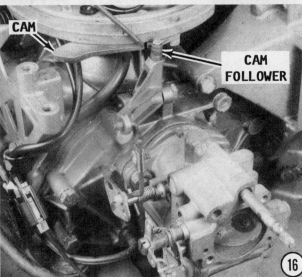

CAM

CAM FOLLOWER

16

low-speed idle by turning the low-speed needle valve **CLOCKWISE** until the engine begins to misfire or the rpm drops noticeably. From this point, rotate the needle valve **COUNTERCLOCKWISE** until the engine is operating at the highest rpm. Advance the throttle to the wide open position (WOT). Adjust the high-speed by rotating the high-speed needle valve **CLOCKWISE** until the number of rpm begins to drop, then rotate the high-speed valve **COUNTERCLOCKWISE** until the highest rpm is reached. Return the throttle to idle speed. Adjust the idle speed a second time as described earlier in this step. Again, advance the throttle to the WOT position and check the high-speed adjustment. Return the engine to idle speed. If the engine coughs and operates as if the fuel is too lean, but the idle and high-speed adjustments have been correctly made, then recheck the synchronization between the fuel and ignition systems. Now, shut off the fuel supply and allow the engine to run until it first begins to misfire from lack of fuel. Retard the spark and shut the engine down. Tighten the sleeve nut securely to prevent the needle valves from changing position through engine vibration while it is operating, but still allow the needle valves to be adjusted by hand using the knob on the end of the valve.

18- The idle stop is located on the port side of the engine, on the outside of the cowling. Adjust the nylon screw inward or outward to obtain the desired idle speed.

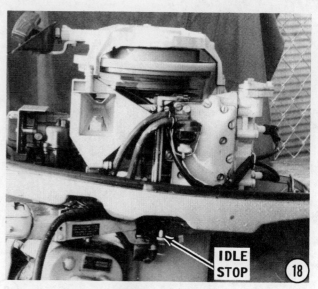

4-7 CHOKE SYSTEM SERVICE

FIRST THESE WORDS

Two different choke systems are used on the Type II carburetors covered in this manual. One system utilizes a manual choke; the other system has an electric choke. These choke systems should not be confused with the primer choke system covered in this chapter beginning on Page 4-55.

The electric choke system is equipped with a manual override to permit operation of the choke without electrical power (a dead battery). Detailed instructions are outlined in this section for removal of the electric choke mounted on the bottom of the carburetor. The following section, 4-8, provides procedures for rebuilding the Type II carburetors. Instructions to install the electric choke system to the carburetor begin on Page 4-30.

The larger powerheads covered in this manual (25 hp thru 60 hp) are equipped with electric starter motors and a generator system. In most cases, these items must be

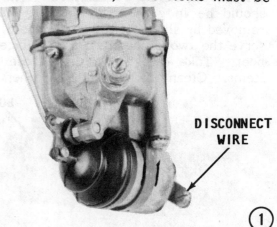

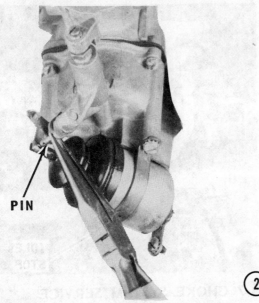

PIN

②

shifted out of the way to provide clearance for the carburetor. Only attaching hardware need be removed in order to shift the item to provide carburetor clearance. Electrical connections may be left intact.

ALL ELECTRIC CHOKE — REMOVAL

The electric choke is mounted on the bottom side of the carburetor.

1- Disconnect the electrical wire from the choke solenoid. Disconnect the fuel line from the carburetor. Remove the two nuts securing the carburetor to the intake manifold. Lift the choke from the engine. Scribe a mark on the side of the bracket to ensure the choke solenoid will be installed in its orginal position.

2- Remove the cotter key, washers, and pin from the choke plunger.

3- Remove the two screws securing the clamp to the carburetor. Remove the choke solenoid from the carburetor. The choke solenoid **CANNOT** be serviced. The boot should be in good condition. It may be removed by sliding it off the solenoid. Observe the two washers and spring under the boot. Take care not to lose these three items. Clean the solenoid, and then slide

SPRING   BOOT

③

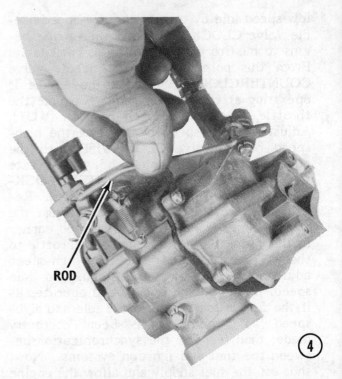

ROD

④

the boot back in place. The solenoid is then ready to be installed.

4- Remove the cotter pins from each end of the rod extending from the upper section to the lower section of the carburetor, and remove the rod. This rod works the choke in the upper body of the carburetor when the electric choke is activated. Further disassembly of the all electric choke is not necessary. If the choke is to be assembled without rebuilding the carburetor, proceed directly to Page 4-30. If the carburetor is to be rebuilt at this time, proceed to Section 4-8.

ADJUSTING KNOB

①

4-8 TYPE II CARBURETOR

DISASSEMBLING

The following procedures outline service procedures for the carburetor with an electric choke system installed. Except for the choke system, this carburetor is very similar to the Type I carburetors installed on Johnson/Evinrude Outboards and covered in the previous section.

1- Remove the knobs from the low-speed needle valve. This is accomplished by holding the knob firmly and at the same time backing out the retaining screw from the center of the knob. Once the screw is removed, the knob may be pulled free of the needle valve. If the carburetor being serviced has a front shield installed, remove the top screw.

2- Remove the two screws on the front side of the carburetor cover. Remove the cover from the front of the carburetor.

3- Remove the small screen and gasket.

4- Loosen the packing nut and rotate the low-speed needle valve counterclockwise to remove it from the carburetor. After the needle valve has been removed, back-out the packing nut.

5- The low-speed needle valve has a removable sleeve. To remove this sleeve, use an **OLD** needle valve as follows: First screw the old valve into the sleeve. Next clamp the needle valve in a vise. Now, tap on the carburetor and remove the packing and sleeve. The packing is installed in front of the sleeve.

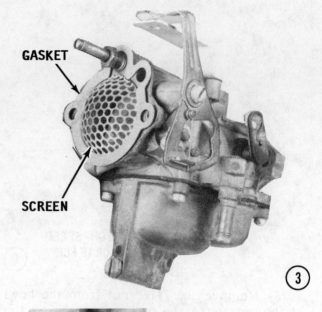

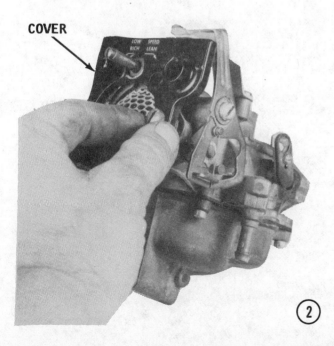

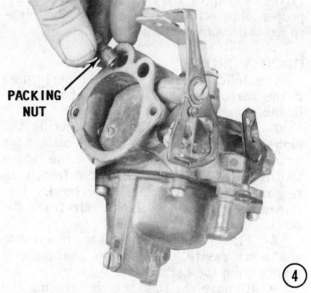

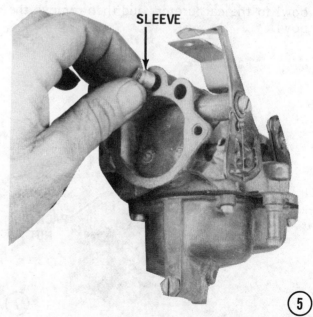

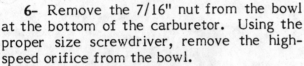

HIGH-SPEED
ORIFICE   ⑥

6- Remove the 7/16" nut from the bowl at the bottom of the carburetor. Using the proper size screwdriver, remove the high-speed orifice from the bowl.

GOOD WORDS

The following steps are to be performed if the carburetor being serviced has an adjustable high-speed needle valve.

7- Loosen the packing nut securing the high-speed needle valve to the carburetor bowl. Turn the high-speed needle valve **COUNTERCLOCKWISE** until it is free, then remove it from the carburetor bowl. Remove the carburetor flange nut from the bowl.

8- With a small screwdriver, reach into the bowl cavity and remove the packing glands from the carburetor.

9- Remove the four screws securing the bowl to the carburetor, and then remove the bowl.

PACKING
NUT   ⑦

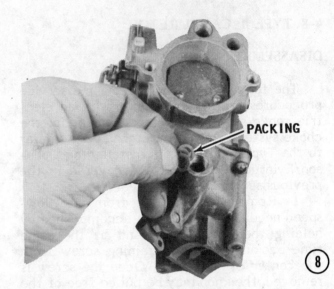

PACKING   ⑧

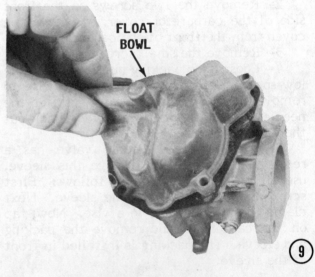

FLOAT
BOWL   ⑨

HIGH-SPEED
NOZZLE   ⑩

10- Remove the bowl gasket from the carburetor. Remove the gasket from the high-speed nozzle. Remove the high-speed nozzle from the carburetor using the proper size screwdriver to prevent possible damage to the nozzle. If difficulty is experienced in removing the high-speed nozzle, leave it in place. When the carburetor is immersed in the carburetor cleaner the high-speed nozzle will be cleaned suitable for further service.

11- Work the hinge pin free of the float and then remove the float.

12- Remove the inlet needle from the seat. Remove the seat and gasket from the carburetor.

Further disassembly of the carburetor is not necessary.

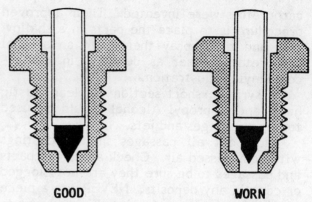

GOOD   WORN

Cross-section drawing to allow comparison of a new needle and seat with one badly worn. Notice how the edges of the worn valve and the seat have become beveled.

CLEANING AND INSPECTING

NEVER dip rubber parts, plastic parts, diaphragms, or pump plungers in carburetor cleaner. These parts should be cleaned **ONLY** in solvent, and then blown dry with compressed air.

Place all metal parts in a screen-type tray and dip them in carburetor cleaner until they appear completely clean, then blow them dry with compressed air.

GOOD WORDS

Since 1983 OMC states that carburetor parts should **NOT** be submerged in carburetor cleaner, as has been the practice since

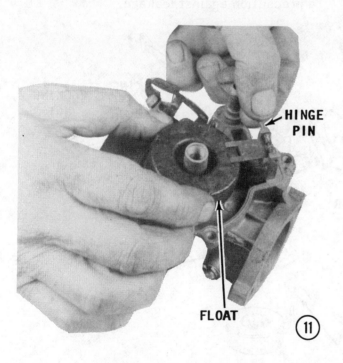

HINGE PIN

FLOAT

⑪

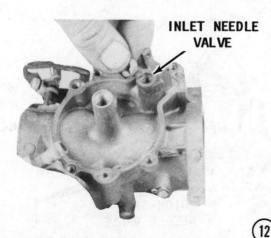

INLET NEEDLE VALVE

⑫

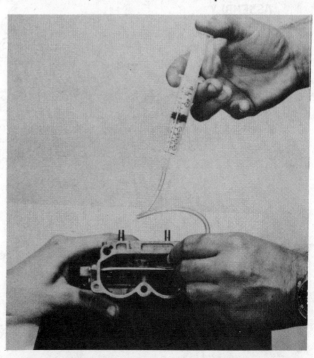

Using a Syringe, piece of clear plastic hose, and Isopropyl Alcohol to clean the idle air bleed system.

carburetors were invented. Their approved procedure is to place the parts in a shallow tray and then spray them with an aerosol carburetor cleaner as depicted in the accompanying illustration.

A syringe, short section of clear plastic hose, and Isopropyl Alcohol should be used to clear passages and jets.

Blow out all passages in the castings with compressed air. Check all of the parts and passages to be sure they are not clogged or contain any deposits. **NEVER** use a piece of wire or any type of pointed instrument to clean drilled passages or calibrated holes in a carburetor.

Move the throttle shaft back-and-forth to check for wear. If the shaft appears to be too loose, replace the complete throttle body because individual replacement parts are **NOT** available.

If any part of the float is damaged, the unit must be replaced. Check the float arm needle contacting surface and replace the float if this surface has a groove worn in it.

Inspect the tapered section of the idle and high-speed adjusting needles and replace any that have developed a groove.

If a high-speed orifice is installed on the carburetor being serviced, check the orifice for cleanliness. The orifice has a stamped number. This number represents a drill size. Check the orifice with the shank of the proper size drill to verify the proper orifice is used. The local OMC dealer will be able to provide the correct size orifice for the engine and carburetor being serviced.

Most parts that should be replaced during a carburetor overhaul are included in overhaul kits. These kits are available from your local marine dealer. One of these kits will contain a matched fuel inlet needle and seat. This combination should be replaced each time the carburetor is disassembled as a precaution against leakage.

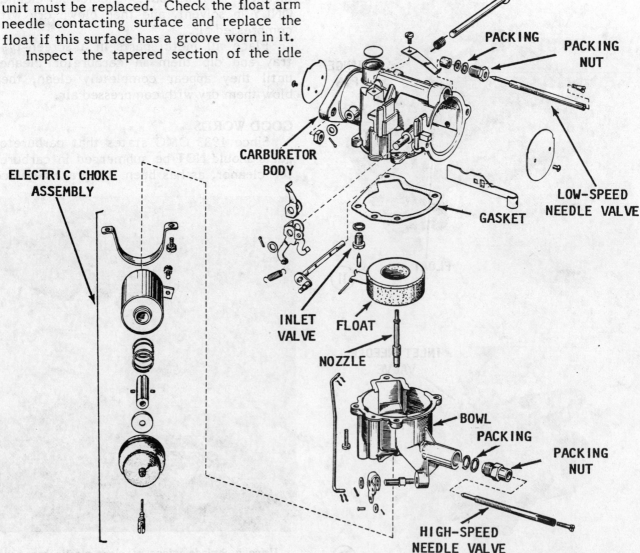

Exploded view of a Type II carburetor installed on larger horsepower engines equipped with the all-electric choke.

ASSEMBLING TYPE II CARBURETOR

Purchase of a carburetor repair kit is almost a must when servicing this unit. All parts required for a complete rebuild, including the necessary gaskets, are contained in the kit. Most repair kits contain more parts and gaskets than are needed because the kit may be used to service a wide range of carburetor models.

1- Install the **NEW** inlet seat and gasket into the carburetor base. Take care to use the proper size screwdriver as a precaution against damaging the inside surface of the the seat. Apply just a drop of oil into the seat, to prevent the needle from sticking when it is installed. Insert the inlet needle into the seat.

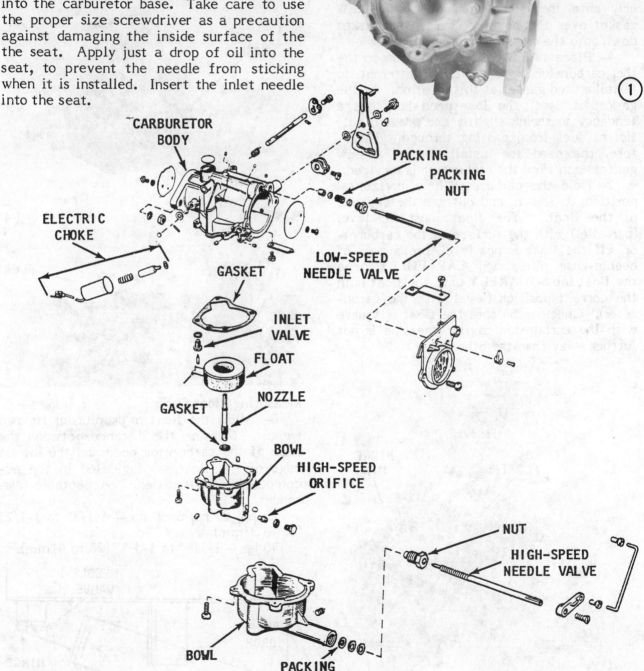

INLET NEEDLE VALVE

①

Exploded view of the Type II carburetor with electric choke mounted on the side of the carburetor instead of on the bottom. The illustration shows the two types of fuel bowls. The upper bowl has the high-speed orifice, and the lower bowl, the high-speed needle valve. The needle valve may be adjusted, the orifice is non-adjustable.

2- Install the **NEW** float and hinge pin from the repair kit with the tang of the float facing **DOWNWARD** toward the inside of the carburetor.

3- If the inlet nozzle was removed, install the nozzle into the center of the carburetor and secure it tightly with the proper size screwdriver. Exercise care not to burr the edges of the nozzle. The end of the nozzle seats in the bowl of the carburetor and burrs on the nozzle will result in fuel leakage, or the bowl will **NOT** fit properly onto the carburetor. Install a **NEW** gasket over the nozzle. Force the gasket down onto the carburetor stem.

4- Place a **NEW** gasket in position on the the carburetor base. **NEVER** attempt to install a used gasket at this location. As the gasket is used, the low-speed hole has a tendancy to shrink slightly and prevent sufficient fuel from passing through. Therefore, the need for installation of a new gasket each time the carburetor is serviced.

5- Hold the carburetor in a horizontal position, as shown, and observe the attitude of the float. The float must be level (parrallel) with the surface of the carburetor. If the float is not level, use a pair of needle-nose pliers and **CAREFULLY** bend the float tab **SQUARELY** until the float is in the correct position (level with the carburetor). Check to be sure the float is square with the carburetor cavity (one side is not further away than the other).

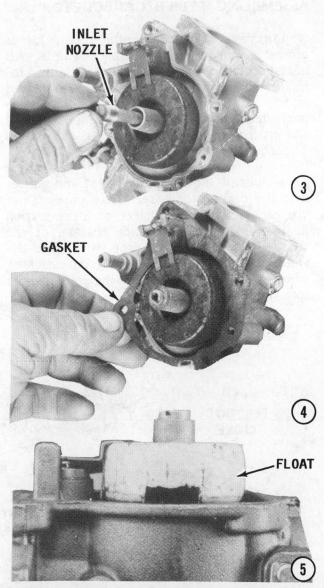

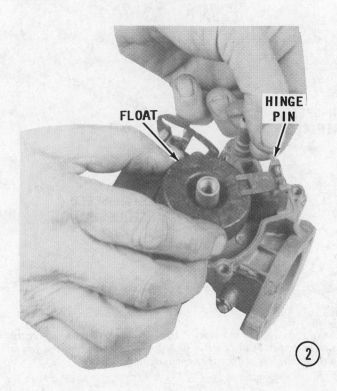

Measuring Float Drop

6- Allow the float to drop under its own weight. Measure the distance between the base of the carburetor body and the lowest edge of the float, as indicated in the accompanying illustration. Acceptable distance:

20 hp, 25 hp, & 30 hp -- 1-1/8" to 1-1/2" (28 to 33mm).

40 hp -- 1-1/8" to 1-5/8" (28 to 41mm).

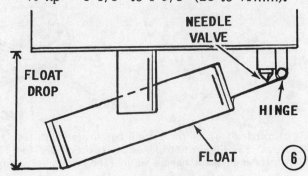

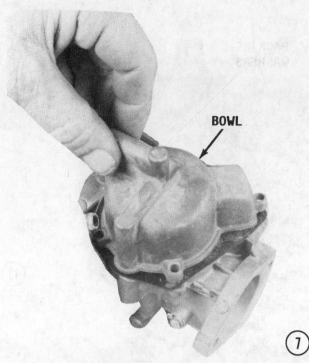

BOWL

⑦

7- Lower the bowl onto the carburetor and secure it with the four screws.

8- Install the bushing into the low-speed needle valve opening.

9- Insert the two fiber washers and the one nylon washer into the low-speed cavity.

10- Start the low-speed needle valve packing nut into the opening. **DO NOT** tighten the nut at this time. Thread the low-speed needle valve into the low-speed opening. Continue threading it into the opening until it barely seats. After the needle valve seats, back it out 1-1/2 turns **COUNTERCLOCKWISE.** Now, tighten the packing nut until there is drag on the needle

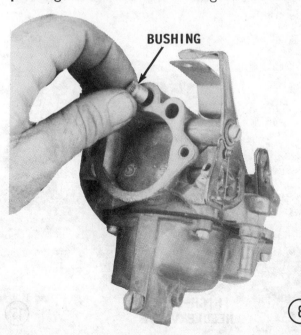

BUSHING

⑧

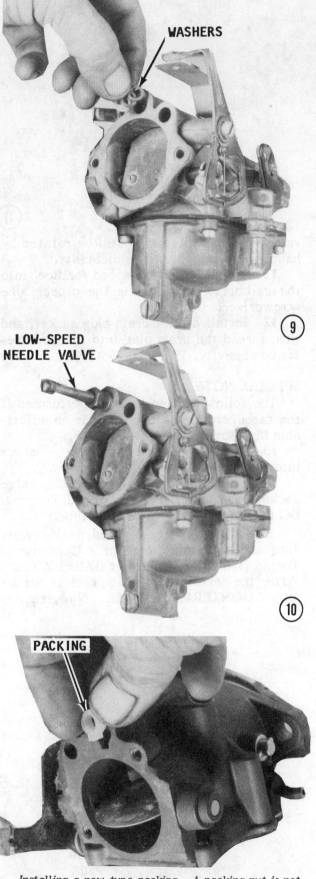

WASHERS

⑨

LOW-SPEED
NEEDLE VALVE

⑩

PACKING

Installing a new type packing. A packing nut is not used to hold the needle adjustment.

HIGH SPEED
ORIFICE

(11)

PACKING
WASHERS

(13)

valve, but the valve may still be rotated by hand, but with just a little difficulty.

11- Install the high-speed orifice into the carburetor bowl, using the proper size screwdriver.

12- Install a **NEW** drain plug gasket, and then thread the drain plug into the carburetor bowl cavity.

SPECIAL NOTE

The following step is to be performed if the carburetor being serviced has an adjustable high-speed needle valve.

13- Install the three packing washers into the carburetor bowl.

14- Start the high-speed needle valve packing nut into the high-speed opening. **DO NOT** tighten the nut at this time.

15- Thread the high-speed needle valve into the high-speed opening. Continue to thread it into place until it **BARELY** seats. After the needle valve seats, back it out 3/4 turn **COUNTERCLOCKWISE**. Now, tighten

PACKING
NUT

(14)

DRAIN
PLUG

(12)

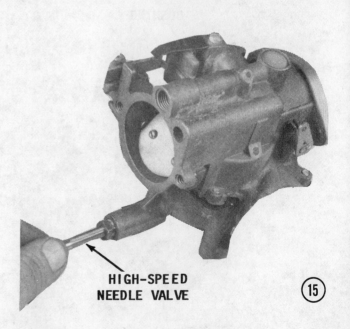

HIGH-SPEED
NEEDLE VALVE

(15)

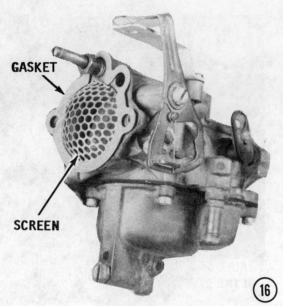

GASKET

SCREEN

16

the packing nut until there is drag on the needle valve, but the valve may still be rotated by hand, but with just a little difficulty.

16– Install a **NEW** gasket and screen onto the front of the carburetor.

17– If a front cover is used on the unit being serviced, install the front cover and secure it in place with the two attaching screws.

18– Install the spring steel lever into the indent of the manual choke on the bottom side of the carburetor. This lever holds the choke in the neutral position and prevents loose movement of the handle.

19– Install the low-speed needle valve (adjusting) knob by sliding it onto the valve stem, and then secure it in place with the screw.

FRONT COVER

17

CHOKE LEVER

18

ADJUSTING KNOB

19

ROD

1

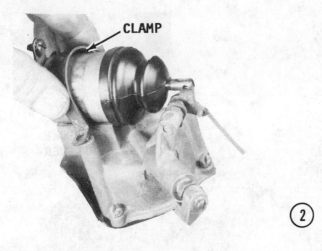

CLAMP

②

All Electric Choke Installation

1- Install the rod extending from the upper choke assembly to the lower hinge of the choke. Secure the rod in place with a cotter pin at each end.

2- Insert the choke assembly into the cavity of the float bowl with the electrical connector facing **DOWN-WARD** to permit installation of the electrical wires after the carburetor is installed. Now, bring the clamp over the top of the choke to secure the choke assembly in place.

3- Install the pin through the end of the shaft of the choke solenoid and into the lever attached to the carburetor. Install the washer and cotter pin.

4- Clean the surface of the intake manifold thoroughly. Check to be sure all old gasket material has been removed. Place a **NEW** gasket over the studs and into place on the manifold. Slide the carburetor onto the intake manifold studs and secure it in place with the attaching nuts. Tighten the nuts **EVENLY** and **ALTERNATELY**. Connect the

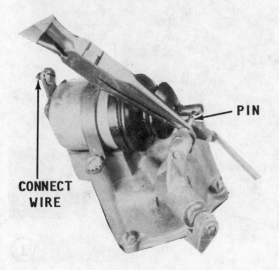

CONNECT WIRE

PIN

③

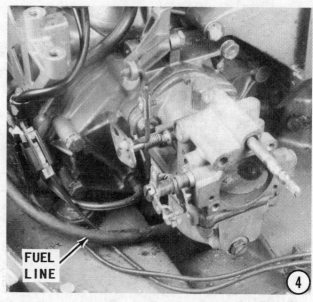

FUEL LINE

④

fuel line to the carburetor. Attach the electrical wires to the choke solenoid. Install the starter motor and generator if the engine being serviced is equipped with these two units. Now, proceed make the carburetor adjustments under a load condition, as outlined in the following paragraphs.

TYPE II CARBURETOR ADJUSTMENTS

GOOD WORD

Under all conditions, the ignition and fuel system **MUST** be synchronized before the fine adjustments to the carburetor are made. See Chapter 5. After the synchronization has been completed, proceed with the following work.

IDLE STOP

FUEL LINE

⑤

5- Mount the engine in a test tank or body of water. If this is not possible, connect a flush attachment and garden hose to the lower unit. **ONLY** the low-speed adjustment may be made using the flush attachment. If the engine is operated above idle speed with no load on the propeller, the engine could **RUNAWAY** resulting in serious damage or destruction of the unit.

CAUTION: Water must circulate through the lower unit to the engine any time the engine is run to prevent damage to the water pump in the lower unit. Just five seconds without water will damage the water pump.

Start the engine and allow it to warm to operating temperature. Adjust the low-speed idle by turning the low-speed needle valve **CLOCKWISE** until the engine begins to misfire or the rpm drops noticeablly. From this point, rotate the needle vavle **COUNTERCLOCKWISE** until the engine is operating at the highest rpm.

If the engine is equipped with an adjustable high-speed needle valve, shift the engine into **FORWARD** gear, and then advance the throttle to the wide open position (WOT). **NEVER** attempt to make this adjustment with a flush attachment and garden hose attached to the lower unit. Adjust the high-speed by rotating the high-speed needle valve **CLOCKWISE** until the number of rpm begins to drop, then rotate the high-speed needle valve **COUNTERCLOCKWISE** until the highest rpm is reached. Return the throttle to idle speed. Adjust the idle speed a second time as described earlier in this step. Again, advance the throttle to the WOT position and check the high-speed adjustment. Return the engine to idle speed. If the engine coughs and operates as if the fuel is too lean, but the idle and high-speed adjustments have been correctly made, then recheck the synchronization between the fuel and ignition systems. Now, shut off the fuel supply and allow the engine to run until it first begins to misfire from lack of fuel. Retard the spark and shut the engine down. Tighten the sleeve nut securely to prevent the needle valves from changing position through engine vibration while it is operating, but still allow the needle valves to be adjusted by hand using the knob on the end of the valve.

The idle stop is located on the port side of the engine, on the outside of the cowling. Adjust the nylon screw inward or outward to obtain the desired idle speed.

4-9 TYPE III CARBURETOR

This carburetor is installed on the 9.5hp engine. The unit has an adjustable low-speed needle valve and fixed high-speed orifice. The only changes that have been made to this particular model carburetor over the years was a redesign of the idle needle valves. On early model 9.5hp engines, the reeds set directly below the carburetor. On later models, the reeds were removed from this location and installed between the engine block and the intake manifold.

The early model idle needle valves extended out of the carburetor with an O-ring, spring, and E-clip. The needle valve was controlled from the front of the engine by means of a flexible cable. On this model carburetor, the cable and control knob must be removed before the carburetor is removed from the engine.

Later model carburetors are equipped with a long packing nut and linkage. An adjustable knob located on the front of the engine controls movement of the valve. When servicing the late model carburetors, the low-speed needle valve adjustment knob and linkage must be removed before the carburetor is removed from the engine.

REMOVAL

1- Remove the choke rod that extends over the top of the carburetor by snapping the choke rod out of the nylon snap. Disconnect the fuel line from the carburetor.

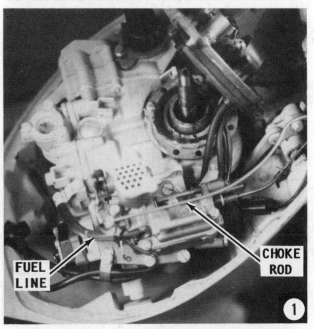

FUEL LINE   CHOKE ROD

①

PACKING
NUT

②

2- If the carburetor has a packing nut
with the needle through the nut and a nylon
adjustment knob, remove the knob or remove the linkage from the knob.

3- If the carburetor has a flexible line to
the front of the engine, remove the knob on
the front of the engine and then turn the
flexible cable COUNTERCLOCKWISE until
the needle valve is removed from the carburetor.

SPECIAL NOTE

As the needle valve is being removed,
take care to retain the washer, O-ring, and
spring installed between the E-ring on the
valve and the carburetor. The washer and
spring will be used again. The O-ring may
be discarded.

4- Remove the five screws securing the
carburetor to the intake manifold. Notice
that four of the screws have slots and one is
a countersunk screw. Lift the carburetor
from the engine.

5- Remove the four screws securing the
float bowl, and then remove the float bowl.

6- Remove and DISCARD the bowl gasket.

CABLE

③

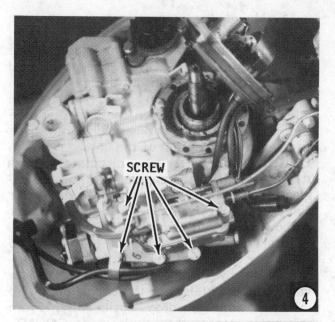

SCREW

④

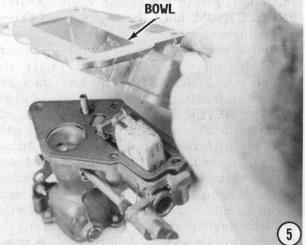

BOWL

⑤

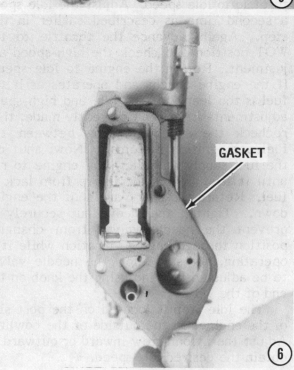

GASKET

⑥

HINGE
PIN

⑦

7- Remove the hinge pin, and then lift the float assembly from the carburetor body.

8- Reach inside the inlet seat and remove the inlet needle. Remove the inlet needle seat.

9- Remove the drain plug from the bottom of the float bowl. Use the proper size screwdriver and remove the high-speed orifice from the float bowl. Loosen the low-speed needle valve packing nut by turning it **COUNTERCLOCKWISE**. Now, remove the low-speed needle valve. Remove the packing nut and washers.

CLEANING AND INSPECTING

NEVER dip rubber parts, plastic parts, nylon parts, diaphragms, or pump plungers in carburetor cleaner. These parts should be cleaned **ONLY** in solvent, and then blown dry with compressed air.

Place all metal parts in a screen-type tray and dip them in carburetor cleaner until they appear completely clean, then blow them dry with compressed air.

Blow out all passages in the castings with compressed air. Check all parts and

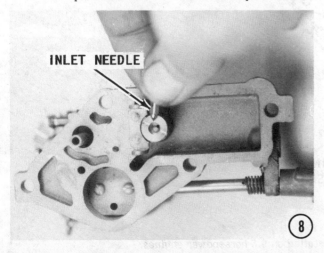

INLET NEEDLE

⑧

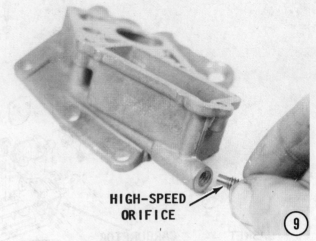

HIGH-SPEED
ORIFICE

⑨

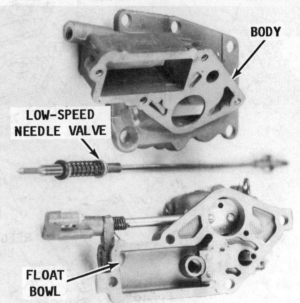

BODY

LOW-SPEED
NEEDLE VALVE

FLOAT
BOWL

Major parts of carburetor installed on a Johnson/ Evinrude 9.5 horsepower engine.

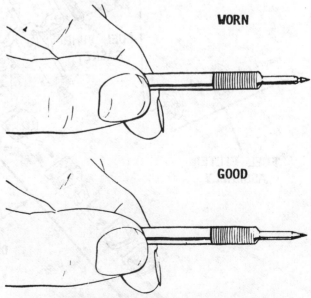

WORN

GOOD

Comparison of worn and new carburetor adjustment screws. The upper screw is unfit for further service.

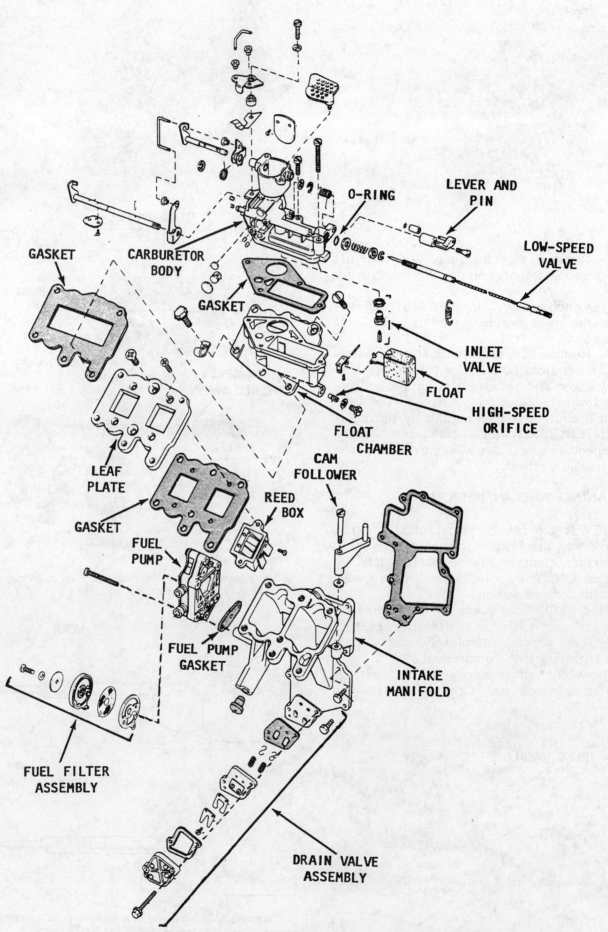

GASKET

CARBURETOR BODY

GASKET

LEVER AND PIN

O-RING

LOW-SPEED VALVE

INLET VALVE

FLOAT

HIGH-SPEED ORIFICE

FLOAT CHAMBER

CAM FOLLOWER

LEAF PLATE

REED BOX

GASKET

FUEL PUMP

INTAKE MANIFOLD

FUEL PUMP GASKET

FUEL FILTER ASSEMBLY

DRAIN VALVE ASSEMBLY

Exploded view of a Type III carburetor installed on 9.5 horsepower engines.

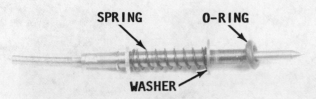

Low-speed needle valve and associated parts used in the carburetor of a 9.5 horsepower engine.

passages to be sure they are not clogged or contain any deposits. **NEVER** use a piece of wire or any type of pointed instrument to clean drilled passages or calibrated holes in a carburetor.

Move the throttle shaft back-and-forth to check for wear. If the shaft appears to be too loose, replace the complete throttle body because individual replacement parts are **NOT** available.

Inspect the main body, airhorn, and venturi cluster gasket surfaces for cracks and burrs which might cause a leak. Check the float for deterioration. Check to be sure the float spring has not been stretched. If any part of the float is damaged, the unit must be replaced. Check the float arm needle contacting surface and replace the float if this surface has a groove worn in it.

Inspect the tapered section of the idle adjusting needles and replace any that have developed a groove.

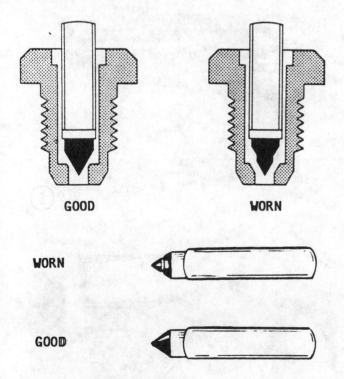

Needle and seat arrangement, showing a worn and new needle for comparison.

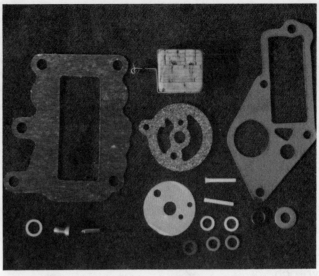

Parts included in a carburetor repair kit for the Johnson/Evinrude 9.5 horsepower engine.

If a high-speed orifice is installed on the carburetor being serviced, check the orifice for cleanliness. The orifice has a stamped number. This number represents a drill size. Check the orifice with the shank of the proper size drill to verify the proper orifice is used. The local OMC dealer will be able to provide the correct size orifice for the engine and carburetor being serviced.

Most of the parts that should be replaced during a carburetor overhaul are included in overhaul kits available from your local marine dealer. One of these kits will contain a matched fuel inlet needle and seat. This combination should be replaced each time the carburetor is disassembled as a precaution against leakage.

ASSEMBLING TYPE III CARBURETOR

1- Install the high-speed orifice into the float body using the proper size screwdriver to prevent burring the edges of the orifice.

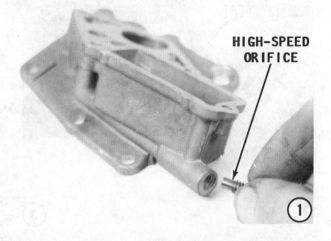

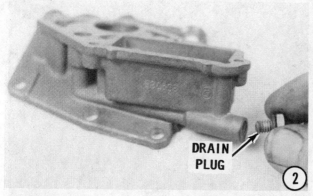

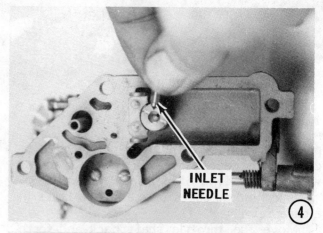

Any damage to the orifice will result fuel leakage and poor engine performance.

2- Install the drain plug with a **NEW** gasket and tighten it securely with a 7/16" wrench.

3- Install the inlet needle seat and gasket. **TAKE CARE** to use the proper size screwdriver to install the seat. If the inside diameter of the seat is damaged the needle valve will leak fuel causing a flooding condition in the carburetor.

4- Apply just a drop of oil into the seat, and then insert the inlet needle into seat.

5- Position a **NEW** float over the needle, and then slide a **NEW** hinge pin into place.

6- Hold the carburetor in a vertical position (up-and-down), and observe the float. The float **MUST** be parallel (align evenly), with the surface of the carburetor body. If the float is not parallel **CAREFULLY** bend the float ever so slightly, as shown, until the correct positioning is obtained.

7- Slide the float bowl gasket over the float and nozzle into position on the carburetor base.

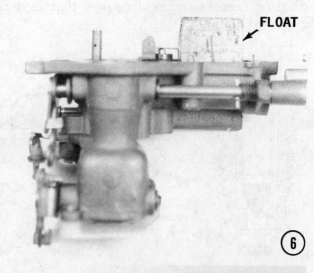

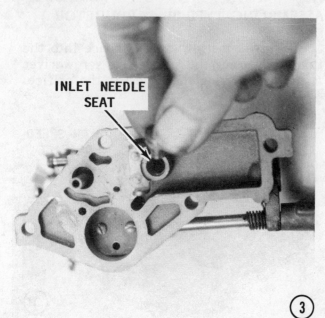

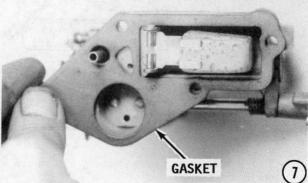

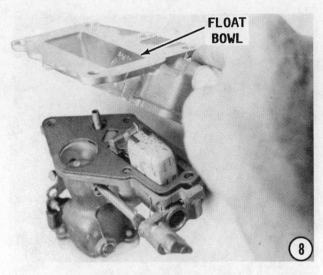

FLOAT BOWL

8

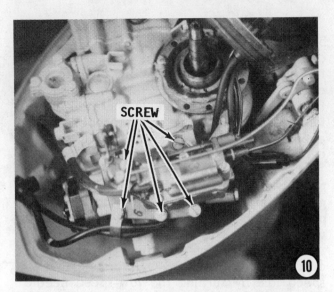

SCREW

10

8- Lower the float bowl down over the top of the carburetor body and secure it in place with the four attaching screws.

INSTALLATION

SPECIAL WORDS

If the carburetor being serviced has packing nuts and washers for the needle valves, perform Step 9. If the carburetor has the flexible line extending from the valve to the front of the engine, the needle valve will be installed AFTER the carburetor is in place on the engine.

9- Install the packing nut washers into the needle valve openings. Thread the packing nut into the opening but DO NOT tighten them at this time. Thread the low-speed needle valve into place until it just BARELY seats. From this position, back it out (COUNTERCLOCKWISE) 1-1/2 turns as a preliminary rough adjustment.

10- Check the surface of the intake manifold to be sure it has been thoroughly cleaned and is free of any old gasket material. Place a NEW gasket in position on the manifold. Set the carburetor into place on the manifold and secure it with the five attaching screws. OBSERVE that one of the screws is a countersunk type. This screw MUST be installed into the countersunk hole.

11- If the carburetor being serviced has the flexible low-speed needle valve arrangement, check to be sure the snap ring is in place and then install the spring, washer and NEW O-ring onto the needle. Apply just a drop of oil onto the O-ring to ease installation of the needle valve. Thread the low-speed needle valve into the carburetor until it just BARELY seats. From this poistion, back it out (COUNTERCLOCKWISE) 1-1/2 turns as a preliminary rough adjustment. Install the choke rod by snapping it into place in the nylon retainer. Connect the fuel line to the carburetor.

GOOD WORDS

It is best to synchronize the fuel and ignition systems at this time. See Chapter 5. After the synchronization has been completed, proceed with the following work.

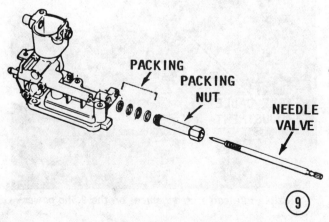

PACKING
PACKING NUT
NEEDLE VALVE

9

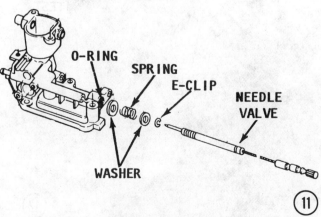

O-RING
SPRING
E-CLIP
NEEDLE VALVE
WASHER

11

12- Mount the engine in a test tank or body of water. If this is not possible, connect a flush attachment and garden hose to the lower unit. **NEVER** operate the engine above idle speed using the flush attachment. If the engine is operated above idle speed with no load on the propeller, the engine could **RUNAWAY** resulting in serious damage or destruction of the unit.

CAUTION: Water must circulate through the lower unit to the engine any time the engine is run to prevent damage to the water pump in the lower unit. Just five seconds without water will damage the water pump.

Start the engine and allow it to warm to operating temperature. Adjust the low-speed idle by turning the low-speed needle valve **CLOCKWISE** until the engine begins to misfire or the rpm drops noticeablly. From this point, rotate the needle valve **COUNTERCLOCKWISE** until the engine is operating at the highest rpm. If the engine coughs and operates as if the fuel is too lean, but the idle and high-speed adjustments have been correctly made, then recheck the synchronization between the fuel and ignition systems.

On engines with the flexible low-speed extension to the front of the engine, the spring maintains tension on the needle and adjustment will not be lost because of vibration during operation. On engines with the packing nut arrangement, the nut must be tightened securely to hold the adjustment. **HOWEVER,** do not tighten it to the point where an adjustment cannot be made by hand.

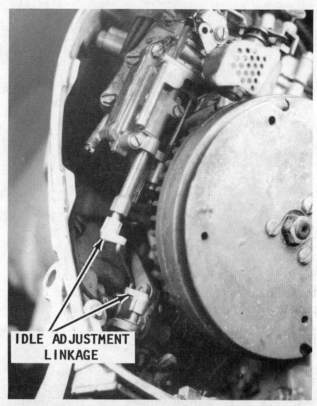

Idle adjustment linkage on a 9.5hp powerhead.

Flexible idle cable arrangement on the 9.5hp powerhead.

4-10 TYPE IV CARBURETOR

This carburetor is used on the 40hp (since 1984), 50hp, 55hp, and the 60hp powerheads covered in this manual.

Two carburetors are installed on each powerhead, one for each cylinder. Three different models of this type carburetor have been used since 1971, however, they are very similar. The major differences are in the idle adjustments and other minor engineering refinements.

The first model is the single barrel, front draft, with an adjustable low-speed needle valve, and special packing for the valve. It has a fixed high-speed orifice in the float bowl. This is usually considered the standard carburetor installation for the smaller horsepower engines.

The second model is also a single barrel, front draft, but with a fixed low-speed, intermediate, and high-speed orifice. Each of these orifices play an important role in metering the amount of fuel to the cylinder as the throttle is advanced from idle to the wide open throttle (WOT) position.

The third model of the Type IV carburetor is a single barrel, front draft, with a fixed low-speed and fixed high-speed orifice. This model is almost identical to the second except for the absence of the intermediate orifice.

All of the Type IV carburetors utilize an electric choke mounted on the side of the engine. On most models, the electric choke does not have to be removed in order to remove and service the carburetor.

REMOVAL

Preliminary Tasks

1- Disconnect the battery cables at the battery as a precaution against an accidental spark igniting the fuel or fuel fumes present during the service work. Disconnect the fuel line from the junction on the port side of the engine. This is the line from the fuel pump to the carburetors. Remove the cowling. Remove the low-speed adjustment knobs at the bottom of the air silencer and remove the silencer cover.

2- Disconnect the hose at the bottom of the air silencer. This hose is connected to a fitting at the bottom of the crankcase, and then the air silencer.

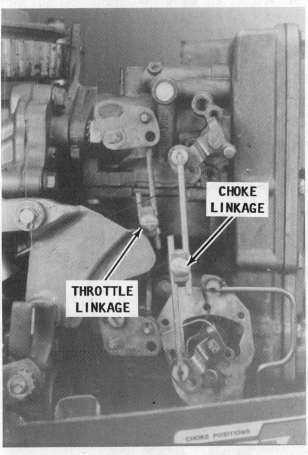

Starboard side view showing the throttle linkage and the choke linkage.

One method of disconnecting the fuel line is to remove the fuel pump cover.

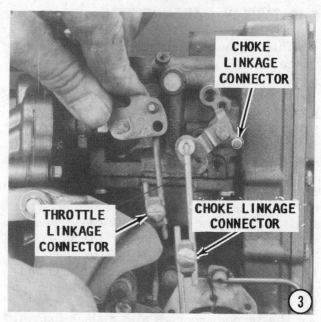

4- Remove the four nuts securing the carburetor to the intake manifold. Identify the top carburetor as an aid to installation. The carburetors **MUST** be installed in their original positions. The fuel connections and the choke arrangement is different for each carburetor. Both carburetors can now be removed as an assembly. The carburetor parts should be kept with the individual unit to ensure all items are installed back in their original positions.

Only one carburetor will be rebuilt in the following procedures. Service of the second unit is to be performed in the same manner.

3- Disconnect the choke and throttle linkage between the two carburetors on the port side of the engine. The linkage will snap out of a retainer on the carburetor cams. The nylon retainers **MUST** be removed before the carburetor is immersed in any type of cleaning solution. Disconnect the choke wire from the electric choke, by first removing the O-ring and coil spring.

DISASSEMBLING

5- Disconnect the fuel hoses between the two carburetors. This is accomplished by simply cutting the tie strap or working the clip securing the hose to the carburetor fitting. Remove the drain plug or the high-speed orifice plug from the bottom of the carburetor bowl.

6- Remove the orifice from the float bowl, using the proper size screwdriver.

7- Remove the low-speed needle valve, if used. After removing the needle valve, observe and remove the retainer that accepts the needle valve. Good shop practice is to replace the retainer, because the retainer is actually the component holding the needle valve in adjustment. If the retainer has become worn, it is not possible to hold the needle valve in an accurate adjustment.

SPECIAL WORDS

The low-speed needle valve has a bearing deep inside the carburetor. Removal of this bearing is no small task. A special tool is not available. However, a paper clip with a hook on the end, can be used to reach inside the bore and remove the bearing. Actually this bearing is made of plastic and the needle valve rotates inside. The bearing centers the needle valve in the carburetor.

8- Remove the low-speed needle valve bearing using a paper clip as described in the previous paragraph.

If the carburetor being serviced does not have the low-speed needle valve, remove the screw plug and washer from the top of

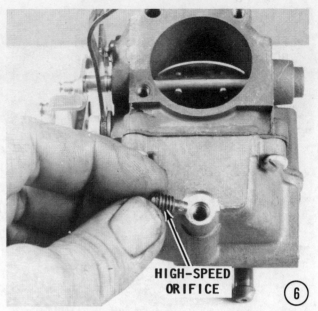

HIGH-SPEED ORIFICE ⑥

the carburetor. Remove the orifice using the proper size screwdriver.

If the carburetor being serviced has the intermediate orifice, remove the plug and gasket from the starboard side of the carburetor. Remove the intermediate orifice using the proper size screwdriver.

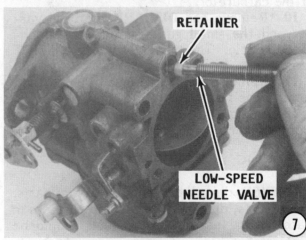

RETAINER

LOW-SPEED NEEDLE VALVE ⑦

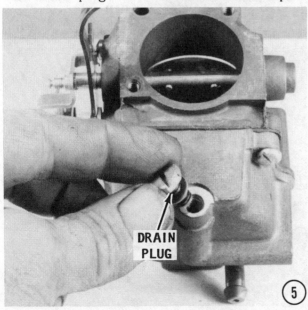

DRAIN PLUG ⑤

BEARING ⑧

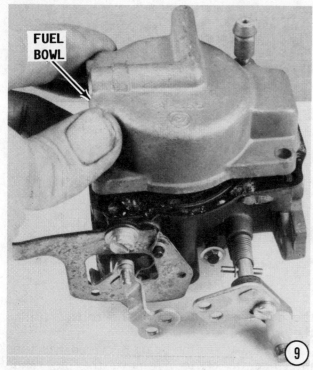

9- Turn the carburetor upside-down and remove the screws securing the float bowl to the carburetor body. Lift the bowl free of the carburetor.

10- Remove and **DISCARD** the bowl gasket and the small gasket around the high-speed nozzle. The high-speed nozzle is **NOT** removable.

11- Remove the float hinge pin, and then remove the float.

12- Remove the inlet needle and seat from the carburetor body.

CLEANING AND INSPECTING

NEVER dip rubber parts, plastic parts, nylon parts, diaphragms, or pump plungers in carburetor cleaner. These parts should be

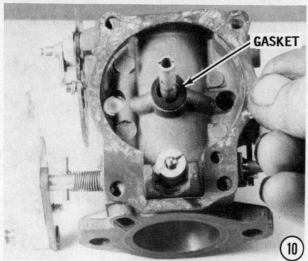

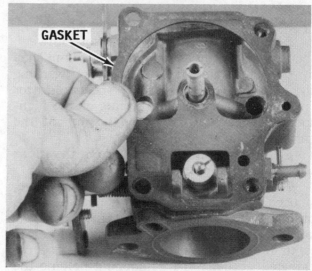

Obsolete one-piece gasket. This type gasket will only be found on an older carburetor that has never been rebuilt. A new two-piece gasket, one for the bowl and other for the nozzle, replaces the one shown.

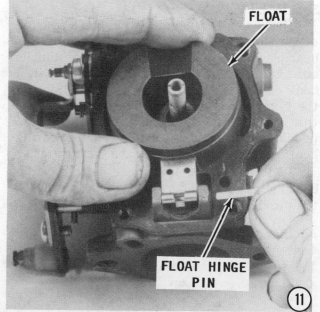

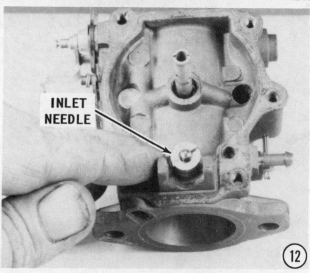

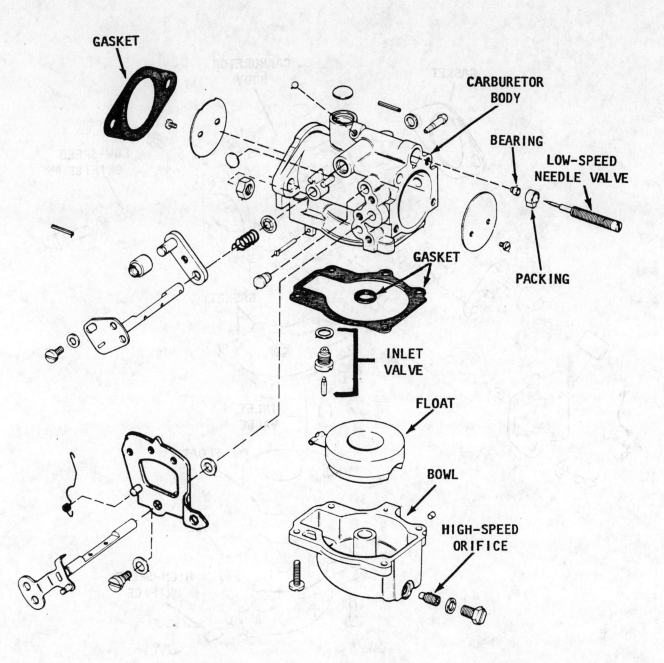

Exploded view of Type IV carburetor with fixed high-speed orifice and adjustable low-speed needle valve. These carburetors were installed on 50 hp models, 1971-75 and 1978. They were also used on the 1976, 55 hp models.

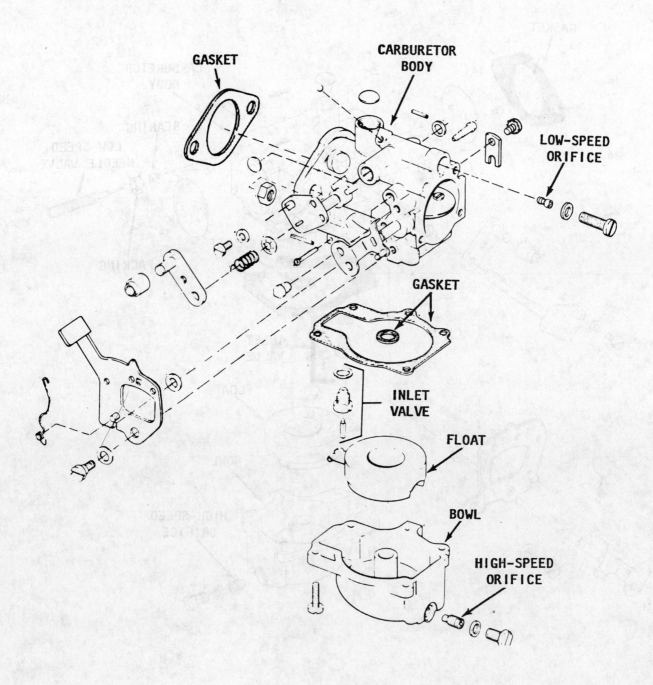

Exploded view of Type IV carburetor with fixed low-speed orifice and fixed high-speed orifice. These carburetors were installed on the 50 hp, 1979; the 55 hp, 1979-81; and on the 60 hp, 1980-81.

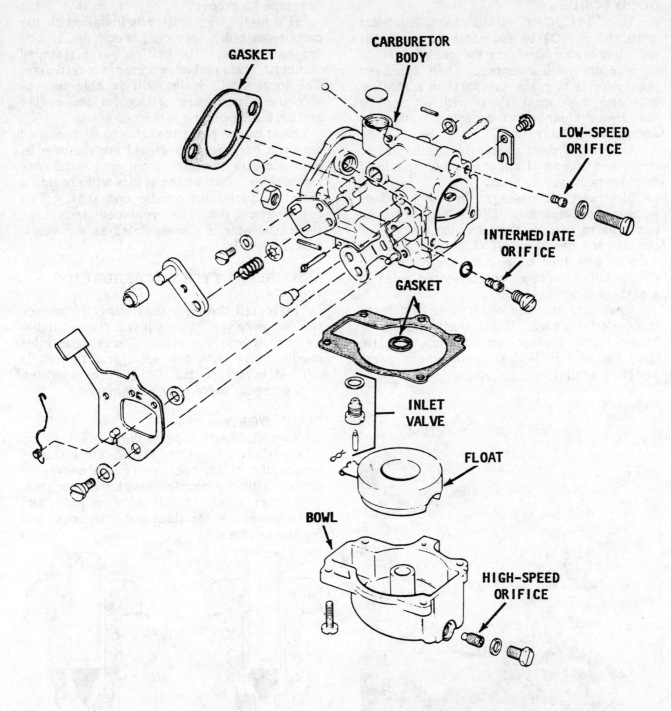

GASKET

CARBURETOR BODY

LOW-SPEED ORIFICE

INTERMEDIATE ORIFICE

GASKET

INLET VALVE

FLOAT

BOWL

HIGH-SPEED ORIFICE

Exploded drawing of a Type IV carburetor with fixed low-speed, fixed intermediate, and fixed high-speed orifices. These carburetors are installed on the 40hp — 1984 and on; 55hp — 1977-78 and the 50hp — 1980 and on.

cleaned **ONLY** in solvent, and then blown dry with compressed air.

Place all metal parts in a screen-type tray and dip them in carburetor cleaner until they appear completely clean, then blow them dry with compressed air.

GOOD WORDS

Since 1983 OMC states that carburetor parts should **NOT** be submerged in carburetor cleaner, as has been the practice since carburetors were invented. Their approved procedure is to place the parts in a shallow tray and then spray them with an aerosol carburetor cleaner as depicted in the accompanying illustration.

A syringe, short section of clear plastic hose, and Isopropyl Alcohol should be used to clear passages and jets.

Blow out all passages in the castings with compressed air. Check all parts and passages to be sure they are not clogged or contain any deposits. **NEVER** use a piece of wire or any type of pointed instrument to clean drilled passages or calibrated holes in a carburetor.

Move the throttle shaft back-and-forth to check for wear. If the shaft appears to be too loose, replace the complete throttle body because individual replacement parts are **NOT** available.

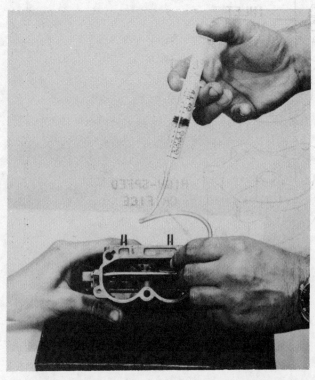

Using a Syringe, piece of clear plastic hose, and Isopropyl Alcohol to clean the idle air bleed system.

If any part of the float or spring is damaged, the unit must be replaced. Check the float arm needle contacting surface and replace the float if this surface has a groove worn in it.

Inspect the tapered section of the idle adjusting needles and replace any that have developed a groove.

If a high-speed orifice is installed on the carburetor being serviced, check the orifice for cleanliness. The orifice has a stamped number. This number represents a drill size. The local OMC dealer will be able to provide the correct size orifice for the engine and carburetor being serviced.

Most of the parts that should be replaced during a carburetor overhaul are included in overhaul kits available from your local marine dealer. One of these kits will contain a matched fuel inlet needle and seat. This combination should be replaced each time the carburetor is disassembled as a precaution against leakage.

ASSEMBLING TYPE IV CARBURETOR

1- Install the inlet seat using the proper size screwdriver. Inject just a drop of lightweight oil into the seat. Thread the inlet needle valve into the seat, and tighten it until it barely seats. **DO NOT** overtighten or the needle valve may be damaged.

GOOD WORD

Two different type gaskets are used on this carburetor. One **MUST** be installed before the float, because it is a one-piece gasket with the nozzle gasket incorporated. The other type consists of two individual gaskets, one for the float and a separate one for the nozzle.

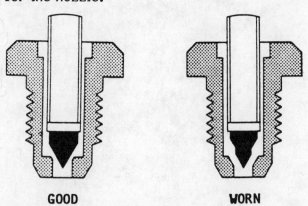

GOOD   WORN

Cross-section drawing to allow comparison of a new needle and seat with one badly worn. Notice how the edges of the worn valve and the seat have become beveled.

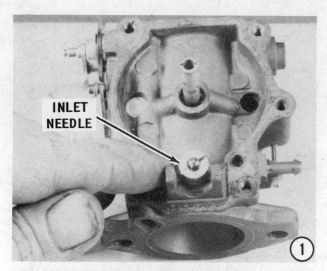

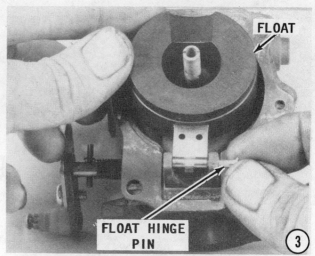

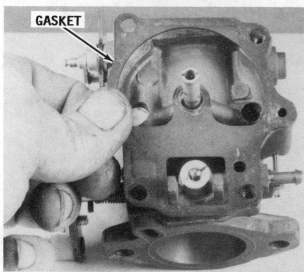

2- Postion **NEW** gaskets in place on the carburetor body and on the nozzle.

3- Lower the float down over the nozzle, and then slide the hinge pin into place.

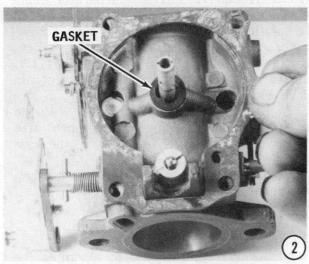

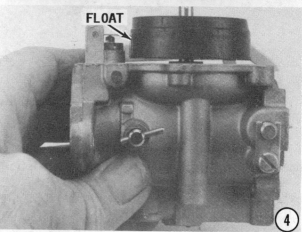

4- Hold the carburetor body in a horizontal position and check to be sure the float is parallel to the carburetor surface, as shown. If the float is not parallel, **CAREFULLY** bend the float tang until the float is in a parallel position and both sides are equal distance from the carburetor surface.

Measuring Float Drop

5- Allow the float to drop under its own weight. Measure the distance between the base of the carburetor body and the lowest edge of the float, as indicated in the accompanying illustration. The measurement should be 1-1/8" to 1-5/8" (28-41mm). Place the float bowl in position, and then secure it in place with the attaching screws.

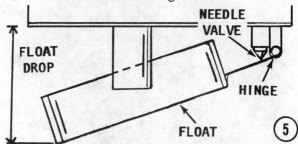

Obsolete one-piece gasket. This type gasket will only be found on an older carburetor that has never been rebuilt. A new two-piece gasket, one for the bowl and other for the nozzle, replaces the one shown.

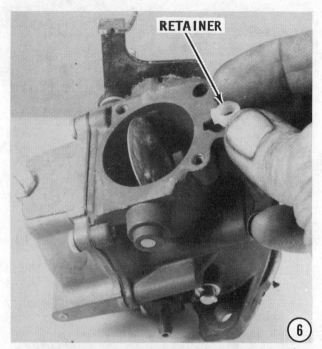

6- Install the low-speed retainer in the carburetor body.

7- Slide the nylon bearing onto the end of the low-speed needle valve, and then thread the valve into the retainer. Continue threading the valve into the retainer until it seats **LIGHTLY**, and then back it out **COUNTERCLOCKWISE** 5/8 turn.

WORDS OF CAUTION

Take time to use the proper size screwdriver to install an orifice. If the orifice is damaged and the edge of the opening burred, because the wrong size screwdriver was used, the flow of fuel will be restricted and the orifice will not function properly. Damage to the orifice will also make it very difficult to remove during the next carburetor service.

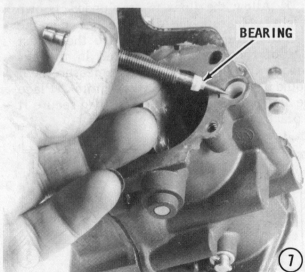

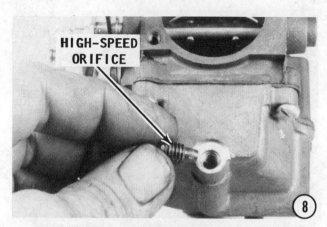

If an orifice is used instead of the needle valve, install the low-speed orifice into its recess, using the proper size screwdriver to prevent damaging the orifice. Tighten the orifice until it seats **LIGHTLY**. Thread the plug, with a **NEW** gasket, into place and tighten it snugly. If the carburetor being serviced has the intermediate orifice, install the orifice, using the proper size screwdriver to prevent damaging the orifice. Tighten the orifice until it seats **LIGHTLY**. Thread the plug, with a **NEW** gasket, in place and tighten it snugly.

8- Install the high-speed orifice into the carburetor bowl, using the proper size screwdriver to prevent damaging the orifice.

9- Thread the plug, with a new gasket, into place and tighten it snugly.

SECOND CARBURETOR

Perform Steps 1 thru 9 to assemble the second carburetor.

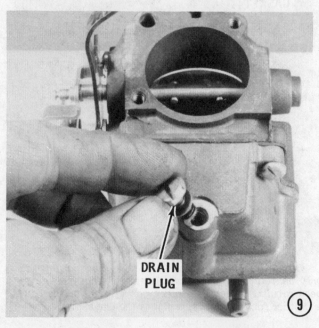

INSTALLATION

10- Connect the fuel line between the top and bottom carburetor. Clean the mating surface of the intake manifold. Check to be sure all old gasket material has been removed. Slide a **NEW** gasket down over the studs into place on the manifold. Install both carburetors at the same time onto the studs and secure them in place with the nuts. Tighten the nuts **ALTERNATELY** and **EVENLY**. Connect the fuel line between the fuel pump and the carburetors.

11- On the starboard side of the engine: Install the four choke and throttle retainers, two for each carburetor. Connect the choke and throttle linkage between the two carburetors. Connect the choke coil solenoid wire onto the linkage stud, and then slide the **O**-ring and coil spring onto the stud to secure the wire in place. Adjust the choke butterflys by loosening the screw between the top and bottom linkage. Make the adjustment to close **BOTH** choke butterflys, then tighten the screw.

12- Adjust the throttle butterflys in the same manner. Loosen the screw between the top and bottom linkage. Make the adjustment to close **BOTH** throttle butterflys, and then tighten the screw.

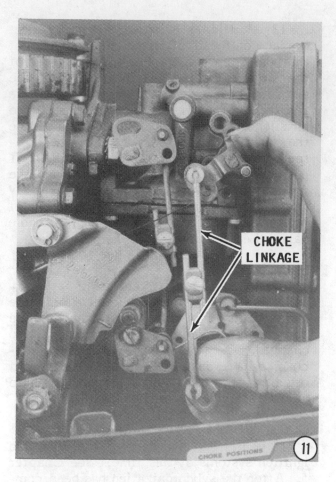

CHOKE LINKAGE

⑪

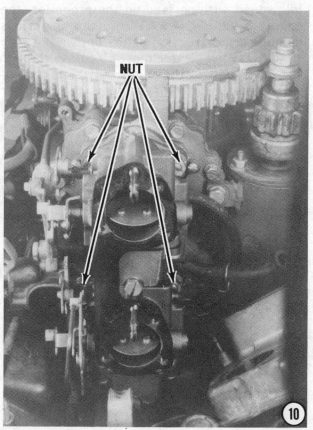

NUT

⑩

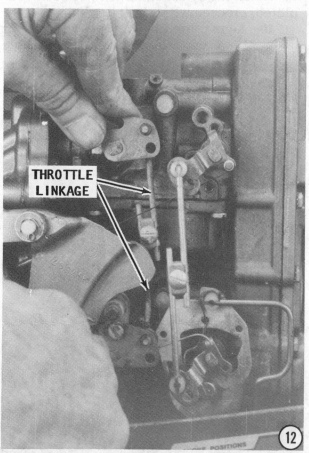

THROTTLE LINKAGE

⑫

13- Position a **NEW** air silencer gasket in place on the front of the engine. Connect the hose from the crankcase to the back of the air silencer. Coat the threads of the air silencer retaining screws with Loctite and then install the air silencer. Install the air silencer cover and at the same time slide the plastic adjusting knob onto the low-speed needle for each carburetor. **DO NOT** push the knobs all the way home onto the needle at this time. You may seriously consider **NOT** to install and use the piece of linkage connecting the low-speed needle valve of each carburetor.

EXPLANATION

This linkage permits adjusting both carburetors simultaneously while using only one knob. Sounds great! But when the one carburetor is adjusted, the other is also changed. A great many professional mechanics have discovered the linkage is not required and it is far more efficient to adjust each carburetor individually.

GOOD WORDS

It is best to synchronize the fuel and ignition systems at this time. See Chapter 5. After the synchronization has been completed, proceed with the following work.

14- Mount the engine in a test tank or body of water. If this is not possible,

Installing the outside cover of the air silencer. As the cover is worked into place, slide the plastic adjustments over the low-speed needle but do not lock them in place until after the adjustments have been made.

connect a flush attachment and garden hose to the lower unit. **NEVER** operate the engine above idle speed using the flush attachment. If the engine is operated above idle speed with no load on the propeller, the engine could **RUNAWAY** resulting in serious damage or destruction of the unit.

CAUTION: Water must circulate through the lower unit to the engine any time the engine is run to prevent damage to the water

pump in the lower unit. Just five seconds without water will damage the water pump.

Start the engine and allow it to warm to operating temperature. Pop the two rubber caps, one for each carburetor, out of the front cover of the air silencer. Adjust the low-speed idle by turning the low-speed needle valve **CLOCKWISE** until the engine begins to misfire or the rpm drops noticeably. From this point, rotate the needle vavle **COUNTERCLOCKWISE** until the engine is operating at the highest rpm. If the engine coughs and operates as if the fuel is too lean, but the idle and high-speed adjustments have been correctly made, then recheck the synchronization between the fuel and ignition systems.

On engines with the flexible low-speed extension to the front of the engine, the retainer maintains tension on the needle and adjustment will not be lost because of vibration during operation. After the idle has been adjusted, push the idle knob retainers onto the needle until it engages the spline on the low-speed needle. Repeat the procedure for the second carburetor. Replace the rubber caps into the front cover of the air silencer.

4-11 FUEL PUMP SERVICE

To test the non-serviceable pump, remove the pump from the engine, operate the squeeze bulb until it is firm, and then carefully observe the vacuum hole in the back side of the pump for any indication of fuel. The smallest amount of fuel indicates a damaged diaphragm. In this case, the pump **MUST** be replaced.

Side of the engine with the vacuum opening shown. The fuel pump is mounted over this opening to receive the vacuum/pressure from the crankcase for operation.

Different fuel pumps installed on Johnson/Evinrude engines. Service on these two models is limited to removing the cover and cleaning the screen.

PUMP REMOVAL AND INSTALLATION

Identify each hose and its location, then disconnect the vacuum hose and two fuel hoses from the fuel pump. Remove the attaching screws securing the pump to the engine. Two screws are visible on top of the pump and the third is hidden behind the fuel inlet nipple.

CLEANING AND INSPECTING

The pump cannot be disassmbled, therefore, the only maintenance is to remove the cover; clean the filter screen; and install new gaskets during installation. If the pump is defective, it must be replaced as a unit.

Install the cover, and then place the pump in position on the engine with a **NEW** gasket. Secure the pump with the attaching screws. Connect the vacuum hose, if used,

The same two fuel pumps shown at the top of this page, with the covers removed exposing the screen. The screen and the gasket are the only replaceable items.

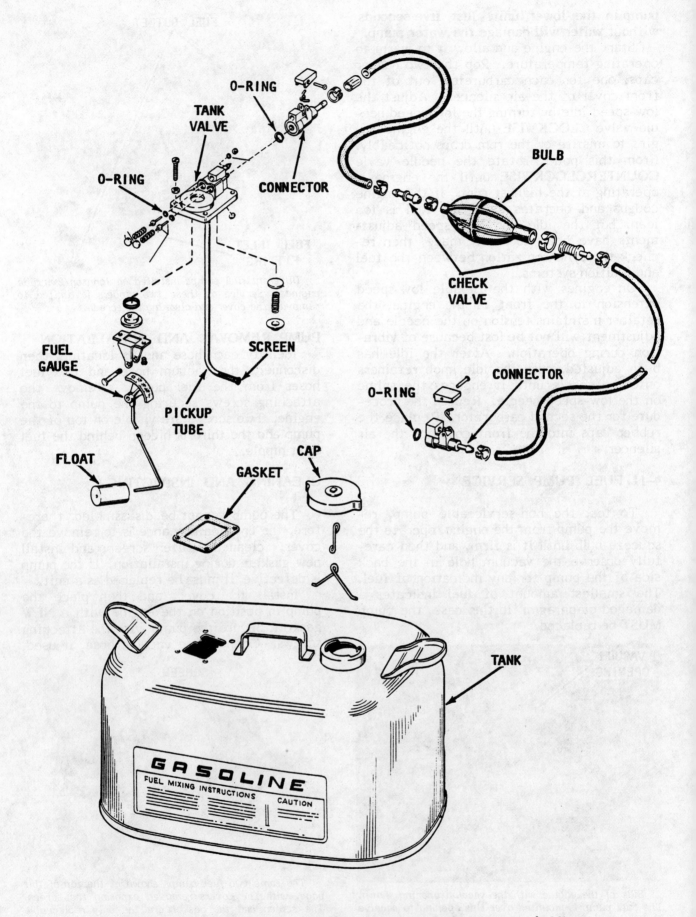

Exploded view of a modern non-pressure type fuel tank using a squeeze bulb in the fuel line.

Typical Johnson/Evinrude non-pressurized fuel tank.

and the two fuel hoses to the pump in the same position from which they were removed. Actually, each fitting is identified by word designation embossed on the pump. The vacuum line is connected to the vacuum fitting; the inlet fuel line from the fuel tank to the inlet fitting on the cover; and the outlet line to the remaining fitting.

4-12 FUEL TANK SERVICE

Late model fuel tanks (since about 1959), are not pressurized. A squeeze bulb is used to move fuel from the tank to the carburetor until the engine is operating. Once the engine starts, the fuel pump, mounted on the engine, transfers fuel from the tank to the carburetor. The pickup unit in the tank is sold as a complete unit, but without the gauge and float.

1- To replace the pickup unit, first remove the four screws securing the unit in the tank. Next, lift the pickup unit up out of the tank.

2- Remove the two Phillips screws securing the fuel gauge to the bottom of the pickup unit and set the gauge aside for

installation onto the new pickup unit.

If the pickup unit is not to be replaced, clean and check the screen for damage. It is possible to bend a new piece of screen material around the pickup and solder it in place without purchasing a complete new unit. Attach the fuel gauge to the new pickup unit and secure it in place with the two Phillips screws.

3- Clean the old gasket material from fuel tank and old pickup unit (if the old pickup unit is to be installed for further service). Work the float arm down through the fuel tank opening, and at the same time the fuel pickup tube into the tank. It will probably be necessary to exert a little force on the float arm in order to feed it all into the hole. The fuel pickup arm should spring into place once it is through the hole. Secure the pickup and float unit in place with the four attaching screws.

Fuel pump pickup assembly. The fuel gauge assembly is not sold as a part of the pickup unit.

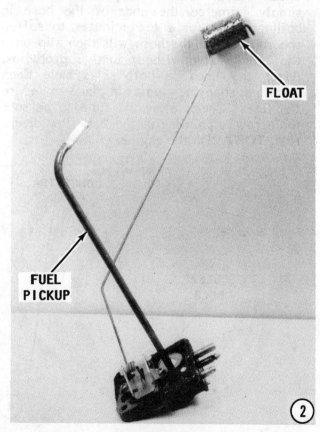

FLOAT

FUEL PICKUP

4- The primer squeeze bulb can be replaced in a short time. A squeeze bulb assembly, complete with the check valves installed, may be obtained from the local OMC dealer.

An arrow is clearly visible on the squeeze bulb to indicate the direction of fuel flow. The sqeeze bulb **MUST** be installed correctly in the line because the check valves in each end of the bulb will allow fuel to flow in **ONLY** one direction. Therefore, if the squeeze bulb should be installed backwards (in a moment of haste to get the job done), fuel will not reach the carburetor.

5- To replace the bulb, first unsnap the clamps on the hose at each end of the bulb. Next, pull the hose out of the check valves at each end of the bulb. New clamps are included with a new squeeze bulb. If the fuel line has been exposed to considerable sunlight, it may have become hardened, causing difficulty in working it over the check valve. To remedy this situation, simply immerse the ends of the hose in boiling water for a few minutes to soften the rubber and the hose will then slip onto the check valve without further problems. After the lines on both sides have been installed, snap the clamps in place to secure the line. Check a second time to be sure the arrow is pointing in the fuel flow direction, **TOWARDS** the engine.

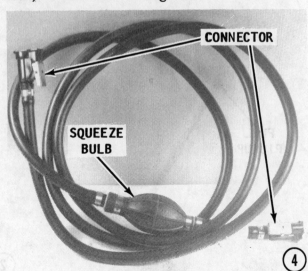

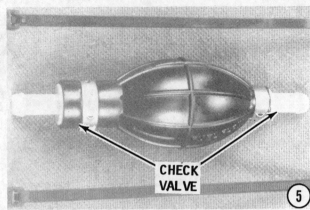

6- Use two ice picks or similar tool, and push down the check valve of the connector and work the O-ring out of the hole.

7- Apply just a drop of oil into the hole of the connector. Apply a thin coating of oil to the surface of the O-ring. Pinch the O-ring together and work it into the hole while simultaneously using a punch to depress the check valve inside the connector.

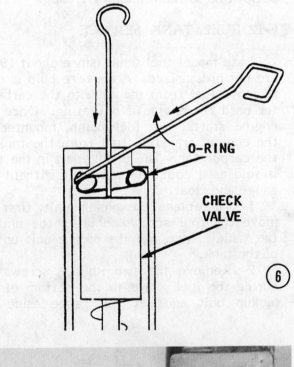

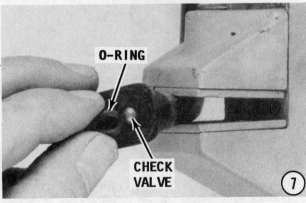

4-13 ELECTRIC PRIMER CHOKE SYSTEM

The electric primer system consists of a solenoid valve, distribution lines, and injection nozzles. The nozzles are tapped into the bypass covers. During engine cranking when the choke system is operating, fuel is injected through metered holes in the nozzles directly into the cylinders, instead of being routed in the usual manner through the crankcase.

During engine operation, from the fuel tank, the fuel passes through the fuel line, to the fuel pump, and into the carburetor. From the carburetor the fuel and air mixture passes through the crankcase and into the cylinder.

The primer system injects fuel directly into the cylinder. The system is controlled by the "push-in" type key switch. As the key is pushed in, the solenoid is activated, moving a small plunger which acts as a pump, injecting fuel through the nozzles directly into the cylinder to assist powerhead startup.

If the battery is dead and the choking effect is desired, a lever on the solenoid may be moved to the **MANUAL** position opening the seat in the valve. When the squeeze bulb is activated, fuel will pass through the nozzles directly into the cylinders. The lever is then returned to the **RUN** position during actual cranking of the powerhead.

WORD OF CAUTION

If the fuel tank has been exposed to direct sunlight, pressure may have developed inside the tank. Therefore, when the solenoid lever is moved to the **MANUAL** position, an excessive amount of fuel may be forced into the cylinders. As a safety precaution, under possible fuel tank pressure conditions, the fuel tank cap should be opened slightly to allow the pressure to escape before attempting to start the engine.

SOLENOID TESTING

Connect an ohmmeter to the solenoid between the blue/white stripe lead and the black (ground) wire. Observe the reading. The ohmmeter should indicate 5.5 + 1.5 ohms. If the reading is not within the prescribed range, the solenoid is defective and must be replaced.

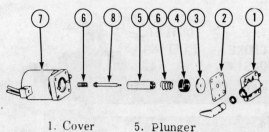

1. Cover
2. Gasket
3. Seal
4. Filler
5. Plunger
6. Spring
7. Solenoid body
8. Plunger valve

Exploded drawing of a primer solenoid valve.

The accompanying illustration will be helpful in ordering and replacing parts of the primer choke system.

The fuel hoses should be checked to ensure they remain flexible and are clear to permit an adequate fuel supply to pass through. Pay particular attention to any evidence of a crack in a fuel line which may permit fuel to escape and cause a very hazardous condition.

OMC tool No. 326623, is available to clean the metered holes in the nozzles.

4-14 MANUAL PRIMER SYSTEM

Description of Operation

During powerhead operation: from the fuel tank, the fuel passes through the fuel line, to the fuel pump, and into the carburetor where it is mixed with ambient air. From the carburetor, the fuel/air mixture passes through the crankcase and into the cylinder.

The manual primer system injects fuel directly into the cylinder. The system is activated by the "push-in" type choke lever. As the lever is pushed in, a small plunger moves inside the cylinder and acts as a pump, injecting fuel through the nozzles directly into the cylinder to assist powerhead startup.

TROUBLESHOOTING

If the manual primer system is suspected of not functioning correctly, remove the fuel line from the primer at the carburetor fitting.

Place the end of the fuel line just removed into a suitable container. Squeeze the fuel tank primer bulb to fill the carburetor bowl with fuel.

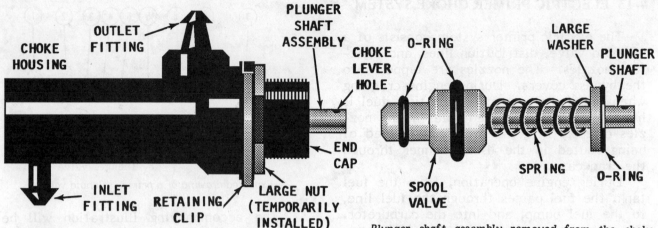

The primer choke valve removed from the powerhead. The large nut is temporarily installed onto the threads of the end cap for safe keeping.

Plunger shaft assembly removed from the choke housing with major parts identified.

Operate the primer choke lever twice. If fuel squirts from the disconnected fuel line into the container, the manual primer system is functioning correctly. If not, a kinked or restricted fuel line may be the problem.

The most probable cause of a malfunctioning primer system is internal leakage past the O-rings. Therefore if the primer itself is still suspected, procede to the following paragraph to service the primer system.

SERVICING MANUAL PRIMER

Removal

Disconnect and plug the inlet and outlet fuel lines to prevent loss of fuel and contamination. Remove the choke lever from the plunger. Back off the large nut securing the choke assembly to the lower cowling and lift the assembly free of the powerhead.

Pry the retaining clip from the choke body housing. Pull out the end cap, plunger, and spool valve assembly. Slide the end cap from the plunger. Remove and discard the O-ring around the end cap.

SPECIAL WORDS

Observe the three small O-rings, two on the spool valve and one around the plunger shaft. These three O-rings are made from a special material and **MUST** be replaced with a genuine OMC replacement part. Matching O-rings will **NOT** work!

Remove and discard the three O-rings.

Remove the large washer and spring from the plunger shaft.

Cleaning and Inspecting

Inspect the grooves of the spool valve and the shaft of the plunger for any scratches or burrs. Polish away any imperfections using crocus cloth. If a smooth finish cannot be obtained without removing excessive material, replace the spool valve and plunger assembly.

Inspect the condition of the plunger spring, replace as required.

Two one-way valves, one at each fuel fitting, can be tested by blowing through them in turn. Each valve is functioning correctly if it allows air to pass one direction, but not in the other direction. If a valve allows air to be drawn both in and out, the valve is defective. Individual valves are not servicable. The primer body must be replaced.

Assembling

Install the two new O-rings around the spool valve. Slide the spring, followed by the large washer and the third O-ring, over the plunger. Install a new O-ring over the end cap and place the end cap over the plunger end. Insert the assembly into the primer housing and install the retaining clip to secure everything together.

Slide the assembled primer into the opening in the lower cowling and thread the large nut over the protruding threads. Tighten the nut securely.

Install the fuel lines to the appropriate fittings and snap the choke lever into the vertical hole in the plunger.

4-15 OIL INJECTION SYSTEMS

INTRODUCTION

The purpose of an oil injection system is to mix oil with the fuel in the proper ratio at all powerhead speeds to ensure adequate lubrication. The system replaces the age-old method of manually adding a quantity of oil to the fuel tank.

Since 1985, all 40hp, and some 25hp and 30hp units have been equipped with a Variable Ratio Oil System, commonly referred to as simply VRO.

Since 1986, all non-electric start 25hp and 30hp units and all 9.9hp and 15hp units have been equipped with an oil injection system known as AutoBlend. In 1987 the name was changed to AccuMix. This AccuMix system is available as an optional accessory on all smaller models down to the 4hp.

ACCUMIX (AUTOBLEND) DESCRIPTION

The AccuMix (AutoBlend) system is located entirely inside the portable fuel tank. A 1-1/2 quart reservoir cannister contains enough oil for almost five tank fulls of fuel. An oil metering pump is located at the base of the cannister. This oil pump is activated by pulses from the fuel pump installed on the powerhead. The oil metering pump automatically blends fuel from the fuel pickup in the portable tank with oil in the reservoir cannister. The oil/fuel mixture passes through a built-in filter also located inside the cannister.

A low-oil warning indicator activates a warning horn when the level of oil falls below one pint. If the operator sustains powerhead operation after the warning horn sounds, the fuel supply is automatically cut off to shutdown the powerhead. The powerhead cannot be restarted until oil has been added to the cannister.

Procedures to service the AccuMix (AutoBlend) oil injection system begin Page 4-62.

VARIABLE RATIO OIL SYSTEM DESCRIPTION

The VRO system consists of an oil reservoir (tank), a VRO oil line primer, a pump to move the oil from the tank to the powerhead, a warning horn, a spark arrestor in the pulse hose to the VRO pump, an oil inlet filter, a vacuum switch in the fuel line and the necessary hoses and fittings to connect the various items for efficient operation. All connections in the system **MUST** be airtight to prevent serious damage to the powerhead.

As the name implies, the VRO pump moves oil from the oil reservoir to the powerhead. However, it is a **dual** pump and also pumps fuel. Pumping action of the pump stops automatically if fuel is not available at the pump for any reason. This automatic pump shutdown feature prevents the carburetors from filling with oil.

The warning horn, located in the control box serves two functions.

First, as a low oil level warning; The horn will sound for 1/2 second every 20 seconds if the oil tank level reaches 1/4 of the tank's capacity. The low oil warning circuit consists of a sending unit in the oil reservoir, a ground wire to the engine, and a wire to the warning horn through the key switch.

Secondly, the warning horn will sound for 1/2 second every 1/2 second to produce a very urgent warning signal. To continue powerhead operation after the no oil warning horn sounds would almost certainly invite serious damage to internal moving parts and powerhead seizure!

The spark arrestor on all models is installed in the pulse hose to the VRO pump. This flame arrestor prevents a backfire flame from entering the VRO pump. A clamp positioned on the hose prevents the spark arrestor from migrating up the hose to the pump.

Variable Ratio Oil (VRO) system installed on a late model powerhead.

GOOD WORDS

Anytime the pulse hose is disconnected at the powerhead, **TAKE CARE** to be sure the spark arrestor remains in the hose. The spark arrestor **MUST** be properly positioned in the pulse hose to prevent serious and permanent damage to the VRO pump.

The oil inlet filter, located in the VRO reservoir oil pickup line, prevents any dirt or foreign material from entering the pump. If the filter should need cleaning, the hose assembly should be removed and reverse flushed using clean solvent. **DO NOT** attempt to remove the filter because the filter and hose are serviced and replaced as an assembly.

NEW POWERHEAD BREAK-IN PROCEDURE

A complete new outboard unit, a new powerhead, or a rebuilt powerhead, must have oil mixed in the fuel tank **IN ADDITION** to the VRO system. The mixture should be 50:1 (1 pint oil to 6 gallons of fuel) and the unit **MUST** be operated with this mixture during the first 10 hours of service.

CRITICAL WORDS

To be convinced the VRO system is working properly, the operator should observe a drop in the oil supply in the VRO oil reservoir during the 10-hour break-in period.

At the end of the 10 hour period, the powerhead mounted fuel filter should be inspected. Remove any dirt or foreign matter collected in the filter.

TROUBLESHOOTING VRO SYSTEM

This short section list a few of the probable problems that might occur in the system with suggested corrective action.

The next section -- **SERVICING** outlines in detail how the tests and service work is to be performed.

Warning Horn Sounds

a- Oil level in the oil reservoir is below 1/4 full. Add oil to the reservoir.

b- Disc on the pickup unit may not be positioned properly. Remove the pickup unit from the reservoir and correct as required.

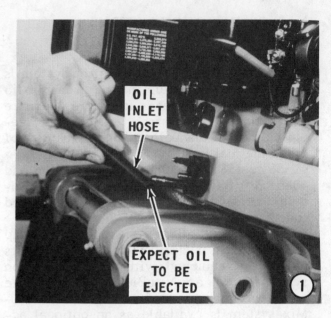

SERVICING VRO SYSTEM

Servicing consists of making simple tests and checks. A vacuum gauge, pressure gauge, a "T" fitting, a short section of clear plastic hose, a source of low-pressure compressed air, and a couple of normal shop tools are all that is required to service the VRO system.

Check Fuel and Oil Circuits

1- Verify there is more than 1/4 tank of oil in the reservoir. Disconnect the oil inlet hose from the VRO pump. Be prepared to catch oil as it is ejected from the end of the hose. Squeeze the bulb and verify oil is ejected from the open end of the hose. The

presence of oil verifies a clear line from the tank to this point. If no oil is ejected, clean the line from the tank and repeat the test.

2- Disconnect the mixed fuel outlet hose from the VRO pump.

3- Insert a "T" fitting into the end of the hose. Insert a drop of oil into each end of a short section of clear plastic hose. Connect one end of the clear piece of plastic hose to one arm of the "T" and the other end of the hose to the VRO pump. Connect a 0-15 psi pressure gauge to the leg of the "T". Secure the connections with tie straps or hose clamps. Mount the engine in an adequate size test tank or move the boat to a body of water.

NEVER operate the engine using a flush attachment for this test. If the engine is operated above idle speed with no load on the propeller, the engine could **RUNAWAY** resulting in serious damage or destruction of the unit.

CAUTION: Water must circulate through the lower unit to the engine any time the engine is run to prevent damage to the water pump in the lower unit. Just five seconds without water will damage the water pump.

Start the engine and shift the unit into gear. Advance the throttle to the near wide open position. Check the pressure gauge. The gauge should indicate 34 kPa (5 psi) to 103 kPa (15 psi) pressure at near full throttle. Each time the pump pulses a small squirt of oil, in addition to the fuel passing through, should be observed in the clear plastic hose discharging from the pump.

Also, the fuel pressure will drop approximately 6.8 kPa (1 psi) to 13.7 kPa (2 psi) and a "click" sound may be heard each time the pump pulses and discharges oil.

Results

Fuel pressure satisfactory and oil is observed discharging from the pump through the clear plastic hose: Fuel and oil systems are verified satisfactory.

No fuel pressure: Check quantity of fuel in the tank to be sure fuel level reaches the pickup. Add fuel if required. Check fuel line to be sure it is not pinched or kinked restricting fuel flow. Check engine pulse hose to be sure it is not pinched, leaking or disconnected. If all above conditions are satisfactory, the VRO pump is defective and **MUST** be replaced as a unit.

Low fuel pressure: Check for restricted fuel filter at the engine. Check the engine pulse hose to be sure it is not pinched, kinked, leaking, or restricting fuel flow. Squeeze the fuel primer bulb a few times to force a possible fuel vapor-lock out of the system.

Warning Horn Check

4- Slide the "boot" down the wire to clear the knife disconnect between the temperature switch and the horn lead. This is not an easy task, but with a pair of needle nose pliers and some patience, it can be done. After the "boot" is clear, disconnect

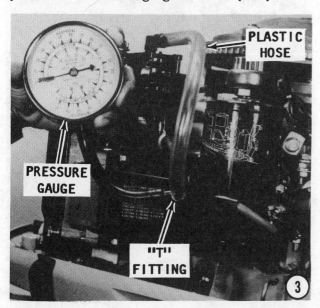

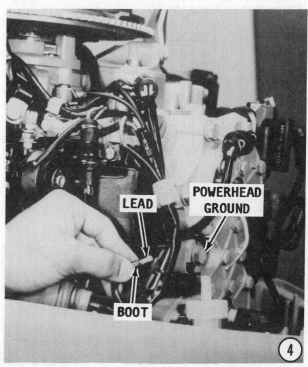

the fitting. Turn the key switch to the **ON** position. Now, make contact with the lead onto a "clean" place on the powerhead. The horn should sound. If the horn does not sound, there is a problem between the knife disconnect and the remote control box.

Clean Vacuum Pulse Hose

5- Remove the vacuum pulse hose. The hose may be cleaned by back-flushing with

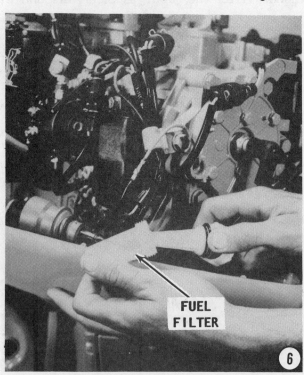

clean solvent. **TAKE CARE** to ensure the spark arrestor installed in the hose stays in the hose. A clamp is positioned on the hose to prevent the spark arrestor from migrating up the hose to the pump. If the hose is damaged and requires replacement, the hose and flame arrestor are purchased as an assembly. The flame arrestor cannot be purchased separately.

VRO Pump

If troubleshooting and service work indicates the VRO pump to be defective, it must be replaced. The pump cannot be serviced or repaired. The pump is removed by first disconnecting the hoses and then removing the three mounting bolts. Lift the VRO pump free.

Engine Mounted Fuel Filter

6- If a fuel inlet filter is mounted at the engine, this filter can be separated and inspected without removing the hoses. The filter should be inspected at the end of the 10-hour break-in period and at regular intervals as part of normal engine maintenance and service.

GOOD WORDS

Use **ONLY** OMC approved clamps on the connections on the inlet side of the VRO

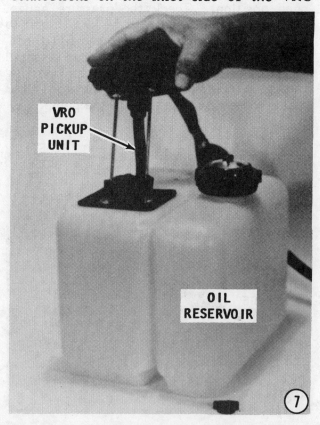

pump. A screw type hose clamp will very likely pinch or break the hose causing a suction leak. A tie-wrap will not tighten down to the degree required to prevent a vacuum leak.

Vacuum Hose Check

This check will verify the system in good condition from the oil pickup to the end of the hose.

7- Remove the VRO pickup unit from the oil reservoir by first removing the four mounting screws and then lifting the pickup straight up and out of the reservoir. Inspect the oil pickup filter, and clean it, if necessary.

8- Purge the system of oil and any small particles of foreign matter by using low-pressure compressed air through the VRO pickup end.

9- Connect a vacuum gauge to the end of the hose. Insert the plug-type nipple that is shipped with the engine (snapped to the fuel hose at the engine) into the VRO pickup. If the plug has been lost, one can be easily made out of suitable material. Secure the plug with a clamp. Pump the gauge until 17.7 cm (7") of mercury is indicated. The system should hold the vacuum reading.

10- If the system fails to hold the required vacuum reading, check each connection by applying a small amount of oil at each fitting. The oil will **MOMENTARILY** stop the leak and the vacuum reading will stop dropping. Carefully inspect the hose for damage.

GOOD WORDS

OMC **STRONGLY** recommends that the hose from the primer bulb to the VRO pump be one continuous hose with no fittings between. Therefore, if the hose is damaged, the entire length should be replaced. Also,

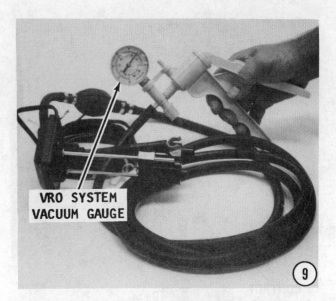

VRO SYSTEM
VACUUM GAUGE

9

if a dual engine installation is used, **DO NOT** "T" into the line from the primer bulb. Use a separate oil reservoir, with separate primer bulb and hose to the VRO pump on the second engine -- a completely separate system for each engine.

11- If the horn sounds indicating low oil or no oil, the contacts on the pickup may not be positioned properly. A disc rises slightly when oil is added into the reservoir and lowers slightly as the oil level drops. Check to be sure the contact surface on both sides of the disc is riding equally with the other side to prevent sounding the horn prematurely. If the disc fails to rise, the horn will sound continuously. Clean the float chamber in solvent and the disc should then rise clear of the contacts.

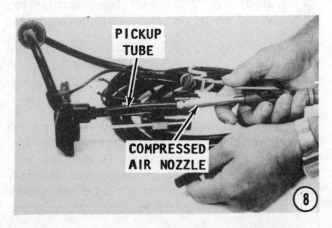

PICKUP
TUBE

COMPRESSED
AIR NOZZLE

8

OIL
CAN

10

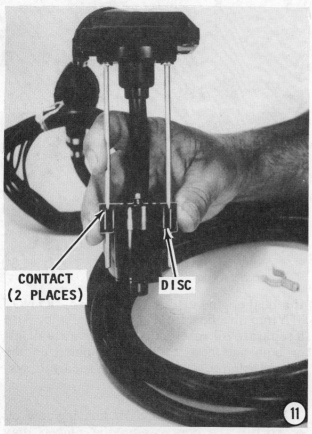

CONTACT (2 PLACES)

DISC

Fuel Line Vacuum Test

12- Connect a "T" fitting and vacuum gauge to the outlet hose from the fuel tank, as shown. Check to be sure all fittings and connections are tight.

NEVER operate the engine above idle speed using a flush attachment for this test.

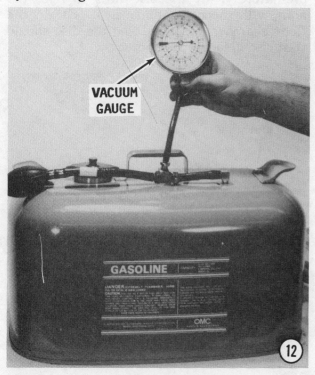

VACUUM GAUGE

GASOLINE

If the engine is operated above idle speed with no load on the propeller, the engine could **RUNAWAY** resulting in serious damage or destruction of the unit.

CAUTION: Water must circulate through the lower unit to the engine any time the engine is run to prevent damage to the water pump in the lower unit. Just five seconds without water will damage the water pump.

Start the powerhead and operate it at idle speed. Observe the vacuum gauge. The gauge should indicate no more than 17.7 cm (7") of mercury.

ACCUMIX (AUTOBLEND) SYSTEM DESCRIPTION

The AccuMix (AutoBlend) oil injection system is designed to provide a fuel/oil ratio of 100/1 regardless of powerhead rpm.

The AccuMix (AutoBlend) system is located entirely inside the portable fuel tank. A 1-1/2 quart reservoir cannister contains enough oil for almost five tankfulls of fuel. An oil metering pump is located at the base of the cannister. This oil pump is activated by pulses from the fuel pump installed on the powerhead. The oil metering pump automatically blends fuel from the fuel pickup in the portable tank with oil in the reservoir cannister. The oil/fuel mixture passes through a built-in filter also located inside the cannister.

A low-oil warning indicator activates a warning horn when the level of oil falls below one pint. If the operator sustains powerhead operation after the warning horn sounds, the fuel supply is automatically cut off to shutdown the powerhead.

A visual low oil level indicator is located on the reservoir cannister cover. This indicator is of the "glass eye" type, similar to those found on newer style automotive batteries.

SPECIAL WORDS ON TROUBLESHOOTING

Due to the inherent design of this oil injection system, very few individual parts can be repaired or replaced, if found to be defective. As an example: if the low oil level float is found to be defective, unfortunately, the entire system must be replaced. Therefore, no "troubleshooting" procedures are given, because no evaluating tests have been provided by the manufacturer.

If any problem is encountered with the delivery of oil, follow the procedures outlined in the following paragraphs. The instructions deal mostly with cleaning out the system. If the problem persists, the only remedy is a new system.

GENERAL MAINTENANCE

Maintenace of this type oil injection system is limited to draining and flushing the oil reservoir cannister each season. To properly clean the cannister and the integral oil filter, it should be removed from the fuel tank and flushed with fresh gasoline or solvent.

Procedures for reservoir cannister service will be found in the following paragraphs.

When the reservoir cannister has been removed from the fuel tank, the fuel line pickup screen can also be serviced.

Clean the vent screw and the area around the screw each time the fuel tank is filled. This vent screw **MUST** be **FULLY OPEN** to allow powerhead operation and to permit air to enter the tank and take the place of the consumed fuel. The vent screw and fuel tank cap cannot be serviced, other than cleaning. If defective for any reason, the assembly must be replaced as a unit.

OIL RESERVOIR CANNISTER SERVICE

Removal

1- Disconnect the electrical harness and the fuel line connector from the top of the cannister. Remove the eight slotted head retaining screws, securing the cover and cannister to the fuel tank. Observe the

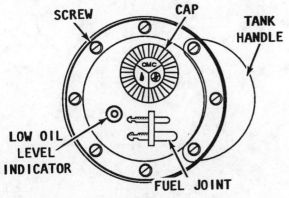

A top view of the AccuMix oil injection system. Eight screws secure the handle and the cannister to the fuel tank.

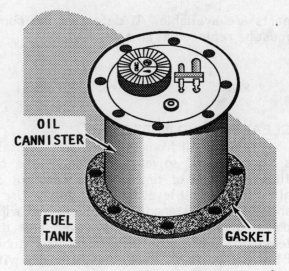

A cork gasket is used between the cannister flange and the fuel tank. Take care to align all holes when installing the cannister to avoid damage to this gasket.

three additional washers under the tank handle bracket, when removing the handle. Lift out the cannister and cover together from the tank, taking care not to spill any oil remaining in the cannister. Remove and discard the gasket between the cannister flange and the tank.

2- Ease the cover from the cannister. The fuel tube must be disengaged from the cavity in the cover. Remove and discard the O-ring between the cover and the cannister.

Good Words

Do not attempt to disassemble the oil pump at the cannister base. No replacement

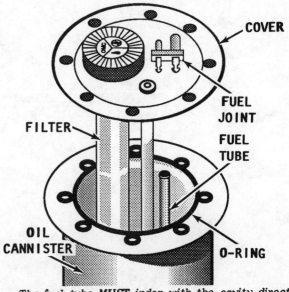

*The fuel tube **MUST** index with the cavity directly under the fuel joint and the eight holes in the O-ring **MUST** align with the cannister and cover holes prior to installing the cannister back into the fuel tank.*

parts are available. If defective the pump must be replaced as an assembly.

CLEANING AND INSPECTING

Drain any oil from the cannister. Obtain a container of solvent and "dunk" the oil filter a couple of times. Take care not to alter the low oil level indicator float. Use a shop towel moistened with solvent and wipe down the surfaces of the cover and the length of the low oil level indicator tube. Blow the cover assembly and filter dry with compressed air.

Pour some solvent into the cannister to rinse any residue from the cannister walls. Make certain no debris obstructs the oil pump pickup or the low oil cutoff float mounted on the pump.

Inspect and service the fuel line pickup screen, as necessary.

Assembling

1- Secure the fuel pickup line into the clip at the base of the cannister. Place a new gasket on the fuel tank surface and lower the cannister into the tank. Align the cannister flange holes with the gasket holes.

2- Apply a light coat of OMC Triple-Guard grease on both sides of the O-ring, and then position the ring onto the cannister flange. Align the holes in the O-ring with the flange holes.

3- Lower the cover over the cannister and make sure the fuel tube indexes with the cavity on the underside of the cover.

4- Position one washer at each of the three holes for the fuel tank handle bracket, and then hold the handle in place while these screws are started. Install the remaining screws fingertight. Tighten all eight screws alternately and evenly to a torque value of 10 in lb (1.1Nm).

Connect the electrical harness and the fuel connector to the fittings on top of the cannister.

5
IGNITION

5-1 INTRODUCTION

The less an outboard engine is operated, the more care it needs. Allowing an outboard engine to remain idle will do more harm than if it is used regularly. To maintain the engine in top shape and always ready for efficient operation at any time, the engine should be operating every 3 to 4 weeks throughout the year.

The carburetion and ignition principles of two-cycle engine operation **MUST** be understood in order to perform a proper tune-up on an outboard motor.

If you have any doubts concerning your understanding of two-cyle engine operation, it would be best to study the operation theory section in the first portion of Chapter 3, before tackling any work on the ignition system.

Three ignition systems are used on the Johnson/Evinrude engines covered in this manual: the flywheel magneto system, the low tension flywheel magneto system, and two types of capacitor discharge (CD) magneto systems. The first sections of this chapter will be devoted to an explanation of the systems and their theory of operation. The latter sections will provide trouble-shooting and repair instructions. For synchronizing procedures, see Section 5-9.

An outboard engine should not be left idle for long periods of time. Lack of use or operation is its worst enemy.

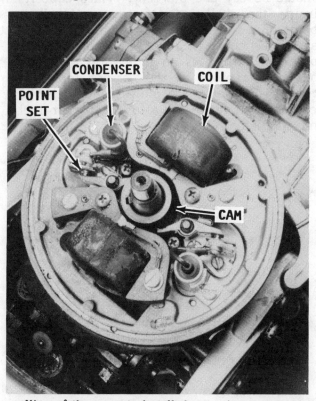

View of the magneto installed on engines covered in this manual. A one-cylinder engine will have only half the parts.

5-2 SPARK PLUG EVALUATION

Removal: Remove the spark plug wires by pulling and twisting on only the molded cap. **NEVER** pull on the wire or the connection inside the cap may become separated or the boot damaged. Remove the spark plugs and keep them in order. **TAKE CARE** not to tilt the socket as you remove the plug or the insulator may be cracked.

Examine: Line the plugs in order of removal and carefully examine them to determine the firing conditions in each cylinder. If the side electrode is bent down onto the center electrode, the piston is traveling too far upward in the cylinder and striking the spark plug. Such damage indicates the wrist pin or the rod bearing is worn excessively. In all cases, an engine overhaul is required to correct the condition. To verify the cause of the problem, turn the engine over by hand. As the piston moves to the full up position, push on the piston crown with a screwdriver inserted through the spark plug hole, and at the same time rock the flywheel back-and-forth. If any play in the piston is detected, the engine must be rebuilt.

Correct Color: A proper firing plug should be dry and powdery. Hard deposits inside the shell indicate too much oil is being mixed with the fuel. The most important evidence is the light gray color of the porcelain, which is an indication this plug has been running at the correct temperature. This means the plug is one with the correct heat range and also that the air-fuel mixture is correct.

*Damaged spark plugs. Notice the broken electrode on the left plug. The broken part **MUST** be found and removed before returning the engine to service.*

Rich Mixture: A black, sooty condition on both the spark plug shell and the porcelain is caused by an excessively rich air-fuel mixture, both at low and high speeds. The rich mixture lowers the combustion temperature so the spark plug does not run hot enough to burn off the deposits.

Deposits formed only on the shell is an indication the low-speed air-fuel mixture is too rich. At high speeds with the correct mixture, the temperature in the combustion chamber is high enough to burn off the deposits on the insulator.

Too Cool: A dark insulator, with very few deposits, indicates the plug is running too cool. This condition can be caused by low compression or by using a spark plug of an incorrect heat range. If this condition shows on only one plug it is most usually caused by low compression in that cylinder. If all of the plugs have this appearance, then it is probably due to the plugs having a too-low heat range.

This spark plug is foul from operating with an over-rich condition, possibly an improper carburetor adjustment.

This spark plug has been operating too-cool, because it is rated with a too-low heat range for the engine.

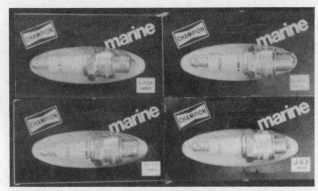

Today, numerous type spark plugs are available for service. ALWAYS check with your local marine dealer to be sure you are purchasing the proper plug for the engine being serviced.

Fouled: A fouled spark plug may be caused by the wet oily deposits on the insulator shorting the high-tension current to ground inside the shell. The condition may also be caused by ignition problems which prevent a high-tension pulse being delivered to the spark plug.

Carbon Deposits: Heavy carbon-like deposits are an indication of excessive oil in the fuel. This condition may be the result of poor oil grade, (automotive-type instead of a marine-type); improper oil-fuel mixture in the fuel tank; or by worn piston rings.

Overheating: A dead white or gray insulator, which is generally blistered, is an indication of overheating and pre-ignition. The electrode gap wear rate will be more than normal and in the case of pre-ignition, will actually cause the electrodes to melt as

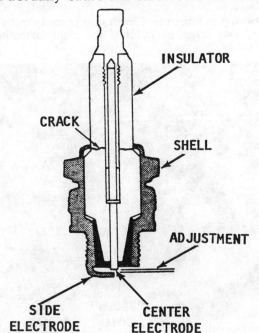

Cut-a-way drawing showing major spark plug parts.

shown in this illustration. Overheating and pre-ignition are usually caused by improper point gap adjustment; detonation from using too-low an octane rating fuel; an excessively lean air-fuel mixture; or problems in the cooling system.

Electrode Wear: Electrode wear results in a wide gap and if the electrode becomes carbonized it will form a high-resistance path for the spark to jump across. Such a condition will cause the engine to misfire during acceleration. If all plugs are in this condition, it can cause an increase in fuel consumption and very poor performance during high-speed operation. The solution is to replace the spark plugs with a rating in the proper heat range and gapped to specification.

Red rust-colored deposits on the entire firing end of a spark plug can be caused by water in the cylinder combustion chamber. This can be the first evidence of water entering the cylinders through the exhaust manifold because of scale accumulation. This condition **MUST** be corrected at the first opportunity. Refer to Chapter 3, Powerhead Service.

5-3 POLARITY CHECK

Coil polarity is extremely important for proper battery ignition system operation. If a coil is connected with reverse polarity, the spark plugs may demand from 30 to 40 percent more voltage to fire. Under such demand conditions, in a very short time the coil would be unable to supply enough voltage to fire the plugs. Any one of the following three methods may be used to quickly determine coil polarity.

1- The polarity of the coil can be checked using an ordinary D.C. voltmeter. Connect the positive lead to a good ground. With the engine running, momentarily touch the negative lead to a spark plug terminal.

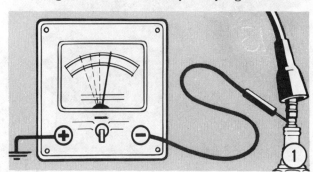

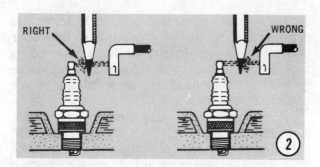

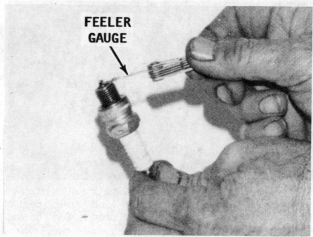

The needle should swing upscale. If the needle swings downscale, the polarity is reversed.

2- If a voltmeter is not available, a pencil may be used in the following manner: Disconnect a spark plug wire and hold the metal connector at the end of the cable about 1/4" from the spark plug terminal. Now, insert an ordinary pencil tip between the terminal and the connector. Crank the engine with the ignition switch ON. If the spark feathers on the plug side and has a slight orange tinge, the polarity is correct. If the spark feathers on the cable connector side, the polarity is reversed.

3- The firing end of a used spark plug can give a clue to coil polarity. If the ground electrode is "dished", it may mean polarity is reversed.

5-4 WIRING HARNESS

CRITICAL WORDS: These next two paragraphs may well be the most important words in this chapter. Misuse of the wiring harness is the most single cause of electrical problems with outboard power plants.

The spark plug gap should always be checked before installing new or used spark plugs.

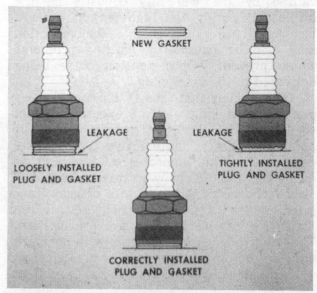

Drawing to illustrate a spark plug properly installed, center, and other plugs, left and right, improperly installed.

This coil was destroyed because 12-volts was connected to the key switch.

A wiring harness is used between the key switch and the engine. This harness seldom contains wire of sufficient size to allow connecting accessories. Therefore, anytime a new accessory is installed, **NEW** wiring should be used between the battery and the accessory. A separate fuse panel **MUST** be installed on the dash. To connect the fuse panel, use one red and one black No. 10 gauge wire from the battery. If a small amount of 12-volt current should be accidently attached to the magneto system, the coil may be damaged or **DESTROYED**. Such a mistake in wiring can easily happen if the source for the 12-volt accessory is taken from the key switch. Therefore, again let it be said, **NEVER** connect accessories through the key switch.

5-5 FLYWHEEL MAGNETO IGNITION

1.25 hp	1986 & on
2 hp	1971-85
4 hp	1971-84*
6 hp	1971-76
9.5 hp	1971-73
18 hp	1971-73
20 hp	1971-72
25 hp	1971-72
40 hp	1971-73

*The 4 hp Deluxe, 1984 and on, has Type II, CD Flywheel Magneto Ignition.

DESCRIPTION

READ AND BELIEVE. A battery installed to crank the engine **DOES NOT** mean the engine is equipped with a battery-type ignition system. A magneto system uses the battery only to crank the engine. Once the engine is running, the battery has absolutely no effect on engine operation. Therefore, if the battery is low and fails to crank the engine properly for starting, the engine may be cranked manually, started, and operated. Under these conditions, the key switch must be turned to the **ON** position or the engine will not start by hand cranking.

A magneto system is a self-contained unit. The unit does not require assistance from an outside source for starting or continued operation. Therefore, as previously mentioned, if the battery is dead, the engine may be cranked manually and the engine started.

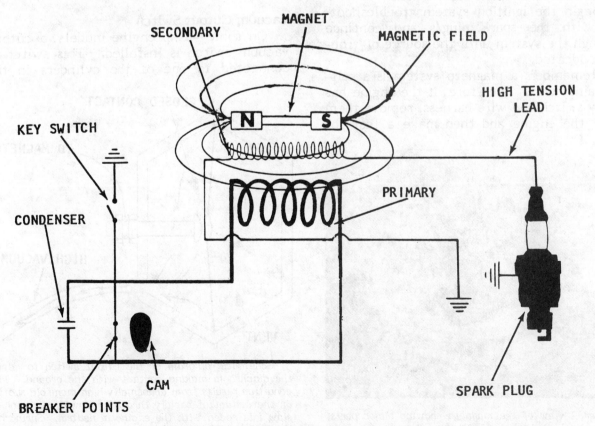

Schematic diagram of a magneto ignition system.

The flywheel-type magneto unit consists of an armature plate and a permanent magnet built into the flywheel. The ignition coil, condenser and breaker points are mounted on the armature plate.

As the pole pieces of the magnet pass over the heels of the coil, a magnetic field is built up about the coil, causing a current to flow through the primary winding.

Now, at the proper time, the breaker points are separated by action of a cam, and the primary circuit is broken. When the circuit is broken, the flow of primary current stops and causes the magnetic field about the coil to break down instantly. At this precise moment, an electrical current of extremely high voltage is induced in the fine secondary windings of the coil. This high voltage is conducted to the spark plug where it jumps the gap between the points of the plug to ignite the compressed charge of air-fuel mixture in the cylinder.

TROUBLESHOOTING

Always attempt to proceed with the troubleshooting in an orderly manner. The shotgun approach will only result in wasted time, incorrect diagnosis, replacement of unnecessary parts, and frustration.

Begin the ignition system troubleshooting with the spark plug/s and continue through the system until the source of trouble is located.

Remember, a magneto system is a self-contained unit. Therefore, if the engine has a key switch and wire harness, remove them from the engine and then make a test for

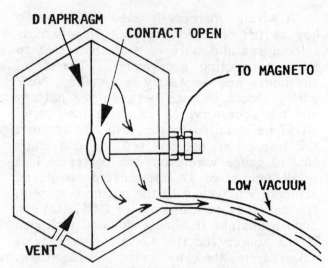

Schematic diagram of a vacuum cutout switch used on early model small horsepower engine. This illustration depicts the position of the diaphragm in relation to the ground contact when operating at normal manifold pressure. Spring omitted for clarity.

spark. If a good spark is obtained with these two items disconnected, but no spark is available at the plug when they are connected, then the trouble is in the harness or the key switch. If a test is made for spark at the plug with the harness and switch connected, check to be sure the key switch is turned to the **ON** position.

Vacuum Cutout Switch

On some 40 hp engine models, a cutout vacuum switch is installed. This switch is connected to one of the cylinders in the

Overall view of the magneto showing major parts, coils, points, condenser and armature plate.

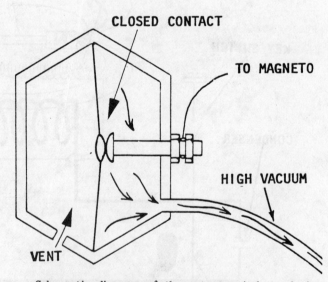

Schematic diagram of the cutout switch to depict the diaphragm making contact with the ground. This condition results from abnormally high manifold suction on the instant of rapidly throttling down from high to slow idle speed with the engine in neutral. Spring not shown for clarity.

ignition system. The switch is actuated by vacuum from the cylinder. When a high vacuum pull is exerted against the switch, during engine operation in gear without the lower unit in the water, the switch is closed and the engine is shut down. This feature is a safeguard against the engine "running away" while operating with a no-load condition on the propeller. A two-cycle engine will continue to increase rpm under a no-load condition and attempts to shut it down will fail, resulting in serious damage or destruction of the unit.

The vacuum switch also serves as a safety feature when the boat is operating in the water. If the propeller is released from the shaft, because of an accident (striking an underwater object, whatever), the engine would then be operating under a no-load condition. The vacuum switch will shut down the engine and prevent extensive damage, resulting from a "runaway" condition.

This cutout switch arrangement was installed on the 40 hp engines, 1971-76.

Therefore, if spark is not present at the spark plug, disconnect the wires from the vacuum switch and again test for spark at the spark plug. If spark is present with the vacuum switch disconnected, the switch is defective and must be replaced.

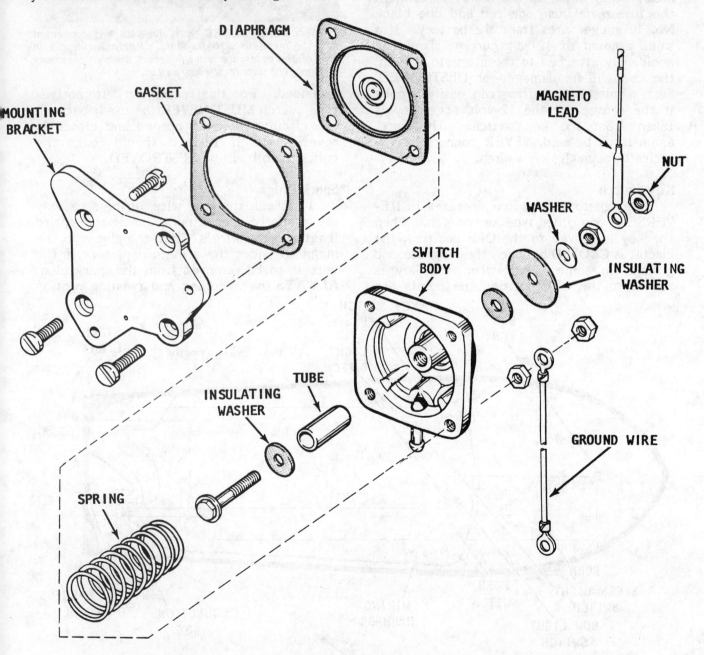

Exploded drawing of a "runaway" cutout switch installed on most late-model powerheads.

WIRING HARNESS

CRITICAL WORDS: These next two paragraphs may well be the most important words in this chapter. Misuse of the wiring harness is the most single cause of electrical problems with outboard power plants.

A wiring harness is used between the key switch and the engine. This harness seldom contains wire of sufficient size to allow connecting accessories. Therefore, anytime a new accessory is installed, **NEW** wiring should be used between the battery and the accessory. A separate fuse panel **MUST** be installed on the dash. To connect the fuse panel, use one red and one black No. 10 gauge wires from the battery. If a small amount of 12-volt current should be accidently attached to the magneto system, the coil will be damaged or **DESTROYED.** Such a mistake in wiring can easily happen if the source for the 12-volt accessory is taken from the key switch. Therefore, again let it be said, **NEVER** connect accessories through the key switch.

Key Switch

A magneto key switch operates in **RE-VERSE** of any other type key switch. When the key is moved to the **OFF** position, the circuit is **CLOSED** between the magneto and ground. In some cases, when the key is turned to the **OFF** position the points are

A coil DESTROYED when 12-volts was connected into the magneto wiring system. Mechanics report in 85% of the cases, the damage occurs when an accessory is connected through the key switch.

grounded. For this reason, an automotive-type switch **MUST NEVER** be used, because the circuit would be opened and closed in reverse, and if 12-volts should reach the coil, the coil will be **DESTROYED.**

Spark Plugs

1- Check the plug wires to be sure they are properly connected. Check the entire length of the wire/s from the plug/s to the magneto under the armature plate. If the wire is to be removed from the spark plug, **ALWAYS** use a pulling and twisting motion

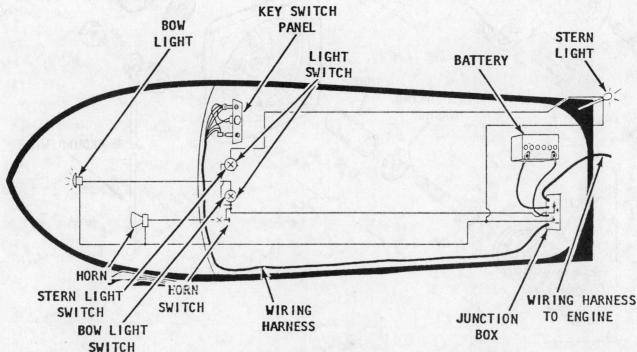

*Functional diagram to illustrate proper hookup of accessories through a junction box. If a junction is not installed on the boat, connect accessories directly to the battery. **NEVER** connect accessories through the key switch.*

as a precaution against damaging the connection.

2- Attempt to remove the spark plug/s by hand. This is a rough test to determine if the plug is tightened properly. You should not be able to remove the plug without using the proper socket size tool. Remove the spark plug/s and keep them in order. Examine each plug and evaluate its condition as described in Section 5-2.

If the spark plugs have been removed and the problem cannot be determined, but the plug appears to be in satisfactory condition, electrodes, etc., then replace the plugs in the spark plug openings.

A conclusive spark plug test should always be performed with the spark plugs installed. A plug may indicate satisfactory spark when it is removed and tested but under a compression condition may fail. An example would be the possibility of a person being able to jump a given distance on the ground, but if a strong wind is blowing, his distance may be reduced by half. The same is true with the spark plug. Under good compression in the cylinder, the spark may be too weak to ignite the fuel properly.

Therefore, to test the spark plug under compression, replace it in the engine and tighten it to the proper torque value. Another reason for testing for spark with the plugs installed is to duplicate actual operat-

ing conditions regarding flywheel speed. If the flywheel is rotated with the pull cord with the plugs removed, the flywheel will rotate much faster because of the no-compression condition in the cylinder, giving the **FALSE** indication of satisfactory spark.

3- Use a spark tester and check for spark at each cylinder. If a spark tester is not available, hold the plug wire about 1/4-inch from the engine. Turn the flywheel with a pull starter or electrical starter and check for spark. A strong spark over a wide gap must be observed when testing in this manner, because under compression a strong spark is necessary in order to ignite the air-fuel mixture in the cylinder. This means it is possible to think you have a strong spark, when in reality the spark will be too weak when the plug is installed. If there is no spark, or if the spark is weak, the trouble is most likely under the flywheel in the magneto.

ONE MORE WORD: Each cylinder has its own ignition system in a flywheel-type ignition system. This means if a strong spark is observed on any one cylinder and not at another, only the weak system is at fault. However, it is always a good idea to check and service all systems while the flywheel is removed.

Compression

A compression check is extremely important, because an engine with low or uneven compression between cylinders CAN-

NOT be tuned to operate satisfactorily. Therefore, it is essential that any compression problem be corrected before proceeding with the tune-up procedure.

If the powerhead shows any indication of overheating, such as discolored or scorched paint, especially in the area of the top (No. 1) cylinder, inspect the cylinders visually thru the transfer ports for possible scoring. A more thorough inspection can be made if the head is removed. It is possible for a cylinder with satisfactory compression to be scored slightly. Also, check the water pump. The overheating condition may be caused by a faulty water pump.

An overheating condition may also be caused by running the engine out of the water. For unknown reasons, many operators have formed a bad habit of running a small engine without the lower unit being submerged. Such a practice will result in an overheated condition in a matter of seconds. It is interesting to note, the same operator would never operate or allow anyone else to run a large horsepower engine without water circulating through the lower unit for cooling. Bear in mind, the laws governing operation and damage to a large unit **ALL** apply equally as well to the small engine.

Checking the rings and cylinder walls through the opening on the exhaust side of the engine to be sure the walls are not scored and the rings are not stuck in the piston (fail to expand properly).

The preferred method of checking the cylinder walls and rings is to pull the head and make an inspection. This method will also reveal the piston condition in each cylinder.

Checking Compression

4- Remove the spark plug wires. AL-WAYS grasp the molded cap and pull it loose with a twisting motion to prevent damage to the connection. Remove the spark plugs and keep them in ORDER by cylinder for evaluation later. Ground the spark plug leads to the engine to render the ignition system inoperative while performing the compression check.

Insert a compression gauge into the No. 1 (top), spark plug opening. Crank the engine with the starter, or pull on the starter cord, through at least 4 complete piston strokes with the throttle at the wide-open position, or until the highest possible reading is observed on the gauge. Record the reading.

Repeat the test and record the compression for each cylinder. A variation between cylinders is far more important than the actual readings. A variation of more than 5 psi between cylinders indicates the lower compression cylinder may be defective. The problem may be worn, broken, or sticking piston rings, scored pistons or worn cylinders. These problems may only be determined after the head has been removed. Removing the head on an outboard engine is not that big a deal, and may save many hours of frustration and the cost of purchasing unnecessary parts to correct a faulty condition.

Condenser

In simple terms, a condenser is composed of two sheets of tin or aluminum foil laid

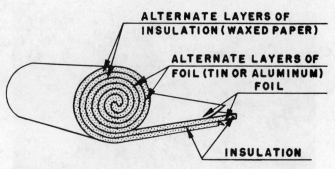

Rough sketch to illustrate how the waxed paper, aluminum foil, and insulation are rolled in a typical condenser.

one on top of the other, but separated by a sheet of insulating material such as waxed paper, etc. The sheets are rolled into a cylinder to conserve space and then inserted into a metal case for protection and to permit easy assembly.

The purpose of the condenser is to absorb or store the secondary current built up in the primary winding at the instant the breaker points are separated. By absorbing or storing this current, the condenser prevents excessive arcing and the useful life of the breaker points is extended. The condenser also gives added force to the charge produced in the secondary winding as the condenser discharges.

Modern condensers seldom cause problems, therefore, it is not necessary to install a new one each time the points are replaced. However, if the points show evidence of arcing, the condenser may be at fault and should be replaced. A faulty condenser may not be detected without the use of special test equipment. The modest cost of a new condenser justifies its purchase and installation to eliminate this item as a source of trouble.

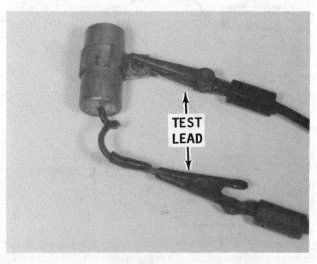

Proper hookup to test a condenser.

Worn and corroded breaker points unfit for further service.

Breaker Points

The breaker points in an outboard motor are an extremely important part of the ignition system. A set of points may appear to be in good condition, but they may be the source of hard starting, misfiring, or poor engine performance. The rules and knowledge gained from association with 4-cycle engines does not necessarily apply to a 2-cycle engine. The points should be replaced every 100 hours of operation or at least once a year. **REMEMBER**, the less an outboard engine is operated, the more care it needs. Allowing an outboard engine to remain idle will do more harm than if it is used regularly.

A breaker point set consists of two points. One is attached to a stationary bracket and does not move. The other point is attached to a moveable mount. A spring is used to keep the points in contact with each other, except when they are separated by the action of a cam built into the flywheel or machined on the crankshaft. Both points are constructed with a steel base and a tungsten cap fused to the base.

To properly diagnose magneto (spark) problems, the theory of electricity flow must be understood. The flow of electricity through a wire may be compared with the flow of water through a pipe. Consider the voltage in the wire as the water pressure in the pipe and the amperes as the volume of water. Now, if the water pipe is broken, the water does not reach the end of the pipe. In a similar manner if the wire is broken the flow of electricity is broken. If the pipe springs a leak, the amount of water reaching the end of the pipe is reduced. Same with the wire. If the installation is defective or the wire becomes grounded, the amount of electricity (amperes) reaching the end of the wire is reduced.

Check the wiring carefully, inspect the points closely and adjust them accurately. The point setting for **ALL** engines covered in this section is 0.020". An added item of useful information simplifying purchase of new points is that **ALL** point sets for the engines covered in this section have the same part number, No. 580148.

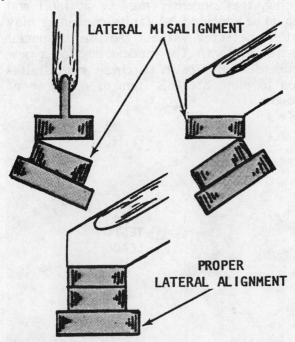

Drawing to illustrate proper point alignment, bottom set, compared with exaggerated misalignment of the other two.

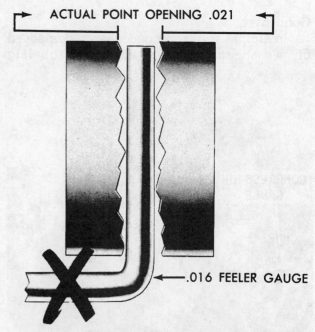

Drawing to depict how a 0.016" feeler gauge may be inserted between a badly worn set of points and the actual opening is 0.021". The point set must be in good condition to obtain an accurate adjustment.

SERVICING FLYWHEEL MAGNETO IGNITION SYSTEM

General Information

Magnetos installed on outboard engines will usually operate over extremely long periods of time without requiring adjustment or repair. However, if ignition system problems are encountered, and the usual corrective actions such as replacement of spark plugs does not correct the problem, the magneto output should be checked to determine if the unit is functioning properly.

Magneto overhaul procedures may differ slightly on various outboard models, but the following general basic instructions will apply to all Johnson/Evinrude high speed flywheel-type magnetos.

REMOVAL

1- Remove the hood or enough of the engine cover to expose the flywheel. Disconnect the battery connections from the battery terminals, if a battery is used to crank the engine. If a hand starter is installed, remove the attaching hardware from the legs of the starter assembly and lift the starter free.

2- On hand started models, a round ratchet plate is attached to the flywheel to allow the hand starter to engage in the ratchet and thus turn the flywheel. This plate must be removed before the flywheel nut is removed.

3- Remove the nut securing the flywheel to the crankshaft. It may be necessary to use some type of flywheel strap to prevent the flywheel from turning as the nut is loosened.

4- Install the proper flywheel puller using the same screw holes in the flywheel

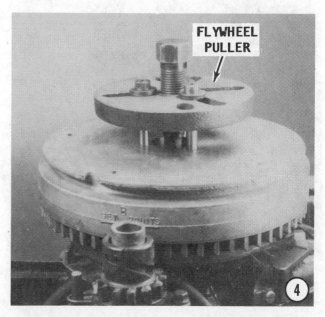

that are used to secure the ratchet plate removed in Step 2. **NEVER** attempt to use a puller which pulls on the outside edge of the flywheel or the flywheel may be damaged. After the puller is installed, tighten the center screw onto the end of the crankshaft. Continue tightening the screw until the flywheel is released from the crankshaft. Remove the flywheel. **DO NOT** strike the puller center bolt with a hammer in an attempt to dislodge the flywheel. Such action could seriously damage the lower seal and/or lower bearing.

5- **STOP**, and carefully observe the magneto and associated wiring layout. Study how the magneto is assembled. **TAKE TIME** to make notes on the wire routing. Observe how the heels of the laminated core, with the coil attached, is flush with the boss on the armature plate. These items must be replaced in their proper positions. You may elect to follow the practice of many professional mechanics by taking a series of photographs of the engine with the flywheel removed: one from the top, and a couple from the sides showing the wiring and arrangement of parts.

Breaker Points/Condenser Service

The armature plate does not have to be removed to service the magneto. If it is necessary to remove the plate for other service work, such as to replace the coil or

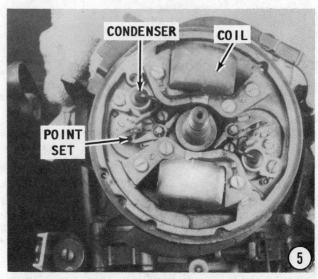

to replace the top seal, see Step 12.

For simplicity and clarity, the following procedures and accompanying illustrations cover a one-cylinder ignition system. If larger than one-cylinder is being serviced, repeat the procedures for each coil and breaker point assembly.

6- Remove the screw attaching the wires from the coil and condenser to one set of points. On engines equipped with a key switch, "kill" button, or "runaway" switch, a ground wire is also connected to this screw.

7- Using a pair of needle-nose pliers remove the wire clip from the post protruding through the center of the points.

8- Again, with the needle-nose pliers, remove the flat retainer holding the set of points together.

This particular engine differs from the text procedures because a washer is installed under the flywheel nut.

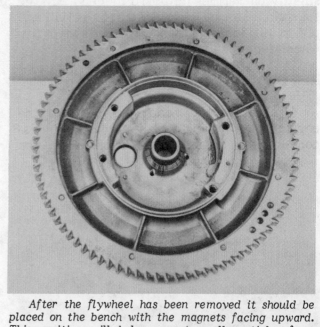

After the flywheel has been removed it should be placed on the bench with the magnets facing upward. This position will help prevent small particles from becoming attached to the magnets.

9- Lift the moveable side of the points free of the other half of the set.

10- Remove the hold-down screw securing the non-moveable half of the point set to the armature plate.

11- Remove the hold down screw securing the condenser to the armature plate. Observe how the condenser sets into a recess in the armature plate.

Repeat the procedure for the other set of points.

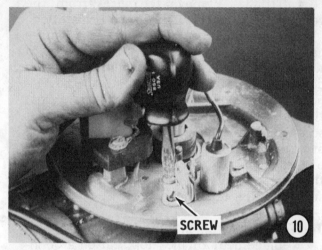

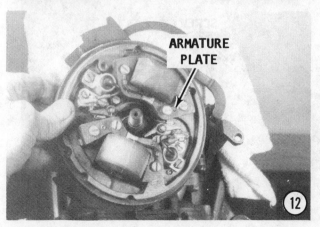

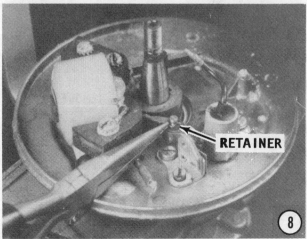

Armature Plate Removal

First, These Words: It is not necessary to remove the armature plate unless the top seal or the coil is to be replaced.

12- Disconnect the advance arm connecting the armature plate with the power shaft on the side of the engine. Next, remove the wires connecting the underside of the armature plate with the "kill" switch. If a "kill" switch is not installed, these wires are connected to the wiring harness plug. The wires of most units have a quick-disconnect fitting. Remove the wires from the vacuum (runaway) switch, if one is installed.

13- Observe the four screws in a square pattern through the armature plate. Two of these screws pass through the laminated core and the armature plate into the powerhead retainer. The other two pass just through the plate. Loosen these four screws. After the screws are loose, lift the armature plate up the crankshaft and clear of the engine. If any oil is present on top of the armature plate, or on the points, the top seal **MUST** be replaced.

Top Seal Replacement

Replacement of the top seal on a Johnson/Evinrude engine is **NOT** a difficult task, with the proper tools: a seal remover and seal installer. **NEVER** attempt to remove the seal with screwdrivers, punch, pick, or other similar tool. Such action will most likely damage the collars in the powerhead.

Obtain OMC Seal Remover P/N 387780. A 1-1/8" open end wrench is needed to hold the remover portion of the tool, while a 3/8" open end wrench is used on the top bolt.

14- To remove the seal, first, work the point cam up and free of the driveshaft. Next, remove the Woodruff key from the crankshaft. A pair of side-cutters is a handy tool for this job. Grasp the Woodruff key with the side-cutters and use the leverage of the pliers against the crankshaft to remove the key.

15- Work the special tool into the seal. Observe how the special tool is tapered and has threads. Continue working and turning the tool until it has a firm grip on the inside of the seal. Now, tighten the center screw of the puller against the end of the crankshaft and the seal will begin to lift from the collars. Continue turning this center screw until the seal can be raised manually from the crankshaft.

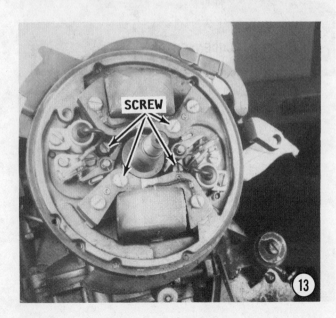

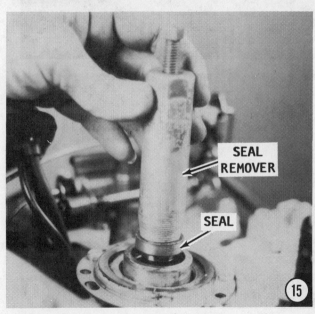

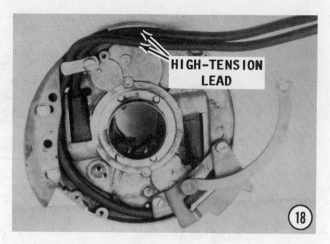

16- To install the new seal: Coat the inside diameter of the seal with a thin layer of oil. Apply OMC sealer to the outside diameter of the seal. Slide the seal down the crankshaft and start it into the recess of the powerhead. Use the special tool and work the seal completely into place in the recess.

17- Install the Woodruff key into the crankshaft. On some models, a pin was used to locate the cam for the points. If the pin was used, install it at this time. Oberve the difference to the sides of the cam. On almost all cams, the word TOP is stamped on one side. Also, on some cams, the groove does not go all the way through. Therefore, it is very difficult to install the cam incorrectly, with the wrong side up.

Slide the cam down the crankshaft with the word TOP facing upward. Continue working the cam down the crankshaft until it is in place over the Woodruff key or pin.

If the coil is not to be removed, proceed directly to Step 7. To remove the coil, perform the procedures in the following section.

Coil Removal from the Armature Plate

The armature plate must be removed as described earlier in this section, Step 12 and Step 13. Notice how the coil has a laminated core. The coil cannot be separated, that is, the laminations from the core.

18- Turn the armature plate over and notice how the high-tension leads are installed on the plate in a recess. The routing of the wires is misleading. The wire to the No. 1 spark plug is NOT connected to the No. 1 coil as might be expected.

19- Remove the three screws attaching the coils to the armature plate.

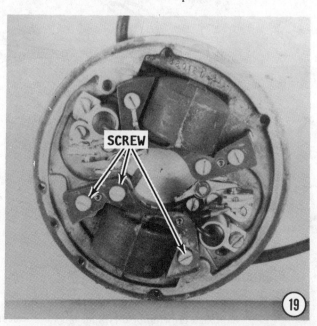

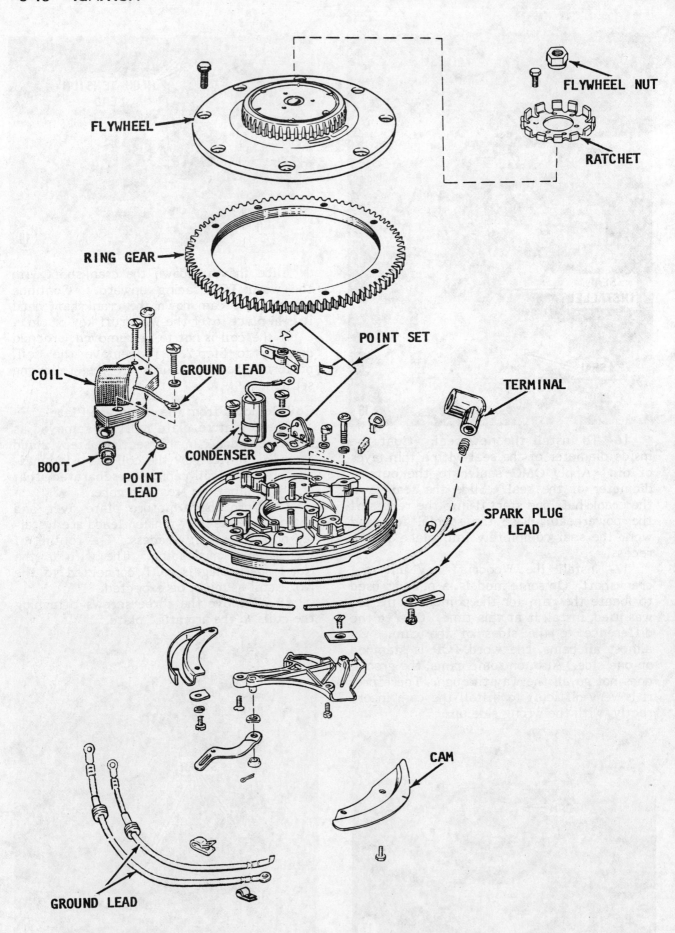

FLYWHEEL NUT

FLYWHEEL

RATCHET

RING GEAR

POINT SET

COIL

GROUND LEAD

TERMINAL

BOOT

CONDENSER

POINT LEAD

SPARK PLUG LEAD

CAM

GROUND LEAD

Exploded drawing of a typical magneto system. Only one coil and set of points are shown.

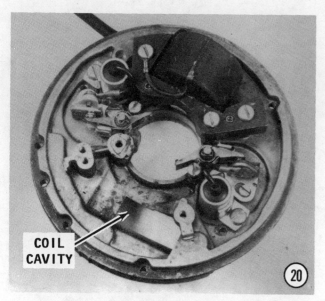

20- Hold the armature plate and separate the coils from the plate. As the coil is separated from the plate, observe the high-tension lead to the spark plug inside the coil. Work the small boot, if used, and the high-tension lead from the coil.

CLEANING AND INSPECTING

Inspect the flywheel for cracks or other damage, especially around the inside of the center hub. Check to be sure metal parts have not become attached to the magnets. Verify each magnet has good magnetism by using a screwdriver or other tool.

Thoroughly clean the inside taper of the flywheel and the taper on the crankshaft to prevent the flywheel from "walking" on the crankshaft while the engine is running.

Check the top seal around the crankshaft to be sure no oil has been leaking onto the armature plate. If there is **ANY** evidence the seal has been leaking, it **MUST** be replaced, as outlined earlier in this section.

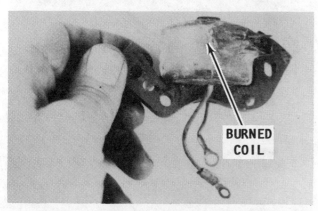

A coil burned where the high-tension lead enters the coil on the bottom side. Arcing caused the damage.

A coil destroyed when the side blew out. This damage was caused when 12-volts was connected to the magneto circuit at the key switch.

Test the armature plate to verify it is not loose. Attempt to lift each side of the plate. There should be little or no evidence of movement.

Clean the surface of the armature plate where the points and condenser attach. Install a new condenser into the recess and secure it with the hold-down screw.

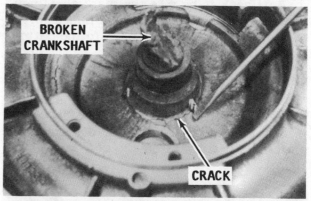

A broken crankshaft and cracked flywheel damaged when the engine was operated at a high rpm with a flush attachment and garden hose connected to the lower unit.

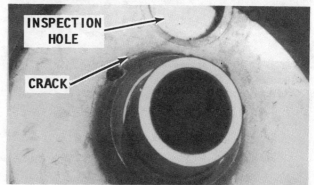

Cracks in the flywheel hub caused by metal fatigue due to flywheel construction and the inspection hole. This hole is no longer incorporated in late-model flywheels.

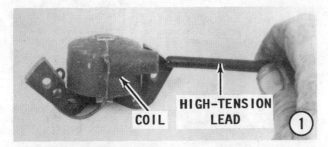

COIL   HIGH-TENSION LEAD ①

ASSEMBLING

Coil to Armature Plate

1- To install a new coil, first turn the armature plate over, and loosen the spark plug lead wires, and push them through the armature plate. Now, work the leads into the coil.

2- After the leads are into the coil, work the small boot up onto the coil. Apply a coating of rubber seal material underneath the boot, if a boot is used.

3- Start the three screws through the laminated core into the armature plate, but **DO NOT** tighten them. If the engine being serviced has a second coil, install the other coil in the same manner.

4- Check to be sure the spark plug (high-tension) leads are properly positioned in the coil and are securely attached to the bottom side of the armature plate.

5- To adjust the coils: A special ring tool is required that fits down over the armature plate. This tool will properly locate the coil in relation to the flywheel. Install this special tool over the armature plate. Push outward on the coil and secure the two outer screws.

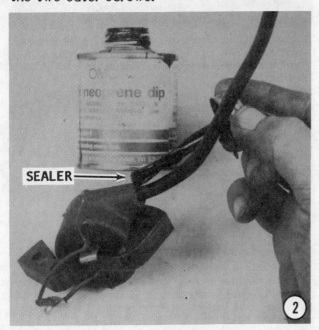

SEALER ②

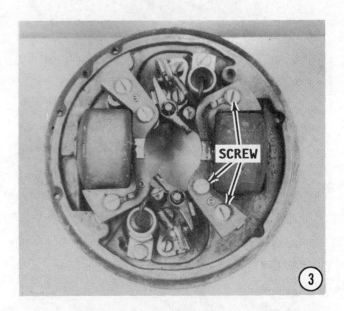

SCREW ③

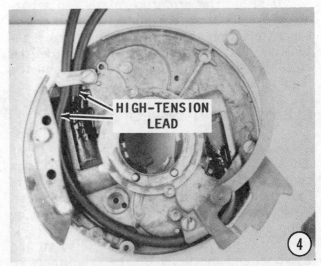

HIGH-TENSION LEAD ④

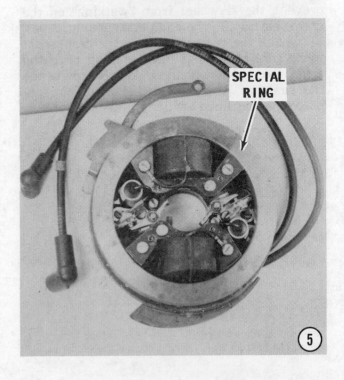

SPECIAL RING ⑤

BOSS

(6)

6- If a special ring tool is not available, and in an emergency, hold a straight edge against the boss on the armature plate and bring the heel of the laminated core out square against the edge of the boss on the armature plate. The ground wire for the coil should be attached under the head of the top screw passing through the laminated core.

Wick Replacement

7- The wick, mounted in a bracket under the coil, can be replaced without removing the armature plate. The wick **SHOULD** be replaced each and every time the breaker points are replaced. To replace the wick, simply loosen all three coil retaining screws and remove the one screw through the wick holder. Lift the coil slightly and remove the wick and wick holder. Slide the new wick into the holder; install the holder and wick under the coil; and secure it in place with the retaining screw. Adjust the coil as described earlier in this section, Steps 5 and 6, and tighten the three screws.

Armature Plate Installation

8- Slide the armature plate down over the crankshaft and onto the engine. Align the screw holes in the armature plate with the holes in the powerhead retainer. After the armature plate is in place, install and

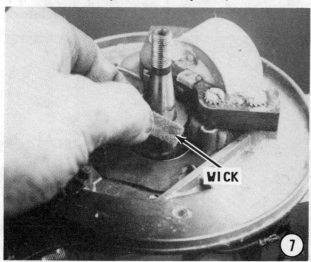

WICK

(7)

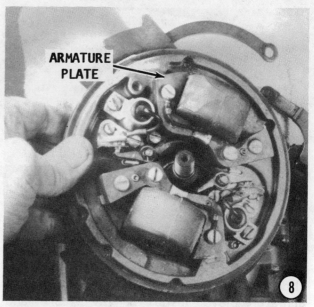

ARMATURE PLATE

(8)

tighten the two screws securing the armature plate to the retainer. Now, take up on the three screws through the laminated core closest to the crankshaft. Tighten the screws securely. Attach the advance arm from the magneto to the tower shaft arm.

GOOD WORDS

All engines covered in this section use the same set of points (Part No. 580148). The points **MUST** be assembled as they are installed. One side of each point set has the base and is non-moveable. The other side of the set has a moveable arm. A small wire clip and a flat retainer are included in each point set package.

9- Hold the base side of the points and the flat retainer. Notice how the base has a bar at right angle to the points. Observe the hole in the bar. Observe the flat retainer. Notice that one side has a slight indentation. When the points are installed, this indentation will slip into the hole in the base bar.

CLIP

(9)

CONDENSER   10

Point Set

10- Install the condensers and secure them in place with their hold-down screws.

11- Hold the base side of the points and slide it down over the anchor pin onto the armature plate. Install the wavy washer and hold-down screw to secure the point base to the armature plate. Tighten the hold-down screw securely.

12- Hold the moveable arm and slide the points down over post, and at the same time, hold back on the points and work the spring arm to the inside of the post of the base points. Continue to work the points on down into the base.

13- Observe the points. The points should be together and the spring part of the moveable arm on the inside of the flat post.

14- Install the flat retainer onto the flat bar of the base points. Check to be sure the flat spring from the other side of the points is on the inside of the retainer. Push the retainer inward until the indentation slips into the hole in the base. The retainer **MUST** be horizontal with the armature plate.

15- Install the wire clip into the groove of the post.

Repeat Steps No. 10 thru 15 for the second set of points.

SCREW

POINT SET
BASE SIDE   11

POINT
SET   12

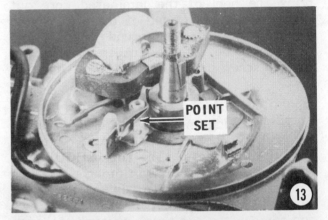

POINT
SET   13

RETAINER   14

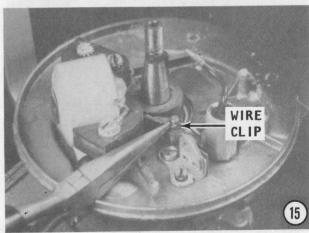

WIRE
CLIP   15

CRITICAL WORDS

As the coil, condenser, and "kill" switch wire are being attached to the point set, take the following precautions and adjustments:

a- The wire between the coil and the points should be tucked back under the coil and as far away from the crankshaft as possible.

b- The condenser wire leaving the top of the condenser and connected to the point set, should be bent downward to prevent the flywheel from making contact with the wire. A countless number of installations have been made only to have the flywheel rub against the condenser wire and cause failure of the ignition system.

c- Check to be sure all wires connected to the point set are bent downward toward the armature plate. The wires **MUST NOT** touch the plate. If any of the wires make contact with the armature plate, the ignition system will be grounded and the engine will fail to start.

16- Connect the wire leads to the set of points, with the attaching screw.

Repeat all of these **Critical Words** for the second set of points.

GOOD WORDS

The point spring tension is predetermined at the factory and does not require adjustment. Once the point set is properly installed, all should be well. In most cases, breaker contact and alignment will not be necessary. If a slight alignment adjustment should be required, **CAREFULLY** bend the insulated part of the point set.

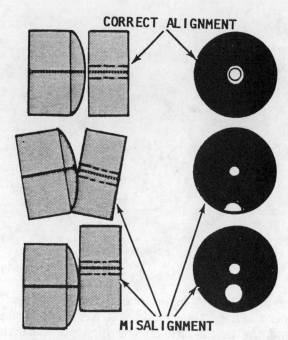

Before setting the breaker point gap, the points must be properly aligned (top). **ALWAYS** *bend the stationary point,* **NEVER** *the breaker lever. Attempting to adjust an old worn set of points is not practical because oxidation and pitting of the points will always give a false reading.*

Point Adjustment

17- Install the flywheel nut onto the end of the crankshaft. Now, turn the crankshaft clockwise and at the same time observe the cam on the crankshaft. Continue turning the crankshaft until the rubbing block of the point set is at the high point of the cam. At this position, use a wire gauge or feeler gauge and set the points at 0.020" for all models covered in this section. A wire gauge will always give a more accurate adjustment than a feeler gauge. Work the gauge between the points and, at the same time, turn the eccentric on the armature plate until the proper adjustment (0.020") is obtained. Rotate the crankshaft a complete

SCREW

16

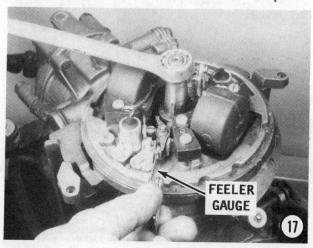

FEELER GAUGE

17

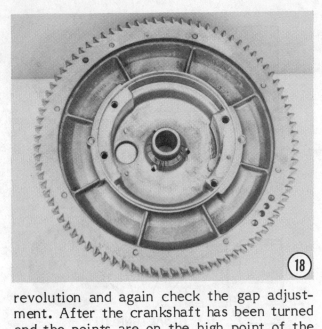

20- Rotate the flywheel clockwise and check to be sure the flywheel does not contact any part of the magneto or the wiring.

21- Place the ratchet for the starter on top of the flywheel and install the three 7/16" screws. On some model engines, a plate retainer covers these screws.

22- Thread the flywheel nut onto the crankshaft and tighten it to the torque value given in the Appendix.

23- After the ratchet and flywheel nut have been installed, install the hand starter over the flywheel, if one is used. Check to be sure the ratchet engages the flywheel properly.

revolution and again check the gap adjustment. After the crankshaft has been turned and the points are on the high point of the cam, check to be sure the hold-down screw is tight against the base. There is enough clearance to allow the eccentric on the base points to turn. If the hold-down screw is tightened **AFTER** the point adjustment has been made, it is very likely the adjustment will be changed. Follow the same procedure and adjust the other set of points. Remove the nut from the crankshaft.

Flywheel Installation

18- Check to be sure the flywheel magnets are free of any metal parts.

19- Place the key in the crankshaft keyway. Check to be sure the inside taper of the flywheel and the taper on the crankshaft are clean of dirt or oil, to prevent the flywheel from "walking" on the crankshaft while the engine is operating. Slide the flywheel down over the crankshaft with the keyway in the flywheel aligned with the key on the crankshaft.

This particular engine differs from the text procedures because a washer is installed under the flywheel nut.

24- Set the gap on each spark plug at 0.030".

25- Install the spark plugs and tighten them to the torque value of 210-246 in lbs. Connect the battery leads to the battery terminals, if a battery is used with a starter motor to crank the powerhead.

To synchronize the powerhead, proceed directly to Section 5-9.

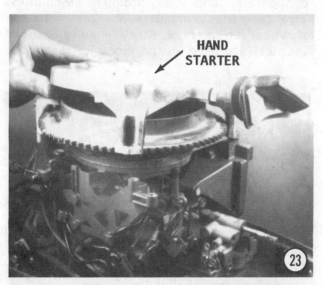

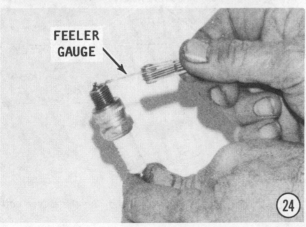

A good grade of OMC approved oil should always be used.

Operating the engine in a test tank to verify proper water circulation and engine performance.

5-6 LOW TENSION FLYWHEEL MAGNETO IGNITION SYSTEM

9.9 hp	1974-76
15 hp	1974-76
18 hp	1973
20 hp	1973
25 hp	1973-76
35 hp	1976
40 hp	1974-76
40 hp Comm.	1981-83*

*Some 40 hp Comm., 1983 models have the Type II, CD Flywheel Magneto Ignition.

DESCRIPTION

READ AND BELIEVE. A battery installed to crank the engine **DOES NOT** mean the engine is equipped with a battery-type ignition system. A low tension magneto system uses the battery only to crank the engine. Once the engine is running, the battery has absolutely no effect on engine operation. Therefore, if the battery is low and fails to crank the engine properly for starting, the engine may be cranked manually, started, and operated. Under these conditions, the key switch must be turned to the **ON** position or the engine will not start by hand cranking.

A low tension magneto system is a self-contained unit. The unit does not require assistance from an outside source for starting or continued operation. Therefore, as previously mentioned, if the battery is dead, the engine may be cranked manually and the engine started.

The low tension magneto of this system

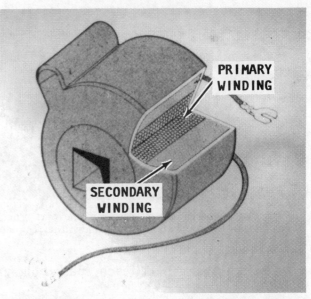

Cut-a-way view of a coil showing the primary and secondary windings.

is essentially a self-contained electrical generating unit. The low tension magneto consists of an armature plate with one driver coil and lamination assembly, two condensers, and two breaker assemblies. Two permanent magnets are cast into the flywheel on manual start engines. Four magnets are used on the electric start models.

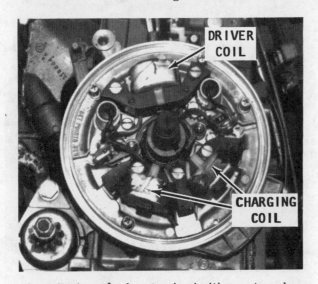

Overall view of a low-tension ignition system showing the driver coil and two charging coils.

Low-tension ignition system showing the driver coil, point set and condenser.

Individual separate coils are installed for each cylinder. These coils are installed on the side of the engine, not under the flywheel as in other systems. Each coil may be considered as two coils. One, called the primary coil, consists of a relatively few turns of heavy gauge copper wire. The other coil, called the secondary, consists of many turns of fine gauge wire. Insulation separates the two coils.

One end of primary coil is grounded and the other end is connected to the stationary breaker point. One end of the secondary coil is also grounded and the other end is connected to the spark plug. The movable side of the breaker point set is grounded to the armature plate.

The condenser can be likened to a storage tank. The condenser consists of thin sheets of metal foil separated by insulation. The sheet and insulation are rolled to save space. One sheet of the foil is grounded to the case and the other sheet is connected to the stationary side of the breaker point set.

THEORY OF OPERATION

On a manual start unit, consider one of the magnets cast into the flywheel as No. 1. Consider the opposite magnet 180° from the first as No. 2. Consider one of the breaker point sets as No. 1 and the other as No. 2. Consider the matching ignition coil as No. 1 and the other as No. 2.

Now, as the flywheel rotates, No. 1 magnet passes the driver coil inducing current flow from the driver coil through point set No. 1 (closed) across the armature plate up through the No. 2 set of points (also closed), and then to the other side of the driver coil.

The cam opens the No. 1 point set and the voltage rises rapidly across the primary of ignition coil No. 1. The condenser in the primary absorbs the current which would otherwise arc across the No. 1 set of open points. The ignition coil, being a transformer, steps up the voltage into the secondary. The current moves on to the No. 1 spark plug and No. 1 cylinder fires, reference illustration "A".

The No. 2 magnet, 180° away and of opposite polarity, passes the driver coil, inducing current flow from the driver coil through the No. 2 point set (closed) across the armature plate up through the No. 1 point set (also closed) and then to the other side of the driver coil. The cam now opens the No. 2 point set, and the voltage rises rapidly across the primary of the No. 2 ignition coil. The condenser in the primary absorbs the current which would otherwise arc across the No. 2 point set opening. The ignition coil steps up the voltage. The current moves to the No. 2 spark plug and No. 2 cylinder fires, reference illustration "B".

TROUBLESHOOTING

Always attempt to proceed with the troubleshooting in an orderly manner. The shotgun approach will only result in wasted time, incorrect diagnosis, replacement of unnecessary parts, and frustration.

Begin the ignition system troubleshooting with the spark plug/s and continue through the system until the source of trouble is located.

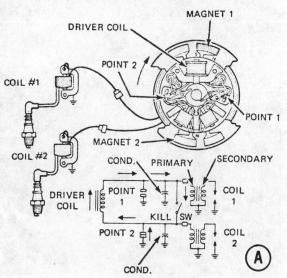

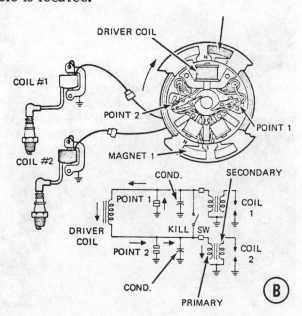

Remember, a low tension magneto system is a self-contained unit. Therefore, if the engine has a key switch and wire harness, remove them from the engine and then make a test for spark. If a good spark is obtained with these two items disconnected, but no spark is available at the plug when they are connected, then the trouble is in the harness or the key switch. If a test is made for spark at the plug with the harness and switch connected, check to be sure the key switch is turned to the **ON** position.

CRITICAL WORDS

A very high voltage exists with this type of ignition system. **NEVER** physically touch a spark plug while attempting to determine if current exists or if there is spark at the plug. The voltage generated in the system will give a person one helluva **JOLT**.

Key Switch

A low tension magneto key switch operates in **REVERSE** of any other type key switch. When the key is moved to the **OFF** position, the circuit is **CLOSED** between the low tension magneto and ground. In some cases, when the key is turned to the **OFF** position the points are grounded. For this reason, an automotive-type switch **MUST NEVER** be used, because the circuit would be opened and closed in reverse, and if 12-volts should reach the coil, the coil will be **DESTROYED.**

Spark Plugs

1- Check the plug wires to be sure they are properly connected. Check the entire length of the wire/s from the plug/s to the coil mounted on the side of the engine. If the wire is to be removed from the spark plug, **ALWAYS** use a pulling and twisting motion as a precaution against damaging the connection.

2- Attempt to remove the spark plug/s by hand. This is a rough test to determine if the plug is tightened properly. You should not be able to remove the plug without using the proper socket size tool. Remove the spark plug/s and keep them in order. Examine each plug and evaluate its condition as described in Section 5-2.

If the spark plugs have been removed and the problem cannot be determined, but the plug appears to be in satisfactory condition, electrodes, etc., then replace the plugs in the spark plug openings.

A conclusive spark plug test should always be performed with the spark plugs installed. A plug may indicate satisfactory spark when it is removed and tested, but under a compression condition may fail. An example would be the possibility of a person being able to jump a given distance on the ground, but if a strong wind is blowing, his distance may be reduced by half. The same is true with the spark plug. Under good compression in the cylinder, the spark may be too weak to ignite the fuel properly.

Therefore, to test the spark plug under compression, replace it in the engine and tighten it to the proper torque value. Another reason for testing for spark with the plugs installed is to duplicate actual operating conditions regarding flywheel speed. If

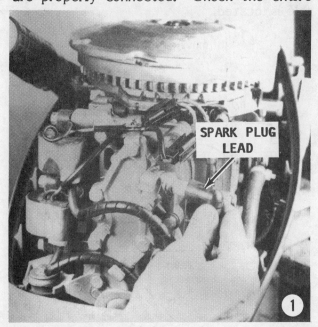

SPARK PLUG LEAD

SPARK PLUG

the flywheel is rotated with the pull cord with the plugs removed, the flywheel will rotate much faster because of the no-compression condition in the cylinder, giving the **FALSE** indication of satisfactory spark.

A spark tester, reference illustration "C", capable of testing for spark while cranking and also while the engine is operating, can be purchased from almost any good automotive supply store. This type tester uses a small neon bulb which flashes under firing voltage.

3- Use a spark tester and check for spark at each cylinder. If a spark tester is not available, hold the plug wire about 1/4-inch from the engine. Turn the flywheel with a pull starter or electrical starter and check for spark. A strong spark over a wide gap must be observed when testing in this manner, because under compression a strong spark is necessary in order to ignite the air-fuel mixture in the cylinder. This means it is possible to think you have a strong spark, when in reality the spark will be too weak when the plug is installed. If there is no spark, or if the spark is weak, the trouble is most likely under the flywheel in the magneto.

ONE MORE WORD:
Each cylinder has its own ignition system in a flywheel-type ignition system. This means if a strong spark is observed on any one cylinder and not at another, only the weak system is at fault. A single driver coil is used for both cylinders. Therefore, it is always a good idea to check and service all systems while the flywheel is removed.

Coil Testing
The following steps outline procedures to be followed to test the driver coil, point set, and condenser.

Proper testing is almost impossible without special equipment. In a marine shop, an S-80 or M-80 tester may be used. If the necessary test equipment is unavailable to the outboard owner, the driver coil may be removed and taken to a marine shop for testing.

4- Disconnect the blue wire from the No. 1 ignition coil (the top coil). Connect one lead of the neon light to the blue wire coming up from under the armature plate. Connect the other test lead to a good ground on the engine. Crank the engine through several revolutions, and at the same time observe the light. If the service work is being performed in the out-of-doors, it may be necessary to provide some type of shade in order to observe the light with confidence. A steady bright light indicates satisfactory output from under the flywheel,

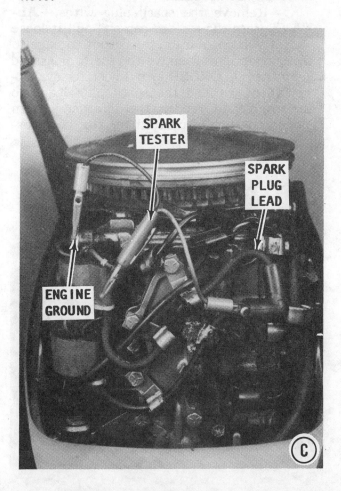

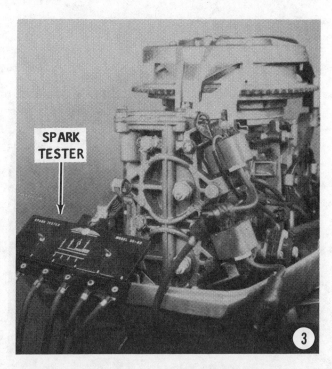

and the ignition problem may very well be the No. 1 coil is unfit for service. A dim light indicates an open condenser. No light requires a check of the driver coil. The points should also be checked for satisfactory condition and proper adjustment. Connect the blue wire back to the No. 1 coil. Disconnect the blue wire with the white stripe from the No. 2 coil. Connect the tester and repeat the test given in Step 1.

Replacement

If the previous tests indicate the coil is unfit for further service, disconnect the coil, remove the attaching hardware, and replace the coil. Bear in mind that the spark plug lead attached to the coil is **NOT** removeable. A new coil will have the spark plug lead with terminal end attached. After the new coil has been installed and the ground wire properly secured with one of the bolts, coat the connection with some type of neoprene rubber sealer. The sealer will prevent moisture, especially in a salt water atmosphere, from causing corrosion and the ground being lost.

Compression

A compression check is extremely important, because an engine with low or uneven compression between cylinders **CANNOT** be tuned to operate satisfactorily. Therefore, it is essential that any compression problem be corrected before proceeding with the tune-up procedure.

If the powerhead shows any indication of overheating, such as discolored or scorched paint, especially in the area of the top (No. 1) cylinder, inspect the cylinders visually thru the transfer ports for possible scoring. A more thorough inspection can be made if the head is removed. It is possible for a cylinder with satisfactory compression to be scored slightly. Also, check the water pump. The overheating condition may be caused by a faulty water pump.

An overheating condition may also be caused by running the engine out of the water. For unknown reasons, many operators have formed a bad habit of running a small engine without the lower unit being submerged. Such a practice will result in an overheated condition in a matter of seconds. It is interesting to note, the same operator would never operate or allow anyone else to run a large horsepower engine without water circulating through the lower unit for cooling. Bear in mind, the laws governing operation and damage to a large unit **ALL** apply equally as well to the small engine.

Checking Compression

5- Remove the spark plug wires. **ALWAYS** grasp the molded cap and pull it loose with a twisting motion to prevent damage to the connection. Remove the spark plugs and keep them in **ORDER** by cylinder for evaluation later. Ground the spark plug leads to the engine to render the ignition system inoperative while performing the compression check.

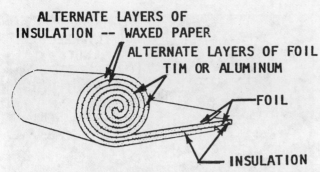

ALTERNATE LAYERS OF
INSULATION -- WAXED PAPER

ALTERNATE LAYERS OF FOIL
TIM OR ALUMINUM

FOIL

INSULATION

Rough sketch to illustrate how the waxed paper, aluminum foil, and insulation are rolled in a typical condenser.

BURNED
POINTS

Worn and corroded breaker points unfit for further service.

Insert a compression gauge into the No. 1, top, spark plug opening. Crank the engine with the starter, or pull on the starter cord, through at least 4 complete piston strokes with the throttle at the wide-open position, or until the highest possible reading is observed on the gauge. Record the reading.

Repeat the test and record the compression for each cylinder. A variation between cylinders is far more important than the actual readings. A variation of more than 5 psi between cylinders indicates the lower compression cylinder may be defective. The problem may be worn, broken, or sticking piston rings, scored pistons, or worn cylinders. These problems may only be determined after the head has been removed. Removing the head on an outboard engine is not that big a deal, and may save many hours of frustration and the cost of purchasing unnecessary parts to correct a faulty condition.

Condenser

In simple terms, a condenser is composed of two sheets of tin or aluminum foil laid one on top of the other, but separated by a sheet of insulating material such as waxed paper, etc. The sheets are rolled into a cylinder to conserve space and then inserted into a metal case for protection and to permit easy assembly.

The purpose of the condenser is to absorb or store the secondary current built up in the primary winding at the instant the breaker points are separated. By absorbing or storing this current, the condenser prevents excessive arcing and the useful life of the breaker points is extended. The condenser also gives added force to the charge produced in the secondary winding as the condenser discharges.

Modern condensers seldom cause problems, therefore, it is not necessary to install a new one each time the points are replaced. However, if the points show evidence of arcing, the condenser may be at fault and should be replaced. A faulty condenser may not be detected without the use of special test equipment. The modest cost of a new condenser justifies its purchase and installation to eliminate this item as a source of trouble.

Breaker Points

The breaker points in an outboard motor are an extremely important part of the ignition system. A set of points may appear to be in good condition, but they may be the source of hard starting, misfiring, or poor engine performance. The rules and knowledge gained from association with 4-cycle engines does not necessarily apply to a 2-cycle engine. The points should be replaced every 100 hours of operation or at least once a year. **REMEMBER,** the less an outboard engine is operated, the more care it needs. Allowing an outboard engine to remain idle will do more harm than if it is used regularly.

A breaker point set consists of two points. One is attached to a stationary bracket and does not move. The other point is attached to a movable mount. A spring is used to keep the points in contact with each other, except when they are separated by the action of a cam built into the flywheel or machined on the crankshaft. Both points are constructed with a steel base and a tungsten cap fused to the base.

To properly diagnose low tension magneto (spark) problems, the theory of electricity flow must be understood. The flow of electricity through a wire may be compared with the flow of water through a pipe. Consider the voltage in the wire as the water pressure in the pipe and the amperes as the volume of water. Now, if the water pipe is broken, the water does not reach the end of the pipe. In a similar manner if the wire is broken the flow of electricity is broken. If the pipe springs a leak, the amount of water reaching the end of the pipe is reduced. Same with the wire. If the installation is defective or the wire becomes grounded, the amount of electricity (amperes) reaching the end of the wire is reduced.

Check the wiring carefully, inspect the points closely and adjust them accurately. The point setting for **ALL** engines covered in this section is 0.020" for a used set of points and 0.022" for a new set. **NEVER** file the points. **DO NOT** change the breaker arm spring tension.

SERVICING LOW TENSION MAGNETO IGNITION SYSTEM

General Information

Magnetos installed on outboard engines will usually operate over extremely long periods of time without requiring adjustment or repair. However, if ignition system problems are encountered, and the usual corrective actions such as replacement of spark plugs does not correct the problem, the low-tension magneto output should be checked to determine if the unit is functioning properly.

Magneto overhaul procedures may differ slightly on various outboard models, but the following general basic instructions will apply to all Johnson/Evinrude high speed flywheel-type low-tension magnetos.

REMOVAL

1- Remove the hood or enough of the engine cover to expose the flywheel. Disconnect the battery connections from the battery terminals, if a battery is used to crank the engine. If a hand starter is

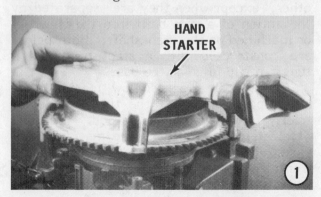

installed, remove the attaching hardware from the legs of the starter assembly and lift the starter free. On hand started models, a round ratchet plate is attached to the flywheel to allow the hand starter to engage in the ratchet and thus turn the flywheel. This plate must be removed before the flywheel nut is removed.

2- Remove the nut securing the flywheel to the crankshaft. It may be necessary to use some type of flywheel strap to prevent the flywheel from turning as the nut is loosened.

3- Install the proper flywheel puller using the same screw holes in the flywheel that are used to secure the ratchet plate (if installed). **NEVER** attempt to use a puller which pulls on the outside edge of the

flywheel or the flywheel may be damaged. After the puller is installed, tighten the center screw onto the end of the crankshaft. Continue tightening the screw until the flywheel is released from the crankshaft. Remove the flywheel. **DO NOT** strike the puller center bolt with a hammer in an attempt to dislodge the flywheel. Such action could seriously damage the lower seal and/or lower bearing.

4- STOP, and carefully observe the magneto and associated wiring layout. Study how the low-tension magneto is assembled. **TAKE TIME** to make notes on the wire routing. Observe how the heels of the laminated core, with the coil attached, is flush with the boss on the armature plate. These items must be replaced in their proper positions. You may elect to follow the practice of many professional mechanics by taking a series of photographs of the engine with the flywheel removed: one from the top, and a couple from the sides showing the wiring and arrangement of parts.

Breaker Points/Condenser Service

The armature plate does not have to be removed to service the low-tension magneto. If it is necessary to remove the plate for other service work, such as to replace the top seal, see Step 12.

For simplicity and clarity, the following procedures and accompanying illustrations cover a two-cylinder ignition system.

5- Remove the screw attaching the wires from the driver coil and condenser to

one set of points. On engines equipped with a key switch, or "kill" button switch, a ground wire is also connected to this screw.

6- Using a pair of needle-nose pliers remove the wire clip from the post protruding through the center of the points.

7- Remove the hold-down screw securing the point set to the armature plate.

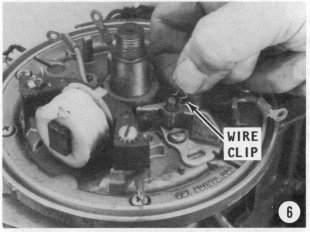

8- Lift the point set free of the armature plate.

9- Remove the hold down screw securing the condenser to the armature plate. Observe how the condenser sets into a recess in the armature plate.

Repeat the procedure for the other set of points.

10- Remove the two screws through the laminated core of the driver coil and into to the armature plate. Lift the driver coil free.

11- Use a pair of needle-nose pliers and remove the oil wick from the armature plate retainer.

Armature Plate Removal

First, These Words: It is not necessary to remove the armature plate unless the top seal, or the driver coil leads are to be replaced.

12- Disconnect the armature plate leads and the Packard connectors. The connectors are located at the back of the engine just above the head. Lift the locking tabs upward, and then slide the two halves apart. Disconnect the armature ground wire between the plate and the coil. If the engine being serviced has charge coils, disconnect

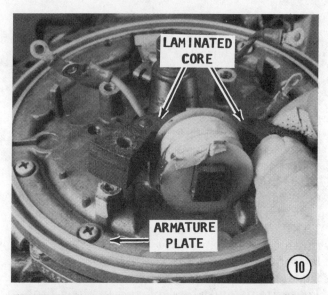

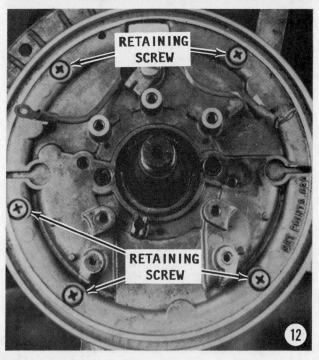

the yellow, yellow/grey, and yellow/blue stripe leads at the terminal board. Loosen the five screws on the outside edge of the armature plate. These screws are "captive" with the armature plate and need not be removed. Lift the plate free of the engine.

13- Check the nylon retainer to be sure it is clean. Good shop practice dictates to replace the nylon retainer any time the armature plate has been removed.

Top Seal Replacement

Replacement of the top seal on a Johnson/Evinrude engine is **NOT** a difficult task, with the proper tools: a seal remover and seal installer. **NEVER** attempt to remove the seal with screwdrivers, punch, pick, or other similar tool. Such action will most likely damage the collars in the powerhead.

Obtain OMC Seal Remover P/N 387780. On newer versions of this tool, a 1-1/8" open end wrench is needed to hold the tool. Older versions are equipped with handles. A 3/8" wrench is also needed for the top bolt.

14- To remove the seal, first, work the point cam up and free of the driveshaft. Next, remove the Woodruff key from the crankshaft. A pair of side-cutters is a handy tool for this job. Grasp the Woodruff key with the side-cutters and use the leverage of the pliers against the crankshaft to remove the key.

15- Work the special tool into the seal. Observe how the special tool is tapered and has threads. Continue working and turning the tool until it has a firm grip on the inside of the seal. Now, tighten the center screw of the puller against the end of the crankshaft and the seal will begin to lift from the collars. Continue turning this center screw

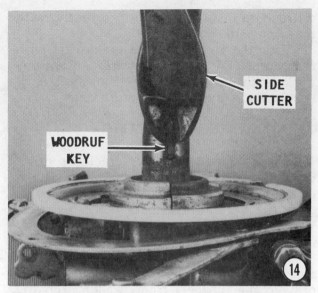

until the seal can be raised manually from the crankshaft.

16- To install the new seal: Coat the inside diameter of the seal with a thin layer of oil. Apply OMC sealer to the outside diameter of the seal. Slide the seal down the crankshaft and start it into the recess of the powerhead. Use the special tool and work the seal completely into place in the recess. Install the Woodruff key into the crankshaft keyway. Observe the difference to the sides of the cam. On almost all cams, the word **TOP** is stamped on one side. Also, on some cams, the groove does not go all the way through. Therefore, it is very

difficult to install the cam incorrectly, with the wrong side up.

Slide the cam down the crankshaft with the word **TOP** facing upward. Continue working the cam down the crankshaft until it is in place over the Woodruff key.

CLEANING AND INSPECTING

Inspect the flywheel for cracks or other damage, especially around the inside of the center hub. Check to be sure metal parts have not become attached to the magnets. Verify each magnet has good magnetism by using a screwdriver or other tool.

Thoroughly clean the inside taper of the flywheel and the taper on the crankshaft to prevent the flywheel from "walking" on the crankshaft while the engine is running.

Check the top seal around the crankshaft to be sure no oil has been leaking onto the armature plate. If there is **ANY** evidence the seal has been leaking, it **MUST** be replaced, as outlined earlier in this section.

Clean the surface of the armature plate where the points and condenser attach. Install a new condenser into the recess and secure it with the hold-down screw.

ASSEMBLING

Armature Plate Installation

1- By volume, mix one part molybdenum sulfide to forty parts OMC Sea-Lube (Trade Mark) Anti-Corrosion Lube. Coat the crankcase boss (the armature plate pilot) with the mixture. Coat the Delrin ring with OMC outboard lubricant, and then install the ring over the edge of the retainer plate. Place the support plate in position on the crankcase. Position the retainer plate over the support plate and align the holes with the holes in the crankcase. Secure the

retainer plate with the attaching screws. Tighten the screws **EVENLY** and **ALTERNATELY**.

2- Carefully slide the armature plate down over the crankshaft. **TAKE CARE** not to damage the breaker arms. Compress the Delrin ring with a pair of needle nose pliers and locate the armature plate over the Delrin ring. Align the holes in the armature plate with the holes in the support. Secure the support plate with the attaching screws. Tighten the screws **EVENLY** and **ALTERNATELY**. Connect the electrical wires, connectors, and the retainer plate to the control shaft arm with the washer beneath the retainer plate link. Tighten the Phillips screws on the outside perimeter of the armature plate. Tighten the screws **ALTERNATELY** and **EVENLY**.

OIL
WICK

3

3- Slide a **NEW** oil wick down into the armature plate retainer. Check to be sure at least half of the wick makes contact with the cam.

4- Position the driver coil in place on the armature plate. Secure the coil in place with the two screws through the laminated core and into the armature plate. Do not tighten the two screws at this time.

5- To adjust the driver coil and charge coils on electric start models, a special coil locating ring tool is required that fits down over the armature plate. This tool will locate the driver coil properly in relation to the flywheel. Install this special tool over the armature plate. Push outward on the coil and secure the two screws.

6- If a special ring tool is not available, and in an emergency, hold a straight edge against the boss on the armature plate and bring the heel of the laminated core out

SPECIAL
TOOL

5

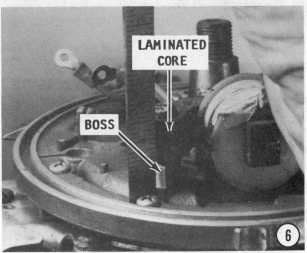

LAMINATED
CORE

BOSS

6

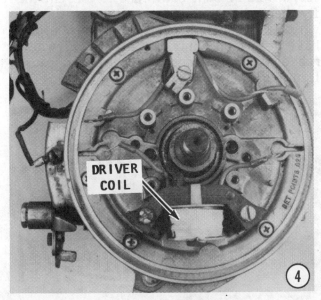

DRIVER
COIL

4

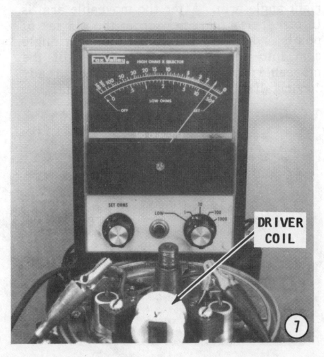

DRIVER
COIL

7

square against the edge of the boss on the armature plate. Tighten the two screws through the laminated core of the driver coil.

Testing the Driver Coil

7- Check to be sure the driver coil leads are disconnected from the point set. Connect the driver coil leads to an ohmmeter. Set the meter to the low ohms scale. The coil is fit for service if the resistance indicated is 1.45 ± 0.4 ohm.

GOOD WORDS

The points **MUST** be assembled **BEFORE** they are installed onto the armature plate. One side of each point set has the base and is non-moveable. The other side of the set has a moveable arm. A small wire clip is included in each point set package. This clip is used on top of the post.

8- Slide the moveable portion of the point set down over the post and onto the base. Check to be sure the spring tension of the moveable arm is on the inside of the right angle arm. After the moveable arm is in place, use the screw from the old set of points and thread it through the right angle arm into the spring tension lever of the moveable arm. The points are now ready for installation onto the armature plate.

9- Position the assembled point set in place on the armature plate. Install the screw with the wavy washer under the head. Tighten the screw securely into the armature plate.

10- Install the wire clip into the groove of the post.

Condenser Installation

11- Position the condenser into place in the armature plate recess. Secure the condenser in place with the attaching screw.

CRITICAL WORDS

As the driver coil, condenser, and "kill" switch wire are being attached to the point set, take the following precautions and make these adjustments:

a- The wire between the driver coil and the points should be tucked onto the armature plate and as far away from the crankshaft as possible.

b- The condenser wire leaving the top of the condenser and connected to the point set, should be bent downward to prevent the flywheel from making contact with the wire. A countless number of installations have been made only to have the flywheel rub against the condenser wire and cause failure of the ignition system.

c- Check to be sure all wires connected to the point set are bent downward toward the armature plate. The wires **MUST NOT** touch the plate. If any of the wires make contact with the armature plate, the ignition system will be grounded and the engine will fail to start.

12- Connect the wire leads to the set of points.

For the second set of points and condenser, repeat Steps 8 thru 12, including the Good Words prior to Step 8 and the Critcal Words following Step 11.

GOOD WORDS

The point spring tension is predetermined at the factory and does not require adjustment. Once the point set is properly installed, all should be well. In most cases, breaker contact and alignment will not be necessary. If a slight alignment adjustment should be required, **CAREFULLY** bend the insulated part of the point set.

Point Adjustment

13- Install the flywheel nut onto the end of the crankshaft. Now, turn the crankshaft clockwise and at the same time observe the cam on the crankshaft. Continue turning the crankshaft until the rubbing block of the point set is at the high point of the cam. At this position, use a wire gauge or feeler gauge and set the points at 0.020" for a used set of points and at 0.022" for a new set. This setting is good for all models covered in this section. A wire gauge will always give a more accurate adjustment than a feeler gauge. Work the gauge between the points and, at the same time, turn the eccentric on the armature plate until the proper adjustment is obtained. Rotate the crankshaft a complete revolution and again check the gap adjustment. After the crank-

shaft has been turned and the points are on the high point of the cam, check to be sure the hold-down screw is tight against the base. There is enough clearance to allow the eccentric on the base points to turn. If the hold-down screw is tightened **AFTER** the point adjustment has been made, it is very likely the adjustment will be changed. Follow the same procedure and adjust the other set of points. Remove the nut from the crankshaft.

Flywheel Installation

14- Check to be sure the flywheel magnets are free of any metal parts.

15- If the key is not already in place, position the key in the crankshaft keyway. Check to be sure the inside taper of the flywheel and the taper on the crankshaft are clean of dirt or oil, to prevent the flywheel

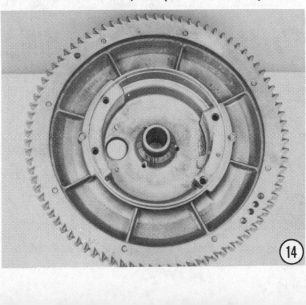

from "walking" on the crankshaft while the engine is operating. Slide the flywheel down over the crankshaft with the keyway in the flywheel aligned with the key on the crankshaft. Rotate the flywheel clockwise and check to be sure the flywheel does not contact any part of the low-tension magneto or the wiring.

16- If servicing a model with the hand starter mounted on top, place the ratchet for the starter on top of the flywheel and install the three 7/16" screws. On some model engines, a plate retainer covers these screws. Thread the flywheel nut onto the crankshaft and tighten it to the torque value given in the Appendix.

After the ratchet plate and flywheel nut have been installed, install the hand starter over the flywheel. Check to be sure the ratchet engages the flywheel properly.

17- Set the gap on each spark plug at 0.030". Some model engines do not require a gap setting. Install the spark plugs and tighten them to the correct torque value.

Spark plugs, 1971-81: 210-246 in lbs.
Spark plugs, 1982 and on: 216-252 in lbs.

Connect the battery leads to the battery terminals, if a battery is used with a starter motor to crank the powerhead.

To synchronize the ignition system with the fuel system, see Section 5-9.

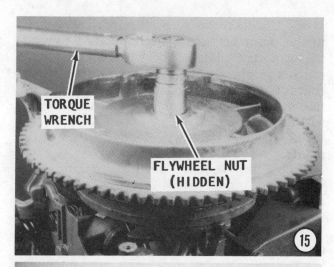

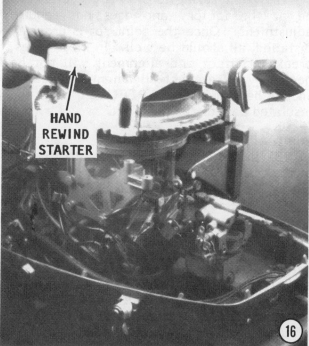

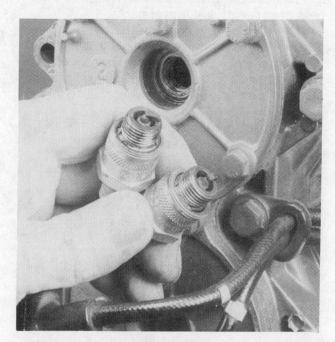

Replace worn spark plugs with new ones and tighten them to the required torque value.

5-7 TYPE I CAPACITOR DISCHARGE (CD) FLYWHEEL MAGNETO WITH TIMER BASE

This section covers the following horse-power units and model years:

50 hp -- 1971-75
55 hp -- 1976-77

DESCRIPTION

READ AND BELIEVE. A battery installed to crank the engine **DOES NOT** mean the engine is equipped with a battery-type ignition system. A magneto system uses the battery only to crank the engine. Once the engine is running, the battery has absolutely no affect on engine operation. Therefore, if the battery is low and fails to crank the engine properly for starting, the engine may be cranked manually, started, and operated. Under these conditions, the key switch must be turned to the **ON** position or the engine will not start by hand cranking.

A magneto system is a self-contained unit. The unit does not require assistance from an outside source for starting or continued operation. Therefore, as previously mentioned, if the battery is dead, the engine may be cranked manually and the engine started.

The capacitor discharge (CD) magneto ignition system consists of the flywheel and ring gear assembly; timer base and sensor assembly installed under the flywheel; a Power Pack installed on the starboard side of the powerhead; and two ignition coils mounted at the rear of the powerhead. On some models, an alternator stator and charge coils assembly is installed directly under the flywheel. The spark plugs might be considered a part of the ignition system.

Repair of these components is not possible. Therefore, if troubleshooting indicates a part unfit for further service, the entire assembly must be removed and replaced in order to restore the outboard to satisfactory performance. As an example the coil and coil wire leading to the spark plug is one assembly. If the coil or wire is found to be faulty the coil and wire must be replaced as an assembly.

Before performing maintenance work on the system, it would be well to take time to read and understand the introduction information presented in Section 5-1 at the beginning of this chapter, the Description at the start of this section, and the Theory of Operation in the following paragraphs.

THEORY OF OPERATION

This system generates approximately 30,000 volts which is fed to the spark plugs without the use of a point set or an outside voltage source.

To understand how high voltage current is generated and reaches a spark plug, imagine the flywheel turning very slowly. As the flywheel rotates, flywheel magnets induce current in the alternator stator and also generate about 300 volts AC in the charge coils. Therefore, no external voltage source is required. The 300 volts AC is converted to DC in the Power Pack, and is stored in the Power Pack capacitor. Sensor magnets are a part of the flywheel hub. Gaps exist between the sensor magnets. As one gap passes the sensor coil, voltage is generated. This small voltage generated in the sensor coil activates one of two electronic switches in the Power Pack. The switch discharges the 300 volts stored in the capacitor into one of the ignition coils. The ignition coil

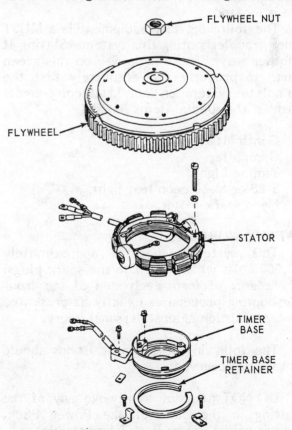

Exploded drawing of a typical flywheel, stator, timer base, and timer base retainer arrangement.

steps the voltage up to approximately 30,000 volts. This high voltage is fed to the spark plug igniting the fuel/air mixture in the cylinder.

Now, as the flywheel continues to rotate, the next sensor magnet gap on the flywheel hub, which is opposite in polarity from the first, generates a reverse polarity voltage in the sensor coil. This voltage activates the second electronic switch in the Power Pack, and discharges the capacitor into the other ignition coil. The voltage is stepped up to approximately 30,000 volts and fed to the next spark plug. The cycle is repeated as the flywheel continues to rotate.

TROUBLESHOOTING

Always attempt to proceed with the troubleshooting in an orderly manner. The shotgun approach will only result in wasted time, incorrect diagnosis, replacement of unnecessary parts, and frustration.

Begin the ignition system troubleshooting with the spark plug/s and continue through the system until the source of trouble is located.

The following test equipment is a **MUST** when troubleshooting this system. Stating it another way, "There is no way on this green earth to properly and accurately test the complete system or individual components without the special items listed."

Continuity Meter
Ohmmeter
Timing Light
S-80 or M-80 neon test light.
Neon spark tester.

SAFETY WORDS

This system generates approximately 30,000 volts which is fed to the spark plugs. Therefore, perform each step of the troubleshooting procedures exactly as presented as a precaution against personal injury.

The following safety precautions should always be observed:

DO NOT attempt to remove any of the potting in the back of the Power Pack. Repair of the Power Pack is impossible.

DO NOT attempt to remove the high tension leads from the ignition coil.

DO NOT open or close any plug-in connectors, or attempt to connect or disconnect any electrical leads while the engine is being cranked or is running.

DO NOT set the timing advanced any further than as specified.

DO NOT hold a high tension lead with your hand while the engine is being cranked or is running. Remember, the system can develop approximately 30,000 volts which will result in a severe shock if the high tension lead is held. **ALWAYS** use a pair of approved insulated pliers to hold the leads.

DO NOT attempt any tests except those listed in this troubleshooting section.

DO NOT connect an electric tachometer to the system unless it is a type which has been approved for such use.

DO NOT connect this system to any voltage source other than given in this troubleshooting section.

ONE MORE WORD: Each cylinder has its own ignition system in a flywheel-type ignition system. This means if a strong spark is observed on any one cylinder and

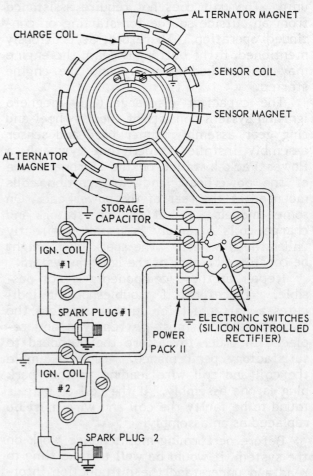

Functional diagram of a typical Type I CD flywheel magneto ignition system with timer base.

not at another, only the weak system is at fault. However, it is always a good idea to check and service all systems while the flywheel is removed.

Preliminary Test

The first area to check on a CD flywheel magneto ignition system is the system ground. Connect one lead of a continuity meter to the No. 4 Power Pack terminal and the other lead to a good ground. The meter should indicate continuity.

Also check the battery charge. If the battery is below a full charge it will not be possible to obtain full cranking speed during the tests.

Compression Check

A compression check is extremely important, because an engine with low or uneven compression between cylinders **CANNOT** be tuned to operate satisfactorily.

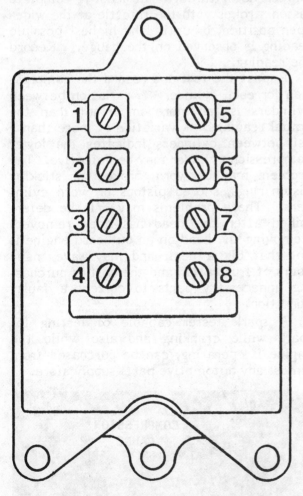

Power pack terminal identification: 1- Charge coil, brown; 2- Ignition coil No. 1, orange; 3- Ignition coil No. 2, orange; 4- Ground lead, black; 5- Key switch, black/yellow; 6- Sensor, white/black; 7- Sensor, black/white; 8- Vacant.

Therefore, it is essential that any compression problem be corrected before proceeding with the tune-up procedure. See Chapter 3.

If the powerhead shows any indication of overheating, such as discolored or scorched paint, especially in the area of the top (No. 1) cylinder, inspect the cylinders visually thru the transfer ports for possible scoring. A more thorough inspection can be made if the head is removed. It is possible for a cylinder with satisfactory compression to be scored slightly. Also, check the water pump. The overheating condition may be caused by a faulty water pump.

An overheating condition may also be caused by running the engine out of the water. For unknown reasons, many operators have formed a bad habit of running a small engine without the lower unit being submerged. Such a practice will result in an overheated condition in a matter of seconds. It is interesting to note, the same operator would never operate or allow anyone else to run a large horsepower engine without water circulating through the lower unit for cooling. Bear in mind, the laws governing operation and damage to a large unit **ALL** apply equally as well to the small engine.

Spark Plugs

1- Check the plug wires to be sure they are properly connected. Check the entire length of the wire/s from the plug/s to the coils. If the wire is to be removed from the spark plug, **ALWAYS** use a pulling and twisting motion as a precaution against damaging the connection.

2- Attempt to remove the spark plug/s by hand. This is a rough test to determine if the plug is tightened properly. You should not be able to remove the plug without using the proper socket size tool. Remove the spark plug/s and keep them in order. Ex-

SPARK PLUG (HIGH-TENSION) LEAD

*Damaged spark plugs. Notice the broken electrode on the left plug. The broken part **MUST** be found and removed before returning the powerhead to service.*

amine each plug and evaluate its condition as described in Section 5-2.

If the spark plugs have been removed and the problem cannot be determined, but the plug appears to be in satisfactory condition, electrodes, etc., then replace the plugs in the spark plug openings.

3- A conclusive spark plug test should always be performed with the spark plugs installed. A plug may indicate satisfactory spark when it is removed and tested, but under a compression condition may fail. An example would be the possibility of a person being able to jump a given distance on the ground, but if a strong wind is blowing, his distance may be reduced by half. The same is true with the spark plug. Under good compression in the cylinder, the spark may be too weak to ignite the fuel properly.

Therefore, to test the spark plug under compression, replace it in the engine and tighten it to the proper torque value. Another reason for testing for spark with the plugs installed is to duplicate actual operating conditions regarding flywheel speed. If the flywheel is rotated with the pull cord with the plugs removed, the flywheel will rotate much faster because of the no-compression condition in the cylinder, giving the **FALSE** indication of satisfactory spark.

Checking Compression

4- Remove all the spark plugs. Ground the spark plug leads to the engine to render the ignition system inoperative while performing the compression check.

Insert a compression gauge into the No. 1, top, spark plug opening. Crank the engine with the starter, or pull on the starter cord, through at least 4 complete piston strokes with the throttle at the wide-open position, or until the highest possible reading is observed on the gauge. Record the reading.

Repeat the test and record the compression for each cylinder. A variation between cylinders is far more important than the actual readings. A variation of more than 5 psi between cylinders indicates the lower compression cylinder may be defective. The problem may be worn, broken, or sticking piston rings, scored pistons or worn cylinders. These problems may only be determined after the head has been removed. Removing the head on an outboard engine is not that big a deal, and may save many hours of frustration and the cost of purchasing unnecessary parts to correct a faulty condition.

A spark tester capable of testing for spark while cranking and also while the engine is operating, can be purchased from almost any automotive parts supply store.

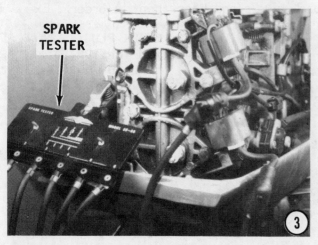

Test No. 1
Ignition Coil Output Check

Use a spark tester and check for spark at each cylinder. If a spark tester is not available, use a pair of insulated pliers and hold the plug wire about 1/4-inch from the engine. Turn the flywheel with a pull starter or electrical starter and check for spark. A strong spark over a wide gap must be observed when testing in this manner, because under compression a strong spark is necessary in order to ignite the air/fuel mixture in the cylinder. This means it is possible to think you have a strong spark, when in reality the spark will be too weak when the plug is installed. If there is no spark, or if the spark is weak, from one coil, proceed directly to Test No. 2, Trigger Coil Input Check. If there is no spark or the spark is weak from both coils, proceed directly to Test No. 5, Charge Coil Output Check. If the spark is strong and steady, across the 1/4-inch gap indicating it is satisfactory, then the problem is in the spark plugs, the fuel system, or the compression is weak.

Test No. 2
Trigger Coil Input Check

Remove the Power Pack cover, and then disconnect the sensor leads from the No. 6 and No. 7 terminals. Obtain a neon tester, No. S-80 or M-80. If one of these testers is not available, a 1-1/2 volt battery with a short lead soldered to each terminal may be used. If neither of these items are available, an ohmmeter may be used. Connect the black lead of the neon tester to the No. 6 terminal and the blue lead to the No. 7 terminal. Set the neon light selector to the No. 3 position.

Crank the engine with the electric starter motor, and at the same time depress the "B" load button rapidly. Stop, reverse the lead at the No. 6 and No. 7 terminals. Repeat the cranking and depressing procedure. Observe the spark across the tester. If both coils fired at the same time, during this test, the Power Pack is at fault. Only one coil should fire when the button is depressed.

If the neon tester is not available, an assistant is required. Have the assistant make contact with the two leads from the small 1-1/2 volt battery to the No. 6 and No. 7 terminals while the engine is being cranked. Reverse the leads at the No. 6 and No. 7 terminals and repeat the test. One coil should fire when the leads are connected one way and the other coil fire when the leads are reversed.

a- If adequate spark was observed from both coils in the previous tests, the problem is in the sensor. To test the sensor assembly, see Test No. 3 and Test No. 4.

b- If no spark was observed from either coil during the previous tests, the charge coil must be checked as outlined in Test No. 5.

c- If spark was observed on only one coil, the Power Pack must be checked, see Test No. 6.

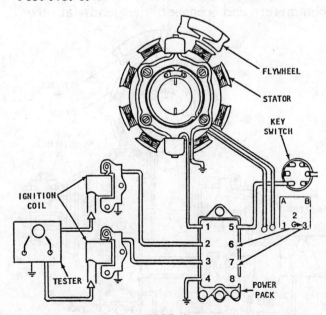

TEST 1   **TEST 2**

Test No. 3
Sensor Coil Low Ohm Check

Disconnect the sensor leads from the Power Pack trerminals No. 6 and No. 7. Connect the ohmmeter leads to the sensor coil leads (the white lead with the black stripes and the black lead with white stripes). Use the low ohm scale and observe the reading. The meter should indicate 15.0 $^{+}_{-}$5.0 ohms at room temperature (70° F). If the ohmmeter reading is not satisfactory, proceed as follows:

Turn the ohmmeter to the High Ohm scale. The meter should indicate zero ohms -- no reading.

If the test fails, the sensor coil and timer base must be replaced as a unit.

If the test is successful disconnect the ohmmeter and connect the lead at the No. 6 and No. 7 Power Pack terminals. If the test is not successful, proceed with the next test, Sensor Coil High Ohm Test.

Test No. 4
Sensor Coil High Ohm Check

Move the ohmmeter to the High Ohm scale. With the sensor leads still disconnected from the Power Pack, as in Test No. 3, connect the red ohmmeter test lead to either one of the sensor leads. Connect the black ohmmeter lead to a good ground. The meter should indicate infinity. Any resistance indicates a short to ground.

If the test fails, the sensor coil and timer base must be replaced as a unit.

If the test is successful disconnect the ohmmeter and connect the leads at the

No. 6 and No. 7 Power Pack terminals.

Test No. 5
Charge Coil Output Check

Connect the black lead from the S-80 or M-80 neon tester to the Power Pack No. 1 terminal. Connect the tester blue lead to the No. 4 Power Pack terminal, or to a good ground on the engine. (The No. 4 terminal is ground.) Turn the neon tester to position No. 2. If a neon test light is being used, connect one lead to the Power Pack No. 1 terminal and the other lead to the No. 4 Power Pack terminal or a good ground on the engine.

Crank the engine and at the same time, depress load button "B" on the tester. Observe the light. If the light glows steadily, the charge coils are good. Check the Power Pack output. Disconnect the coil lead at terminal No. 1 on the Power Pack. Connect the black tester lead to the wire just removed from the Power Pack terminal. Crank the engine. If the neon light glows intermittently or does not glow at all, the charge coil and stator assembly must be replaced.

If a neon tester is not available, obtain an ohmmeter. Disconnect the lead at the No. 1 Power Pack terminal. Connect one ohmmeter lead to the lead just disconnected from the terminal and the other ohmmeter lead to a good ground. Use the high ohmmeter scale and observe the reading. The meter should indicate 750 $^{+}_{-}$75 ohms at room temperature (70°F). If the proper reading cannot be obtained, the charge coil and stator must be replaced as an assembly.

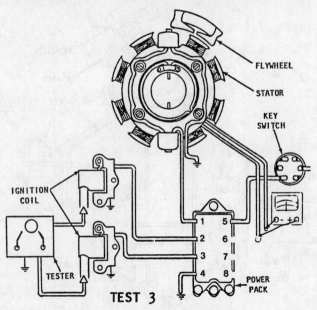

TEST 3

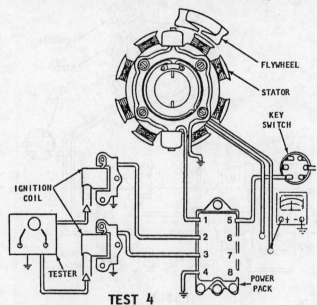

TEST 4

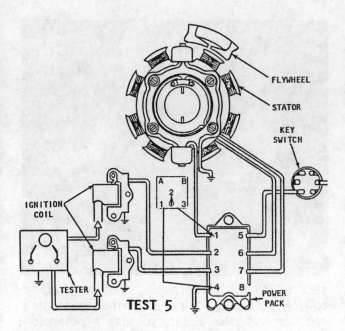

TEST 5

terminal. Crank the engine again and observe the light.

If the light is strong and steady on both outputs, replace the faulty coil or coils. The No. 2 terminal connection is the top coil and the No. 3 terminal is the bottom coil.

If no light is visible on both outputs, check the key switch as outlined in Key - Switch Check, Test No. 7.

If a steady light was observed on one output but no light on the other, replace the Power Pack.

If the light was very dim or intermittent on one or both outputs during the tests, the Power Pack must be replaced.

Connect the wires disconnected at the start of the tests.

Test No. 6
Power Pack Output Check

Check all connections at the Power Pack to be sure they are clean and secure. Disconnect the lead from the No. 2 and No. 3 terminals. If the S-80 or M-80 neon tester is used, connect the black lead to the No. 2 Power Pack terminal and the blue lead to a good ground or to the No. 4 terminal. If a neon test light is used, connect one lead to the No. 2 terminal and the other lead to a good ground or the No. 4 terminal. Move the neon tester switch to the No. 1 positon, and then depress load button "A" and crank the engine with the electric starter motor. Observe the light. If the test light is used, crank the engine with the electric starter motor and observe the light. Now, move the lead from the No. 2 terminal to the No. 3

Test No. 7
Key Switch Check

A magneto key switch operates in REVERSE of any other type key switch. When the key is moved to the OFF position, the circuit is CLOSED between the magneto and ground. In some cases, when the key is turned to the OFF position the points are grounded. For this reason, an automotive-type switch MUST NEVER be used, because the circuit would be opened and closed in reverse, and if 12-volts should reach the coil, the coil will be DESTROYED.

Connect the spark tester to the high tension leads. Disconnect the lead at the No. 5 Power Pack terminal. This is the wire from the key. Crank the engine with the key and observe the spark.

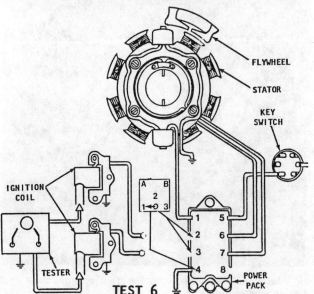

TEST 6

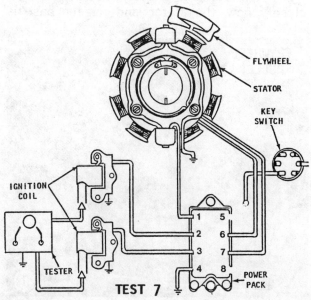

TEST 7

If there is no indication of spark on either coil or on only one, replace the Power Pack.

If spark is indicated on both coils, the problem is most likely in the lead from the key.

If the key switch leads appear to be in good condition, replace the key switch.

SERVICING THE TYPE I CD FLYWHEEL MAGNETO IGNITION WITH TIMER BASE

General Information

CD magneto systems installed on outboard engines will usually operate over extremely long periods of time without requiring adjustment or repair. However, if ignition system problems are encountered, and the usual corrective actions, such as replacement of spark plugs, does not correct the problem, the magneto output should be checked to determine if the unit is functioning properly.

CD magneto overhaul procedures may differ slightly on various outboard models but the following general basic instructions will apply to all Johnson/Evinrude high speed flywheel-type CD magnetos.

STATOR AND CHARGE COIL REPLACEMENT

Removal

1- Hold the flywheel with a proper tool and remove the flywheel nut. Obtain special tool OMC No. 378103 or an equivalent puller. **NEVER** use a puller that exerts a force on the rim or ring gear of the flywheel. Remove the flywheel. Observe closely how the stator and trigger base is

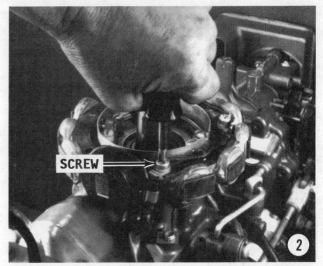

secured to the powerhead. You may elect to follow the practice of professional mechanics and take a picture with a polaroid-type camera as an aid during the assembling and installation work.

2- Disconnect the wires at the terminal block and Power Pack. The leads from the charge coil are connected to the No. 1 terminal of the Power Pack and the leads from the stator are connected to the No. 1 and No. 2 terminals of the terminal board. Remove the four screws securing the stator to the powerhead. Lift the stator and charge coil assembly free of the powerhead. The stator **CANNOT** be repaired. Therefore, if troubleshooting and testing indicates the stator or the charge coil are unfit for service, the complete unit must be replaced as an assembly.

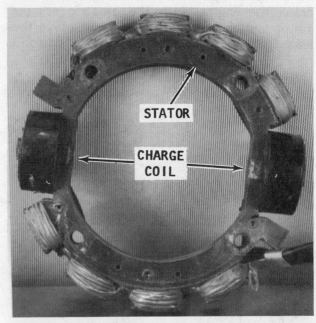

Stator assembly showing the two charge coils and the generating coils.

Installation

3- Slide the stator assembly down the crankshaft and into place on the powerhead. Apply just a drop of Loctite to the threads of the attaching screws, and then install and tighten them **ALTERNATELY** and **EVENLY** to the torque value given in the Appendix.

4- Check the crankshaft and flywheel tapers for any traces of oil, burrs, nicks, or other damage. Clean the tapered surfaces with solvent, and then blow them dry with compressed air. These two surfaces **MUST** be absolutely dry. Slide the flywheel onto the crankshaft with the slot in the flywheel indexed over the Woodruff key in the crankshaft. Thread the flywheel nut onto the crankshaft. Hold the flywheel from rotating with a proper tool and tighten the nut to the torque value given in the Appendix.

TIMER BASE AND SENSOR ASSEMBLY REPLACEMENT

Removal

1- Remove the stator and charge coil assembly as outlined elsewhere in this section. Remove the four retaining clips and screws. The clips engage in a Delrin ring which fits around the timer base. Lift the timer base and Delrin free of the powerhead.

GOOD WORDS

A brass bushing is an integral part of the timer base. This bushing has a very close tolerance with the upper bearing and seal assembly. The bushing rotates as the spark is advanced or retarded. After the assembly has been removed, make a careful check for dirt, chips, or damage which might prevent the timer base from rotating freely, reference illustration **"A"** and **"B"**.

RETAINER

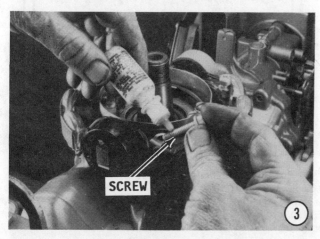

SCREW

BRASS BUSHING

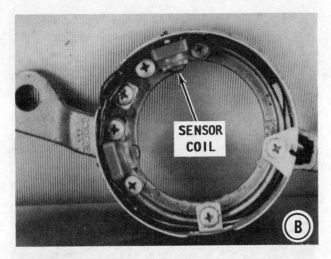

Installation

2- Coat the upper bearing and seal assembly with OMC Sea-Lube (Trade Mark) or equivalent. Apply a coating of light-weight oil to the Delrin ring. Slip the ring into place on the timer base. Position the timer base assembly into position on the powerhead and secure it with the four retaining clips and screws. If the Woodruff key on the crankshaft was removed, insert it into the keyway with the outer edge of the key parallel to the centerline of the crankshaft.

Install the stator assembly and flywheel.

POWER PACK REPLACEMENT

Removal

3- Disconnect the battery from the engine. Disconnect all leads at the Power Pack terminal board. Remove the attaching hardware and lift the Power Pack free of the powerhead.

Installation

Place the new Power Pack in place on the powerhead. Secure it with the attaching hardware. Connect the electrical leads to the terminal board following the color code designations given on the Power Pack cover. Install and secure the cover in place.

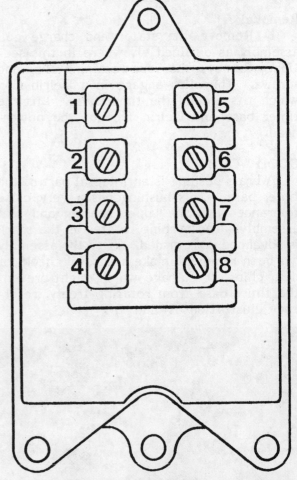

Power pack terminal identification: 1- Charge coil, brown; 2- Ignition coil No. 1, orange; 3- Ignition coil No. 2, orange; 4- Ground lead, black; 5- Key switch, black/yellow; 6- Sensor, white/black; 7- Sensor, black/-white; 8- Vacant.

TIMING CHECK AND ADJUSTMENT

GOOD WORDS

Under normal operating conditions, the timing should not change. If the spark advance stop screw has been moved, or if the Power Pack assembly has been replaced, the timing should be checked. The timing **CANNOT** be properly or accurately adjusted without a timing light.

The engine **MUST** be mounted in a test tank or body of water to adjust the timing. **NEVER** attempt to make this adjustment with a flush attachment connected to the lower unit. The no-load condition on the propeller would cause the engine to **RUN-AWAY** resulting in serious damage or destruction of the unit.

CAUTION: Water must circulate through the lower unit to the engine any time the engine is run to prevent damage to the water pump in the lower unit. Just five seconds without water will damage the water pump.

1- As a preliminary adjustment, loosen the locknut on the spark advance screw and rotate the screw inward or outward until the exposed portion of the screw outside the bracket is 1/2" (12.7 mm).

2- Connect the timing light to the No. 1 cylinder. Start the engine. With the unit in neutral or in gear, set engine speed to a minimum of 3500 rpm, full spark advance. The spark advance should be as listed in the Appendix. If necessary, advance or retard the spark to obtain the proper degree reading, as follows: Shut down the engine. Loosen the locknut and move the advance stop adjustment screw to obtain the proper

setting. One full turn of the screw will result in approximately 1° of adjustment. Again start the engine and check the degree reading. Tighten the locknut securely after the final adjustment.

Sychronizing

To synchronize the ignition system with the fuel system, see Section 5-9.

5-8 TYPE II CD FLYWHEEL MAGNETO IGNITION WITH SENSOR COIL

2.5 hp	1987 & on	15 hp	1977 & on
4 hp	1984-85	20 hp	1981 & on
4.5 hp	1980-84	25 hp	1977 & on
5 hp	1984-85	30 hp	1985 & on
6 hp	1977-79	35 hp	1977-84
6 hp	1982 & on	40 hp	1984 & on
7.5 hp	1980-83	50 hp	1978 & on
8 hp	1984 & on	55 hp	1978-83
9.9 hp	1977 & on	60hp	1980-85

DESCRIPTION

READ AND BELIEVE. A battery installed to crank the engine **DOES NOT** mean the engine is equipped with a battery-type ignition system. A CD magneto system uses the battery only to crank the engine. Once the engine is running, the battery has absolutely no affect on engine operation. Therefore, if

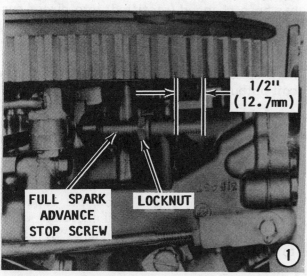

FULL SPARK ADVANCE STOP SCREW   LOCKNUT   1/2" (12.7mm)

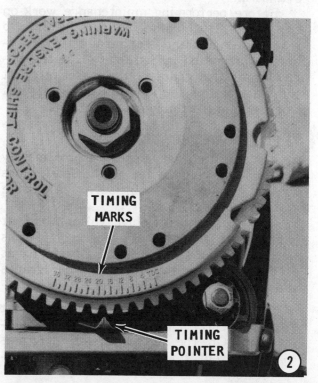

TIMING MARKS   TIMING POINTER

the battery is low and fails to crank the engine properly for starting, the engine may be cranked manually, started, and operated. Under these conditions, the key switch must be turned to the **ON** position or the engine will not start by hand cranking.

A CD magneto system is a self-contained unit. The unit does not require assistance from an outside source for starting or continued operation. Therefore, as previously mentioned, if the battery is dead, the engine may be cranked manually and the engine started.

The Type II capacitor discharge (CD) ignition system consists of the following assemblies: flywheel, charge coil, sensor, power pack, and ignition coils.

The flywheel assembly contains two integral magnets installed 180° apart. These magnets are charged with opposite polarity.

The charge coil assembly is composed of a large coil of wire which generates alternating current (ac). The current is fed into the Power Pack.

The sensor assembly consists of a small coil of wire which generates the triggering voltage necessary for the Power Pack.

In simple terms, the Power Pack contains the electronic circuits required to produce ignition at the proper time.

The ignition coils generate approximately 30,000 volts which is fed to the spark plugs to ignite the fuel/air mixture in the cylinders.

Before performing maintenance work on the system, it would be well to take time to read and understand the introduction information presented in Section 5-1 at the beginning of this chapter, the Description at the start of this section, and the Theory of Operation in the following paragraphs.

THEORY OF OPERATION

This system generates approximately 30,000 volts which is fed to the spark plugs without the use of a point set or an outside voltage source.

To understand how high voltage current is generated and reaches a spark plug, it is well to understand a couple of basic terms used throughout the following paragraphs and troubleshooting procedures.

Diode -- a small electrical part that allows current to flow in only one direction. On the accompanying diagrams the current flow is indicated by the direction of the arrow.

Silicon Controlled Rectifier -- commonly referred to as a SCR -- a small electrical part in which current can flow through in only one direction, and in which the current can flow only when a "Gate" in the part receives a positive (+) input to trigger or open the gate in the SCR. On the accompanying diagrams the direction of current flow is indicated by an arrow.

Capacitor -- a part that stores voltage. One end of the capacitor is positive (+), and the other end is negative (-). The voltage is stored in the capacitor until the trigger circuit is activated.

While studying the following circuit explanations, imagine the flywheel turning very, very slowly.

Capacitor, Charge Circuit

The terms and letter designations are keyed to reference illustration **"A"**, this page.

As the flywheel rotates, the magnetic field of a magnet passes through the charge coil winding. As a result, the charge coil produces a small amount of alternating current (AC). This current flows from the charge coil through wire **"E"** which is positive (+), then enters the Power Pack. It then flows through diode **"B"** which applies a positive (+) charge to the ground side of the capacitor. Current flow is blocked by diodes **"A"** and **"C"**. On the return path, current flows from the capacitor to the charge coil through diode **"D"** and wire **"F"**.

Alternating current from the charge coil has been changed (rectified) into direct current (DC) for capacitor charge by the four diodes. The diodes maintain a positive (+) charge on the ground side of the capacitor, regardless of the constantly changing charge coil output.

Triggering Circuit

The terms and letter designations are keyed to reference illustration **(B)**.

Purpose of the triggering circuit is to control which silicon controlled rectifier (SCR) will turn on and allow the capacitor to discharge. A sensor coil installed under the flywheel produces the signal necessary to turn on the SCR which controls the ignition timing.

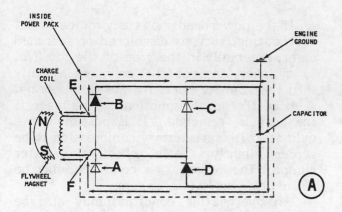

(A)

After the capacitor is charged, the flywheel continues to rotate so the magnetic field of a magnet passes through the sensor coil winding, producing alternating current.

The current leaves the sensor coil through wire "E" which is positive (+) and enters the Power Pack. Current flows through diode "B" to the No. 1 SCR gate, turning on SCR No. 1 and then returns to the other side of the sensor coil through diode "D". Current flow is blocked by diodes "A" and "C".

As the flywheel continues to rotate, the opppsite flywheel magent is passed by the sensor coil, current flow is reversed, and wire "F" is positive (+). Current enters the Power Pack and flows through diode "C" to the gate of SCR No. 2 and returns to the other side of the sensor coil through diode "A". Current flow is blocked by diode "B" and bypasses diode "D".

The four diodes in the triggering circuit direct current flow to the SCR gates and back to the sensor.

Capacitor Discharge (CD) Circuit

The terms and letter designations are keyed to reference illustration "C".

When SCR No. 1 is triggered, the positive (+) charge stored in the capacitor flows to engine ground, enters the No. 1 ignition

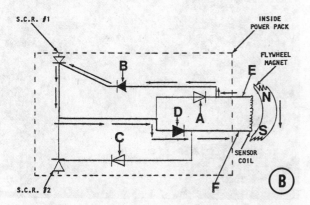

(B)

coil through the grounded primary winding wire, and then flows through SCR No. 1 to the other side of the capacitor.

During capacitor discharge, current flows through the ignition coil primary winding and SCR No. 1 until the capacitor is discharged and SCR No. 1 turns itself off. The capacitor can now be recharged as described in the capacitor charge circuit paragraphs.

Current flowing through the No. 1 ignition coil primary winding produces a large magnetic field surrounding the secondary winding and produces the high voltage to the spark plug to ignite the fuel/air mixture in the cylinder.

When SCR No. 2 is triggered, the capacitor is discharged through the No. 2 ignition coil primary winding. Current flows through the SCR No. 2 and returns to the other side of the capacitor. The No. 2 ignition coil then produces the high voltage to the spark plug to ignite the fuel/air mixture in the next cylinder.

Ignition Stop/Kill Circuit

The terms and letter designations are keyed to reference illustration "D".

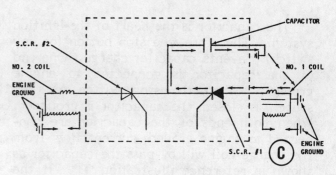

(C)

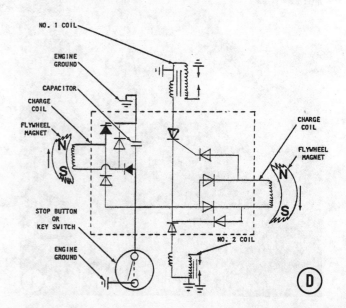

(D)

The capacitor is the heart of the ignition system. Therefore, the stop button (or key switch) prevents capacitor charge. One end of the capacitor is connected to engine ground. When the stop button (or key switch) is closed, the capacitor is grounded. When both ends of the capacitor are connected to engine ground, current flow from the charge coil will by-pass the capacitor as shown in reference illustration "D". If the capacitor is not charged, ignition cannot occur.

TROUBLESHOOTING

Always attempt to proceed with the troubleshooting in an orderly manner. The shotgun approach will only result in wasted time, incorrect diagnosis, replacement of unnecessary parts, and frustration.

Begin the ignition system troubleshooting with the spark plug/s and continue through the system until the source of trouble is located.

Compression

A compression check is extremely important, because an engine with low or uneven compression between cylinders CAN-NOT be tuned to operate satisfactorily. Therefore, it is essential that any compression problem be corrected before proceeding with the tune-up procedure. See Chapter 3.

If the powerhead shows any indication of overheating, such as discolored or scorched paint, especially in the area of the top (No. 1) cylinder, inspect the cylinders visually thru the transfer ports for possible scoring. A more thorough inspection can be made if the head is removed. It is possible for a cylinder with satisfactory compression to be scored slightly. Also, check the water pump. The overheating condition may be caused by a faulty water pump.

An overheating condition may also be caused by running the engine out of the water. For unknown reasons, many operators have formed a bad habit of running a small engine without the lower unit being submerged. Such a practice will result in an overheated condition in a matter of seconds. It is interesting to note, the same operator would never operate or allow anyone else to run a large horsepower engine without water circulating through the lower unit for cooling. Bear in mind, the laws governing operation and damage to a large unit ALL apply equally as well to the small engine.

Checking Compression

1- Remove the spark plug wires. AL-WAYS grasp the molded cap and pull it loose with a twisting motion to prevent damage to the connection.

2- Remove the spark plugs and keep them in ORDER by cylinder for evaluation later. Ground the spark plug leads to the engine to render the ignition system inoperative while performing the compression

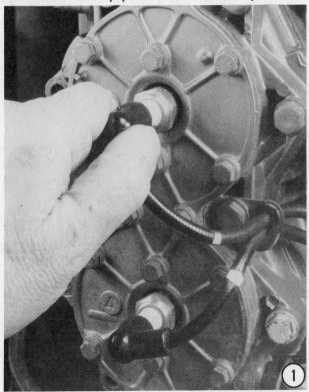

COMPRESSION GAUGE

check and to prevent damage to the ignition coil. If a high tension lead is not grounded, the coil will attempt to match the demand created by the spark trying to jump from the electrode to the nearest ground.

Insert a compression gauge into the No. 1 (top) spark plug opening. Crank the engine with the starter, or pull on the starter cord, through at least 4 complete piston strokes with the throttle at the wide-open position, or until the highest possible reading is observed on the gauge. Record the reading.

Repeat the test and record the compression for each cylinder. A variation between cylinders is far more important than the actual readings. A variation of more than 5 psi between cylinders indicates the lower compression cylinder may be defective. The problem may be worn, broken, or sticking piston rings, scored pistons or worn cylinders. These problems may only be determined after the head has been removed. Removing the head on an outboard engine is not that big a deal, and may save many hours of frustration and the cost of purchasing unnecessary parts to correct a faulty condition.

CHECKING THE TYPE II SYSTEM

The following equipment is required to test the Type II CD Magneto Ignition System.

There is no way on this green earth to properly or accurately check the system without the following text equipment:

Spark Test -- with the air gap adjusted to 1/2 inch (12.7 mm).
Neon Test Light -- Model M-80 or S-80 with a 1-1/2 volt battery adapter. If a Model M-90 test light is available an adapter is not necessary. A Stevens or Electro Specialties C.D. Voltmeter Tester will also do the job.
Ohmmeter -- capable of indicating low ohms (RX1) and high ohms (RX1,000).
Jumper Wires -- four required. These jumper wires may be made using a piece of No. 16 solid wire about 8 inches long with the insulation stripped back about an inch from each end.

Spark Plugs
3- Check the plug wires to be sure they

are properly connected. Check the entire length of the wire/s from the plug/s to the coils. If the wire is to be removed from the spark plug, **ALWAYS** use a pulling and twisting motion as a precaution against damaging the connection.

4- Attempt to remove the spark plug/s by hand. This is a rough test to determine if the plug is tightened properly. You should not be able to remove the plug without using the proper socket size tool. Remove the spark plug/s and keep them in order. Examine each plug and evaluate its condition as described in Section 5-2.

If the spark plugs have been removed and the problem cannot be determined, but the plug appears to be in satisfactory condition, electrodes, etc., then replace the plugs in the spark plug openings.

A conclusive spark plug test should always be performed with the spark plugs installed. A plug may indicate satisfactory spark when it is removed and tested, but under a compression condition may fail. An example would be the possibility of a person being able to jump a given distance on the ground, but if a strong wind is blowing, his distance may be reduced by half. The same is true with the spark plug. Under good compression in the cylinder, the spark may be too weak to ignite the fuel properly.

Therefore, to test the spark plug under compression, replace it in the engine and tighten it to the proper torque value. Another reason for testing for spark with the plugs installed is to duplicate actual operating conditions regarding flywheel speed. If the flywheel is rotated with the pull cord with the plugs removed, the flywheel will rotate much faster because of the no-compression condition in the cylinder, giving the FALSE indication of satisfactory spark.

5- A spark tester capable of testing for spark while cranking and also while the engine is operating, can be purchased from almost any automotive parts supply store, reference illustration "E".

Use a spark tester and check for spark at each cylinder. If the spark tester is used, it should be held at least 2" from any metal part of the engine. If a spark tester is not available, hold the plug wire about 1/4-inch from the engine. Turn the flywheel with a pull starter or electrical starter and check for spark. A strong spark over a wide gap must be observed when testing in this manner, because under compression a strong spark is necessary in order to ignite the air-fuel mixture in the cylinder. This means it is possible to think you have a strong spark, when in reality the spark will be too weak when the plug is installed. If there is no spark, or if the spark is weak, the trouble is most likely under the flywheel in the magneto or in the key switch.

If a satisfactory spark cannot be obtained, proceed with the troubleshooting as outlined in the following paragraphs.

GOOD WORDS

Each of the leads from the Power Pack and the leads from under the flywheel have a connector. The leads from the Power Pack to the coils have three-prong connectors. The other connectors for the charge coil and the sensor to the Power Pack each have four-prong connectors. Numerous references to these connectors will be made in the following tests.

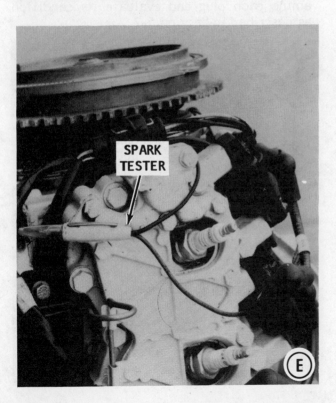

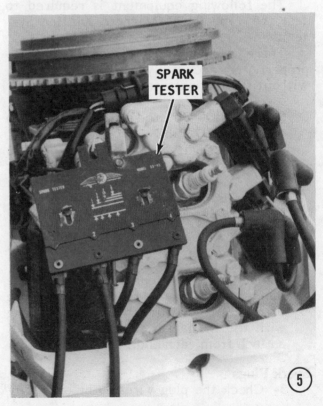

MORE GOOD WORDS

A magneto key switch operates in **RE-VERSE** of any other type key switch. When the key is moved to the **OFF** position, the circuit is **CLOSED** between the magneto and ground. In some cases, when the key is turned to the **OFF** position the points are grounded. For this reason, an automotive-type switch **MUST NEVER** be used, because the circuit would be opened and closed in reverse, and if 12-volts should reach the coil, the coil will be **DESTROYED**.

Test No. 1
Kill Button or Key Switch
Three-prong Connector
Models Prior to 1986

Separate the three prong connector between the Power Pack and the ignition coils.

Set the ohmmeter to the high ohms scale. Connect the ohmmeter red lead to the "A" terminal in the ignition coil end of the connector, and the ohmmeter black lead to a good ground on the engine.

If the engine is equipped with a push stop button, proceed as follows: With the button at rest, there should be no continuity. Depress the button -- the meter should indicate continuity.

If a key switch is used, the meter should indicate continuity with the switch in the **OFF** position. Turn the key switch to the **ON** position and the meter should indicate **NO** continuity.

If there is any needle movement, leave the ohmmeter connected and disconnect the black/yellow stripe wire from the key switch "M" terminal.

Now, if the meter indicates an open circuit with the wire disconnected from the key switch, replace the switch or perform Test No. 2, before replacing the switch.

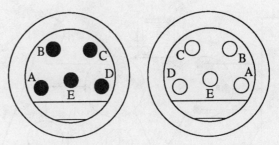

Illustration showing both halves of a five prong quick disconnect fitting used since 1986. Identification letters are embossed on the surface next to each terminal. Use this diagram for reference while performing the tests on the following pages.

If a closed circuit (0 ohms) is indicated, repair or replace the black/yellow strip lead between the key switch and the ignition coil end of the three-prong connector.

Test No. 2
Three-prong Connector
Kill Button or Key Switch
Models Prior to 1986
Five-prong Connector
Models 1986 and after

Separate the three or five prong connector from the Power Pack. Connect jumper wires between the male and female part of the connector, matching color to color, for the four outer terminals. Do not jumper the center terminal marked "E".

Crank the engine with the starter motor. If spark is now present but was not with the connector connected, either the key switch is faulty or the stop button is at fault.

If there is no spark with this test, continue with the troubleshooting.

Test No. 3
Ignition Coil Test
Primary Winding
Three-prong Connector
Models Prior to 1986

Disconnect the spark tester from the engine. Separate the three-prong connector

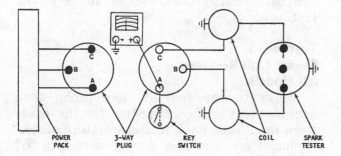

TEST 1

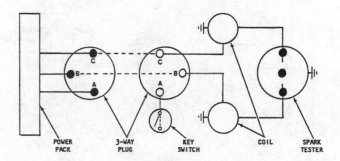

TEST 2

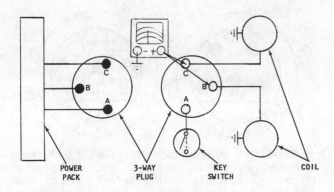

TEST 3

between the Power Pack and the ignition coils.

Connect the Red lead of an ohmmeter to the "B" prong in the connector to the coils. Connect the Black meter lead to a good ground on the engine.

Set the ohmmeter to the low ohm scale. The resistance of the primary winding should be 0.1 +0.05 ohms.

To check the primary winding of the other coil, move the red ohmmeter lead to the "C" prong of plug to the coils. Repeat the tests. The ohmmeter should indicate the same amount of resistance.

Five-prong Connector
Models 1986 and after

Rotate the primary leads (Orange and Orange/Blue) **CLOCKWISE** to remove them from the primary terminals at the ignition coils.

Connect the Red lead of an ohmmeter to the primary terminal at the ignition coil.

Connect the Black meter lead to a good ground on the engine.

Set the ohmmeter to the low ohm scale. The resistance of the primary winding should be 0.1 +0.05 ohms.

To check the primary winding of the other coil, move the Red ohmmeter lead to

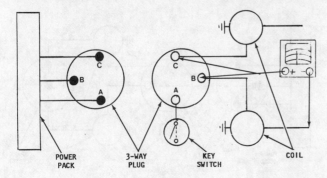

TEST 4

the primary terminal on the other ignition coils. Repeat the tests. The ohmmeter should indicate the same amount of resistance.

Test No. 4
Ignition Coil Test
Secondary Winding
Three-prong Connector
Models Prior to 1986

To check the secondary winding of the coils, connect the Red ohmmeter lead in either "B" or "C", depending on the coil being tested, and the Black test lead to the inside of the high tension spark plug lead.

Set the ohmmeter to the high scale. The meter should indicate 275 ohms +50 ohms.

Test the other coil in the same manner with the red ohmmeter lead connected to the other point "B" or "C" and the Black lead to the high tension lead for that coil.

Five-prong Connector
Models 1986 and after

To check the secondary winding of the coils, make contact with the Red ohmmeter lead to the inside of the high tension spark plug lead and the Black test lead to a good ground on the engine.

Set the ohmmeter to the high scale. The meter should indicate 275 ohms +50 ohms.

Test the other coil in the same manner with the Red ohmmeter lead connected to the other high tension lead and the Black test lead to a good ground on the engine.

If either winding -- primary or secondary -- fails the test, the coil must be replaced.

Coil Power Test

Very special equipment is required to make a power test of the coil. Therefore, either purchase a new coil or take the old coil to a nearby OMC dealer for a power test.

Test No. 5
Sensor Coil Resistance
All Models

Separate the four or five prong connector. This is the connector for the leads from the Power Pack to the armature plate.

Insert one end of a jumper wire into "B" of the female portion of the connector. Insert another jumper wire into "C" of the female portion of the connector.

Connect the test leads of an ohmmeter to each of the jumper wires. Set the ohmmeter to the low scale. The sensor coil resistance is satisfactory if the meter indicates 40 +10 ohms. If the resistance is not within this range, replace the sensor coil.

Test No. 6
Sensor Coil
Testing for a Short
All Models

Connect the black ohmmeter test lead to the armature plate ground. Connect the red meter lead to the jumper wire connected to terminal "C". Turn the ohmmeter to the high ohm scale. Any meter needle movement indicates the sensor coil or the leads are shorted to ground. A shorted sensor coil **MUST** be replaced. A shorted sensor coil lead can and **MUST** be repaired.

Disconnect the two jumper wires from "B" and "C".

Test No. 7
Charge Coil
Resistance Test
All Models

Insert a jumper wire into terminal "A" into the female portion of the armature plate connector. Insert a jumper wire into terminal "D" of the connector.

Now, connect one meter lead to one jumper wire and the other test lead to the other jumper wire.

Set the ohmmeter to the high ohm scale. The meter should indicate 575 ohms +75 oms.

If servicing an electric start model outboard, the meter should indicate 475 ohms +75 ohms.

If the resistance is not within the limts given, the charge coil **MUST** be replaced.

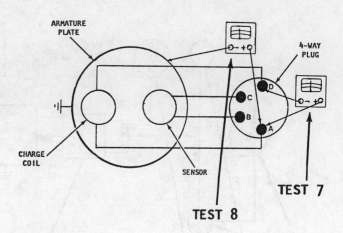

TEST 7

TEST 8

Test No. 8
Charge Coil
Testing for a Short

Set the ohmmeter to the High Ohm scale. Connect the black ohmmeter lead to the armature plate ground. Connect a jumper wire to the "A" terminal of the connector. Connect the red meter lead to the jumper wire. Any meter needle movement indicates the charge coil or the charge coil leads are shorted to ground.

A shorted charge coil **MUST** be replaced. A shorted charge coil lead can and **MUST** be repaired or replaced.

Test No. 9
Charge Coil
Output Test

An M-80 or S-80 tester can be used for this test. A neon test light available from almost any automotive parts supply store may also be used.

Insert a jumper wire into the "A" and "D" pin holes in the armature plate side of the connector. Connect the neon test leads or the neon test light leads to the jumper wires.

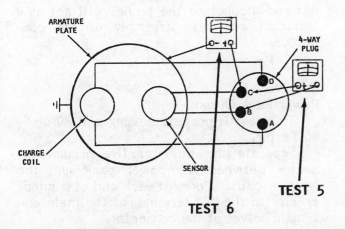

TEST 5

TEST 6

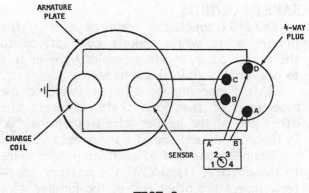

TEST 9

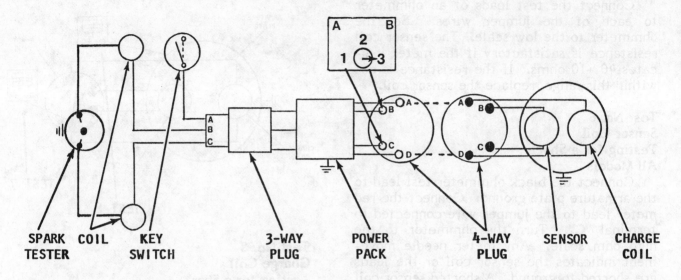

TEST 10

If the neon tester is being used, turn the switch to position **"2"**. Depress load button **"B"** and at the same time crank the engine using the hand starter rope and observe the light of the neon tester. If the neon light flashes, the charge coil is satisfactory for further service. If the neon light fails to flash, reverse the jumper leads of either tester and repeat the test. If the light fails to flash during either test, the charge coil **MUST** be replaced and the test repeated. After a new coil has been installed, the test should be successful.

Test No. 10
Trigger Input

This test is presented using a M-80 or S-80 Tester.

Insert a jumper wire between the **"A"** terminals of both parts of the four or five prong connector. Insert a jumper wire between the **"D"** terminals of the connector.

SAFETY WORDS

DO NOT touch the terminal ends of the jumper wires while making connections to the power pack, or when cranking the engine to prevent possible shock hazards.

Attach one end of a jumper wire to the blue lead of the neon tester. Insert the other end of the jumper wire to terminal **"B"** in the power pack end of the connector.

Connect one end of another jumper wire to the positive (+) side of the battery adapter. Insert the other end of the jumper wire into terminal **"C"** in the power pack end of

the connector. Move the switch to position No. 3.

Now, crank the engine with the hand starter rope, and tap the load button **"B"**. When the load button **"B"** is depressed, a spark should jump at the gap of the spark tester connected to the No. 2 ignition coil only. If spark is visible at both gaps of the spark tester during this test, replace the power pack.

Reverse the jumper wires on terminals **"B"** and **"C"** in the power pack end of the connector. Repeat the test to fire No. 1 ignition coil. If a spark occurred during each of these tests, the sensor coil is defective and must be replaced. If there was no spark, or a spark occurred on one of the tests, but not on the other, continue with the troubleshooting procedures.

Remove the jumper wires. Carefully align the connector halves and assemble the connector. Just a drop of acetone or denatured alcohol on the prongs will act as a lubricant and ease assembly of the connector.

Test No. 11
Power Pack Output

This test is presented using a M-80 or S-80 Tester.

Separate the three or five prong connector between the power pack and the ignition coils. Connect each end of a jumper wire to the **"C"** terminal of the male and female halves of the connector.

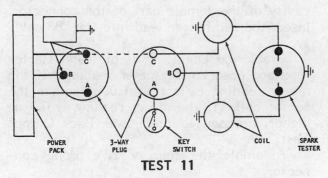

TEST 11

Connect the blue lead of the neon tester to a good ground on the engine. Connect one end of another jumper wire to the black lead of the tester. Insert the other end of the jumper wire to terminal "B" in the power pack end of the connector.

Set the neon tester switch to position No. 1. Now, hold load button "A" depressed, and at the same time crank the engine with the hand starter rope, and observe the light in the tester.

Reverse the jumper wires between terminals "B" and "C" in the power pack end of the connector.

Hold the load button "A" depressed and at the same time crank the engine with the starter rope and again observe the light in the tester.

If the neon light flashes on each output test, the power is satisfactory. Check the ignition coils. If the neon light does not flash, or flashes on only during one of the tests, replace the power pack and repeat this power pack output test.

Power Pack Output
Using Neon Test Light

Connect one lead of the tester to either the "B" or "C" terminals of the power pack end of the connector. Connect the other lead to a good ground on the engine. Crank the engine with the starter rope and observe the light. If the light flashes during both tests, the power is in good condition. If the light flashes during only one test or does not flash at all, the power pack is unfit for further service and must be replaced.

Remove all jumper wires and test equipment. Assemble the connector.

Test No. 12
Ignition System
Using CD Voltmeter Tester

This is a special tester available only through OMC. These test can only be performed using this piece of equipment.

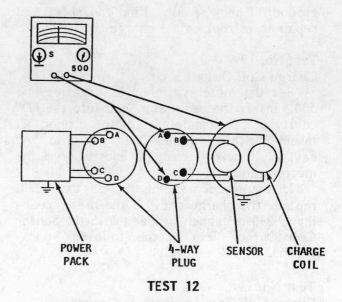

TEST 12

BEFORE conducting the following test, the Spark Test, Stop Button Test, and the Ohmmeter Tests should have been performed.

Checking for Shorts to Ground

Set the meter switches to Negative and 500. Disconnect the four or five prong connector. Insert the red meter lead into the "A" cavity of the female half. Connect the black meter lead to the armature plate or to a good ground on the engine.

Crank the engine with the hand starter rope and observe the meter reading. Remove the meter lead from the "A" cavity and insert it into cavity "D", same half of the connector. Again, crank the engine with the starter rope and observe the meter.

ANY meter reading during **EITHER** test indicates the charge coil is shorted to

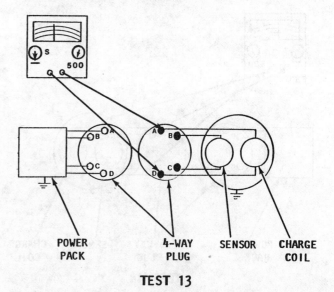

TEST 13

ground. Check it out. Find the short and repair it, or replace the charge coil.

Test No. 13
Charge Coil Output

Set the meter switches to Negative and 500. Insert the red meter lead into the "D" cavity of the female part of the connector. Insert the black meter lead into the "A" cavity. Crank the engine and observe the meter reading.

If the meter reading is less than 230, replace the charge coil. If the meter reading is 230 or higher, proceed with the Sensor Coil Output Test in the following paragraphs.

Test No. 14
Sensor Coil Test for Short to Ground

Set the meter switches to "S" and "5". Disconnect the four or five prong connector. Insert the red meter lead into the "C" cavity of the female part of the connector. Connect the black meter lead to the armature plate or to a good ground on the engine.

Crank the engine with the hand starter rope and observe the meter reading. Remove the lead from the "C" cavity and insert it into the "B" cavity. Again, crank the engine and observe the meter reading.

ANY reading during **EITHER** test indicates the sensor coil is shorted to ground. Check it out. Find the short and repair it, or replace the sensor coil.

Test No. 15
Sensor Coil Output Test

Set the meter switches to "S" and "5". Insert the black tester lead into the "C"

cavity of the female part of the connector. Insert the red meter lead into the "B" cavity.

Crank the engine with the hand starter rope and observe the meter reading. If the meter reading is less than 0.3, replace the sensor coil. If the meter reading is 0.3 or higher, proceed to the Power Pack Output Test.

Assemble the four or five prong connector.

SPECIAL WORDS

If testing a unit with a five prong connector, proceed to Test No. 18.

Test No. 16
Power Pack Output Test

Units with three prong connector: Set the meter switches to **NEGATIVE** and **500**. Disconnect the three-prong connector between the power pack and the ignition coils. Insert jumper leads between the "B" and "C" terminals of both male and female halves of the connector -- "B" to "B" and "C" to "C". Connect the black meter lead to a good ground on the engine. Connect the red meter lead to the metal part of the jumper lead coming from the power pack cavity "B" --the female connector. Crank the engine and observe the meter reading.

Now, connect the red meter lead to the metal part of the jumper lead in the "C" cavity. Again, crank the engine. The meter reading should be 180 or higher.

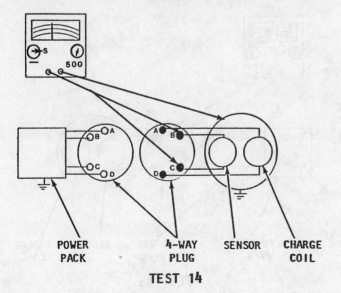

POWER PACK   4-WAY PLUG   SENSOR   CHARGE COIL

TEST 14

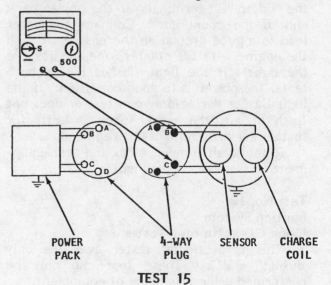

POWER PACK   4-WAY PLUG   SENSOR   CHARGE COIL

TEST 15

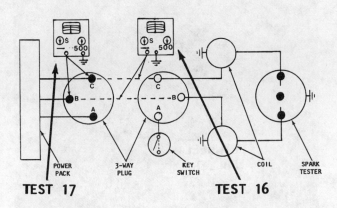

TEST 17   TEST 16

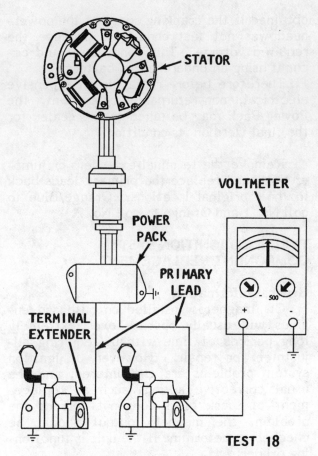

TEST 18

Test No. 17
Another Power Pack Output Test

If the meter reading during the previous two checks was not 180 or higher, disconnect the jumper lead from the cavity on the Power Pack half of the connector which had the low reading (either "B" or "C"). Insert the red meter lead into the cavity from which the jumper lead was just removed (either "B" or "C"), on the Power Pack half of the connector.

Again, crank the engine and observe the meter reading. If the meter now reads 180 or higher, the ignition coil is defective and **MUST** be replaced. If the meter does **NOT** read 180 or higher, the Power Pack is defective.

If the meter does not read 180 or higher on both outputs, the power pack is defective.

Test No. 18
Power Pack Output Test

Units with five prong connector: Rotate the primary leads (Orange and Orange/Blue) **CLOCKWISE** to remove them from the primary terminals at the ignition coils. Obtain a pair of terminal extenders from the dealer, or a pair of very short jumper wires and connect them in series with the coil terminals and the primary leads back in their original locations.

Make contact with the Red voltmeter lead to a metal portion of the terminal extender or jumper lead aligator clip of the No. 1 ignition coil. Make contact with the Black voltmeter lead to a suitable ground on the powerhead. Crank the powerhead with the starter motor while observing the voltmeter reading. The meter should register a minimum of 200 volts.

If the reading is less than indicated, disconnect the primary lead from the terminal extender or jumper lead and connect the Red voltmeter lead directly to the spring clip inside the boot of the primary lead. Crank the powerhead once more with the starter motor while observing the voltmeter reading. If the meter registers 200 volts or higher, test the ignition coil as described in Test 3 and 4 in this testing section. If the voltmeter registers less than 200 volts, check the condition of the spring clip inside the boot and the primary lead. Replace these items as necessary and then retest Power Pack output.

Repeat this test for the No. 2 ignition coil. If one test is good but the other test is bad, there is a high probability the Power Pack is defective and must be replaced.

If both tests prove bad, the problem could still be with the Power Pack, but could be with the charge coil, repeat Test No. 13, to check the input signal to the Power Pack. If this test proves good, replace the Power Pack.

GOOD WORDS

To take testing the Power Pack one step further: An erroneous reading could be obtained if the cranking speed of the power-head was not fast enough to produce the required voltage. This condition would occur if using an undercharged battery.

Therefore before replacing an expensive electrical, non-returnable component, the Power Pack may be taken to the dealer for the final word on its condition.

Remove the terminal extenders or jumper leads and replace the primary leads back in their original locations: Orange/Blue to coil No. 1 and Orange to coil No. 2.

TYPE II CD IGNITION SYSTEM COMPONENT REPLACEMENT

General Information

CD magnetos installed on outboard engines will usually operate over extremely long periods of time without requiring adjustment or repair. However, if ignition system problems are encountered, and the usual corrective actions such as replacement of spark plugs does not correct the problem, the magneto output should be checked to determine if the unit is functioning properly.

CD magneto overhaul procedures may differ slightly on various outboard models, but the following general basic instructions will apply to all Johnson/Evinrude high speed flywheel-type CD magnetos.

FLYWHEEL REMOVAL

1- Remove the hood or enough of the engine cover to expose the flywheel. Disconnect the battery connections from the battery terminals, if a battery is used to crank the engine. If a hand starter is installed, remove the attaching hardware from the legs of the starter assembly and

FLYWHEEL NUT

FLYWHEEL PULLER

lift the starter free. Remove the nut securing the flywheel to the crankshaft. It may be necessary to use some type of flywheel strap to prevent the flywheel from turning as the nut is loosened.

2- Install the proper flywheel puller using the screw holes in the flywheel. **NEVER** attempt to use a puller which pulls on the outside edge of the flywheel or the flywheel may be damaged. After the puller is installed, tighten the center screw onto the end of the crankshaft. Continue tightening the screw until the flywheel is released from the crankshaft. Remove the flywheel. **DO NOT** strike the puller center bolt with a hammer in an attempt to dislodge the flywheel. Such action could seriously damage the lower seal and/or lower bearing.

3- **STOP,** and carefully observe the magneto and associated wiring layout. Study how the magneto is assembled. **TAKE TIME** to make notes on the wire routing. Observe how the heels of the laminated core, with the coil attached, is flush with the boss on

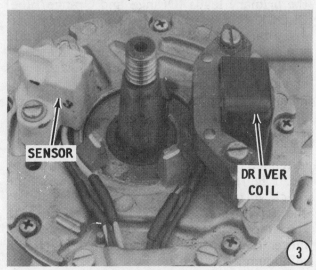

SENSOR

DRIVER COIL

the armature plate. These items must be replaced in their proper positions. You may elect to follow the practice of many professional mechanics by taking a series of photographs of the engine with the flywheel removed: one from the top, and a couple from the sides showing the wiring and arrangement of parts.

ARMATURE PLATE — REMOVAL

4- Release the armature leads from the support clamps. Disconnect the four-prong connector. This is the connector for the leads from the armature plate to the Power Pack. If servicing an electric start model, disconnect the wires from the terminal board on the starboard side of the engine.

Remove the five screws along the outside rim of the armature plate, and then lift the plate free of the powerhead. The retainer plate under the armature plate does not have to be removed, unless the plate is damaged or further work on the powerhead is required.

ARMATURE PLATE INSTALLATION

5- Install the retainer plate onto the powerhead, if it was removed. Position the support plate over the retainer plate and align the holes through the plate with the holes in the crankcase. Apply a thin coat of OMC nut lock to the threads of the attaching screws. Install and tighten the screws.

Apply OMC Molylube to the crankcase boss -- the pilot for the armature plate.

Lubricate the armature plate bearing with a light coating of OMC lubricant.

Compress the armature plate bearing with a pair of needle nose pliers in the slots provided, and at the same time, guide the armature plate into position.

6- Position the armature plate with the center of the charge coil in line with the beginning of the throttle cam. Tighten the armature attaching screws to the torque value given in the Appendix.

If servicing an electric start model, position the cutouts in the armature plate in line with the beginning of the throttle cam.

Secure the armature plate lead under the bracket with enough slack to allow for full advance.

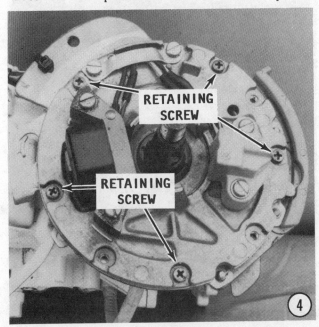

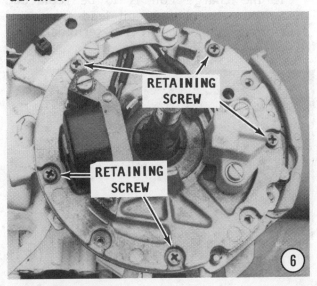

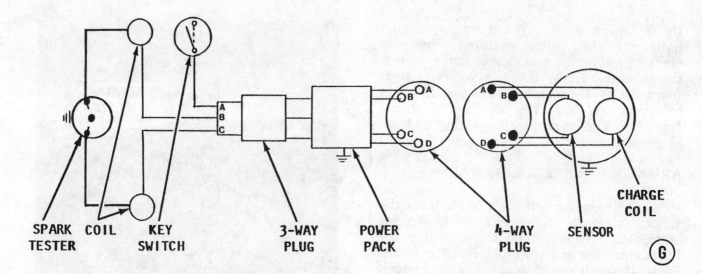

SPARK COIL KEY   3-WAY   POWER   4-WAY SENSOR   CHARGE
TESTER   SWITCH   PLUG   PACK   PLUG   COIL

G

If servicing an electric start model, route the stator assembly lead behind the starboard bracket, and then connect the three wires to the terminal board on the starboard side of the engine. Match the wire color codes of the stator lead with those from the rectifier.

CHARGE OR SENSOR COIL REMOVAL

First, These Words

A special tool is required to disconnect leads from the connectors. The special tool numbers are OMC 322697, 322698, and 322699, reference illustration "F". On the electric start model engines, the wires are connected to a terminal board on the starboard side of the engine. Therefore, the special tools are **NOT** required. Each terminal has a letter designation embossed into the end of the plug for positive identification.

If the charge coil is to be replaced, disconnect the charge coil lead from the four-prong connector at the terminals "A" and "D".

If the sensor coil is to be replaced, disconnect the sensor coil lead from the four-prong connector at terminals "B" and "C", reference illustration "G".

GOOD WORD

The armature plate need not be removed in order to remove either the charge coil or the sensor coil.

1- Rotate the armature plate in either direction until the retainer plate and attaching screw are exposed. Remove the screw and then the retainer plate from the armature plate.

2- Remove the attaching screws for the coil to be removed. Pull the defective coil and leads free of the armature plate. If

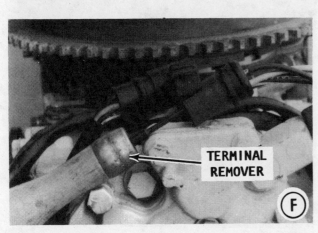

TERMINAL REMOVER

F

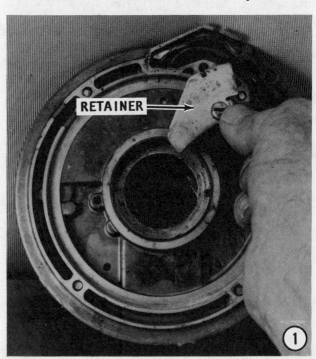

RETAINER

1

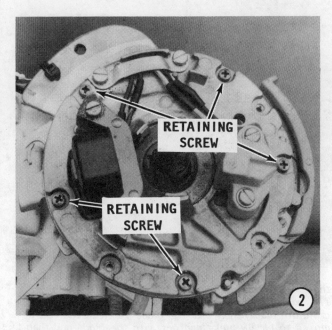

RETAINING SCREW

RETAINING SCREW

servicing an electric start model, remove the screws from the stator assembly, and then pull the stator free of the insulated sleeve and armature plate.

CHARGE COIL OR SENSOR COIL INSTALLATION

3- Lubricate the insulating sleeve with acetone or denatured alcohol. Insert the coil leads through the armature plate and sleeve into the four-prong connector.

Lay the lead wires down into the recess on the underside of the armature plate, as shown.

Start the screw through the retainer plate and into the armature plate. Now,

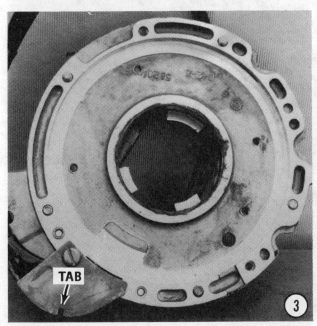

TAB

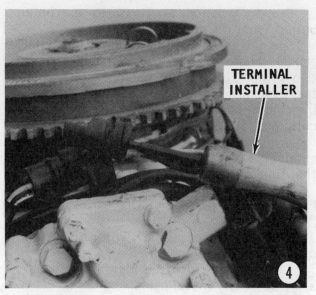

TERMINAL INSTALLER

rotate the retainer plate clockwise to cover the lead wires. The small tab on the retainer plate must index with the small notch on the edge of the armature plate. After the tab is properly indexed with the notch, tighten the screw securely.

4- Using a terminal installer, install the leads into the four-prong connector. Position the charge coil, or sensor coil onto the armature plate and start the attaching screws just finger tight.

CHARGE COIL AND SENSOR COIL ADJUSTMENT

5- The machined surfaces of the armature plate governs location of the coil assembly. The outside edge of the charge coil and the sensor coil MUST be flush with the machined surfaces on the armature plate to prevent contact with the flywheel magnet.

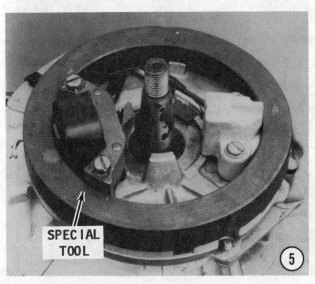

SPECIAL TOOL

Proper location also assures maximum output from the coils.

Special tool, Coil Locating Ring, OMC No. 317001, will simplify alignment of the two coils. This tool is machined to fit over the armature plate bosses. Place the locating ring in position, push the ring toward the coil, and at the same time, pull the coil toward the ring, and then tighten the coil retaining screws to the torque value given in the Appendix.

FLYWHEEL INSTALLATION

6- Clean the crankshaft with solvent and blow it dry with compressed air.

Check to be sure the flywheel magnets are free of any metal parts.

If the key is not already in place, position the key in the crankshaft keyway. Check to be sure the inside taper of the flywheel and the taper on the crankshaft are clean of dirt or oil, to prevent the flywheel from "walking" on the crankshaft while the engine is operating. Slide the flywheel down over the crankshaft with the keyway in the flywheel aligned with the key on the crankshaft.

Rotate the flywheel clockwise and check to be sure the flywheel does not contact any part of the low-tension magneto or the wiring.

Using a hacksaw blade to set the sensor coil when the special tool, called out in the text, is not available. The hacksaw blade is placed against the boss on the armature plate, then with the blade vertical, the coil is moved to barely make contact with the blade and then secured in place.

Thread the flywheel nut onto the crankshaft and tighten it to the torque value given in the Appendix.

Carefully align the four-prong connector. Assemble the two halves. Just a drop of acetone or denatured alcohol will permit the pins of the male half to index with the cavities in the other half. Secure the connector with the retainer.

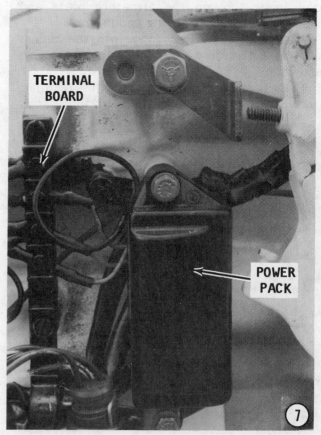

Using a hacksaw blade to set the heels of the charge coil to the armature plate when the special tool, called out in the text, is not available. The adjustment is made in the same manner as described in the illustration directly above.

If servicing a manual start model, install the starter assembly onto the top of the powerhead, see Chapter 9.

POWER PACK REPLACEMENT

Removal
7- Separate the three-prong connector between the power pack and the ignition coils, and the four-prong connector between the armature plate and the power pack.

GOOD WORDS
Bear in mind and **REMEMBER** the routing of the power pack leads and the location of the clamps to ensure proper lead routing and positioning during the installation.

Remove the attaching hardware securing the power pack to the powerhead.

MORE GOOD WORDS
Except on rare occasions, when noted, always install lockwashers on electrical connectors. Place the washer on the engine side of the connector to provide a good ground.

Installation
The leads and half of the connectors will be installed on a new power pack. The assembly is completely ready for installation when purchased.

Connect the power pack ground wire, and route the wire to prevent any pull when the pack is installed. Pull on this ground wire could very easily result in ignition failure.

Mount the power pack and tighten the attaching screws to the torque value given in the Appendix.

Carefully align the three-prong and four-prong connector halves. Assemble the two halves. Just a drop of acetone or denatured alcohol will permit the pins of the male half to index more easily in the cavities of the other half. Secure each connector with the retainer.

TIMING CHECK AND ADJUSTMENT

GOOD WORDS
Under normal operating conditions, the timing should not change. If the spark advance stop screw has been moved, or if the Power Pack assembly has been replaced, the timing should be checked. The timing **CANNOT** be properly or accurately adjusted without a timing light.

The engine **MUST** be mounted in a test tank or body of water to adjust the timing. **NEVER** attempt to make this adjustment with a flush attachment connected to the lower unit. The no-load condition on the propeller would cause the engine to **RUN-AWAY** resulting in serious damage or destruction of the unit.

CAUTION: Water must circulate through the lower unit to the engine any time the engine is run to prevent damage to the water

pump in the lower unit. Just five seconds without water will damage the water pump.

1- Connect the timing light to the No. 1 cylinder. Start the engine. With the unit in **FORWARD** gear, set engine speed to a minimum of 3500 rpm, full spark advance. The spark advance should be as listed in the Appendix.

2- If necessary, advance or retard the spark to obtain the proper degree reading, as follows: Shut down the engine. Loosen the locknut and move the advance stop adjustment screw to obtain the proper setting. One full turn of the screw will result in approximately 1° of adjustment. Again start the engine and check the degree reading. Tighten the locknut securely after the final adjustment.

5-9 SYNCHRONIZATION FUEL AND IGNITION SYSTEMS

Timing is **NOT** adjustable with a flywheel CD magneto system. The timing is controlled by the point setting. The correct point setting for **ALL** engines covered in this manual is 0.020". The fuel and ignition systems **MUST** be carefully synchronized to achieve maximum performance from the engine. In simple terms, synchronization is timing the carburetion to the ignition. This means, as the throttle is advanced to increase engine rpm, the carburetor and ignition systems are both advanced equally and at the same rate.

Therefore, any time the fuel system or the ignition system is serviced to replace a faulty part, or any adjustments are made for

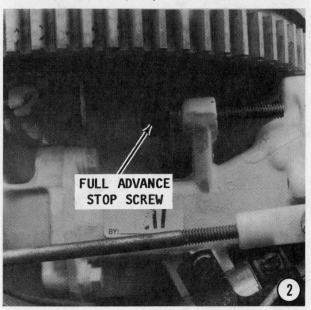

FULL ADVANCE STOP SCREW

any reason, the engine synchronization must be carefully checked and verified.

Before making any adjustments with the synchronization, the ignition system should be thoroughly checked according to the procedures outlined in this chapter and the fuel checked according to the procedures outlined in Chapter 4.

PRIMARY PICKUP ADJUSTMENTS AND LOCATIONS

To properly adjust the synchronization, locations are assigned letters and primary adjustments are assigned numbers for every powerhead covered in this manual. These letters and numbers are referenced in the Appendix under Tune-up Specifications. The locations are lettered A thru E and the adjustments are numbered from 1 thru 7. The information, a letter and a number, to be used is taken from the table by following across from the first column for the engine being serviced to the column titled Primary-Pickup Location and the column titles Primary Adjustment Note. Therefore, from the Appendix, a letter will be obtained for the location of the primary adjustment and a number, indicating the method of making the adjustment.

GOOD WORDS

When making the synchronization adjustment, it is well to know and understand exactly what to look for and why. The critical time when the throttle shaft in the carburetor begins to move is of the utmost importance. First, realize that the time the cam follower makes contact with the cam is not the time the throttle shaft starts to move. Instead, the critical time is when the follower hits the designated position (as described in the next five paragraphs) and the throttle shaft **AT THE CARBURETOR** begins to move.

A considerable amount of play exists between the follower at the top of the carburetor through the linkage to the actual throttle shaft. Therefore, the most important consideration is to watch for movement of the **THROTTLE SHAFT**, and not the follower. Movement of the shaft can be exaggerated by attaching a short piece of stiff wire to an alligator clip; grinding down the teeth on one side of the clip; and then attaching the clip to the throttle shaft, as shown. The wire jiggling will instantly indicate movement of the shaft.

Almost all of the photographs were taken with the flywheel removed for clarity. Normally the synchronization is set with the flywheel installed.

EXAMPLE

If the powerhead being serviced is a 25 hp, 1971, then the primary pickup location letter obtained from the Appendix is **"D"** and the primary pickup adjustment note number is **"3"**. In this case:

Perform Steps **D1** and **D2**, valid for all powerheads with the **"D"** location.
DO NOT perform D3, D5, or D6.
Therefore, perform Step **D4** which describes note **"3"**.

The following paragraphs describe the location in detail and give specific instructions as to exactly how the synchronization is to be made.

The "A" Location

A1– The pickup for the **"A"** location is **PORT** side of the mark. This means that the pickup on the carburetor arm should be just to the **PORT** side of the mark on the cam. To obtain this position, the cam, or the cam follower, is to be adjusted until pickup is at the proper location and the throttle shaft just begins to move.

A2– Movement of the shaft can be exaggerated by attaching a short piece of stiff wire to an alligator clip; grinding down the teeth on one side of the clip; and then attaching the clip to the throttle shaft, as shown. The wire jiggling will instantly indicate movement of the shaft. The actual adjustment is accomplished by **ONLY ONE** method, depending on the engine being ser-

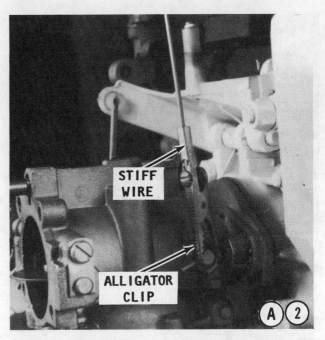

viced. Check the Appendix for the adjustment procedure to be performed. The reference numbers are listed in the Appendix under Primary Pickup Adjustment Note.

A3– Per Note 1 in the Appendix, loosen the two screws under the armature plate and move the primary pickup inward or outward to meet the follower.

The "B" Location

B1– Powerheads referenced in the Appendix to the **"B"** Primary Pickup Location are to be adjusted to the **STARBOARD** side of the mark. To obtain this position, the cam, or the cam follower, (depending the type of powerhead being serviced), is to be adjusted until pickup is at the proper location and the throttle shaft just begins to move. Movement of the shaft can be exaggerated by attaching a short piece of stiff

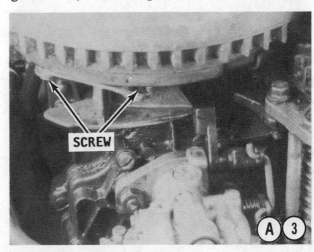

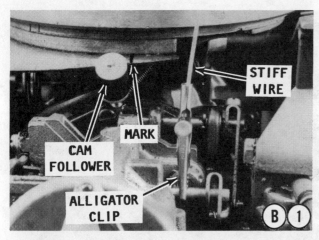

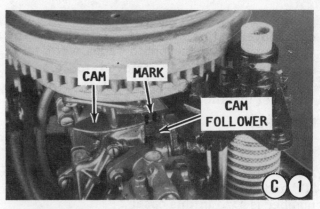

wire to an alligator clip; grinding down the teeth on one side of the clip; and then attaching the clip to the throttle shaft, as shown. The wire jiggling will instantly indicate movement of the shaft.

GOOD WORDS

The actual adjustment is accomplished by **ONLY ONE** method, regardless of the powerhead being serviced. Check the Appendix for the adjustment procedure to be performed. The reference numbers are listed in the Appendix under Primary Pickup Adjustment Note.

B2- Per Note 1 in the Appendix, loosen the two screws under the armature plate and move the primary pickup inward or outward to meet the follower.

The "C" Location

C1- Powerheads referenced in the Appendix to the "C" Primary Pickup Location are to be adjusted to the **CENTER** of the mark. To obtain this position, the cam, or the cam follower, (depending the type of engine being serviced), is to be adjusted

until pickup is at the proper location and the throttle shaft just begins to move.

C2- Movement of the shaft can be exaggerated by attaching a short piece of stiff wire to an alligator clip; grinding down the teeth on one side of the clip; and then attaching the clip to the throttle shaft, as shown. The wire jiggling will instantly indicate movement of the shaft.

GOOD WORDS

The actual adjustment is accomplished by **ONLY ONE** of two means, depending on the powerhead being serviced. Check the Appendix for the adjustment procedures to be performed. The reference numbers are listed in the Appendix under Primary Pickup Adjustment Note. **AFTER** the pickup adjustment note has been obtained from the Appendix, perform **ONE** of the following numbered procedures with the matching note number from the Appendix.

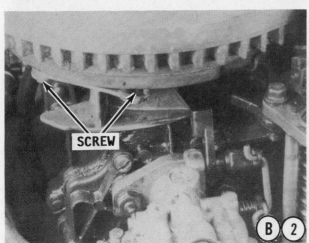

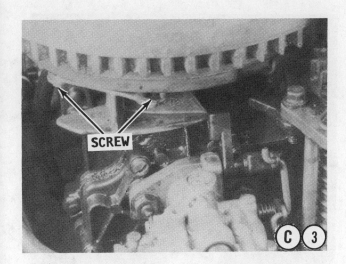

C3- Per Note 1 in the Appendix, loosen the two screws under the armature plate and move the primary pickup inward or outward to meet the follower.

OR

C4- Per Note 2 in the Appendix, loosen the center screw on the throttle lever and move the lever inward or outward to match the line on the cam.

OR

C5- Per Note 3 in the Appendix, on the starboard side of the engine, loosen the clamps on the throttle shaft and move the roller to make contact with the armature cam.

OR

C6- Per Note 4 in the Appendix, loosen the clamp on the throttle shaft and move the roller to meet the cam.

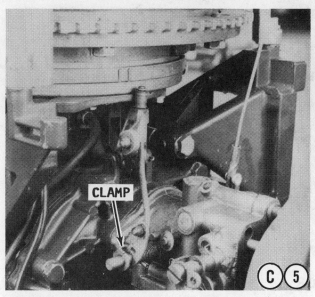

OR

C7- Per Note 6 in the Appendix, loosen the throttle linkage screw on the **STAR-BOARD** and move the roller to make contact with the cam.

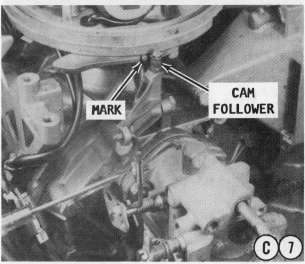

OR

C8- Per Note 7 in the Appendix, advance the throttle control with the adjusting screw at the base of the follower, until the roller barely makes contact with the cam at the mark.

The "D" Location

D1- Powerheads referenced in the Appendix to the **"D"** Primary Pickup Location are to be adjusted with the cam follower midway between the two marks on the cam, at the moment the throttle shaft at the carburetor begins to move. The marks on the cam are about 1/4" apart.

D2- Movement of the shaft can be exaggerated by attaching a short piece of stiff wire to an alligator clip; grinding down the teeth on one side of the clip; and then attaching the clip to the throttle shaft, as shown. The wire jiggling will instantly indicate movement of the shaft.

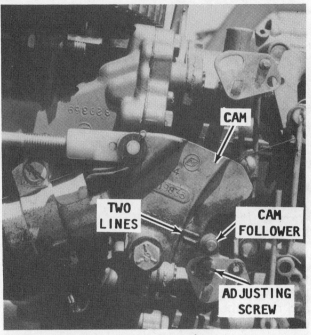

Even though the cam and cam follower are mounted vertically on the larger size powerheads covered in this manual, the procedures given the text are still valid.

GOOD WORDS

The actual adjustment is accomplished by **ONLY ONE** of two means, depending on the powerhead being serviced. Check the Appendix for the adjustment procedure to be performed. The reference numbers are listed in the Appendix under Primary Pickup Adjustment Note. **AFTER** the pickup adjustment note has been obtained from the Appendix, perform **ONE** of the following numbered procedures with the matching note number from the Appendix.

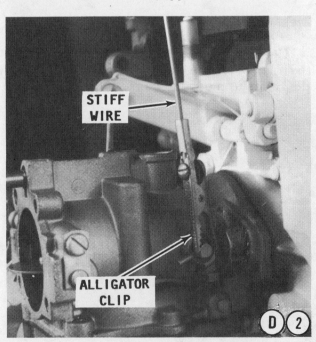

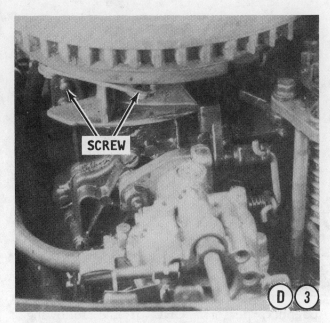

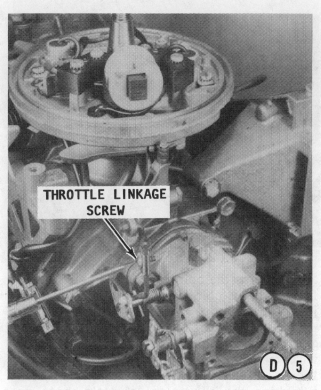

D3– Per Note 1 in the Appendix, loosen the two screws under the armature plate and move the primary pickup inward or outward to meet the follower.

OR

D4– Per Note 3 in the Appendix, on the starboard side of the engine, loosen the clamps on the throttle shaft and move the roller to make contact with the armature cam.

OR

D5– Per Note 6 in the Appendix, loosen the throttle linkage screw on the **STARBOARD** and move the roller to make contact with the cam.

OR

D6– Per Note 7 in the Appendix, advance the throttle control with the adjusting screw at the base of the follower, until the roller barely makes contact with the cam at the mark.

The "E" Location

E1– Powerheads referenced in the Appendix to the **"E"** Primary Pickup Location have a pointer attached to the intake manifold. Synchronization is made by advancing

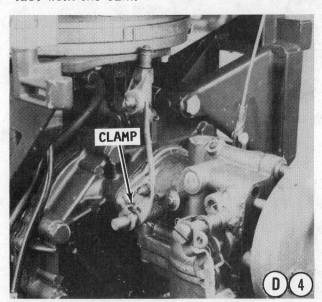

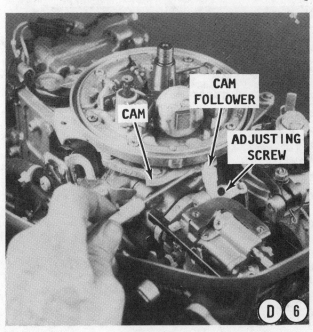

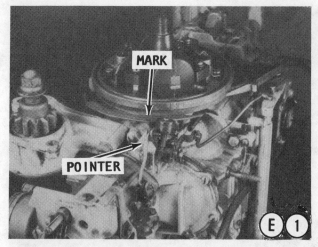

the magneto until the mark on the cam is aligned with the pointer on the intake manifold. At this point the throttle shaft should just begin to move.

E2- Movement of the shaft can be exaggerated by attaching a short piece of stiff wire to an alligator clip; grinding down the teeth on one side of the clip; and then attaching the clip to the throttle shaft, as shown. The wire jiggling will instantly indicate movement of the shaft.

GOOD WORDS

The actual adjustment is accomplished by **ONLY ONE** of three means, depending on the powerhead being serviced. Check the Appendix for the adjustment procedure to be performed. The reference numbers are listed in the Appendix under Primary Pickup Adjustment Note. **AFTER** the pickup adjustment note has been obtained from the

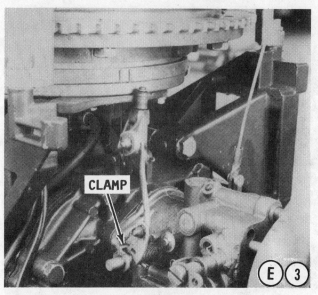

Appendix, perform **ONE** of the following numbered procedures with the matching note number from the Appendix.

E3- Per Note 3 in the Appendix, on the starboard side of the engine, loosen the clamps on the throttle shaft and move the roller to make contact with the armature cam.

OR

E4- Per Note 5 in the Appendix, loosen the eccentirc lock screw on the **PORT SIDE** on the throttle shaft and turn the eccentric to move the roller to make contact with the armature cam.

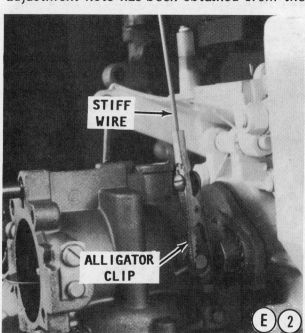

6
ELECTRICAL

6-1 INTRODUCTION

The battery, gauges and horns, charging system, and the cranking system are all considered subsystems of the electrical system. Each of these units or subsystems will be covered in detail in this chapter beginning with the battery.

The engines covered in this manual use one of four ignition systems: A conventional magneto; a low-tension magneto; a Type I magneto capacitor discharge (CD) system; and a Type II CD system.

A battery is not required to operate the engine. Most of the larger horsepower units use a cranking motor for starting and the battery is only used to supply power for this motor.

The starting circuit consists of a cranking motor and a starter-engaging mechanism. A solenoid is used as a heavy-duty switch to carry the heavy current from the battery to the starter motor. The solenoid is actuated by turning the ignition key to the **START** position. On some models, a pushbutton is used to actuate the solenoid.

These engines are also equipped with a hand starter for use when the electric starter motor system is inoperative.

6-2 BATTERIES

The battery is one of the most important parts of the electrical system. In addition to providing electrical power to start the engine, it also provides power for operation of the the running lights, radio, electrical accessories, and possibly the pump for a bait tank.

Because of its job and the consequences, (failure to perform in an emergency) the best advice is to purchase a well-known brand, with an extended warranty period, from a reputable dealer.

The usual warranty covers a prorated replacement policy, which means you would be entitled to a consideration for the time left on the warranty period if the battery should prove defective before its time.

Do not consider a battery of less than 70-ampere hour capacity. If in doubt as to how large your boat requires, make a liberal estimate and then purchase the one with the next higher ampere rating.

MARINE BATTERIES

Because marine batteries are required to perform under much more rigorous conditions than automotive batteries, they are constructed much differently than those used in automobiles or trucks. Therefore, a marine battery should always be the No. 1 unit for the boat and other types of batteries used only in an emergency.

Marine batteries have a much heavier exterior case to withstand the violent pounding and shocks imposed on it as the

A fully charged battery, filled to the proper level with electrolyte, is the heart of the ignition system. Engine starting and efficient performance can never be obtained if the battery is below a fully charged rating.

boat moves through rough water and in extremely tight turns.

The plates in marine batteries are thicker than in automotive batteries and each plate is securely anchored within the battery case to ensure extended life.

The caps of marine batteries are "spill proof" to prevent acid from spilling into the bilges when the boat heels to one side in a tight turn, or is moving through rough water.

Because of these features, the marine battery will recover from a low charge condition and give satisfactory service over a much longer period of time than any type of automotive-type unit.

BATTERY CONSTRUCTION

A battery consists of a number of positive and negative plates immersed in a solution of diluted sulfuric acid. The plates contain dissimilar active materials and are kept apart by separators. The plates are grouped into what are termed elements. Plate straps on top of each element connect all of the positive plates and all of the negative plates into groups.

The battery is divided into cells which hold a number of the elements apart from the others. The entire arrangement is contained within a hard-rubber case. The top is a one-piece cover and contains the filler caps for each cell. The terminal posts protrude through the top where the battery connections for the boat are made. Each of the cells is connected to its neighbor in a positive-to-negative manner with a heavy strap called the cell connector.

The battery MUST be located near the engine in a well-ventilated area. It must be secured in such a manner that absolutely no movement is possible in any direction under the most violent actions of the boat.

BATTERY RATINGS

Two ratings are used to classify batteries: One is a 20-hour rating at 80°F and the other is a cold rating at 0°F. This second figure indicates the cranking load capacity and is referred to as the Peak Watt Rating of a battery. This Peak Watt Rating (PWR) has been developed to measure the cold-cranking ability of the battery. The numerical rating is embossed on each battery case at the base and is determined by multiplying the maximum current by the maximum voltage.

The ampere-hour rating of a battery is its capacity to furnish a given amount of amperes over a period of time at a cell voltage of 1.5. Therefore, a battery with a capacity of maintaining 3 amperes for 20 hours at 1.5 volts would be classified as a 60-ampere hour battery.

Do not confuse the ampere-hour rating with the PWR, because they are two unrelated figures used for different purposes.

A replacement battery should have a power rating equal or as close to the old unit as possible.

BATTERY LOCATION

Every battery installed in a boat must be secured in a well-protected ventilated area. If the battery area is not well ventilated, hydrogen gas which is given off during charging could become very explosive if the gas is concentrated and confined. Because of its size, weight, and acid content, the battery must be well-secured. If the battery should break loose during rough boat maneuvers, considerable damage could be done, including damage to the hull.

BATTERY SERVICE

The battery requires periodic servicing and a definite maintenance program to ensure extended life. If the battery should test satisfactorily, but still fails to perform properly, one of five problems could be the cause.

1- An accessory might have accidently been left on overnight or for a long period during the day. Such an oversight would result in a discharged battery.

2- Slow speed engine operation for long periods of time resulting in an undercharged condition.

3- Using more electrical power than the generator can replace will result in an undercharged condition.

4- A defect in the charging system. A faulty generator, defective regulator, defective alternator or diodes on units with an alternator, or high resistance somewhere in the system could cause the battery to become undercharged.

5- Failure to maintain the battery in good order. This might include a low level of electrolyte in the cells; loose or dirty cable connections at the battery terminals; or possibly an excessive dirty battery top.

Electrolyte Level

The most common practice of checking the electrolyte level in a battery is to remove the cell cap and visually observe the level in the vent well. The bottom of each vent well has a split vent which will cause the surface of the electrolyte to appear distorted when it makes contact. When the distortion first appears at the bottom of the split vent, the electrolyte level is correct.

Some late-model batteries have an electrolyte-level indicator installed which operates in the following manner:

A transparent rod extends through the center of one of the cell caps. The lower tip of the rod is immersed in the electrolyte when the level is correct. If the level should drop below normal, the lower tip of the rod is exposed and the upper end glows as a warning to add water. Such a device is only necessary on one cell cap because if the electrolyte is low in one cell it is also low in the other cells. **BE SURE** to replace the cap with the indicator onto the second cell from the positive terminal.

During hot weather and periods of heavy use, the electrolyte level should be checked more often than during normal operation. Add colorless, odorless, drinking water to bring the level of electrolyte in each cell to the proper level. **TAKE CARE** not to overfill, because adding an excessive amount of water will cause loss of electrolyte and any loss will result in poor performance, short battery life, and will contribute quickly to corrosion. **NEVER** add electrolyte from another battery. Use only clean pure water.

Cleaning

Dirt and corrosion should be cleaned from the battery just as soon as it is discovered. Any accumulation of acid film or dirt will permit current to flow between the terminals. Such a current flow will drain the battery over a period of time.

Clean the exterior of the battery with a solution of diluted ammonia or a soda solution to neutralize any acid which may be present. Flush the cleaning solution off with clean water. **TAKE CARE** to prevent any of the neutralizing solution from entering the cells, by keeping the caps tight.

A poor contact at the terminals will add resistance to the charging circuit. This resistance will cause the voltage regulator to register a fully charged battery, and thus

One of the most effective means of cleaning the battery terminals is to use a wire brush designed for this specific purpose.

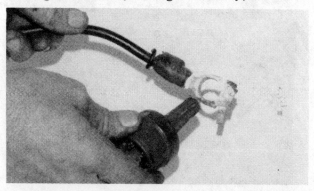

An inexpensive brush can be purchased and used to clean battery lead connectors to ensure a proper connection.

cut down on the alternator output adding to the low battery charge problem.

Scrape the battery posts clean with a suitable tool or with a stiff wire brush. Clean the inside of the cable clamps to be sure they do not cause any resistance in the circuit.

Battery Testing

A hydrometer is a device to measure the percentage of sulfuric acid in the battery electrolyte in terms of specific gravity. When the condition of the battery drops from fully charged to discharged, the acid leaves the solution and enters the plates, causing the specific gravity of the electrolyte to drop:

The following six points should be observed when using a hydrometer.

1- NEVER attempt to take a reading immediately after adding water to the battery. Allow at least 1/4 hour of charging at a high rate to thoroughly mix the electrolyte with the new water and to cause vigorous gassing.

2- ALWAYS be sure the hydrometer is clean inside and out as a precaution against contaminating the electrolyte.

3- If a thermometer is an integral part of the hydrometer, draw liquid into it several times to ensure the correct temperature before taking a reading.

4- BE SURE to hold the hydrometer vertically and suck up liquid only until the float is free and floating.

5- ALWAYS hold the hydrometer at eye level and take the reading at the surface of the liquid with the float free and floating.

Disregard the light curvature appearing where the liquid rises against the float stem. This phenomenon is due to surface tension.

6- DO NOT drop any of the battery fluid on the boat or on your clothing, because it is extremely caustic. Use water and baking soda to neutralize any battery liquid that does accidently drop.

After withdrawing electrolyte from the battery cell until the float is barely free, note the level of the liquid inside the hydrometer. If the level is within the green band range, the condition of the battery is satisfactory. If the level is within the white band, the battery is in fair condition, and if the level is in the red band, it needs charging badly or is dead and should be replaced. If the level fails to rise above the red band after charging, the only answer is to replace the battery.

A check of the electrolyte in the battery should be on the maintenance schedule for any boat. A hydrometer reading of 1.300 or in the green band, indicates the battery is in satisfactory condition. If the reading is 1.150 or in the red band, the battery needs to be charged. Observe the six safety points given in the text when using a hydrometer.

A pair of pliers should be used to tighten the wingnuts, when they are used. Securing the wingnuts by hand is not adequate, the connections will vibrate loose.

JUMPER CABLES

If booster batteries are used for starting an engine the jumper cables must be connected correctly and in the proper sequence to prevent damage to either battery, or to the alternator diodes.

ALWAYS connect a cable from the positive terminal of the dead battery to the positive terminal of the good battery **FIRST.** **NEXT,** connect one end of the other cable to the negative terminal of the good battery and the other end to the **ENGINE** for a good ground. By making the ground connection on the engine, if there is an arc when you make the connection it will not be near the battery. An arc near the battery could cause an explosion, destroying the battery and causing serious personal **INJURY.**

DISCONNECT the battery ground cable before replacing an alternator or before connecting any type of meter to the alternator.

If it is necessary to use a fast-charger on a dead battery, **ALWAYS** disconnect one of the boat cables from the battery **FIRST,** to prevent burning out the diodes in the rectifier. **NEVER** use a fast-charger as a booster to start the engine because the voltage regulator may be **DAMAGED.**

STORAGE

If the boat is to be laid up for the winter or for more than a few weeks, special attention must be given to the battery to prevent complete discharge or possible dam-age to the terminals and wiring. Before putting the boat in storage, disconnect and remove the batteries. Clean them thoroughly of any dirt or corrosion, and then charge them to full specific gravity reading. After they are fully charged, store them in a clean cool dry place where they will not be damaged or knocked over.

NEVER store the battery with anything on top of it or cover the battery in such a manner as to prevent air from circulating around the filler caps. All batteries, new and old, will discharge during periods of storage, more so if they are hot than if they remain cool. Therefore, the electrolyte level and the specific gravity should be checked at regular intervals. A drop in the specific gravity reading is cause to charge them back to a full reading.

In cold climates, care should be exercised in selecting the battery storage area. A fully-charged battery will freeze at about 60° below zero. A discharged battery, almost dead, will have ice forming at about 19 degrees above zero.

DUAL BATTERY INSTALLATION

Three methods are available for utilizing a dual battery hookup.

Someone smoking close to this battery during high charge may have ignited the explosive fumes, blowing a hole in the top surface.

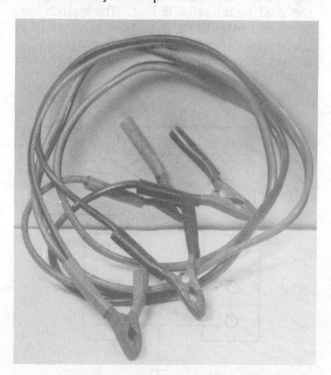

A common set of heavy-duty jumper cables. Observe the safety precautions given in the text when using jumper cables.

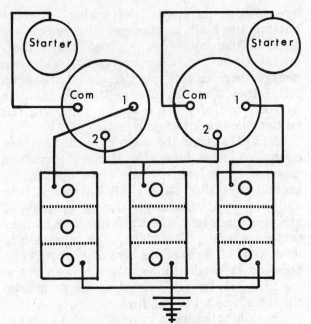

Schematic diagram for a three battery, two engine hookup.

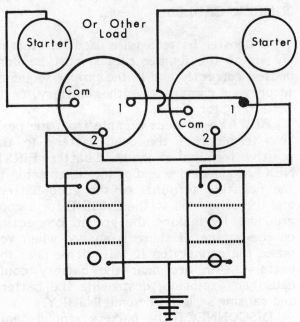

Schematic diagram for a two battery, two engine hookup.

1- A high-capacity switch can be used to connect the two batteries. The accompanying illustration details the connections for installation of such a switch. This type of switch installation has the advantage of being simple, inexpensive, and easy to mount and hookup. However, if the switch is accidently left in the closed position, it will cause the convenience loads to run down both batteries and the advantage of the dual installation is lost. The switch may be closed intentionally to take advantage of the extra capacity of the two batteries, or it may be temporarily closed to help start

the engine under adverse conditions.

2- A relay can be connected into the ignition circuit to enable both batteries to be automatically put in parallel for charging or to isolate them for ignition use during engine cranking and start. By connecting the relay coil to the ignition terminal of the ignition-starting switch, the relay will close during the start to aid the starting battery. If the second battery is allowed to run down, this arrangement can be a disadvantage since it will draw a load from the starting battery while cranking the engine. One way to avoid such a condition is to connect the

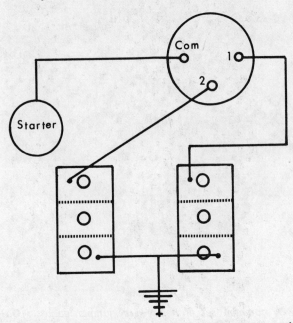

Schematic diagram for a two battery, one engine hookup.

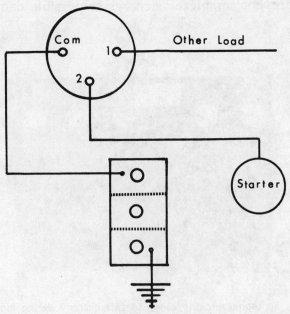

Schematic diagram for a single battery, one engine hookup.

relay coil to the ignition switch accessory terminal. When connected in this manner, while the engine is being cranked, the relay is open. But when the engine is running with the ignition switch in the normal position, the relay is closed, and the second battery is being charged at the same time as the starting battery.

3- A heavy duty switch installed as close to the batteries as possible can be connected between them. If such an arrangement is used, it must meet the standards of the American Boat and Yacht Council, Inc. or the Fire Protection Standard for Motor Craft, N.F.P.A. No. 302.

6-3 GAUGES AND HORNS

Gauges or lights are installed to warn the operator of a condition in the cooling and lubrication systems that may need attention. The fuel gauge gives an indication of the amount of fuel in the tank. If the engine overheats or the oil pressure drops too low for safety, a gauge or warning light reminds the operator to shut down the engine and check the cause of the warning before serious damage is done.

CONSTANT-VOLTAGE SYSTEM

In order for gauges to register properly, they must be supplied with a steady voltage. The voltage variations produced by the engine charging system would cause erratic gauge operation, too high when the alternator voltage is high and too low when the alternator is not charging. To remedy this problem, a constant-voltage system is used to reduce the 12-14 volts of the electrical system to an average of 5 volts. This steady 5 volts ensures the gauges will read accurately under varying conditions from the electrical system.

6-4 SERVICE PROCEDURES

Systems utilizing warning lights do not require a constant-voltage system, therefore, this service is not needed.

Service procedures for checking the gauges and their sending units is detailed in the following sections.

TEMPERATURE GAUGES

The body of temperature gauges must be grounded and they must be supplied with 12 volts. Many gauges have a terminal on the mounting bracket for attaching a ground wire. A tang from the mounting bracket makes contact with the gauge. **CHECK** to be sure the tang does make good contact with the gauge.

Ground the wire to the sending unit and the needle of the gauge should move to the full right position indicating the gauge is in serviceable condition.

See Chapter 3, to test the sender unit.

WARNING LIGHTS

If a problem arises on a boat equipped with water and temperature lights, the first area to check is the light assembly for loose wires or burned-out bulbs.

When the ignition key is turned on, the light assembly is supplied with 12 volts and grounded through the sending unit mounted on the engine. When the sending unit makes contact, because the water temperature is too hot, the circuit to ground is completed and the lamp should light.

Check The Bulb: Turn the ignition switch on. Disconnect the wire at the engine sending unit, and then ground the wire. The lamp on the dash should light. If it does not light, check for a burned-out bulb or a break in the wiring to the light.

The gauges and controls on the dashboard should be kept clean and protected from water spray, especially when operating in a salt water atmosphere.

THERMOMELT STICKS

Thermomelt sticks are an easy method of determining if the engine is running at the proper temperature. Thermomelt sticks are not expensive and are available at your local marine dealer.

Start the engine with the propeller in the water and run it for about 5 minutes at roughly 3000 rpm.

CAUTION: Water must circulate through the lower unit to the engine any time the engine is run to prevent damage to the water pump in the lower unit. Just five seconds without water will damage the water pump.

The 140 degree stick should melt when you touch it to the lower thermostat housing or on the top cylinder. If it does not melt, the thermostat is stuck in the open position and the engine temperature is too low.

Touch the 170 degree stick to the same spot on the lower thermostat housing or on the top cylinder. The stick should not melt. If it does, the thermostat is stuck in the closed position or the water pump is not operating properly because the engine is running too hot. For service procedures on the cooling system, see Chapter 8.

A thermomelt stick is a quick, simple, inexpensive, and fairly accurate method to determine engine running temperature.

6-5 FUEL SYSTEM

FUEL GAUGE

The fuel gauge is intended to indicate the quantity of fuel in the tank. As the experienced boatman has learned, the gauge reading is seldom an accurate report of the fuel available in the tank. The main reason for this false reading is because the boat is rarely on an even keel. A considerable difference in fuel quantity will be indicated by the gauge if the bow or stern is heavy, or if the boat has a list to port or starboard.

Therefore, the reading is usually low. The amount of fuel drawn from the tank is dependent on the location of the fuel pickup tube in the tank. The engine may cutout while cruising because the pickup tube is above the fuel level. Instead of assuming the tank is empty, shift weight in the boat to change the trim and the problem may be solved until you are able to take on more fuel.

FUEL GAUGE HOOKUP

The Boating Industry Association recommends the following color coding be used on all fuel gauge installations:

Black -- for all grounded current-carrying conductors.

Pink -- insulated wire for the fuel gauge sending unit to the gauge.

Red -- insulated wire for a connection from the positive side of the battery to any electrical equipment.

Connect one end of a pink insulated wire to the terminal on the gauge marked **TANK** and the other end to the terminal on top of the tank unit.

Connect one end of a black wire to the terminal on the fuel gauge marked **IGN** and the other end to the ignition switch.

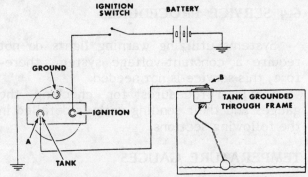

Schematic diagram for a safe fuel tank gauge hookup.

Connect one end of a second black wire to the fuel gauge terminal marked **GRD** and the other end to a good ground. It is important for the fuel gauge case to have a good common ground with the tank unit. Aboard an all-metal boat, this ground wire is not necessary. However, if the dashboard is insulated, or made of wood or plastic, a wire **MUST** be run from the gauge ground terminal to one of the bolts securing the sending unit in the fuel tank, and then from there to the **NEGATIVE** side of the battery.

FUEL GAUGE TROUBLESHOOTING

In order for the fuel gauge to operate properly the sending unit and the receiving unit must be of the same type and preferably of the same make.

The following symptoms and possible corrective actions will be helpful in restoring a faulty fuel gauge circuit to proper operation.

If you suspect the gauge is not operating properly, the first area to check is all electrical connections from one end to the other. Be sure they are clean and tight.

Next, check the common ground wire between the negative side of the battery, the fuel tank, and the gauge on the dash.

If all wires and connections in the circuit are in good condition, remove the sending unit from the tank. Run a wire from the gauge mounting flange on the tank to the flange of the sending unit. Now, move the float up-and-down to determine if the receiving unit operates. If the sending unit does not appear to operate, move the float to the midway point of its travel and see if the receiving unit indicates half full.

If the pointer does not move from the **EMPTY** position one of four faults could be to blame:

1- The dash receiving unit is not properly grounded.

2- No voltage at the dash receiving unit.

3- Negative meter connections are on a positive grounded system.

4- Positive meter connections are on a negative grounded system.

If the pointer fails to move from the **FULL** position, the problem could be one of three faults.

1- The tank sending unit is not properly grounded.

2- Improper connection between the tank sending unit and the receiving unit on

the dash.

3- The wire from the gauge to the ignition switch is connected at the wrong terminal.

If the pointer remains at the 3/4 full mark, it indicates a 6-volt gauge is installed in a 12-volt system.

If the pointer remains at about 3/8 full, it indicates a 12-volt gauge is installed in a 6-volt system.

Preliminary Inspection

Inspect all of the wiring in the circuit for possible damage to the insulation or conductor. Carefully check:

1- Ground connections at the receiving unit on the dash.

2- Harness connector to the dash unit.

3- Body harness connector to the chassis harness.

4- Ground connection from the fuel tank to the tank floor pan.

5- Feed wire connection at the tank sending unit.

GAUGE ALWAYS READS FULL when the ignition switch is **ON:**

1- Check the electrical connections at the receiving unit on the dash; the body harness connector to chassis harness connector; and the tank unit connector in the tank.

2- Make a continuity check of the ground wire from the tank to the tank floor pan.

3- Connect a known good tank unit to the tank feed wire and the ground lead. Raise and lower the float and observe the receiving unit on the dash. If the dash unit follows the arm movement, replace the tank sending unit.

GAUGE ALWAYS READS EMPTY when the ignition switch is **ON:**

Disconnect the tank unit feed wire and do not allow the wire terminal to ground. The gauge on the dash should read **FULL.**

If Gauge Reads Empty:

1- Connect a spare dash unit into the dash unit harness connector and ground the unit. If the spare unit reads **FULL,** the original unit is shorted and must be replaced.

2- A reading of **EMPTY** indicate a short in the harness between the tank sending unit and the gauge on the dash.

If Gauge Reads Full:

1- Connect a known good tank sending unit to the tank feed wire and the ground lead.

2- Raise and lower the float while observing the dash gauge. If the dash gauge follows movement of the float, replace the tank sending unit.

GAUGE NEVER INDICATES FULL

This test requires shop test equipment.

1- Disconnect the feed wire to the tank unit and connect the wire to a good ground through a variable resistor or through a spare tank unit.

2- Observe the dash gauge reading. The reading should be **FULL** when resistance is increased to about 90 ohms. This resistance would simulate a full tank.

3- If the check indicates the dash gauge is operating properly, the trouble is either in the tank sending unit rheostat being shorted, or the float is binding. The arm could be bent, or the tank may be deformed. Inspect and correct the problem.

6-6 TACHOMETER

An accurate tachometer can be installed on any engine. Such an instrument provides an indication of engine speed in revolutions per minute (rpm). This is accomplished by measuring the number of electrical pulses per minute generated in the primary circuit of the ignition system.

The meter readings range from 0 to 6,000 rpm, in increments of 100. Tachometers have solid-state electronic circuits which eliminate the need for relays or batteries and contribute to their accuracy. The

electronic parts of the tachometer, susceptible to moisture, are coated to prolong their life.

6-7 HORNS

The only reason for servicing a horn is because it fails to operate properly or because it is out of tune. In most cases the problem can be traced to an open circuit in the wiring or to a defective relay.

Cleaning

Crocus cloth and carbon tetrachloride should be used to clean the contact points. **NEVER** force the contacts apart or you will bend the contact spring and change the operating tension.

Check Relay and Wiring

Connect a wire from the battery to the horn terminal. If the horn operates, the problem is in the relay or in the horn wiring. If both of these appear satisfactory, the horn is defective and needs to be replaced.

Before replacing the horn however, connect a second jumper wire from the horn frame to ground to check the ground connection.

Test the winding for an open circuit, faulty insulation, or poor ground. Check the

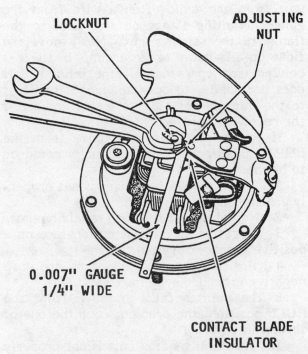

LOCKNUT   ADJUSTING NUT

0.007" GAUGE 1/4" WIDE

CONTACT BLADE INSULATOR

The tone of a horn can be adjusted with a 0.007" feeler gauge, as described in the text. **TAKE CARE** *to prevent the feeler gauge from making contact with the case, or the circuit will be shorted out.*

Maximum engine performance can only be obtained through proper tuning using a tachometer.

resistor with an ohmmeter, or test the condenser for capacity, ground, and leakage. Inspect the diaphragm for cracks.

Adjust Horn Tone

Loosen the locknut, and then rotate the adjusting screw until the desired tone is reached. On a dual horn installation, disconnect one horn and adjust each, one-at-a-time. The contact point adjustment is made by inserting a 0.007" feeler gauge blade between the adjusting nut and the contact blade insulator. **TAKE CARE** not to allow the feeler gauge to touch the metallic parts of the contact points because it would short them out. Now, loosen the locknut and turn the adjusting nut down until the horn fails to sound. Loosen the adjusting nut slowly until the horn barely sounds. The locknut **MUST** be tightened after each test. When the feeler gauge is withdrawn the horn will operate properly and the current draw will be satisfactory.

6-8 ELECTRICAL SYSTEM GENERAL INFORMATION

Probably 75-80% of all Johnson/Evinrude engines covered in this manual are started by pulling on a rope. As the manufacturer increased the size and horsepower of the engines, it was necessary to replace the rope starter with some form of power cranking system. Today, most small engines are still started by pulling on a rope.

On the larger hp engines, an electric starter motor coupled with a mechanical gear mesh between the cranking motor and

Starter motor mounted on the port side of the engine. The attaching hardware for this unit consists of two bolts on the side and one in the front.

the engine flywheel, similar to the method used to crank an automobile engine, was added. This system provided an alternate method to the hand starter rope arrangement. If the electric cranking system is inoperative for any reason, including a dead or weak battery, the engine may still be cranked and started by hand.

Since the starting motor requires a large amount of electrical current, it is necessary to have a fully charged battery available for the starting system. If the boat is equipped with several electrical accessories, such as bait tank with circulating pump, radio, a number of running and accessory lights etc., the charging system must be performing properly to keep the battery charged.

Generator Charging Circuit

The charging circuit consists of a generator driven by a belt connected to the flywheel. The flywheel is equipped with a pulley arrangement to transfer flywheel rotation to the generator pulley through the belt.

Alternator Charging Circuit

The alternator charging system consists of the alternator, rectifiers, and the flywheel. This is a direct charge system. A 5 amp to 7 amp current is constantly produced while the engine is operating. A voltage regulator is used with higher capacity alternator to prevent overcharging the battery.

Small horsepower engines use a hand starter instead of an electric starter motor. The hand starter may be mounted on top of the engine or on the side, depending on the model.

Choke Circuit

The choke is activated by a solenoid. This solenoid attracts a plunger to close the choke valves. The solenoid is energized when the ignition key is turned to the **START** position and the choke button is depressed. When using the electric choke, the manual choke **MUST** be in the **NEUTRAL** position.

Only the electric choke is used on the engines covered in this manual.

On some newer model engines, the choke will be activiated if the key is pushed inward while it is being rotated to the **START** position.

Starting Circuit

The starting circuit consists of a cranking motor and a starter-engaging mechanism. A solenoid is used as a heavy-duty switch to carry the heavy current from the battery to the starter motor. The solenoid is actuated by turning the ignition key to the **START** position. On some models, a pushbutton is used to actuate the solenoid. See Section 6-12 for detailed service procedures on the starter motor circuit.

6-9 GENERATOR CHARGING CIRCUIT SERVICE

The generator has two terminals on the lower end. One terminal is larger than the other and the wires connected have different size connectors to ensure the proper wire is connected to the correct terminal.

If several electrical accessories are used and the engine is operating at idle speed, or below 1500 rpm for extended periods of time, the battery will not receive adequate current to remain in serviceable condition.

The rated capacity of the generator is 10-amps. Therefore, the electrical accessory load should not exceed 10 amps or current will be drawn from the battery at a greater rate than the generator is able to produce. Such a negative draw on the battery will result in a run-down condition and failure of the battery to provide the required current to the starter for cranking the engine.

To calculate the amperage draw of an accessory the following simple formula may be used: Amps equals watts divided by volts. $Amps = \dfrac{Watts}{Volts}$

The volts will always be 12. Accessories will usually be given in watts. If the obsolete measurement of candlepower is used, then one candle power is equal to approximately one watt. Example: A boat has running lights requiring 8 watts; auxiliary lights use 10 watts; and a radio rated at 30 watts.

$Amps = 48 \ Watts/12 \ volts = 4 \ Amps.$

In this case, if all the lights are on and the radio is being used, the total draw on the battery would be 4 amps. If the engine is running at 1500 rpm or higher, and the generator circuit is performing properly by charging the battery with 10 amps, then a net postive gain of 6 amps is being received by the battery.

The battery can be externally charged or the engine can be equipped with a generator to charge the battery while the engine is operating.

A voltage regulator, mounted in a junction box on the rear of the engine, is connected between the generator and the battery to prevent overcharging of the battery while the engine is operating. The junction box also houses a fuse to protect the charging circuit.

The generator circuit requires at least 1500 rpm engine speed to effectively charge the battery. At this speed, the ampere

Front view of a Johnson outboard with the generator mounted on the port side.

meter on the dash will indicate a positive charge to the battery.

If the boat has a twin engine installation, the usual practice is to use only one battery for cranking both units. With such a twin installation, only one engine generator should be used to charge and maintain the battery at its full amperage rating.

Most mechanics have discovered if both generators of a twin installation are connected to charge the battery, one seems to "fight" the other. Instead of having an improved system, this type of hookup causes many serious electrical problems that are unexplainable. See Section 6-9 for detailed service procedures on the generator circuit.

TROUBLESHOOTING

One of three areas may cause problems in the generating circuit and failure of the system to provide sufficient current to maintain the battery at a satisfactory charge. Remember, the generator will only produce approximately 10 amps of current. Most amp-meters have a 20 amp scale. Therefore, it is only necessary for the scale to register in the 10 amp area, while the engine is operating above 1500 rpm, to indicate satisfactory performance.

a- The 4-amp or 20-amp fuse in the junction box may have burned, opening the circuit. If the fuse requires replacement, a check should be made immediately, to determine why the fuse burned protecting the circuit.

b- The voltage regulator may be defective. If the regulator has failed, a thorough check of the circuit must be made to determine the cause. Simply replacing the regulator usually will not solve the problem and the new regulator may be damaged when the engine and generator are operating.

c- The generator may be defective and fail to produce the current necessary to maintain the battery. The problem may simply be worn brushes. Replacement with new brushes may solve the problem. However, if the brushes are in good condition, and testing reveals the generator must be replaced, the conservative mechanic will install a new voltage regulator at the same time.

CRITICAL WORDS

The engine must be operated, in gear, at speeds in excess of 1500 rpm to test the generator circuit. Therefore, the engine **MUST** be mounted in a body of water to prevent a **RUNAWAY** condition and serious damage to internal parts, or destruction of the unit. **NEVER** attempt to operate the engine above idle speed with a flush attachment connected to the lower unit or with the engine mounted in a small test tank, such as a fifty-gallon drum.

CAUTION: Water must circulate through the lower unit to the engine any time the engine is run to prevent damage to the water pump in the lower unit. Just five seconds without water will damage the water pump.

1- Check to be sure all electrical connections in the circuit are secure and free of corrosion. Double check the battery connections and terminals. If the terminals are badly corroded, there is no way on this green earth for the current produced by the generator to reach the battery cells. A special wire brush can be purchased at very modest cost to clean the inside of the wire connectors. A common wire brush may be used to clean the battery terminals. Baking soda and water is a good cleaning agent for the battery suface.

2- Check the wiring in the circuit for broken insulation, or an actual break in the line. Disconnect the **POSITIVE** electrical lead from the battery as a precaution against an accidental short causing damage to the voltage regulator. Remove the junction box cover. Check the condition of the 4-amp or 20-amp fuse with a continuity light, or install a new fuse and check the

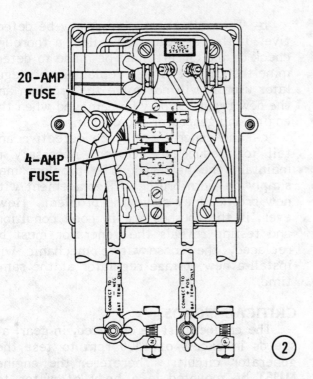

20-AMP FUSE

4-AMP FUSE

② CONNECT TO + POS BAT TERM ONLY / CONNECT TO - NEG BAT TERM ONLY

charging circuit again with the engine operating. The 4-amp and 20-amp fuses can easily be "popped" out their retainers in the panel of the junction box base and replaced after the cover has been removed.

3- Ground the field of the generator very **QUICKLY** and only **MOMENTARILY** with a jumper wire while the engine is operating at approximately 2000 rpm. By **MOMENTARILY** grounding the field, the voltage regulator is actually bypassed and the generator will "run wild". If the amp meter registers a high reading while the generator field is grounded, the circuit has a broken wire, or the voltage regulator is defective. If the amp meter reading does not change while the generator field is grounded, then the indication is a faulty generator.

Voltage Regulator

1- Disconnect the positive lead from the battery terminal to prevent an accidental short causing damage in the circuit. Loosen the two wingnuts on both sides of the junction box cover. These wingnuts are "captive" with the cover and cannot be completely removed, only released from the junction box. (This arrangement prevents loss of the wingnuts). If the regulator is to be replaced, disconnect the wires between the generator and the terminal board in the junction box at the board.

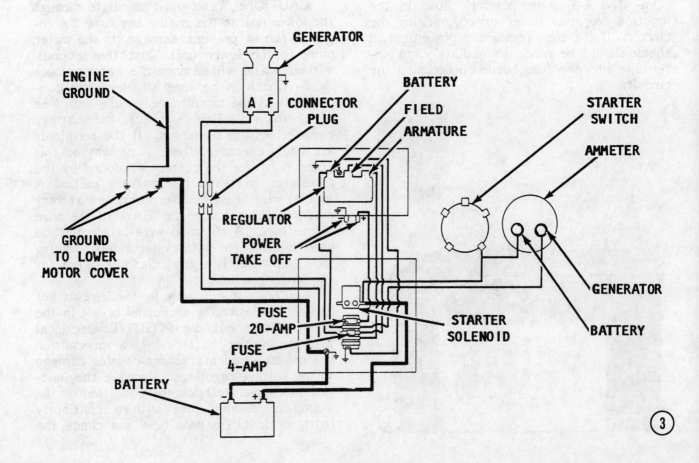

GENERATOR

ENGINE GROUND

A F CONNECTOR PLUG

BATTERY
FIELD
ARMATURE

STARTER SWITCH

AMMETER

GROUND TO LOWER MOTOR COVER

REGULATOR
POWER TAKE OFF

GENERATOR

BATTERY

FUSE 20-AMP

FUSE 4-AMP

STARTER SOLENOID

BATTERY

③

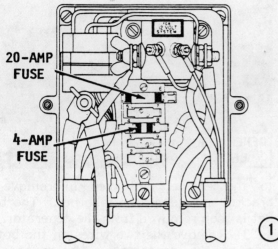

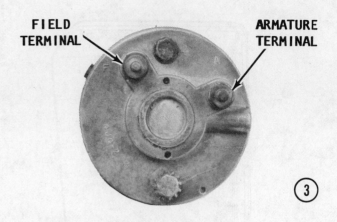

2– Place the junction box cover on its back and remove the five leads to the voltage regulator. Notice how the leads are color-coded, as an assist in connecting them correctly during installation. Remove the attaching hardware securing the voltage regulator to the cover. Remove the regulator. Place the new regulator in position in the junction box cover and secure it with the attaching hardware. Connect the four color-coded wires to the regulator: Yellow, from the generator armature; blue, from the generator field; brown, from the battery; and the black is the ground wire. A second brown wire is connected to the bottom of the junction box to allow additional electrical accessories to be connected. This second wire would not have been disconnected to remove the regulator.

CRITICAL WORDS

When a new voltage regulator is installed, the generator must be "polarized" **BEFORE** the cover is installed.

3– "Polarize" the new regulator by first connecting the positive lead to the battery, and then using a small jumper wire to make a **MOMENTARY** connection between the battery terminal and the armature terminal of the regulator. The generator is now properly "polarized" with the new regulator for service.

4– Now, disconnect the positive battery lead again, before installing the junction box cover. The few moments involved in disconnecting and connecting the positive lead at the battery is well spent. This small task will prevent any possible short from causing damage to the circuit when working with the wires.

5– Install the junction box cover to the box base. As the cover is moved into place, work the wires alongside the regulator. Secure the cover in place with the two "captive" wingnuts.

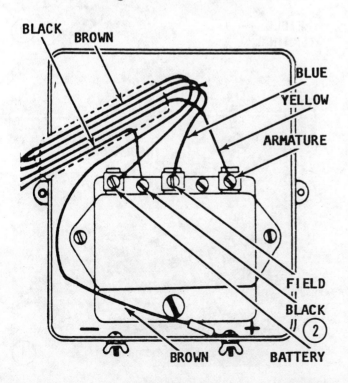

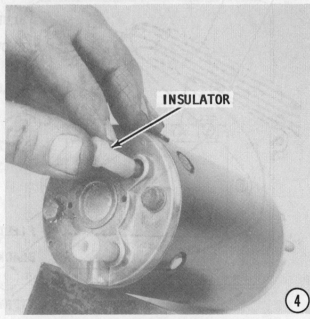

ACCESSORY CONNECTOR ⑤

GENERATOR REMOVAL AND DISASSEMBLING

1- Disconnect the positive lead at the battery terminal. Remove the hood from the engine. If the engine has a hand starter, remove the retaining bolts and lift off the hand starter. Prevent the shaft from turning by engaging an open-end wrench with the flats on the generator shaft underneath the pulley. Now, while continuing to hold the wrench on the shaft, remove the nut from the top of the generator pulley. Remove the cover from the inside of the generator pulley.

2- Remove the generator belt from the pulley. Use a screwdriver or other similar tool and pry the puller up and free of the generator shaft. Hold the generator, and at the same time remove the nuts from the top of the generator support bracket, and the generator is free.

SPECIAL WORDS

In some cases, it is just as easy to remove the 7/16" bolts securing the bracket

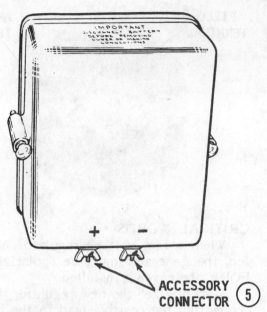

PULLEY ①

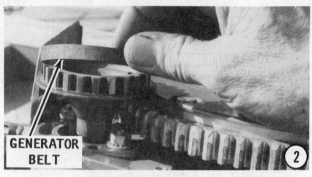

GENERATOR BELT ②

to the engine, and then to remove the bracket and generator together. The bracket is then removed from the generator.

3- Remove the two wires on the bottom side of the generator. Notice how one generator stud is smaller than the other and the electrical connectors are different sizes to match the studs.

4- Remove the two nuts from the wire terminals at the bottom of the generator. Work out the two white insulators from around the studs. Remove the two thrubolts.

ARMATURE TERMINAL

FIELD TERMINAL

③

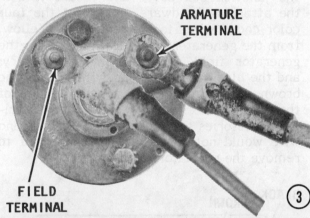

INSULATOR

④

5- Remove the end cap from the generator. After the cap has been removed, take notice of the small dowel in the end of the generator. This dowel ensures the cap will be installed correctly when the dowel is indexed in a matching hole in the cap. Pull the armature and upper cap out of the frame.

ARMATURE TESTING

Testing for a Short

6- Position the armature on a growler, then hold a hacksaw blade over the armature core. Turn the growler switch to the **ON** position. Slowly rotate the armature. If the hacksaw blade vibrates, the armature or commutator has a short. Clean the grooves

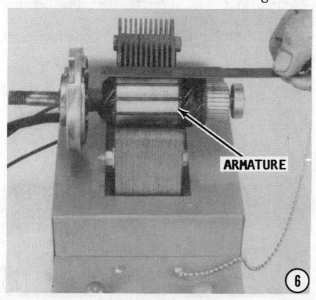

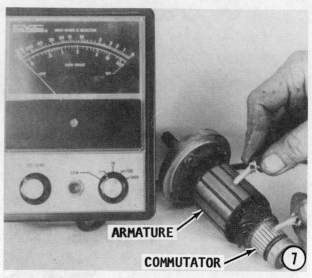

between the commutator bars on the armature. Perform the test again. If the hacksaw blade still vibrates during the test, the armature has a short and **MUST** be replaced.

Testing for a Ground

7- Obtain a test lamp or continuity meter. Make contact with one probe lead on the armature core and the other probe lead on the commutator bar. If the lamp lights, or the meter indicates continuity, the armature is grounded and **MUST** be replaced.

Checking the Commutator Bar

8- Check between or check bar-to-bar as shown in the accompanying illustration. The test light should light, or the meter should indicate continuity. If the commutator fails the test, the armature **MUST** be replaced.

Field Coil Test for Ground

9- Check to be sure the free end of the field wire is not grounded to the frame and

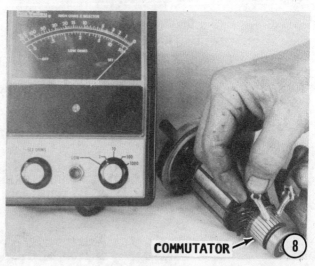

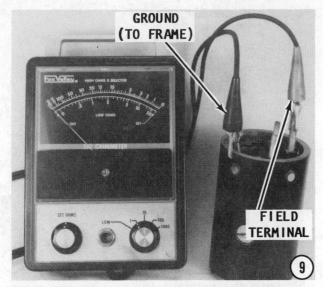

the field insulation is not broken. Using a test lamp or ohmmeter, make contact with one probe lead to the ground of the generator frame. Make contact with the other lead to the field terminal. If the lamp lights or the ohmmeter indicates continuity, the field coils are grounded. If the location of the ground in the field coils cannot be determined, or repaired, the coils **MUST** be replaced.

Armature Terminal Test for Ground

10- Check to be sure the loose end of the armature terminal lead of the generator is **NOT** grounded to the frame. Using a test lamp or ohmmeter, make contact with one probe lead to the armature terminal of the generator. Make contact with the other probe lead to a good ground on the generator frame. If the test lamp lights or the ohmmeter indicates continuity, the positive terminal insulation through the generator frame is broken down and **MUST** be replaced.

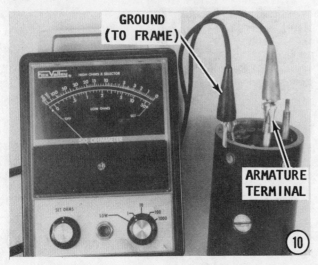

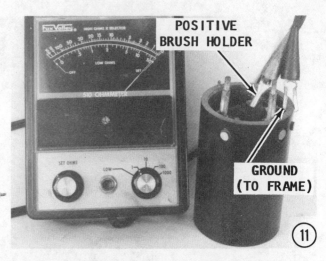

Positive Brush Test for Ground

11- Using a test lamp or ohmmeter, make contact with one probe lead to the positive or insulated brush holder. Make contact with the other probe lead to a good ground on the generator frame. If the lamp lights, or the ohmmeter indicates continuity, the brush holder is grounded due to defective insulation at the frame.

Field Test

12- Using a test lamp or ohmmeter, make contact with one probe lead to the armature stud. Make contact with the other probe lead to the armature brush. The lamp should light or the ohmmeter should indicate continuity. If this test is not successful, check for a poor connection between the stud and the brush.

CLEANING AND INSPECTING

Check the ball bearing at the end of the commutator bar. Verify that the bearing turns free with no sign of "rough spots" or binding. Hold the armature in one hand and

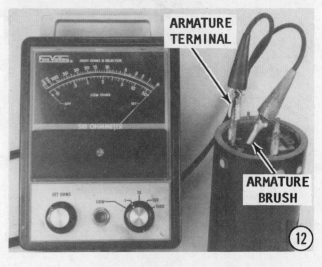

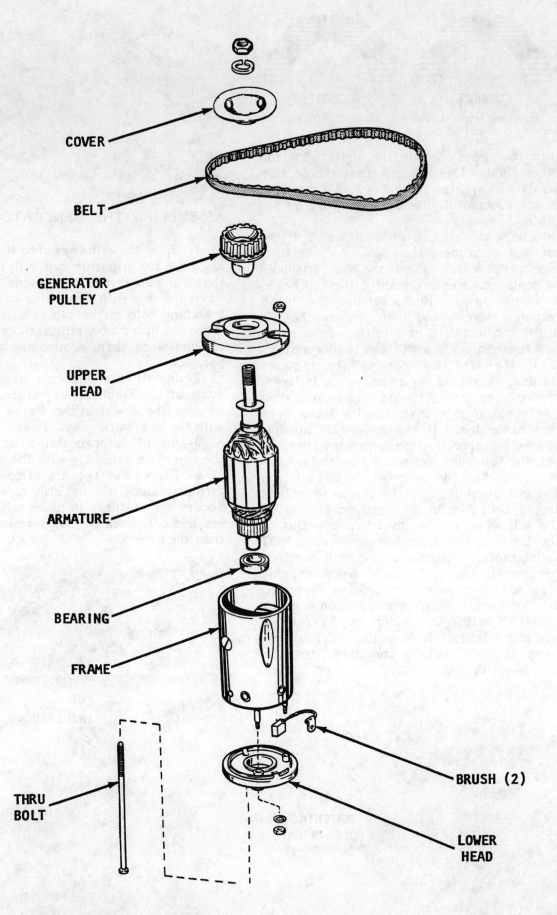

COVER

BELT

GENERATOR
PULLEY

UPPER
HEAD

ARMATURE

BEARING

FRAME

BRUSH (2)

THRU
BOLT

LOWER
HEAD

Exploded drawing showing arrangement of principle parts to a typical generator.

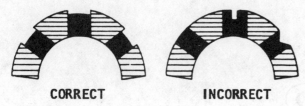

CORRECT   INCORRECT

Armature segments properly cleaned (left) and improperly cleaned (right).

turn the upper cap on the shaft with the other hand. The cap and shaft should turn freely with no sign of binding. If either test is not successful, the bearing **MUST** be replaced.

Check the amount of brush wear. If the brush is worn more than 50% of its original size, or to within 1/4" of the base, it should be replaced. Replacement of the brushes is a simple task. First, remove the brush retaining screw, and then remove the old brush and install a new brush. Secure the new brush in place with the retaining screw.

If the armature commutator requires turning, it should be turned in a lathe to ensure accuracy. The local generator shop can perform this task, usually for a very reasonable fee. If the turning is accomplished by other than generator shop personnel, the following words are necessary. After the turning, an undercut should be made. The insulation between the commutator bars should be 1-3/4". This undercut must be the full width of the insulation and flat at the bottom. A triangular groove is **NOT** satisfactory. After the undercut work is completed, the slot should be thoroughly cleaned to remove any foreign material, dirt, or copper dust. Sand the commutator **LIGHTLY** with "00" sandpaper to remove any slight burrs left from the undercutting. After all work has been completed, test the unit again, on the growler.

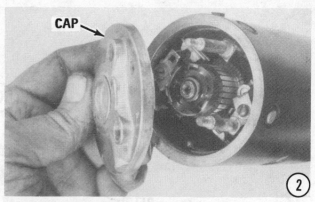

ASSEMBLING THE GENERATOR

1- Slide the armature into the frame and align the top armature cap with the dowel in the frame. Proper alignment is achieved when the dowel in the frame indexes into a matching hole in the cap. As the armature is moved into place, pull back on the brushes, and work them around the commutator bar.

2- Install the end cap down over the studs of the field and armature. Check to be sure the dowel in the frame has indexed with the hole in the cap.

3- Install the two thru-bolts and secure the complete assembly with the nuts.

4- Place the two insulators over the terminal studs of the armature and field. Secure the bushings in place with the washers and proper nuts (one terminal is larger than the other).

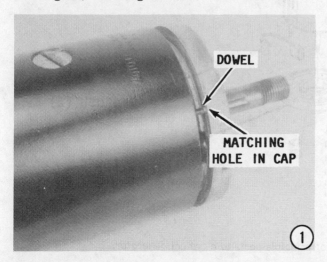

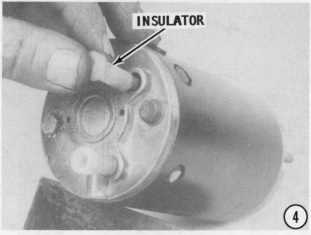

Testing by Rotating the Armature

Performing this test will also "polarize" the new or rebuilt generator. If this test is not performed, the new or rebuilt generator **MUST** still be "polarized" following installation. "Polarization" at that time is accomplished by first connecting the battery to the system in the normal manner, and then connecting a jumper lead to the **BAT** terminal of the voltage regulator. Next, **MOMENTARILY** make contact with the other end of the jumper lead to the **GEN** terminal of the regulator. The generator is now "polarized" for service.

Now, returning to bench testing after rebuilding the generator:

CAUTION: The armature will turn rapidly during this test. Therefore, the generator **MUST** be well **SECURED** before making the test to prevent personal **INJURY** or damage to the generator.

1- Connect a jumper wire between the field terminal and a good ground on the case. Connect a second jumper wire between the positive battery terminal and the ground stud on the armature. **MOMENTARILY** make contact with the positive lead from the battery to the armature lead of the generator. The generator should rotate rapidly. If the generator fails to rotate, the generator must be disassembled again and the service work carefully checked. Sorry

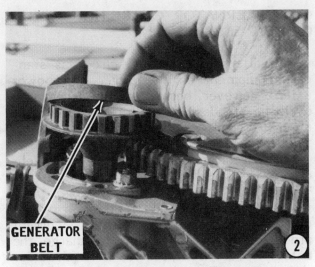

GENERATOR BELT

about that, but some phase of the rebuild task was not performed properly.

2- Install the holding bracket to the generator, or if the bracket remained on the engine, install the generator into the bracket and secure it in place with the attaching hardware, but **DO NOT** tighten the nuts on the generator thru-bolts at this time. Check to be sure the Woodruff key is in place in the generator shaft. Slide the pulley onto the armature shaft with the slot in the pulley indexed with the Woodruff key. Install the drive belt around the pulley on the flywheel and the generator pulley.

3- Install the pulley cap, lockwasher, and nut, to secure the generator pulley in place. Hold the generator shaft from turning with an open wrench on the flats of the shaft underneath the pulley. Adjust tension on the generator pulley by pulling the generator away from the engine, and then tightening the thru-bolt nuts securing the generator in the bracket. The pulley is properly adjusted when it may be depressed approximately 1/4" at a point midway between the two pulleys.

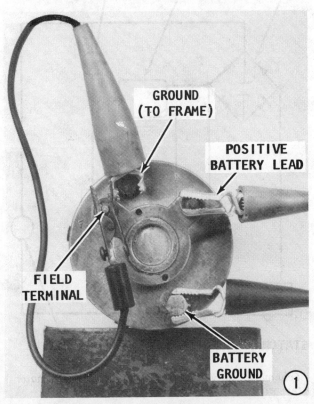

GROUND (TO FRAME)

POSITIVE BATTERY LEAD

FIELD TERMINAL

BATTERY GROUND

GENERATOR PULLEY

6-10 ALTERNATOR CHARGING CIRCUIT SERVICE

An alternator charging circuit is standard equipment on the larger horsepower units. However, because of the advantages over the standard generator, an alternator may be found on all except for the very smallest engines.

The circuit consists of the flywheel, generating coils (commonly referred to as the "stator") and the rectifier, including a positive and negative diode. Function of the diodes is to change the alternating current produced to direct current to charge the battery. Very little service is required for this system. Troubleshooting is confined to work with an ohmmeter. Seldom, if ever, do the magnets in the flywheel lose their magnetism. Problems with the circuit are usually traced to the circuitry, such as a burned coil wire, or defective diodes.

SPECIAL WORDS

A diode in an electrical circuit could be compared with a check valve in a hydraulic system. A check valve will allow fluid to pass in one direction, but close and prevent the fluid from moving in the opposite direction regardless of the pressure buildup. Likewise, a diode will allow electrical current to move in one direction but close the circuit and prevent any current flow in the opposite direction. Therefore, when testing the circuit with an ohmmeter, the meter should indicate current flow in one direction but not in the opposite direction.

TROUBLESHOOTING

Failure in the alternator charging circuit will usually become evident when the battery reaches a run-down condition. To determine why the charging system has failed to maintain the battery in satisfactory condition, first, check the condition of the battery and all electrical connections, especially at the battery terminals. Many times, this visual inspection will reveal the problem area.

Verify the battery cables have been **PROPERLY** connected. If the battery has not been connected properly the diodes in the circuit will be blown instantly. Determine the number of accessories connected, their draw, and the capability of the system to maintain the battery in a fully charged condition.

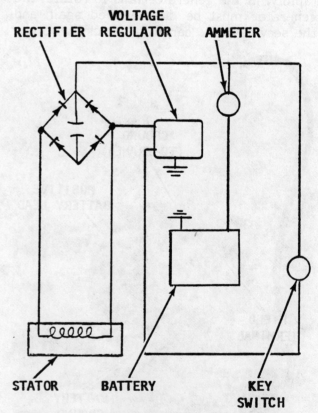

Schematic drawing of the low-output alternator system without a voltage regulator.

Schematic drawing of a high-output alternator system with a voltage regulator installed.

SAFETY WORD

Before conducting any troubleshooting tests or actual service work, the battery cables should be disconnected at the battery terminals. Battery current is **NOT** required for any of the following troubleshooting tests. This safety measure will prevent accidental current from reaching a component resulting in possible damage to the part or personal injury.

Testing Diodes Use High-Ohm Scale

1- Check the diodes by connecting one test lead of an ohmmeter to one of the yellow wires at the diode and the other test lead to a good ground. Now reverse the connections. The ohmmeter should indicate current flow in only one direction. Perform the test on the other yellow lead. If current flow is indicated in both directions, the diode is defective and **MUST** be replaced.

2- Connect one ohmmeter test lead to the yellow diode lead and the other test lead to the red lead of the rectifier. Observe the ohmmeter reading. Now, reverse the test leads and again observe the ohmmeter. Current should be indicated in only one direction. Next, connect the test lead to the other diode yellow lead and perform the same two tests. If current flow is indicated in both directions when the test lead is connected to either of the yellow leads, the diode is defective and **MUST** be replaced.

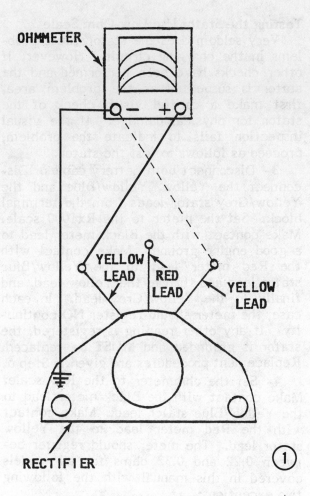

*Tightening the wingnuts of a cable connection at the battery terminal using a pair of pliers. Tightening the wingnuts by hand is not adequate as explained in the text. The battery terminals **MUST** be kept clean with an alternator system.*

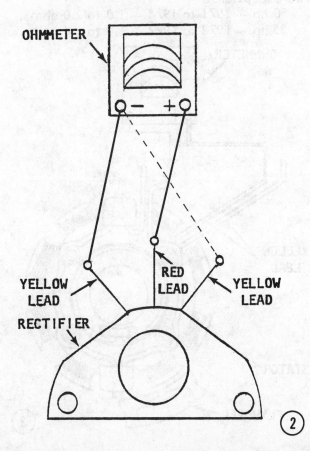

Testing the Stator Use Low-Ohm Scale

Very seldom does the stator cause problems in the charging circuit. However, if other checks have been performed and the stator is suspected as the problem area, first make a careful visual check of the stator for physical damage. If the visual inspection fails to indicate the problem, proceed as follows to test the stator:

3- Disconnect both battery cables. Disconnect the Yellow, Yellow/Blue and the Yellow/Grey stator leads from the terminal block. Set the meter to the Rx1000 scale. Make contact with the Black meter lead to a good engine ground. Make contact with the Red meter lead to the Yellow/Blue stator lead and then to the Yellow lead, and finally to the Yellow/Grey lead. In each case, the meter should register **NO** continuity. If any other reading is registered, the stator is grounded and **MUST** be replaced. Replacement procedures are given in Step 5.

4- Set the ohmmeter to the Rx1 scale. Make contact with the Black meter lead to the Yellow/Blue stator lead. Make contact with the Red meter lead to the Yellow stator lead. The meter should register between 0.22 and 0.32 ohms for all models covered in this manual with the following two exceptions:

50 hp -- 1971 to 1975 -- 1.0 to 2.0 ohms.
55 hp -- 1978 to 1977 -- 1.0 to 1.6 ohms.

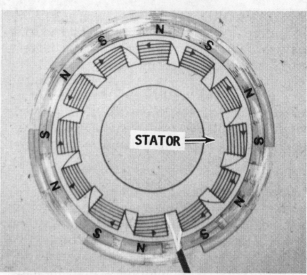

Functional diagram to show the flywheel and stator arrangement for a typical alternator system.

Keep the black meter lead in place and move the Red meter lead to contact the Yellow/Grey stator lead. The meter should register the same reading as in the previous test.

Stator Replacement

5- Because the stator is located under the flywheel, the proper puller must be obtained; the flywheel removed; the two and sometimes three yellow wires disconnected at the junction box; and then the stator assembly removed. The new stator is installed and secured with the attaching hardware; the wires connected to the junction

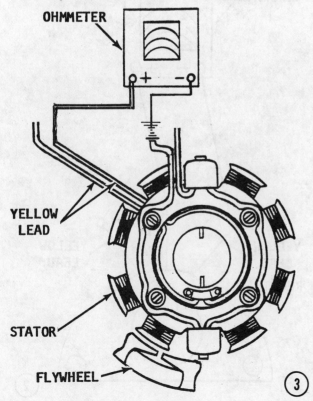

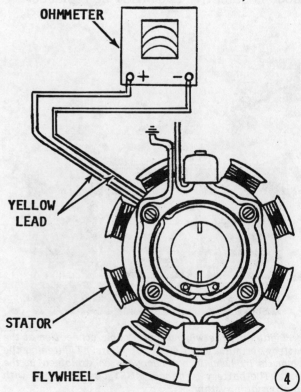

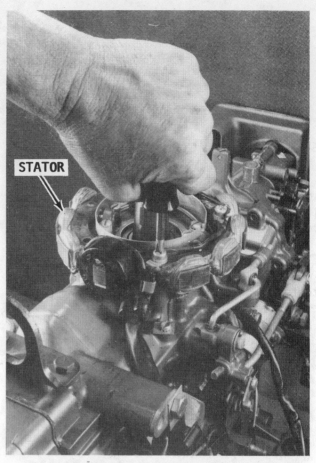

box; and the flywheel installed and secured. Details of flywheel removal and installation will be found in Chapter 5.

6- Use Loctite on threads of the stator attaching screws to prevent them from vibrating loose and being struck by the flywheel.

Diode Replacement

7- The diode assembly is a sealed unit secured to the terminal block with two attaching screws. Take note of the wire color coding and the terminals to which each is connected as an aid during installation of the new diode. Disconnect the two yellow and one red wire from the diode. Remove the two attaching screws, and then the diode assembly. Secure the new diode assembly in place with the two attaching screws, and then connect the two yellow and one red wire to the same assembly terminals from which they were disconnected.

Tightening the screws that extend down through the stator coil assembly using a Phillips screwdriver.

6-11 CHOKE CIRCUIT SERVICE

This short section provides instructions to test the choke circuit. If the system fails the test, the attaching hardware can be removed and the choke assembly replaced.

Choke Circuit Testing

1- The choke circuit may be quickly tested to determine if it is functioning properly as follows:

a- Obtain an ohmmeter.

b- Connect the black meter lead to an unpainted portion of the engine block for a good ground.

c- Connect the red meter lead to the choke terminal.

d- Test the circuit using the Rx1 scale of the ohmmeter. A satisfactory reading is approximately 3 ohms.

e- After the test is completed, check to be sure the choke plunger is pulled into the choke solenoid.

6-12 STARTER MOTOR CIRCUIT SERVICE

CIRCUIT DESCRIPTION

As the name implies, the sole purpose of the starter motor circuit is to control operation of the starter motor to crank the engine until the engine is operating. The circuit includes a solenoid or magnetic switch to connect or disconnect the starter from the battery. The operator controls the switch with a pushbutton or key switch.

A cutout switch is installed in the system to prevent starting the engine if the throttle is advanced too far, beyond idle speed. When the throttle is advanced, the starter solenoid is not grounded and the starter motor will not rotate. On electric shift models, a cutout switch is installed in the circuit to permit operation of the starter motor **ONLY** if the shift control lever is in **NEUTRAL**. This switch is a safety device to prevent accidental engine start when the engine is in gear.

STARTER MOTOR DESCRIPTION

Prestolite and Bosch starter motors are used on the Johnson/Evinrude engines covered in this manual. These two starter motors are used interchangeably. Therefore, two

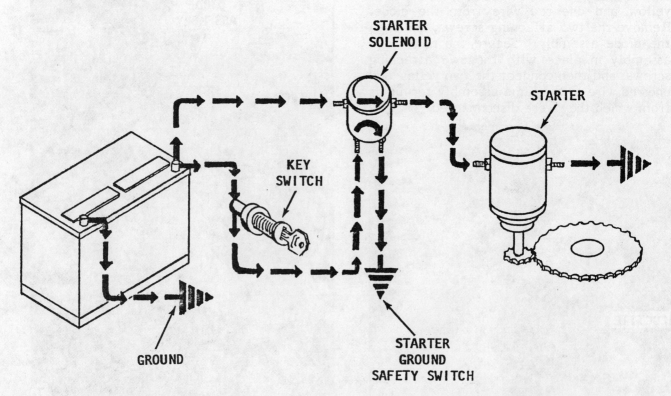

Functional diagram depicting current flow when the key switch is turned to the START position.

engines of the same horsepower and model year may have either the Prestolite or Bosch unit installed.

Marine starter motors are very similar in construction and operation to the units used in the automotive industry. All marine starter motors use the inertia-type drive assembly. This type assembly is mounted on an armature shaft with external spiral splines which mate with the internal splines of the drive assembly.

The starter motor is a series wound electric motor which draws a heavy current from the battery. It is designed to be used only for short periods of time to crank the engine for starting. To prevent overheating the motor, cranking should not be continued for more than 30 seconds without allowing the motor to cool for at least three minutes. Actually, this time can be spent in making preliminary checks to determine why the engine fails to start.

Most starter motors operate in much the same manner and the service work involved in restoring a defective unit to service is almost identical. Therefore, the information in this chapter is grouped together for the major components of the starter under separate headings. Differences, where they occur, between the various manufacturers, are clearly indicated.

Theory of Operation

Power is transmitted from the starter motor to the engine flywheel through a Bendix drive. This drive has a pinion gear mounted on screw threads. When the motor is operated, the pinion gear moves up to mesh with the teeth on the flywheel ring gear.

When the engine starts, the pinion gear is driven faster than the shaft, and as a result, it screws out of mesh with the flywheel. A rubber cushion is built into the Bendix drive to absorb the shock when the pinion meshes with the flywheel ring gear. The parts of the drive **MUST** be properly assembled for efficient operation. If the drive is removed for cleaning, **TAKE CARE** to assemble the parts as shown in the accompanying illustration. If the screw shaft assembly is reversed, it will strike the splines and the rubber cushion will not absorb the shock.

The sound of the motor during cranking is a good indication of whether the starter motor is operating properly or not. Natural-

ly, temperature conditions will affect the speed at which the starter motor is able to crank the engine. The speed of cranking a cold engine will be much slower than when cranking a warm engine. An experienced operator will learn to recognize the favorable sounds of the cranking engine under various conditions.

Faulty Symptoms

If the starter spins, but fails to crank the engine, the cause is usually a corroded or gummy Bendix drive. The drive should be removed, cleaned, and given an inspection.

If the starter motor cranks the engine too slowly, the following are possible causes and the corrective actions that may be taken:

a- Battery charge is low. Charge the battery to full capacity.

b- High resistance connections at the battery, solenoid, or motor. Clean and tighten all connections.

c- Undersize battery cables. Replace cables with sufficient size.

d- Battery cables too long. Relocate the battery to shorten the run to the starter solenoid.

Maintenance

The starter motor does not require periodic maintenance or lubrication **EXCEPT** just a drop of lightweight oil on the starter shaft to ease movement of the Bendix drive. If the motor fails to perform properly, the checks outlined in the previous paragraph should be performed.

The frequency of starts governs how often the motor should be removed and reconditioned. The manufacturer recommends removal and reconditioning every 1000 hours.

Naturally, the motor will have to be removed if the corrective actions outlined under **Faulty Symptoms** above, does not restore the motor to satisfactory operation.

STARTER MOTOR TROUBLESHOOTING

Before wasting too much time troubleshooting the starter circuit, the following checks should be made. Many times, the problem will be corrected.

a- Battery fully charged.

b- Throttle advanced too far (beyond fast idle speed).

c- Shift control lever not in **NEUTRAL** (electric shift models only).

Two handy instruments for use in checking the generating circuit (left) and the starter motor circuit (right). These meters do not require any wire connections. A reading will be obtained by simply placing the meter on the line.

d- All electrical connections clean and tight.

e- Wiring in good condition, insulation not worn or frayed.

f- One of the cutout switches may be defective.

Two more areas may cause the engine to turn over slowly even though the starter motor circuit is in excellent condition: A tight or "frozen" engine; or water in the lower unit causing the bearings to tighten up. The following troubleshooting procedures are presented in a logical sequence, with the most common and easily corrected areas listed first in each problem area. The connection number refers to the numbered positions in the accompanying illustrations.

Typical Bendix spring arrangement on a starter motor. A small amount of oil on the shaft in the spring area will prolong satisfactory operation.

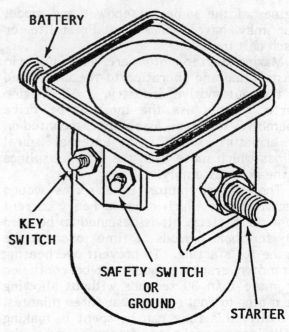

Functional diagram of a starter motor solenoid. Notice the separate terminal for a ground wire. This solenoid is NOT grounded through the mounting bracket.

Perform the following quick checks and corrective actions for following problems:

TESTING

SAFETY WORD

Before making any test of the cranking system, disconnect the spark plug leads at the spark plugs to prevent the engine from possibly starting during the test and causing personal injury.

The following tests are to be performed according to the faulty condition described. The numbers referenced in the steps are correlated with numbers on the accompanying circuit diagram to identify exactly where the connection or test is to be made.

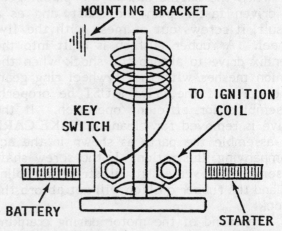

Functional diagram of a "slave-type" starter motor solenoid used on four-cycle engine installations. This solenoid CANNOT be used on a two-cycle engine.

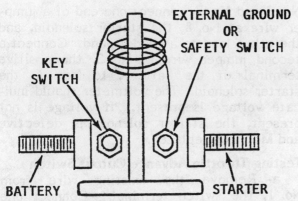

*Functional diagram of a typical two-cycle outboard engine starter motor solenoid. Notice that the right-hand small terminal is connected to ground or a safety switch. The unit is **NOT** grounded through the mounting bracket.*

Starter Motor Turns Slowly

a- Battery charge is low. Charge the battery to full capacity.

b- Electrical connections corroded or loose. Clean and tighten.

c- Defective starter motor. Perform an amp draw test. Lay an amp draw gauge on the cable leading to the starter motor No. 5. Turn the key to the **START** position and attempt to crank the engine. If the gauge indicates an excessive amperage draw, the starter motor **MUST** be replaced or rebuilt.

Starter Motor Fails To Turn Over Voltage Check

a- Check the voltage at No. 2, the battery and ground.

b- If satisfactory voltage is indicated at the battery, check the voltage at No. 3, the positive side of the starter solenoid. Weak, or no voltage at this point indicates corroded battery terminals, poor connection at the solenoid, or defective wiring between the battery and the solenoid.

c- Test the voltage at No. 4, the key. A full 12-volt reading should be registered at the key. Weak or no voltage at the key indicates a poor connection at the solenoid, or a broken wire between the starter solenoid and the key.

d- If satisfactory voltage is indicated during Steps a, b, and c, connect a volt meter at No. 5 and ground, and then turn the key switch to the **START** position. If 9-1/2 or more volts is registered at No. 5 and the starter still fails to operate, the starter is defective and requires service. If voltage

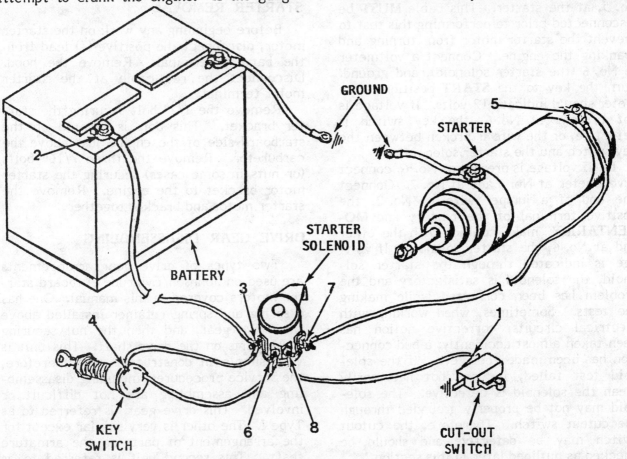

Diagram of hookup for making the various tests outlined in the text. This illustration and the numbers shown are to be used when testing the starter motor components.

is **NOT** present at No. 5, proceed to the next section, Testing Starter Solenoid.

Testing Starter Solenoid

FIRST THESE WORDS

The starter solenoid is actually nothing more than a switch between the battery and the starter motor. Several types of solenoids are used and many appear similar. **NEVER** attempt to use an automotive-type solenoid in a marine installation. Such practice will lead to more problems than can be imagined. An automotive-type solenoid has a completely different internal wiring circuit. If such a solenoid is connected into the starter system, and the system is activated, current will be directed to ground. The wires will be burned and the cutout switch will be burned and rendered useless. Therefore, when installing replacement parts in the starter or other circuits on a marine installation, always take time to obtain parts from a **MARINE** outlet to ensure proper service and to prevent damage to other expensive components.

a- Remove the heavy starter cable at No. 5, at the starter. This cable **MUST** be disconnected prior to performing this test to prevent the starter motor from turning and cranking the engine. Connect a voltmeter to No. 6 (the starter solenoid), and ground. Turn the key to the **START** position. The meter should indicate 12 volts. If voltage is not present at No. 6, the key switch is defective, or the wire is broken between the key switch and the starter solenoid.

b- If voltage is present at No. 6, connect a voltmeter at No. 3 and at No. 7. Connect one end of a jumper wire to No. 2, the positive terminal of the battery and **MOMENTARILY** make contact with the other end at No. 6, the starter solenoid. If voltage is indicated through the starter solenoid, the solenoid is satisfactory and the problem has been corrected while making the tests. Sometimes, when working with electrical circuits, corrective action has been taken almost accidently, a bad connection has been made good, etc. If the solenoid test failed, it does not necessarily mean the solenoid is defective. The solenoid may not be properly grounded through the cutout switch. Therefore, the cutout switch may be defective and should be checked as outlined later in this section.

c- With the voltmeter still connected at No. 3 and No. 7, connect one end of a jumper wire at No. 8, the starter solenoid, and the other lead to a good ground. Connect a second jumper wire at No. 2, the positive terminal of the battery, to No. 6, the starter solenoid. The voltmeter should indicate voltage is present. If voltage is not present, the starter solenoid is defective and **MUST** be replaced.

Testing Throttle Advance Cutout Switch

a- Remove the existing wire from No. 1, the switch terminal. Connect one probe lead of an ohmmeter to the terminal. Connect the other test probe lead to a good ground. Depress the switch button and the ohmmeter should indicate continuity. If continuity is not indicated, the switch is defective and **MUST** be replaced. Connect the heavy cable at No. 5, the starter motor. On 50 hp, 55 hp, 60 hp and 40 hp models, since 1983, the cutout switch is located in the shift box and functions in the same manner as the cutout switch mounted on the engine.

6-13 STARTER DRIVE GEAR SERVICE

STARTER REMOVAL

Before beginning any work on the starter motor, disconnect the positive (+) lead from the battery terminal. Remove the hood. Disconnect the red cable at the starter motor terminal.

Remove the 1/2" bolt securing the starter bracket. This bolt is located on the starboard side of the engine just above the carburetor. Remove the three 7/16" bolts (or nuts in some cases) securing the starter motor bracket to the engine. Remove the starter motor and bracket together.

DRIVE GEAR DISASSEMBLING

Two types of drive gear arrangements are used on Johnson/Evinrude outboard starter motors covered in this manual. One has a spring and spring retainer installed above the drive gear, and then the nut securing these parts on the drive shaft. This unit is very simple in construction and therefore, the service procedures, including disassembling and assembling are not difficult or involved. This drive gear is referred to as Type I. The other is very similar except for the arrangement of parts on the armature shaft. This second unit is referred to as Type II.

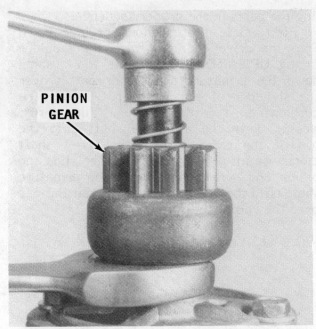

Removing the starter drive using an open-end wrench and a box-end wrench to remove the shaft nut.

To determine which starter motor drive type is being serviced observe the unit and make a comparison with the two exploded illustrations in this section, especially the pinion gear and the screw shaft.

DISASSEMBLING
Type I Drive Gear

Prevent the armature from turning by holding it with the proper size wrench on the hex nut provided for this purpose on the opposite end from the shaft nut. If the hex nut is not provided, hold the drive assembly with a pair of water pump pliers. Remove the shaft nut, spring retainer, spring, and then the drive assembly. The shaft nut

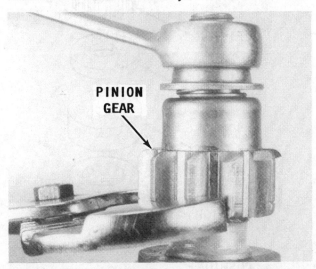

Removing the drive gear using a pair of pliers to hold the gear and a box-end wrench to remove the shaft nut.

Worn teeth on a Bendix drive gear. This gear is no longer fit for service.

should be replaced and **NOT** used a second time. The manufacturer **STRONGLY** recommends against using any type of self-locking nut on the shaft.

The exploded drawing accompanying this section will be helpful in assembling the starter motor in the proper sequence.

CLEANING AND INSPECTING

Inspect the drive gear teeth for chips, cracks, or a broken tooth. Check the spline inside the drive gear for burrs and to be sure the drive gear moves freely on the armature shaft. Check to be sure the return spring is flexible and has not become distorted. Clean the armature shaft with crocus cloth.

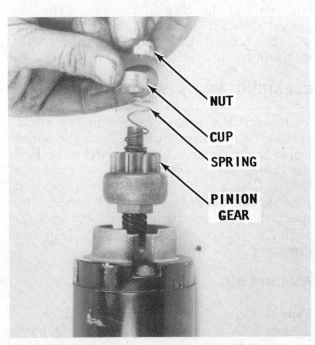

Installation sequence of parts when assembling the drive gear.

ASSEMBLING
Type I Starter Drive

Begin by assembling the following parts in the order given. The accompanying illustration will be most helpful in assembling the parts in the proper sequence. First, slide the drive gear onto the shaft, then the spring, spring retainer, and then a **NEW** locking nut. Prevent the armature shaft from turning by holding it with the proper size wrench on the hex nut provided for this purpose on the opposite end from the shaft nut. If the armature hex nut is not provided, hold the drive assembly with a pair of water pump pliers. Tighten the shaft nut securely.

To test the complete starter motor, proceed directly to Section 6-16.

To install the starter motor onto the engine, if no further work is to be performed, proceed directly to Section 6-17.

DISASSEMBLING

Type II Drive Gear with Rubber Cushion

First, remove the nut from the end of the armature shaft. Scratch a mark on the top of the screw shaft and one on the top of the pinion as an aid during assembling. These marks will identify the top of both parts. Next, remove the following parts in the sequence given: the pinion stop; anti-drift spring; sleeve; screw shaft cup; screw shaft; pinion; thrust washer; cushion cap; cushion; and the cushion retainer. The exploded drawing accompanying this section will be helpful in assembling the starter motor in the proper sequence.

CLEANING AND INSPECTING

Inspect the drive gear teeth for chips, cracks, or a broken tooth. Check the spline inside the drive gear for burrs and to be sure the drive gear moves freely on the armature shaft. Check to be sure the return spring is flexible and has not become distorted. Inspect the rubber cushion for cracks and for signs of oil on the cushion. Clean the armature shaft with crocus cloth.

ASSEMBLING

Type II Drive Gear with Rubber Cushion

Begin by assembling the following parts in the sequence given: The accompanying illustration will be most helpful in assembling the parts in the proper sequence.

First, slide the cushion retainer down onto the drive end cap, with the shoulder facing **UPWARD.** Next, slide the cushion down the armature shaft and seat it over the shoulder of the retainer. Install the cushion cap over the cushion. Slide thrust washer down the armature shaft onto the top of the cap. Rotate the screw shaft clockwise into the pinion, and then slide the pinion and screw shaft down the armature shaft onto the thrust washer. Install the cap over the end of the screw shaft.

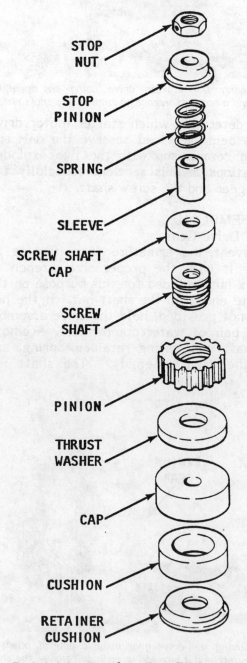

STOP
NUT

STOP
PINION

SPRING

SLEEVE

SCREW SHAFT
CAP

SCREW
SHAFT

PINION

THRUST
WASHER

CAP

CUSHION

RETAINER
CUSHION

Exploded drawing of the Type II drive gear.

Slide the following parts onto the armature shaft in the order given: the sleeve; spring; pinion stop washer; and finally thread the nut onto the end of the shaft. Tighten the nut securely.

To test the complete starter motor, proceed directly to Section 6-16.

To install the starter motor onto the engine, if no further work is to be performed, proceed directly to Section 6-17.

6-14 BOSCH STARTER MOTOR SERVICE

REMOVAL

1- Before beginning any work on the starter motor, disconnect the positive (+) lead from the battery terminal. Remove the engine hood. Disconnect the red cable at the starter motor terminal. Remove the three attaching bolts securing the starter motor to the engine. One bolt is very near the carburetor and the other two bolts are on the side. Remove the starter motor from the engine.

GOOD NEWS

If the only motor repair necessary is replacement of the brushes, the drive gear does not have to be removed. All starter motors have thru-bolts securing the upper and lower cap to the field frame assembly. In all cases both caps have some type of mark or boss. These marks are used to properly align the caps with the field frame assembly.

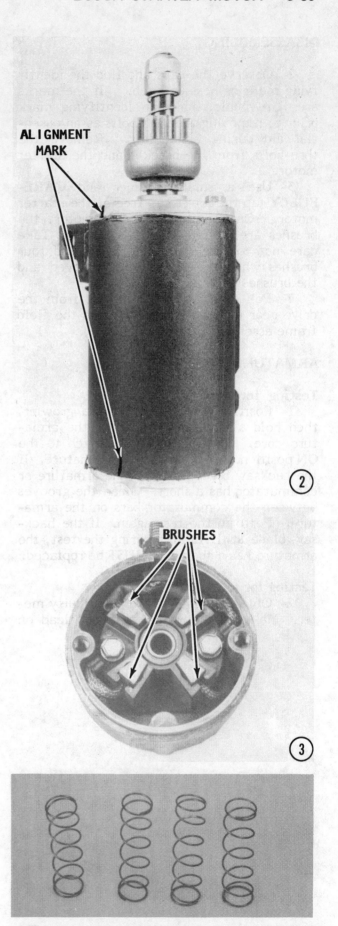

Typical brush spring. If the springs have turned blue in color, they must be replaced.

DISASSEMBLING

2- Observe the caps and find the identifying mark or boss on each.   If the marks are not visible, make an identifying mark prior to removing the thru-bolts as an essential aid during assembling.   Remove the thru-bolts from the bracket and the starter motor.

3- Use a small hammer and **CARE-FULLY** tap the lower cap free of the starter motor.   On the Bosch starter motor, the brushes are mounted in the end cap. Take care not to lose the four springs and four brushes when the end cap is removed and the brushes pop out.

4- Pull on the armature shaft from the drive gear end and remove it from the field frame assembly.

ARMATURE TESTING

Testing for a Short

1- Position the armature on a growler, then hold a hacksaw blade over the armature core. Turn the growler switch to the **ON** position. Slowly rotate the armature. If the hacksaw blade vibrates, the armature or commutator has a short. Clean the grooves between the commutator bars on the armature. Perform the test again. If the hacksaw blade still vibrates during the test, the armature has a short and **MUST** be replaced.

Testing for a Ground

2- Obtain a test lamp or continuity meter. Make contact with one probe lead on

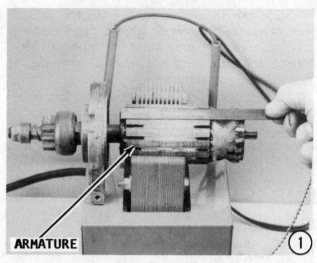

the armature core and the other probe lead on the commutator bar. If the lamp lights, or the meter indicates continuity, the armature is grounded and **MUST** be replaced.

Checking the Commutator Bar

3- Check between or check bar-to-bar as shown in the accompanying illustration. The test light should light, or the meter should indicate continuity. If the commutator fails the test, the armature **MUST** be replaced.

Turning the Commutator

4- True the commutator, if necessary, in a lathe. **NEVER** undercut the mica because the brushes are harder than the insulation. Undercut the insulation between the commutator bars 1/32" to the full width of the insulation and flat at the bottom. A triangular groove is not satisfactory. After the under-cutting work is completed, clean out the slots carefully to remove dirt and copper dust. Sand the commutator lightly with No. "00" sandpaper to remove any burrs left

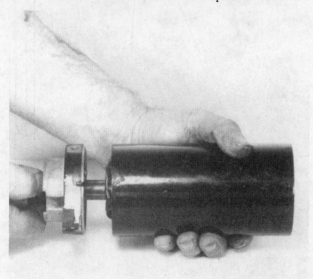

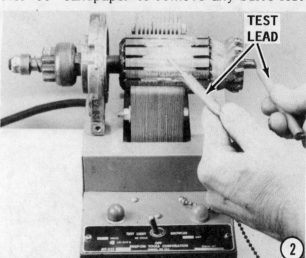

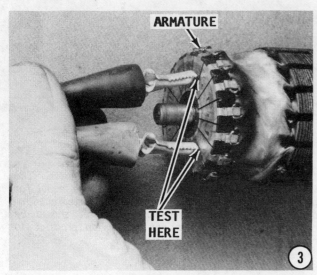

from the undercutting. Check the armature a second time on the growler for possible short circuits.

Positive Brushes

5– The positive brushes can always be identified as the brush with the lead connected to the starter terminal.

Obtain an ohmmeter. Connect one lead of the meter to the positive terminal of the cap and the other lead alternately to the positive brushes. The ohmmeter **MUST** indicate continuity between the brush and the terminal. If the meter indicates any resistance, check the lead to the brush and the lead to the positive terminal solder connection. If the connection cannot be repaired, the brush **MUST** be replaced.

Negative Brush

6– The negative brush can always be identified because the lead is connected to the starter cap.

Obtain an ohmmeter. Make contact with one lead to the starter motor frame and the other lead alternately to the negative brushes. If the meter does not indicate continuity, the field coils open and **MUST** be replaced.

CLEANING AND INSPECTING

Clean the field coils, armature, commutator, armature shaft, brush-end plate and

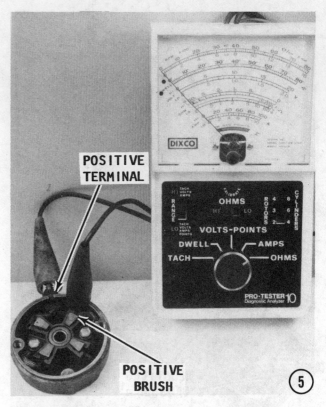

drive-end housing with a brush or compressed air. Wash all other parts in solvent and blow them dry with compressed air.

Inspect the insulation and the unsoldered connections of the armature windings for breaks or burns.

Perform electrical tests on any suspected defective part, according to the procedures outlined in Section 6-14.

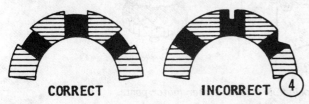

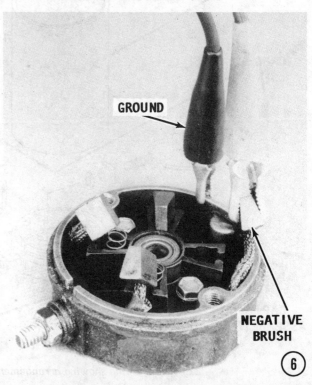

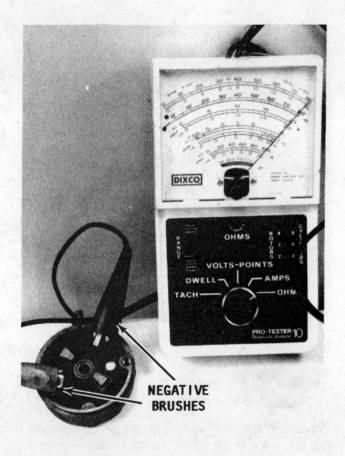

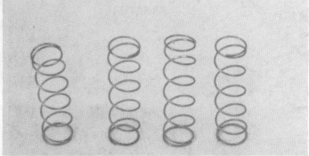

Brush springs removed from a starter motor. The length and condition of all springs must be equal for proper operation. If the springs have been stretched, or appear bluish in color the complete set should be replaced. Usually it is not good shop practice to replace less than a full set of springs.

Hookup to check the continuity between the negative brushes in a Bosch starter motor, as explained in the text.

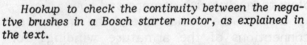

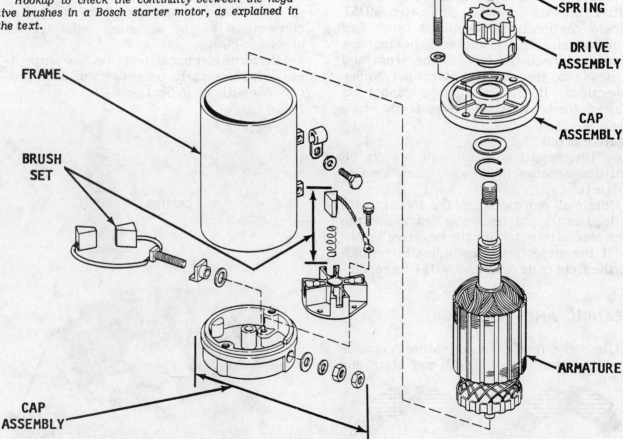

Exploded drawing showing arrangement of principle Bosch starter motor parts.

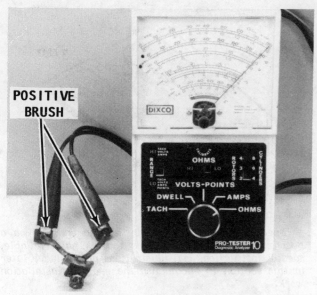

Checking the positive brushes on a Bosch starter motor. Each test lead is connected to a positive brush. Continuity must be indicated between the brushes.

Check the commutator for run-out. Inspect the armature shaft and both bearings for scoring.

Turn the commutator in a lathe if it is out-of-round by more than 0.005".

Check the springs in the brush holder to be sure none are broken. Check the spring tension and replace if the tension is not 32-40 ounces. Check the insulated brush holders for shorts to ground. If the brushes are worn down to 1/4" or less, they must be replaced.

Check the field brush connections and lead insulation. A brush kit and a contact kit are available at your local marine dealer, but all other assemblies must be replaced rather than repaired.

The armature, fields, and brush holders must be checked before assembling the star-

Badly corroded starter motor end cap and brushes. Such damage can only be corrected with replacement parts.

Cracked lower end cap of a Bosch starter motor. This damage was caused by salt water corrosion.

ter motor. See the testing section in this chapter for detailed procedures to test the starter motor.

ASSEMBLING THE BOSCH

Brush Installation

1- Both the positive and negative brushes on a Bosch starter motor are mounted in the lower cap. The positive brushes are attached to the positive terminal and are sold as an assembled set. The negative brushes are attached to the lower cap with a bolt.

To remove the positive brushes, slip the terminal out of the slot in the cap. The negative brushes are removed by simply removing the two bolts attaching the brush lead to the lower cap.

Installation of the new positive brushes is accomplished by sliding the new positive terminal into the slot of the end cap. Install the negative brushes by positioning them in place in the lower cap, and then securing the leads with the attaching bolts.

Assembling a Bosch Using Special Tool

Make a tool as shown in the accompanying illustration to prevent the brushes from being damaged during installation of the commutator end cap. If a special tool is not possible, see the next section, Assembling a Bosch Without a Special Tool.

1- Slide the brush springs into the brush holders, and then install the positive and negative leads. Position the special tool over the cap and brushes, to hold the brushes in place.

2- Clamp the drive gear in a vise equipped with soft jaws and with the drive gear down. Lower the frame assembly over the armature. Align the marks on the frame assembly with the marks on the upper end cap.

3- Position the lower end cap onto the frame assembly. Lower the cap as far as it will go, and then remove the special tool. Now, align the mark on the cap with the mark on the frame, and then install the thru-bolts and tighten them securely.

4- Place the starter motor on the floor. To test the operation of the motor, first connect one lead from a set of jumper cables to the positive terminal of a battery. Connect the other end of the same lead to the positive terminal of the motor. Connect one end of the second lead of the jumper cables to the negative terminal of the battery.

Now hold the motor firmly on the floor with one foot and at the same time, momentarily make contact with the other end of the second jumper lead to the motor case. The pinion gear should spin rapidly.

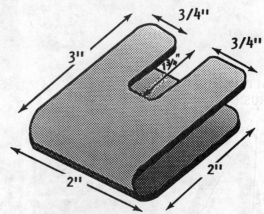

Special tool required to install the end cap on a Bosch starter motor. If this tool is not available, special instructions and illustrations are included later in this section to accomplish the end cap installation work.

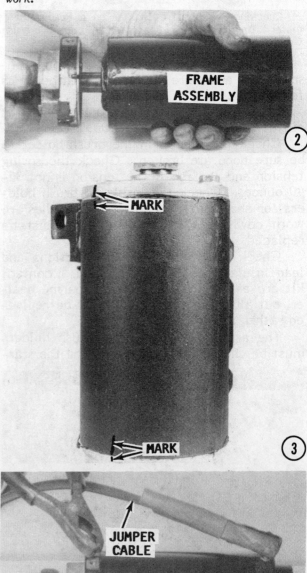

5- Install the starter motor onto the powerhead and secure it in place with the clamps and mounting bolts. If the starter motor has the mounting flanges permanently attached, then position the motor in place and start the three bolts to attach the motor to the powerhead. Now, tighten the three bolts **ALTERNATELY** and **EVENLY** until all bolts are tight. The bolts **MUST** be tightened alternately to prevent binding and possibly bending the bolts.

Assembling a Bosch Starter Without Special Tool

GOOD WORDS

The drive gear assembly must have been removed in order to assemble the starter using this method.

1- Install the brush springs into the brush holder, and then place each brush on top of the springs. Lay the lower cap on the bench with the brush facing up. Pickup the armature and place the commutator on top of the brushes. Lower the armature and at the same time, work each brush into its holder. Continue to lower the armature until the full weight of the armature is on the brushes.

2- Now, very **CAREFULLY** lower the frame assembly down over the armature. **TAKE CARE** because the magnets in the frame assembly will tend to pull against the armature.

3- When the frame makes contact with the lower cap, align the marks on the cap and the frame.

4- Slide the upper cap washer onto the shaft.

5- Install the upper cap with the mark on the cap aligned with the mark on the frame.

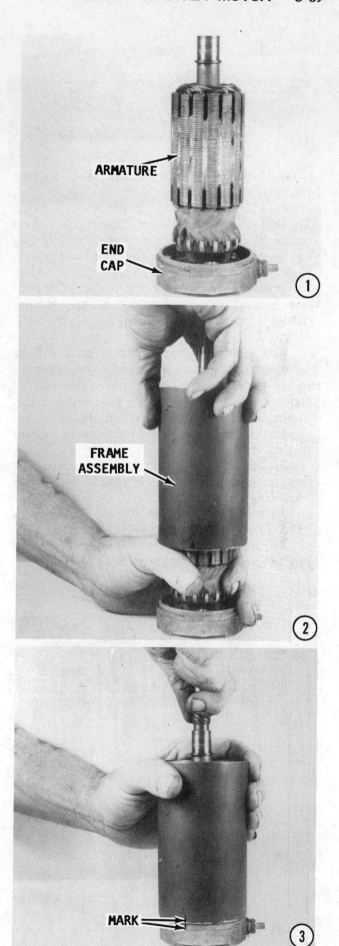

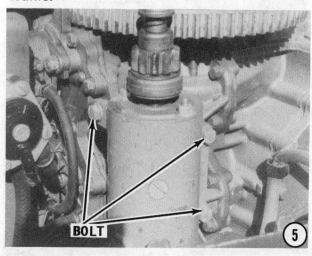

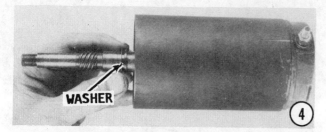

WASHER

(4)

6- Install the thru-bolts and tighten them securely.

7- Place the starter motor on the floor. To test the operation of the motor, first connect one lead from a set of jumper cables to the positive terminal of a battery. Connect the other end of the same lead to the positive terminal of the motor. Connect one end of the second lead of the jumper cables to the negative terminal of the battery.

Now hold the motor firmly on the floor with one foot and at the same time, momentarily make contact with the other end of the second jumper lead to the motor case. The pinion gear should spin rapidly.

Install the drive gear as described earlier in this section.

8- Install the starter motor onto the engine and secure it in place with the mounting bolts.

To test the complete starter motor, proceed directly to Section 6-16.

To install the starter motor onto the engine, proceed directly to Section 6-17.

MARK

(5)

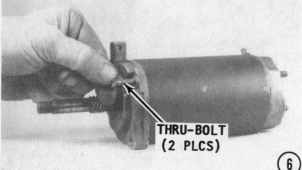

THRU-BOLT (2 PLCS)

(6)

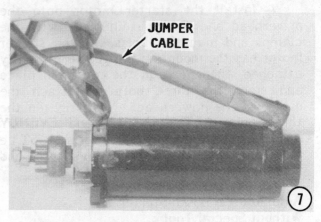

JUMPER CABLE

(7)

6-15 PRESTOLITE SERVICE

REMOVAL

1- Before beginning any work on the starter motor, disconnect the positive (+) lead from the battery terminal. Remove the engine hood. Disconnect the red cable at the starter motor terminal. Remove the 1/2" bolt securing the starter bracket. This bolt is located on the starboard side of the engine just above the carburetor. Remove the three 7/16" bolts (or nuts in some cases) securing the starter motor bracket to the engine. Some starter motors have flanges permanently attached to starter motor and the attachment to the engine is accomplished through these flanges instead of the usual mounting bracket. If the flanges are attached to the starter motor, remove the three bolts attaching the motor to the engine. Remove the starter motor and bracket together.

BOLT

(8)

GOOD NEWS

If the only motor repair necessary is replacement of the brushes, the drive gear does not have to be removed. All starter motors have thru-bolts securing the upper and lower cap to the field frame assembly. In all cases both caps have some type of mark or boss. These marks are used to properly align the caps with the field frame assembly.

DISASSEMBLING

2- Observe the caps and find the identifying mark or boss on each. If the marks are not visible, make an identifying mark prior to removing the thru-bolts as an essential aid during assembling. Remove the thru-bolts from the bracket and the starter motor.

3- Use a small hammer and **CAREFULLY** tap the lower cap free of the starter motor.

4- Pull on the armature shaft from the drive gear end and remove it from the field frame assembly. Remove the brushes from their holders, and then remove the brush springs. Lift the white plastic retainer free from the frame. Observe the location of the notch on the retainer in relation to the frame. The retainer must be installed in the same position.

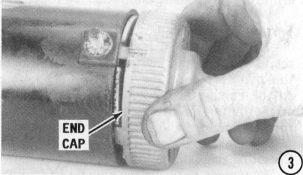

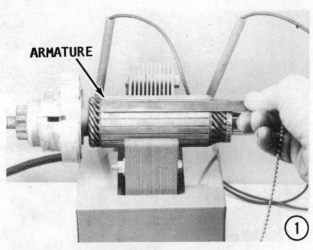

ARMATURE

ARMATURE TESTING

Testing for a Short

1- Position the armature on a growler, then hold a hacksaw blade over the armature core. Turn the growler switch to the **ON** position. Slowly rotate the armature. If the hacksaw blade vibrates, the armature or commutator has a short. Clean the grooves between the commutator bars on the armature. Perform the test again. If the hacksaw blade still vibrates during the test, the armature has a short and **MUST** be replaced.

Testing for a Ground

2- Obtain a test lamp or continuity meter. Make contact with one probe lead on the armature core and the other probe lead on the commutator bar. If the lamp lights, or the meter indicates continuity, the armature is grounded and **MUST** be replaced.

Checking the Commutator Bar

3- Check between or check bar-to-bar as shown in the accompanying illustration.

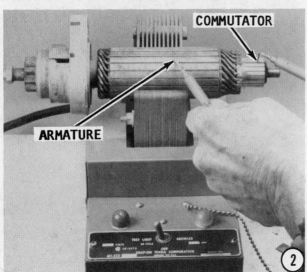

COMMUTATOR

ARMATURE

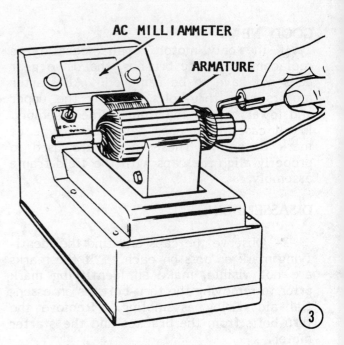

AC MILLIAMMETER

ARMATURE

The test light should light, or the meter should indicate continuity. If the commutator fails the test, the armature **MUST** be replaced.

Turning the Commutator

4- True the commutator, if necessary, in a lathe. **NEVER** undercut the mica because the brushes are harder than the insulation. Undercut the insulation between the commutator bars 1/32" to the full width of the insulation and flat at the bottom. A triangular groove is not satisfactory. After the undercutting work is completed, clean out the slots carefully to remove dirt and copper dust. Sand the commutator lightly with No. "00" sandpaper to remove any burrs left from the undercutting. Check the armature a second time on the growler for possible short circuits.

Positive Brushes

5- Notice how the positive brush lead is attached to the terminal on the end of the frame. This is the same terminal to which the heavy battery cable is attached. The terminal may be removed from the frame. Pull the terminal free of the frame.

Obtain an ohmmeter. Connect one test lead of an ohmmeter to the brush and the other test lead to the terminal. Continuity

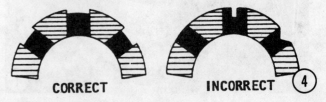

CORRECT   INCORRECT

POSITIVE BRUSH LEAD

⑤

GROUND (TO FRAME)

POSITIVE BRUSH LEAD

⑥

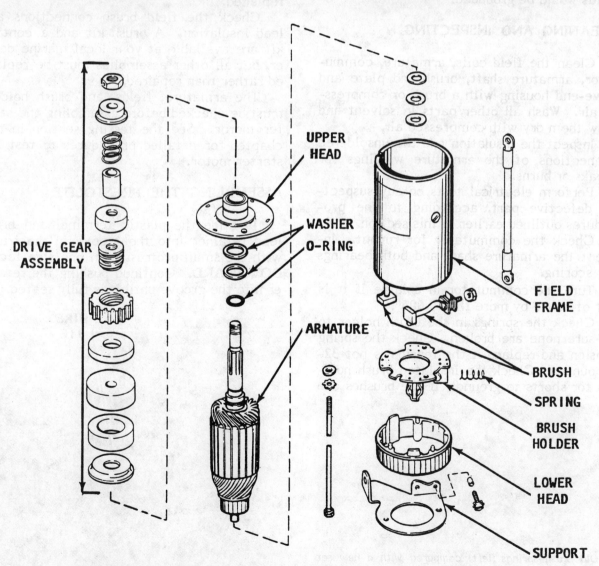

DRIVE GEAR ASSEMBLY

UPPER HEAD

WASHER

O-RING

ARMATURE

FIELD FRAME

BRUSH

SPRING

BRUSH HOLDER

LOWER HEAD

SUPPORT

Exploded drawing showing arrangement of principle Prestolite starter motor parts.

should be indicated on the ohmmeter. If continuity is not indicated, the brush must be replaced. The brush and terminal are sold as an assembly, eliminating the necessity for soldering.

Negative Brushes

6- On Prestolite starters the negative brushes are connected to the field coils inside the starter frame.

Obtain an ohmmeter. Make contact with one test lead to the negative brush and make contact with the other lead to the starter frame. If the meter does not indicate continuity, the field coils are open and **MUST** be replaced.

Check to be sure the soldered connections are **NOT** touching the frame. The fields must not be grounded. If the connections make contact with the frame, the fields would be grounded.

CLEANING AND INSPECTING

Clean the field coils, armature, commutator, armature shaft, brush-end plate and drive-end housing with a brush or compressed air. Wash all other parts in solvent and blow them dry with compressed air.

Inspect the insulation and the unsoldered connections of the armature windings for breaks or burns.

Perform electrical tests on any suspected defective part, according to the procedures outlined earlier in this section.

Check the commutator for runout. Inspect the armature shaft and both bearings for scoring.

Turn the commutator in a lathe if it is out-of-round by more than 0.005".

Check the springs in the brush holder to be sure none are broken. Check the spring tension and replace if the tension is not 32-40 ounces. Check the insulated brush holders for shorts to ground. If the brushes are

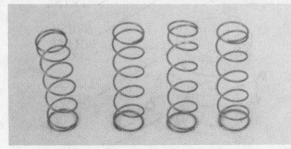

Old brush springs (left) compared with a new set (right). The springs must be in good condition, the same length, and free of any discoloration.

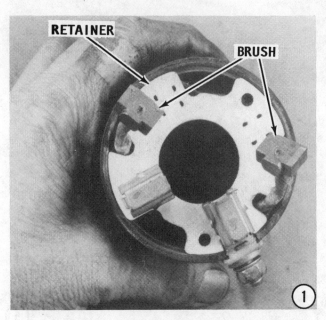

worn down to 1/4" or less, they must be replaced.

Check the field brush connections and lead insulation. A brush kit and a contact kit are available at your local marine dealer, but all other assemblies must be replaced rather than repaired.

The armature, fields, and brush holders must be checked before assembling the starter motor. See the testing section in this chapter for detailed procedures to test the starter motor.

ASSEMBLING THE PRESTOLITE

1- Slide the plastic terminal and brush lead retainer into the groove in the frame with the small protrusion on one side facing **DOWNWARD**. Continue pushing the retainer into the groove until it is fully seated.

Work the brush retainer down on top of the frame with the positive lead through the cutaway in the retainer plate. Check to be sure the field coil negative brush passes through the **cutaway** in the plate.

2- Install the spring into the retainer. Push the negative brush into its retainer and then, wrap a fine piece of wire around the front side of the brush and the back side of the retainer. Tighten the wire snugly. This wire will hold the brush in the retainer. Repeat the procedure for the positive brush.

Check to be sure the plate is secured onto the frame and the **cutaway** is over the protrusion of the positive plastic terminal.

Clamp the armature in a vise equipped with soft jaws with the drive gear facing **DOWNWARD**. Install the thrust washers onto the end of the armature shaft. Lower the frame assembly down over the armature until the brushes are over the commutator.

3- After the armature is in place, cut and remove the wire wrapped around the brushes to hold them in place. The brushes should then make firm contact with the commutator.

4- Install the end cap onto the end of the starter motor. Observe three small nipples on the inside of the end cap. These nipples **MUST** index with matching dimples in the retaining plate. Align the mark on the side of the end cap with the terminal. Lower the cap onto the frame, and seat it **GENTLY**. **NEVER** tap with a hammer or other tool, because the nipples may not be indexed with the dimples and the tapping may cause damage.

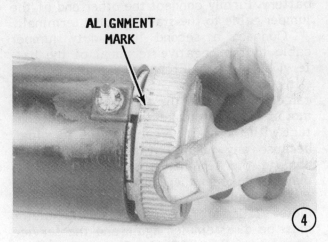

ALIGNMENT MARK

Align the end cap notch or mark with the mark on the frame and the upper cap mark with its mark.

Slide the rubber spacer or collar into the starter bracket, if used. Now, install the starter bracket over the starter motor, if used. Install the thru-bolts through the end cap, the frame, and thread them into the starter bracket. Tighten the thru-bolts securely.

6-16 STARTER MOTOR TESTING

1- Place the starter motor on the floor. Hold the motor firmly with one foot while testing its operation.

CAUTION: The armature will turn rapidly during this test. Therefore, the starter motor **MUST** be well **SECURED** before making the test to prevent personal **INJURY** or damage to the starter motor.

Firmly connect one end of a heavy-duty jumper cable to the **POSITIVE** terminal of a

COMMUTATOR

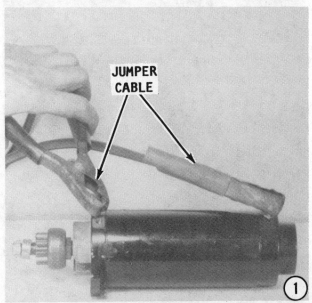

JUMPER CABLE

battery. Firmly connect the other end of the jumper cable to the starter motor terminal.

Connect a second heavy-duty jumper cable to the negative terminal of the battery. Now, **MOMENTARILY** make contact with the other end of the second jumper cable anywhere to the frame of the starter motor. **NEVER** make the momentary contact with the positive cable to the terminal, because any arcing at the terminal may damage the terminal threads and the nut may not take to the damaged threads. The motor should turn rapidly. If the starter motor fails to rotate, the starter motor must be disassembled again and the service work carefully checked. Sorry about that, but some phase of the rebuild task was not performed properly.

6-17 STARTER MOTOR INSTALLATION

With Mounting Bracket

1- Mount the starter motor and bracket onto the engine. Align the top bolt above the carburetor, and then thread it into the block about half-way. Align the other three bolts on the starboard side or start the nuts onto the studs, depending on the model engine being serviced. Tighten the three on the side evenly and alternately until they are secure.

With Flanges

2- If the starter motor has the mounting flanges permanently attached, then position the motor in place and start the three bolts to attach the motor to the engine. Now, tighten the three bolts **ALTERNATELY** and **EVENLY** until all bolts are tight. The bolts **MUST** be tightened alternately to prevent binding and possibly bending the flanges.

Connections

Connect the positive red lead to the starter motor. Connect the electrical lead to the battery. Connect the fuel line to the fuel pump.

Test the completed work by cranking the engine with the starter motor.

DO NOT, under any circumstances, start the engine unless it is mounted in a test tank or body of water.

CAUTION: Water must circulate through the lower unit to the engine any time the engine is run to prevent damage to the water pump in the lower unit. Just five seconds without water will damage the water pump.

7
REMOTE CONTROLS

7-1 INTRODUCTION

Remote controls are seldom obtained from the original equipment manufacturer, except in the case of the electric shift unit. The electric shift box is considered a part of the new engine. Therefore, unless an owner made a change, the electric shift unit with the engine is probably original engine manufacturer equipment. Mechanical shift units are sold and installed separately.

Shift boxes, steering, and other similar equipment may be added after the boat leaves the plant. Because of the wide assortment, styles, and price ranges of such accessories, the distributor, dealer, or customer has a wide selection from which to draw, when outfitting the boat.

Therefore, the procedures and suggestions in this chapter are general in nature in order to cover as many units as possible, but still specific and in enough detail to allow troubleshooting, repair, and adjustment for each of these accessories. Proper operation will do much for maximum comfort, performance, and enjoyment.

Complete procedures for removal, installation, and adjustment of four shift arrangements are covered in this chapter: manual shift; electric shift; push-button shift mechanism; and the single-lever remote control shift box. These shift boxes are all considered original Johnson/Evinrude equipment.

7-2 SHIFT BOXES DESCRIPTION

Undoubtedly, the most used accessory on any boat is the shift control box. This unit is a remote control device for shifting the outboard and at the same time controlling the throttle. Engines equipped with the manual mechanical shift are the only engine sold that do not have a shift box included as part of the complete package. If this engine is to be converted to a shift box operation, a shift box kit must be purchased as an accessory. All other engines will have the shift box included. Therefore, on the engines covered in this manual, only rarely will the installation have other than an OMC installed unit.

Because the cable length requirements cannot be known for each installation, the shift and throttle cables must be purchased

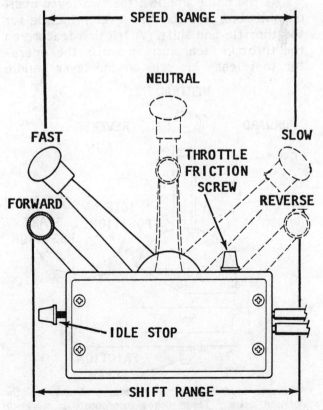

Double lever shift box installed with manual shift lower unit engines. This shift box is still in use and parts are available from OMC dealers.

separately.   OMC equipped boats may be equipped with one of four different type shift boxes: the two-lever manual shift; electric shift; the pushbutton shift arrangement; and the new improved electric shift box incorporating all of the electrical harness, key switch, and the "hot horn".

The mechanical shift box units have two levers, a long lever handle and a short lever handle. The long handle controls the throttle and the short one the shift mechanism. The electric shift units, including the pushbutton models, have only one lever handle for control of the shift and throttle. The shift box installed with Johnson engines has one handle for shifting and throttle control. Another lever, considered a "warmup" lever, is installed at the rear, or at the side of the box. This warmup lever may be adjusted for low and fast idle speeds. The push-button type locksout the shift lever to prevent shifting if the throttle is advanced too far while the engine is in neutral. The single-lever remote control box has one lever for both shift and throttle control, plus the "warmup" lever on the side of the box, and a safety button on the handle to prevent movement into gear unless the button is depressed.

As the name implies, the two-lever manual shift box uses the two lever principle for the throttle and shift. A friction feature on the throttle mechanism permits the operator to release his grip on the lever handle

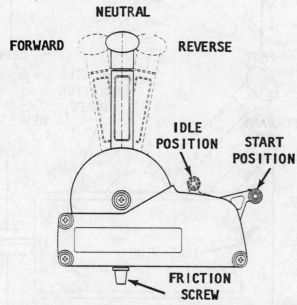

Single lever electric shift box used only on the Johnson units. These boxes incorporate a warm-up throttle lever at the rear and a friction screw on the bottom to hold the throttle position after the operator releases the handle.

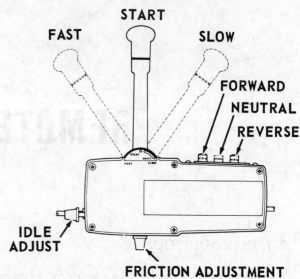

Single lever electric shift box installed with the Evinrude units. This box incorporates pushbuttons for shift control, an idle adjustment at the front, and a friction adjustment on the bottom to hold the throttle position.

without the throttle changing position. An idle stop is also built into the shift box. This feature prevents the throttle from being retracted past normal idle to the point where the engine would shut down.

Outboard models are equipped with a cutout switch in the cranking system to open the circuit to the starter solenoid. This   arrangement prevents the cranking

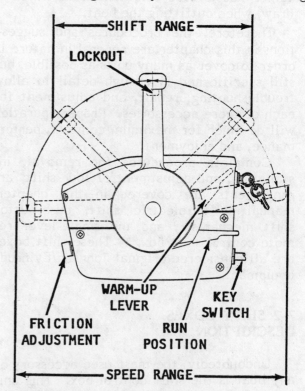

Single lever remote control shift box with key, choke, and "hot horn" incorporated.

system from operating unless the throttle is in the proper idle range. Stating it another way, the throttle **MUST** be in the idle position or the starter system will not operate. The position of the shift lever does not affect the starting motor circuit. All shift box models have a means of advancing the throttle without moving the shift lever into gear. This device is commonly known as the "warm-up" lever and may be adjusted for low and fast idle speeds.

7-3 DOUBLE-LEVER SHIFT BOX SERVICE

TROUBLESHOOTING

The following paragraphs provide a logical sequence of tests, checks, and adjustments, designed to isolate and correct a problem in the shift box operation.

The procedures and suggestions are keyed by number to matching numbered illustrations as an aid in performing the work.

The double-lever shift boxes are fairly simple in construction and operation. Seldom do they fail creating problems requiring service in addition to normal lubrication.

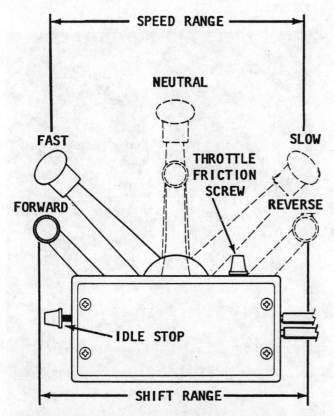

Double lever shift box installed with manual shift lower unit engines. This shift box including replacement parts is still available from OMC dealers.

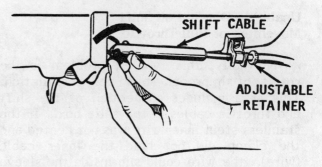

Attaching a shift cable, with adjustable trunnion to the shift arm.

Hard Shifting or Difficult Throttle Advance

Checking Throttle Side

Remove the throttle and shift control at the engine. Now, at the shift box, attempt to move the throttle or shift lever. If the lever moves smoothly, without difficulty, the problem is immediately isolated to the engine. The problem may be in the tower shaft between the connector of the throttle and the armature plate. The armature plate may be "frozen", unable to move properly. Late model engines do not have the tower shaft but a "lever advance arm" located on the starboard side of the engine. This arm may become "frozen" and require disassembly and lubrication.

If the problem with shifting is at the engine, the first place to check is the area where the shift lever extends through the exhaust housing. The bushing may be worn, or corroded. If the bushing requires replacement, the engine powerhead must be removed. Another cause of hard shifting is water entering the lower unit. In this case the lower unit must be disassembled, see Chapter 8.

If hard shifting is still encountered at the shift box when the controls are disconnected from the engine, the cables may be corroded, and require replacement, or lack of lubrication in the shift box has resulted in excessive wear, or corrosion.

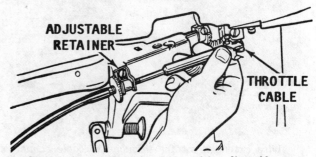

Connecting a throttle cable with adjustable trunnion.

Unable to Obtain Full Shift Movement or Full Throttle

Normally, this type of problem is the result of improper shift box installation. This area includes connection of the shift and throttle cables in the shift box. If the stainless steel inner wire was not heated and the clamp did not hold the inner cable (wire), the wire could slip inside the sleeve and the cable would be shortened. Therefore, if it is not possible to obtain full shift or full throttle, the shift box must be removed, opened, and checked for proper installation work. The inner wire could also slip at the engine end of the control, but problems at that end are very rare. Usually if improper installation work has been done at the engine end, the ability to shift at all is lost, or the throttle cannot be actuated.

REMOVAL

Removing Double Lever Shift Box

1- Remove the attaching hardware securing the shift box to the side of the boat. Once the shift box is free, the service work may be performed in the boat. The cables may remain as routed. Remove the two screws, at the rear side of the shift box, holding the two halves together. Separate the two halves.

OBSERVE

Observe how one side accommodates the throttle and the other side the shift mechanism. Notice the metal plate between the two halves. This plate prevents any contact between the shift parts and those for the

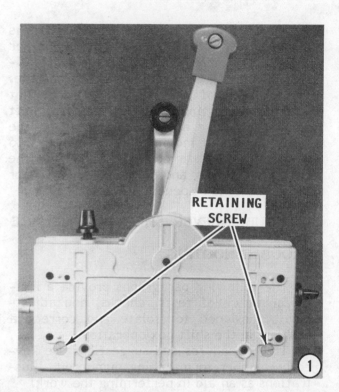

throttle. Notice the friction screw and throttle stop on the throttle side of the box.

The shift side of the box does not have any adjustments, except for the low idle stop. Observe how the shift lever pivots at the bottom and the throttle lever pivots at the top.

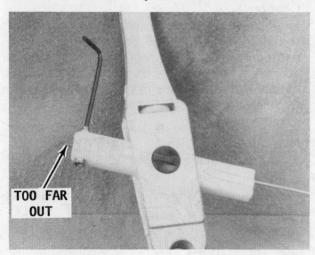

Wire extended too far through the cable connector (slider). The end of the wire should be flush with the slider surface.

Divider plate to separate the shift half components from the throttle half parts inside the control box.

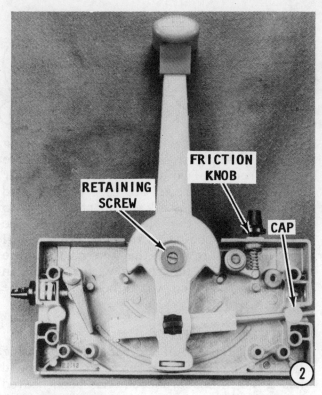

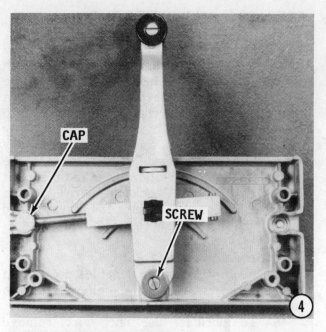

DISASSEMBLING

Throttle Half

2- Remove the screw from the center of the throttle lever. Notice how the washer has a concave side to allow the screw to fit flush with the washer. Loosen the screw on the top side of the shift box to relieve pressure on the anti-friction knob. Lift the lever and throttle cable free of the shift box. Notice how the cable on the trunnion has two small caps --one on the underside and the other on top.

3- Loosen the two screws securing the gear rack to the end of the throttle cable and remove the end of the cable from the throttle lever. Take care not to lose the small sleeve from the end of the rack to which the screws were attached. Push out the center square button, and then remove the rack from the shift lever.

Shift Half Disassembling

4- Remove the screw and washer from the bottom of the shift lever. Notice how this washer also has a concave side to accommodate the screw. Lift the lever and shift cable free of the shift box. Notice how the cable on the trunnion has two small caps -- one on the underside and the other on top.

5- Loosen the two screws securing the gear rack to the end of the shift cable and remove the end of the cable from the throttle lever. Take care not to lose the small sleeve from the end of the rack to which the screws were attached. Push out the center square button, and then remove the rack from the shift lever.

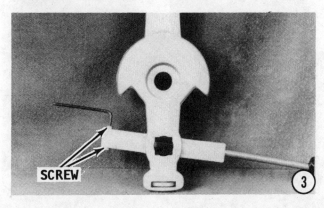

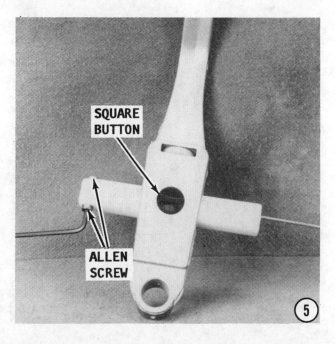

CLEANING AND INSPECTING

Check the nylon wear block on the end of the anti-friction cap. The cap has teeth which index into the inside diameter of the throttle lever. If the teeth are damaged a new block may be purchased and slipped into place. Clean the box halves thoroughly inside and out with solvent, and then dry them with compressed air. Inspect the spring on the anti-friction lever to be sure it is not distorted. Check the screw on the throttle idle stop to ensure it moves in and out freely without any sign of binding.

Throttle Cable Lubrication

If the throttle or shift cables are not to be replaced, now is an excellent time to lubricate the inner wire.

6- To lubricate the inner wire, remove the casing guide from the cable at both ends. Attach an electric drill to one end of the wire. Momentarily turn the drill on and off to rotate the wire and at the same time allow lubricant to flow into the cable, as shown.

ASSEMBLING

Throttle Cable into Shift Box

1- If the slider sleeve was removed from the throttle lever, install the slide rack into the throttle lever with the hole on the end for securing the throttle cable on the opposite end of the hole that accommodates the cable. Position the center of the slide with the center of the throttle lever. Install the

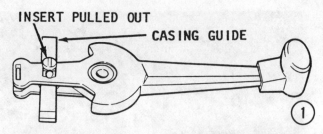

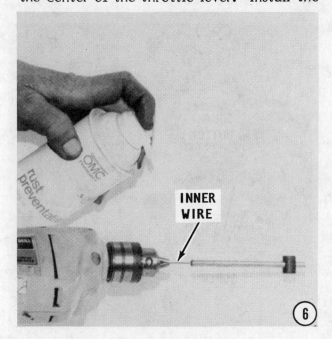

square nylon plug with the holes in the plug in a vertical position to permit the cable to slide through. Two different size screws, or possibly Allen screws, are used on each end of the sleeve. Install the short screw into the bottom of the sleeve to prevent the sleeve from rubbing on the shift box. Install the longer screw on the top part of the sleeve. Install the sleeve with the hole in the sleeve aligned with the hole for the cable.

CRITICAL WORDS

Check the end of the cable to determine if the temper has been removed. If the end has a bluish appearance, it has been heated at an earlier date and the temper removed. The temper **MUST** be removed to permit the holding screw to make a crimp in the wire to hold an adjustment. If the wire has not been tempered, heat the end, but not enough to melt the wire.

2- Slide the cable into the rack. Work the inner wire into the sleeve and out the end of the rack. Push the wire back until

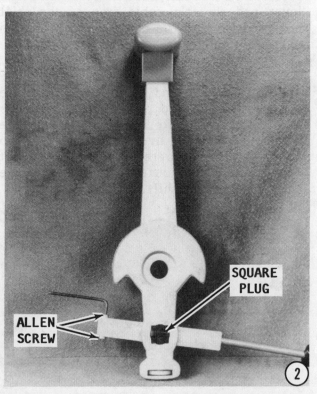

the end is flush with the rack surface. Tighten the **TOP** holding screw enough to make a definite crimp in the wire, as shown. If this screw is not tightened to make the crimp, the wire will slip during operation and the adjustment will be lost. After the top screw has been fully tightened, bring the other screw up tight against the wire. It is not necessary for this second screw to make a crimp in the wire.

3- Work the throttle lever handle down over the friction nylon block and at the same time feed the throttle cable into place in the box half. Check to be sure one of the small caps is on the bottom side of the trunnion. Install the washer and the screw with the concave side of the washer on the same side as the screw. Tighten the screw securely. Install the other trunnion cap on top of the trunnion. Check the throttle lever for ease of movement with no sign of binding.

Shift Cable into Shift Box

4- If the slider sleeve was removed from the shift lever, install the slide rack into the shift lever with the hole on the end for securing the shift cable on the opposite end of the hole that will accommodate the cable. Position the center of the slide with the center of the shift lever. Install the square nylon plug with the holes in the plug in a vertical position to permit the cable to slide through. Two different size screws are used on each end of the sleeve. Install the short screw into the bottom of the sleeve to prevent the sleeve from rubbing on the shift box. Install the longer screw on the top part of the sleeve. Install the sleeve with the hole in the sleeve aligned with the hole for the cable.

CRITICAL WORDS

Check the end of the cable to determine if the temper has been removed. If the end

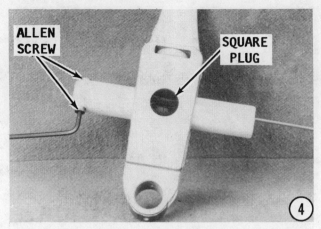

has a bluish appearance, it has been heated at an earlier date and the temper removed. The temper **MUST** be removed to permit the holding screw to make a crimp in the wire to hold an adjustment. If the wire has not been tempered, heat the end, but not enough to melt the wire.

Slide the cable into the rack. Work the inner wire into the sleeve and out the end of the rack. Push the wire back until the end is flush with the rack surface. Tighten the **TOP** holding screw enough to make a definite crimp in the wire, as shown. If this screw is not tightened to make the crimp, the wire will slip during operation and the adjustment will be lost. After the top screw has been fully tightened, bring the other screw up tight against the wire. It is not necessary for this second screw to make a crimp in the wire.

5- Place the wavy washer and regular washer into the shift box, and then work the shift lever handle down into the shift box with one of the small caps under the shift cable trunnion. Install the bushing into the bottom of the shift handle. Install the washer and the screw with the concave side of the washer on the same side as the screw. Tighten the screw securely. Install the other trunnion cap on top of the trunnion.

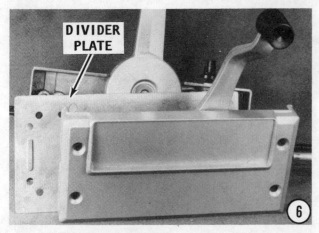

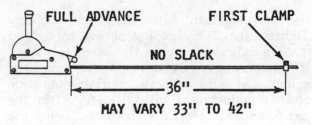

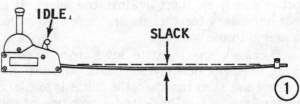

Check the shift lever for ease of movement with no sign of binding.

6- Place the divider plate between the two halves. Bring the two halves together and secure them with the two screws from the back side. Install the box in the boat and secure it in place with the attaching hardware. Again check the levers for ease of movement and no sign of binding.

7-4 ELECTRIC GEAR BOXES AND SINGLE LEVER CONTROL

TROUBLESHOOTING

The following paragraphs provide a logical sequence of tests, checks, and adjustments, designed to isolate and correct a problem in the Johnson single lever shift box with the warm-up lever to the rear and the Evinrude single lever pushbutton shift box operation.

The procedures and suggestions are keyed by number to matching numbered illustrations as an aid in performing the work.

When the lower unit is in FORWARD gear, all electrical systems are at rest. When the shift lever is moved to the NEUTRAL position, current flows through the green wire circuit to the neutral solenoid in the lower unit, which is activated. When the shift lever is moved to the REVERSE position, current flows through the neutral green and reverse blue wire circuits to activate the neutral and reverse solenoids.

1- Difficult Shift Operation

Many times this type of problem is the result of incorrect cable installation -- the cable is not the proper length or there are too many bends or kinks in the routing. Such an installation will cause the inner cable to travel much further than necessary

and therefore, wear on the outer cable. Over a period of time, inner cable wear will result in difficult shifting or throttle operation.

BE SURE to cycle the shift lever to the full position in both directions, when making any test on the shift box. The shift switch may have a dead spot and will not indicate the switch is defective unless the shift lever is fully cycled for each test.

2- Amp Draw Test

Turn the ignition switch to the ON position and note the ammeter reading. If the boat is not equipped with an ampere gauge, then temporarily disconnect the GREEN and BROWN (or RED) wires from the back side of the key switch and temporarily install an amp gauge for the test. Replace the wires after the test is completed. There should be no reading with the shift lever in the FORWARD position.

Now, operate the shift control lever to the NEUTRAL position, and then to the REVERSE position. Note how much the ammeter reading increased each time the shift lever was moved. If the reading was more than 1.5 to 2.0 amperes for either shift positions, continue with the following checks.

Disconnect the shift leads at the rear of the engine. Temporarily lay a piece of cloth or other insulating material under the wires to prevent them from shorting out during the following tests.

Again operate the shift lever and note the current loss. If the current draw is still more than 1.5 to 2.0 amperes, then check for a short in the control box switch or wiring. If the current draw is normal with the leads disconnected from the engine, then check for a short in the gear case solenoids or wiring. If the solenoid leads are shorted to each other, both shift solenoids would be energized, causing the lower unit to move into reverse gear.

3- Solenoid Tests

Obtain an ohmmeter. At the rear of the engine, disconnect the green and blue wires at the knife disconnect. Set the ohmmeter to the low scale. Connect one lead to the green wire to the lower unit, and the other lead to a good ground. The meter should indicate 5 to 7 ohms. Connect the meter to the blue wire to the lower unit and ground. The meter should again indicate from 5 to 7 ohms.

BAD NEWS

If the unit fails the ohmmeter tests just outlined, the only course of action is to disassemble the lower unit to determine and correct the problem.

4- Testing Shift Switch — NEUTRAL

To test the switch for the **NEUTRAL** position, make contact with one probe of a continuity meter (or a test light) to the terminal (purple or red lead) and to the (green lead) terminal with the other probe. Now, move the shift lever to the **NEUTRAL**

SHIFT TERMINAL

position. The meter should indicate continuity (or the test light come on), when the shift lever is in the **NEUTRAL** position.

5- Testing Shift Switch — Reverse

Move the shift lever to the **REVERSE** position. To test the shift switch for reverse, make contact with one probe of a continuity meter (or test light), to the purple or red lead from the plug. Now, with the other probe make contact **ALTERNATELY** to the blue and green wires in the plug. The meter should indicate continuity (or the test light come on), during both of these tests when the lever is in the reverse position.

6- Testing Shift Switch — Forward

After the neutral and reverse tests have been completed check for continuity with the shift lever in the **FORWARD** position. Leave the red lead connected and check the blue lead **(REVERSE)** and the green lead **(NEUTRAL)**. Continuity should not be indicated when the shift handle is in **FORWARD**. If the switch is defective and

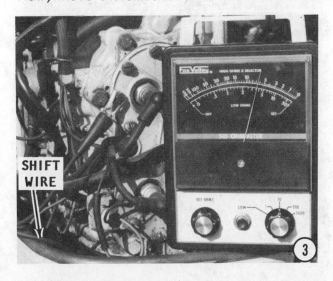

SHIFT WIRE

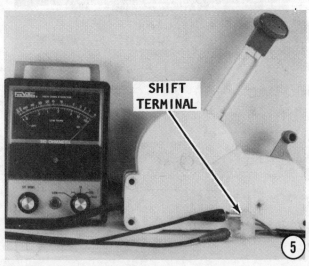

SHIFT TERMINAL

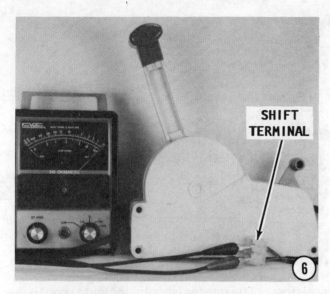

requires replacement, proceed to Section 7-5, Single Lever Shift Box Service.

7- Cranking System Inoperative

If the starter fails to crank the engine, check to be sure the throttle lever is in the idle position. If the throttle is advanced more than 1/4 forward, the cutout switch attached to the armature plate will open the circuit to the starter solenoid. Late models have the cutout switch installed in the shift box. If the cranking system fails to operate the starter properly when the throttle lever is in the IDLE position, check the 20-ampere fuse between the ignition switch BAT terminal and the ammeter GEN terminal.

If the starter operates in full throttle (which it should not do), check for a short between the two white leads in the shift box wiring.

Further problems in the cranking system may indicate more serious problems. See Chapter 6, starter motor sections.

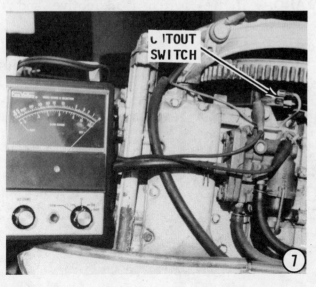

7-5 SINGLE LEVER SHIFT BOX SERVICE JOHNSON UNITS ONLY

GOOD WORDS

A friction screw is installed in the bottom side of the shift box. This screw allows friction adjustment of the throttle handle. This arrangement prevents the throttle handle from "creeping" after the operator releases his grip on the handle. The box has a maximum advance screw. The adjustment is made through movement of a screw in the warmup lever. If the engine shuts down when the throttle lever is moved back, then an adjustment must be made at the engine. This is accomplished through an adjustment knob at the engine. Movement of this knob will actually lengthen or shorten the cable slightly for proper operation.

1- Remove the attaching screws; pull the shift box clear; and then disconnect the shift wire under the dash. Remove the screws from the back side of the box.

2- CAREFULLY separate the two halves. Check the shift box for salt water corrosion, worn bushings, and general condition.

Throttle Cable

Notice how the throttle cable enters the throttle half of the shift box through the idle link. On the top side of the shift box, observe the screw and the concave washer.

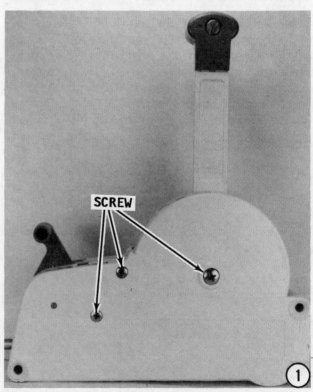

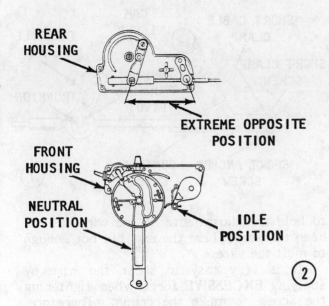

REAR HOUSING

EXTREME OPPOSITE POSITION

FRONT HOUSING

NEUTRAL POSITION

IDLE POSITION

②

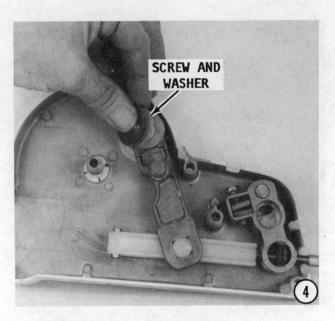

SCREW AND WASHER

④

The washer must be installed with the concave side toward the screw to allow the screw to seat properly.

3- Remove the bushing from the shift rod. Observe the two slots in the bushing and how the flat area without a hole faces toward you. The bushing must be installed in this same position.

4- Remove the screw and washer from the top of the shift box half. This is the screw and washer described in the previous paragraph. Lift out the cam lever and the idle link as an assembly.

5- Remove the screws from the end of the sleeve on the end of the cable, and then pull the throttle cable free of the link and sleeve.

6- If the switch fails to check out, as described in the previous tests, the switch and cable assembly **MUST** be replaced. The switch is easily removed by simply removing the attaching screws and lifting the switch free of the shift box.

CLEANING AND INSPECTING

Clean the box halves thoroughly inside and out with solvent and blow them dry with compressed air. Apply a thin coat of engine oil on all metal parts. The three-position switch installed in the gear box cannot be repaired. Therefore, if a problem is isolated to the switch, it must be replaced.

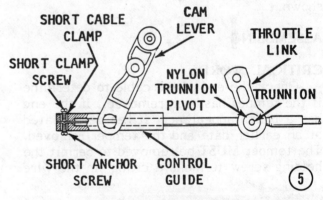

SHORT CABLE CLAMP

CAM LEVER

THROTTLE LINK

SHORT CLAMP SCREW

NYLON TRUNNION PIVOT

TRUNNION

SHORT ANCHOR SCREW

CONTROL GUIDE

⑤

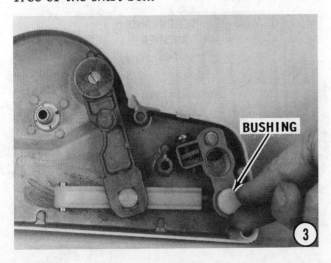

BUSHING

③

SHIFT SWITCH

⑥

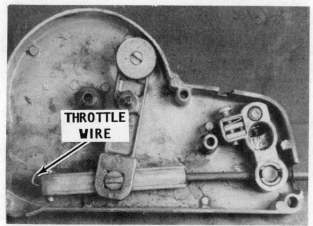

Interior view of a used shift box showing the results of an improper installation. The inner wire was not crimped to hold the adjustment. The wire, therefore, slipped through and was bent as the casing struck the shift box during operation.

Throttle Cable Lubrication

If the throttle cable is **NOT** to be replaced, now is an excellent time to lubricate the inner wire.

7- To lubricate the inner wire, remove the casing guide from the cable at both ends. Attach an electric drill to one end of the wire. Momentarily turn the drill on and off to rotate the wire and at the same time allow lubricant to flow into the cable, as shown.

ASSEMBLING

CRITICAL WORDS

Check the end of the cable to determine if the temper has been removed. If the end has a bluish appearance, it has been heated at an earlier date and the temper removed. The temper **MUST** be removed to permit the holding screw to make a crimp in the wire

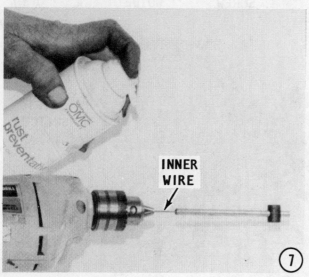

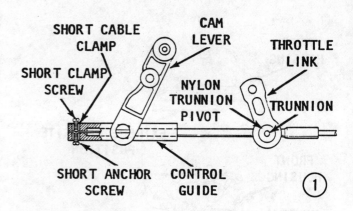

to hold an adjustment. If the wire has not been tempered, heat the end, but not enough to melt the wire.

It is very easy to shear the wire by applying **EXCESSIVE** force when tightening the screw to make the crimp. Therefore, play it cool. Tighten the screw; make one more complete turn to make the crimp; call it good; and then bring the other screw up just tight against the wire.

1- Start the two cable retaining screws into the sleeve. These screws are different sizes. On some models, Allen screws are used. Install the short screw on the bottom to prevent the sleeve from rubbing on the shift box. Slide the sleeve onto the cable with the hole aligned with the hole in the plastic sleeve. Feed the wire on through until the end of the wire is flush with the end of the white plastic sleeve.

2- Lower the shift link and throttle link into the box half and secure the throttle link with the screw and washer. Check to be sure the concave side of the washer is facing toward the screw side to permit the screw to seat properly.

3- Slide the throttle cable through the idle link, and then slide the bushing down over the cable.

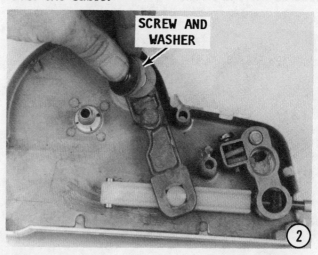

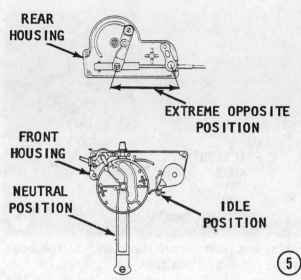

4- ALWAYS TAKE CARE when assembling the shift box, not to damage the remote-control unit. The arm on the switch **MUST** lay in the cut-out portion of the throttle cam.

5- Carefully work the two halves of the box together with the cam lever fitting into the recess of the throttle handle and the throttle link fitting into the warm-up lever.

6- Secure the two halves together with the screws into the side of the box. Secure the shift box to the side of the boat with the attaching hardware. Bolts with self-locking nuts **SHOULD ALWAYS BE USED** because a loose shift box during high speed operation could be extremely dangerous. Connect the shift wire under the dash.

7- The tension of the throttle lever is adjusted by the friction knob under the shift box. Turn the knob clockwise to increase friction and counterclockwise to decrease friction.

8- Remote-Control Cable Installation In the Boat

The remote-control cable must be installed properly for satisfactory operation. The clamp nearest the shift box **MUST** be positioned correctly as follows:

First, move the warm-up lever on the shift box to full advance. Now, measure 36" (actually this measurement could range from 33" to 42") on the cable from the shift box. **BE SURE** there is no slack in the

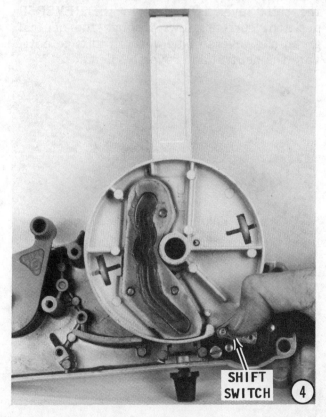

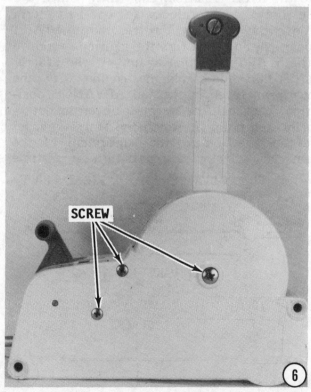

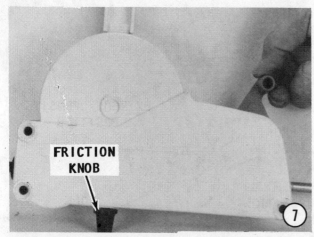

FRICTION KNOB

(7)

cable, and then secure the clamp to the boat at the measured position.

Next, place the warm-up lever in the slow position and observe the amount of slack in the cable between the shift box and the first clamp. The slack should not be more than 1/2 inch. **AVOID SHARP TURNS** in the cable. The radius of any bend **MUST** not be less than 5".

ALWAYS use the correct length of cable when replacing the assembly.

9- Adjusting Starter Lockout Switch

Move the shift lever to the **NEUTRAL** position and the warm-up lever to the **START** position. Now, turn the ignition switch to the **START** position in an attempt to crank the engine. If the starter fails to crank the engine, move the warm-up lever to the **IDLE** position, and then rotate the setscrew counterclockwise one-half turn. Move the auxiliary throttle to the **START** position, and then turn the ignition switch to the **START** position again. If the starter still fails to crank the engine with the warm-up lever in the full **ADVANCE** position, back off the warm-up lever to determine the point at which the starter ceases to operate. Make the adjustment of the setscrew clockwise a half-turn at a time

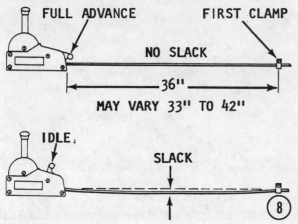

FULL ADVANCE   FIRST CLAMP

NO SLACK

36"

MAY VARY 33" TO 42"

IDLE

SLACK

(8)

until the starter operates only with the warm-up lever in the full **START** position.

7-6 PUSHBUTTON TYPE SHIFT BOX SERVICE — EVINRUDE UNITS ONLY

GOOD WORDS

A friction screw is installed in the bottom side of the shift box. This screw allows friction adjustment of the throttle handle. This arrangement prevents the throttle handle from "creeping" after the operator releases his grip on the handle. A thumbscrew on the front of the box permits adjustment of the throttle handle to prevent movement past a satisfactory idle position and subsequent shutdown of the engine. If adequate adjustment cannot be made at the shift box, and the engine continues to shut down when the throttle lever is moved back, then an adjustment must be made at the engine. This is accomplished through an adjustment knob at the engine. Movement of the engine knob will actually lengthen or shorten the cable slightly for proper operation.

When the lower unit is in **FORWARD** gear, all electrical systems are at rest. When the shift lever is moved to the **NEUTRAL** position, current flows through the green wire circuit to the neutral solenoid in the lower unit, which is activated. When the shift lever is moved to the **REVERSE** position, current flows through the neutral green and reverse blue wire circuits to activate the neutral and reverse solenoids.

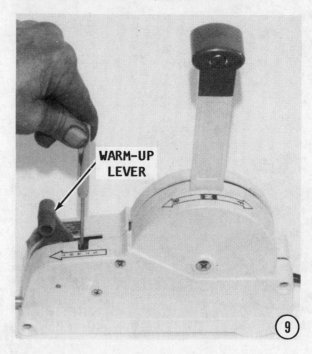

WARM-UP LEVER

(9)

TROUBLESHOOTING

The following paragraphs provide a logical sequence of tests, checks, and adjustments, designed to isolate and correct a problem in the Johnson single lever shift box with the warm-up lever to the rear and the Evinrude single lever pushbutton shift box operation.

The procedures and suggestions are keyed by number to matching numbered illustrations as an aid in performing the work.

Many times this type of problem is the result of incorrect cable installation -- the cable is not the proper length or there are too many bends or kinks in the routing. Such an installation will cause the inner cable to travel much further than necessary and therefore, wear on the outer cable. Over a period of time, inner cable wear will result is difficult shifting or throttle operation.

BE SURE to cycle all three shift pushbuttons to the full shift position in both directions, when making any test on the shift box.   The shift switch may have a dead spot and will not indicate the switch is defective unless the three buttons are fully cycled for each test.

1- Amp Draw Test

Turn the ignition switch to the **ON** position and note the ammeter reading. If the boat is not equipped with an ampere gauge, then temporarily disconnect the **GREEN** and **BROWN (or RED)** wires from the back side of the key switch and temporarily install an amp gauge for the test. Replace the wires after the test is completed. Now, depress the shift button for the **FORWARD** position. There should be **NO** reading. Depress the button for the **REVERSE** and **NEUTRAL** positions.   Note how much the ammeter reading increased each time a shift button was depressed. If the reading was more than 1.5 to 2.0 amperes for either the neutral or reverse positions, continue with the following checks.

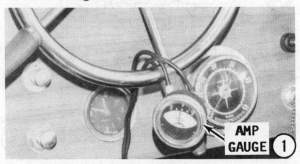

AMP GAUGE ①

Disconnect the shift leads at the rear of the engine. Temporarily lay a piece of cloth or other insulating material under the wires to prevent them from shorting out during the following tests.

Again operate the shift buttons and note the current loss. If the current draw is still more than 1.5 to 2.0 amperes, then check for a short in the control box switch or wiring. If the current draw is normal with the leads disconnected from the engine, then check for a short in the gear case solenoids or wiring. If the solenoid leads are shorted to each other, both shift solenoids would be energized, moving the lower unit into reverse gear.

2- Solenoid Tests

Obtain an ohmmeter. At the rear of the engine, disconnect the green and blue wires at the knife disconnect. Set the ohmmeter to the low scale. Connect one lead to the green wire to the lower unit, and the other lead to a good ground. The meter should indicate 5 to 7 ohms. Connect the meter to the blue wire to the lower unit and ground. The meter should again indicate from 5 to 7 ohms.

BAD NEWS

If the unit fails the ohmmeter tests just outlined, the only course of action is to disassemble the lower unit to determine and correct the problem.

3- Testing Shift Switch — NEUTRAL

To test the switch for the **NEUTRAL** position, make contact with one probe of a continuity meter (or a test light) to the terminal (purple or red lead) and to the (green lead) terminal with the other probe. Now, depress the **NEUTRAL** pushbutton. The meter should indicate continuity (or the test light come on). When the **FORWARD** button is depressed, the meter should indicate **NO** continuity or the test light should **NOT** come on.

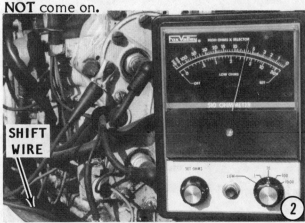

SHIFT WIRE ②

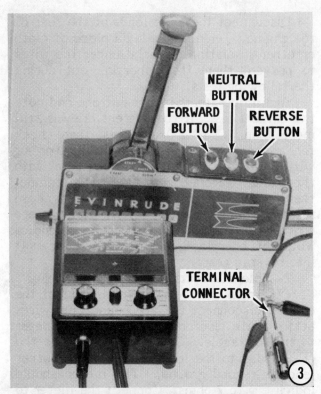

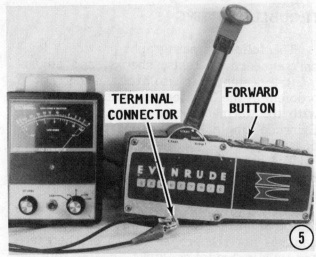

4- Testing Shift Switch — Reverse

Depress the pushbutton for the **RE-VERSE** position. To test the shift switch for reverse, make contact with one probe of a continuity meter (or test light), to the purple or red lead from the plug. Now, with the other probe make contact **ALTERNATE-LY** to the blue and green wires in the plug. The meter should indicate continuity (or the test light come on), during both of these tests when the pushbutton is depressed for the reverse position.

5- Testing Shift Switch — Forward

After the neutral and reverse tests have been completed check for continuity depress the pushbutton for the **FORWARD** position. Leave the red lead connected and check the

blue lead **(REVERSE)** and the green lead **(NEUTRAL)**. Continuity should **NOT** be indicated when the **FORWARD** pushbutton is depressed. If the switch is defective and requires replacement, Page 7-17.

6- Cranking System Inoperative

If the starter fails to crank the engine, check to be sure the throttle lever is in the idle position. If the throttle is advanced more than 1/4 forward, the cutout switch attached to the armature plate will open the circuit to the starter solenoid. On late models, the switch is installed in the shift box. If the cranking system fails to operate the starter properly when the throttle lever is in the **IDLE** position, check the 20-ampere fuse between the ingition switch **BAT** terminal and the ammeter **GEN** terminal.

If the starter operates at full throttle (which it should not do), check for a short between the two white leads in the shift box wiring.

Further problems in the cranking system may indicate more serious problems. See Chapter 6, starter motor sections.

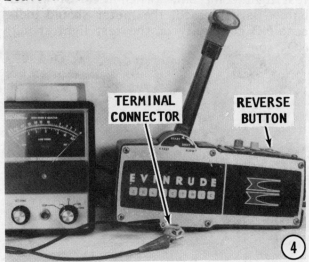

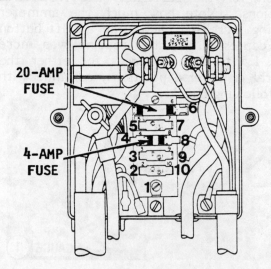

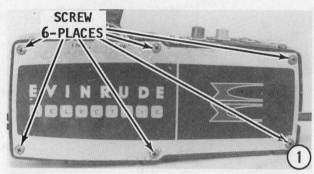

Throttle Cable
GOOD WORDS

The throttle cable and switch box may be replaced without removing the shift box from the boat. If the only service to be performed is replacement of the cable, leave the shift box in place.

1- Remove the Phillips screws on the side plate of the shift box. Remove the front side cover.

2- Notice the screw and retainer at the forward end of the casing guide and just below the throttle lever. Remove the screw and retainer. Pull the throttle cable and casting guide free of the shift box. Take care not loose the trunnion caps, one on the top and another on the bottom.

3- Remove the screws from the end of the casing guide and then pull the throttle cable free. **TAKE CARE** not to lose the screws and sleeve from the end of the guide.

Shift Switch Removal

4- Remove the four Phillips screws from the top of the shift box, and then lift off the shift box cover around the push buttons.

5- Pull upward and remove the three push buttons. New buttons are not supplied with replacement switches. Therefore, **SAVE** the **THREE BUTTONS** for installation with the new switch.

6- Pull the red, green, and blue wires from the bottom of the switch box.

7- Notice the two small Phillips screws on top of the switch box holding the switch to the retainer. Remove these two screws.

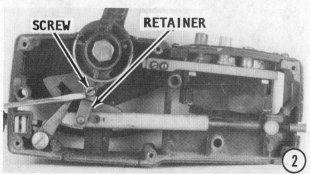

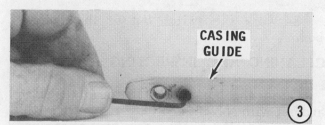

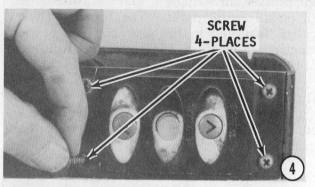

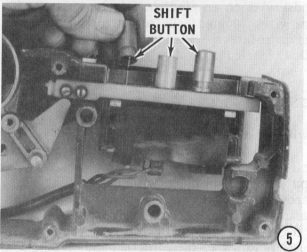

8– Work the switch out of the switch box.

CLEANING AND INSPECTING

Clean the box halves thoroughly inside and out with solvent and blow them dry with compressed air. Apply a thin coat of engine oil on all metal parts. The three-position switch installed in the gear box cannot be repaired. Therefore, if a problem is isolated to the swtich, it must be replaced.

Throttle Cable Lubrication

If the throttle cable is not to be replaced, now is an excellent time to lubricate the inner wire.

9– To lubricate the inner wire, remove the casing guide from the cable at both ends. Attach an electric drill to one end of the wire. Momentarily turn the drill on and off to rotate the wire and at the same time allow lubricant to flow into the cable, as shown.

ASSEMBLING

Switch Installation

1– Position the switch box inside the shift box underneath the retainer and slider and secure it in place with the two Phillips screws. Check to be sure the two terminals on the bottom side of the switch are towards you. This will place the forward button closest to the throttle handle.

2– Install the two small Phillips screws into the top of the shift box. These two screws secure the switch box to the retainer.

3– Connect the wires to the bottom of the switch. Connect the red wire to the **POSITIVE** terminal; the green to the **NEUTRAL** terminal; and the blue wire to the **REVERSE** terminal.

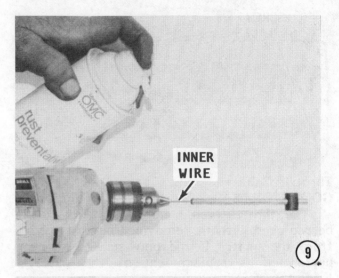

INNER WIRE

SHIFT SWITCH

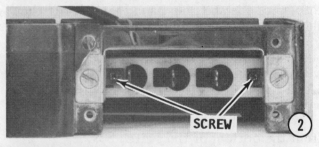

SCREW

SHIFT WIRE

SHIFT SWITCH

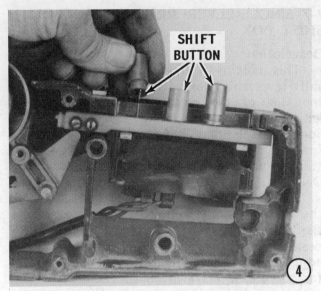

4- Slide the buttons down over the protrusions of the switch and seat them in place.

Adjustment

5- Temporarily install the side plate, and then move the throttle hand forward until the boss mark on the bottom of the throttle hand aligns with the mark on the side of the shift box panel.

6- Remove the panel and depress the three buttons one at-a-time. If it is not possible to depress the buttons, loosen the two screws on the selector bracket. Move the bracket forward or aft until the buttons can be depressed. Tighten the screws to secure the bracket in the proper position.

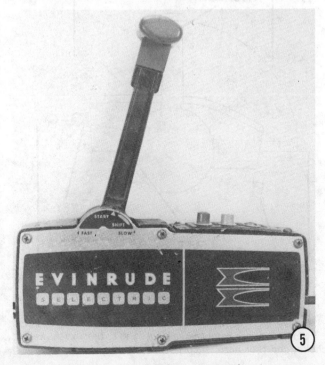

7- Install the shift box cover and secure it in place with the four screws.

CRITICAL WORDS

If the throttle adjustment is not properly performed, the circuit to the starter solenoid will be opened preventing the starter motor from cranking the engine. Adjustment is made by moving the throttle cable adjustment knob in the trunnion on the side of the engine.

Check the end of the cable to determine if the temper has been removed. If the end has a bluish appearance, it has been heated at an earlier date and the temper removed. The temper MUST be removed to permit the holding screw to make a crimp in the wire to hold an adjustment. If the wire has not been tempered, heat the end, but not enough to melt the wire.

8- Feed the inner cable into the casing guide and align it with the hole in the sleeve. Tighten the two Allen screws in the sleeve until one screw makes a crimp in the wire. The screw must be tightened to this degree to prevent the wire from slipping during operation. Bring the other Allen screw up tight against the wire.

9- Install the cable and cable end into the shift box with the trunnion cap on the bottom side. Lower the cable trunnion retainer into the recess and at the same time install the end of the shift cable sleeve over

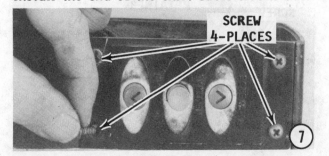

the end of the protrusion. Install the other trunnion cap over the top of the cable retainer.

Slide the retaining clip over the end of the guide. The guide slips over a pin and the retainer has a hole in the end. The retainer fits over the pin and holds the end of the throttle cable onto the pin. Install the side plate with the attaching Phillips screws. Start the engine and run it at 700 rpm.

CAUTION: Water must circulate through the lower unit to the engine any time the engine is run to prevent damage to the water pump in the lower unit. Just five seconds without water will damage the water pump.

Now, adjust the slide yoke to allow the pushbuttons to be depressed at 700 rpm, but not at 750 rpm. If it is not possible to depress the buttons at 700 rpm, remove the side panel and loosen the two screws on the selector bracket. Move the bracket forward or aft until the buttons can be depressed. To adjust the friction knob under the shift box, turn the knob clockwise to increase friction and counterclockwise to decrease friction.

Cable Adjustments

See Chapter 8, Lower Unit, to properly adjust the shift cable and to adjust the throttle cable.

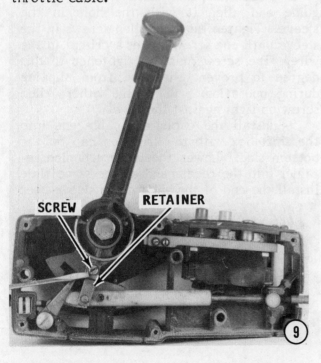

7-7 SINGLE-LEVER REMOTE CONTROL SHIFT BOX

Description

This unit is a single throttle and gear shift lever shift box with a warmup lever on the side of the box. The unit has the ignition key switch, and choke built-in. Some models may have additional built-in features such as a motor overheat horn, a start-in neutral only switch, and an ignition ON light.

The control cables connected between the motor and the remote control lever at the shift box open the throttle after the desired gear is engaged. A throttle friction adjustment is provided to permit the operator to release his grip on the control lever without a change in engine speed.

The warmup lever mounted on the side of the shift box opens the throttle enough to start the engine and to control the fast idle speed for warmup after the engine has started.

On models equipped with the start-in-neutral only switch the starting circuit is completed only when the control lever is in the NEUTRAL position. The switch opens the circuit when the control lever is in either FORWARD or REVERSE position making the ignition key switch inactive.

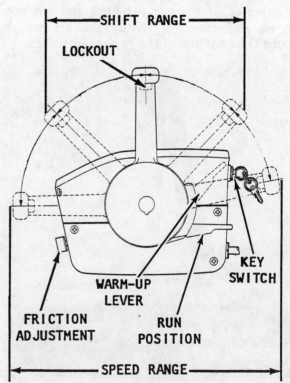

Single lever remote control shift box with key, choke, and "hot horn" incorporated.

To start the engine, the control lever moved to the **NEUTRAL** position and the warmup lever to the **START** position. After engine has started and allowed to warm to normal operating temperature, the warmup lever should be moved to the **RUN** position.

A lockout knob is installed under the control lever handle. This knob **MUST** be depressed to permit the control lever to move to the **FORWARD** or to the **REVERSE** position. The control lever handle must be moved approximately 45° of its total travel for complete shift movement in the lower unit. If the control handle is moved past the 45° point, the throttle is advanced and engine speed increases.

A throttle friction adjustment knob installed on the front of the control box can be adjusted to permit the operator to release his grasp on the handle without the throttle "creeping" and thus changing engine speed. The friction knob should be adjusted only to the point to prevent the throttle from "creeping".

TROUBLESHOOTING

The following paragraphs provide a logical sequence of tests, checks, and adjustments, designed to isolate and correct a problem in the shift box operation.

The procedures and suggestions are keyed by number to matching numbered

illustrations as an aid in performing the work.

The single-lever remote control shift boxes are fairly simple in construction and operation. Seldom do they fail creating problems requiring service in addition to normal lubrication.

Hard Shifting or Difficult Throttle Advance

Checking Throttle Side

Remove the throttle and shift control at the engine. Now, at the shift box, attempt to move the throttle or shift lever. If the lever moves smoothly, without difficulty, the problem is immediately isolated to the engine. The problem may be in the tower shaft between the connector of the throttle and the armature plate. The armature plate may be "frozen", unable to move properly. On the late model units, the "lever advance arm" located on the starboard side of the engine may be "frozen" and require disassembly and lubrication.

If the problem with shifting is at the engine, the first place to check is the area where the shift lever extends through the exhaust housing. The bushing may be worn or corroded. If the bushing requires replacement, the engine powerhead must be removed. Another cause of hard shifting is water entering the lower unit. In this case the lower unit must be disassembled, see Chapter 8.

If hard shifting is still encountered at the shift box when the controls are disconnected from the engine, one of two areas may be causing the problem: the cables may be corroded and require replacment; or, the teeth on the plastic shift lever assembly in the shift box may be worn or broken. This is a common area for problems.

Throttle and shift side lever. This linkage should be checked for corrosion and freedom of movement.

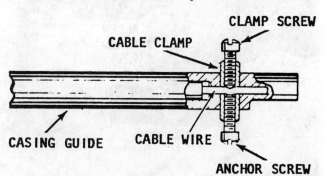

Cut-a-way view of the cable passing through the casing guide. Note the crimp made by the clamp screw. The anchor screw is brought up just tight, as described in the text.

Unable to Obtain Full Shift Movement or Full Throttle

Normally, this type of problem is the result of improper shift box installation. This area includes connection of the shift and throttle cables in the shift box. If the stainless steel inner wire was not heated and the clamp did not hold the inner cable (wire), the wire could slip inside the sleeve and the cable would be shortened. Therefore, if it is not possible to obtain full shift or full throttle, the shift box must be removed, opened, and checked for proper installation work. The inner wire could also slip at the engine end of the control, but problems at that end are very rare. Usually if improper installation work has been done at the engine end, the ability to shift at all is lost, or the throttle cannot be actuated. If the lower unit has previously been removed, the shift rod may not have been adjusted properly.

DISASSEMBLING

Throttle Cable and Shift Cable

Preparation Tasks: Disconnect the leads at the battery terminals and disconnect the spark plug wires at the plugs, as a safety precaution to prevent possible personal injury during the work. A professional mechanic can usually service the cables,

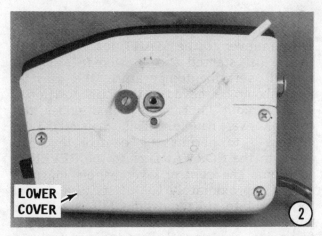

including replacement, without removing the box from the side of the boat. However, the job is made much easier if the box is removed and laid on its side on the boat seat.

1- Use the proper size Allen wrench and remove the screw from the center of the throttle and shift handle. It is not necessary to remove the shift handle, however the work will progress easier if the handle is removed and out of the way.

2- Remove the three Phillips screws on the bottom, the side, or both ends of the panel on the lower section of the shift box. Observe inside the box and notice the attaching screws securing the box to the side of the boat. Remove the screws and lay the shift box on the boat seat.

3- Lift the electrical cable and grommet up out of the slot at the rear of the shift box. Loosen the anchor screws on the end of the casing guides. On the back side of the shift box two holes are provided to permit inserting an Allen wrench to hold the underneath screw while the upper outside Allen screws are removed. After the screws

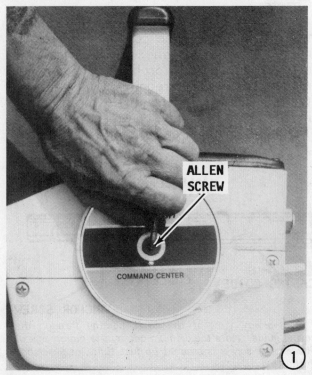

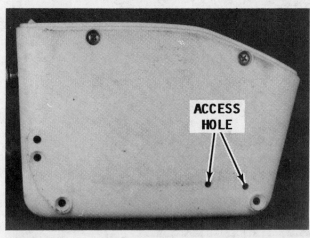

Back side of the shift box showing the two access holes through which an Allen wrench may be passed to loosen the inner cable retaining screws.

have been loosened, pull the shift cable out of the cable casing.

4- To remove the cables at the engine end: Remove the self-locking nuts securing the cables to the engine. Remove the clip on the trunnion. Slip the end of the cable out of the engine retainer, and then remove the cable from the boat. To remove the guide casings at the engine end: Loosen the Allen screws on both sides and pull the casing free.

Throttle Cable Lubrication

If the throttle cable is **NOT** to be replaced, now is an excellent time to lubricate the inner wire.

5- To lubricate the inner wire, remove the casing guide from the cable at both ends. Attach an electric drill to one end of the wire. Momentarily turn the drill on and off to rotate the wire and at the same time allow lubricant to flow into the cable, as shown.

ASSEMBLING

Shift or Throttle Cable Into Shift Box

The following procedures are to be followed to install either the shift cable or the throttle cable.

CRITICAL WORDS

Check the end of the cable to determine if the temper has been removed. If the end has a bluish appearance, it has been heated at an earlier date and the temper removed. The temper **MUST** be removed to permit the holding screw to make a crimp in the wire

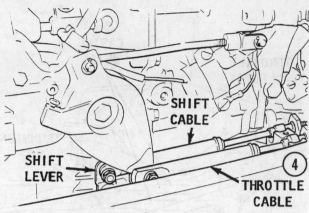

to hold an adjustment. If the wire has not been tempered, heat the end, but not enough to melt the wire.

It is very easy to shear the wire by applying **EXCESSIVE** force when tightening the screw to make the crimp. Therefore, play it cool. Tighten the screw; make one more complete turn to make the crimp; call it good; and then bring the other screw up just tight against the wire.

1- Place the shift control handle in the **NEUTRAL** position and the warmup lever in the **START** position. Coat the cable sleeve with anti-corrosive lubricant, and then slide the casing guide over the cable end. Thread one screw into the control cable clamp. Insert the casing guide into the shift control clevis and align the hole in the casing guide with the hole in the clevis. Lubricate and insert control wire clamp in the clevis hole with the screw toward the back of the control lever.

Align the wire holes in the clamp with wire holes in the casing guide. Feed the control wire through the clamp until the wire is flush to within 1/32" (0.8 mm) recessed with the end of the casing. Secure the

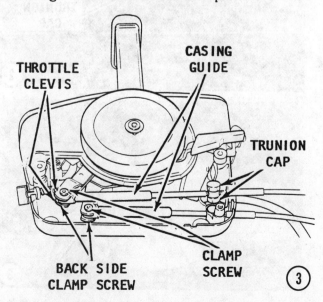

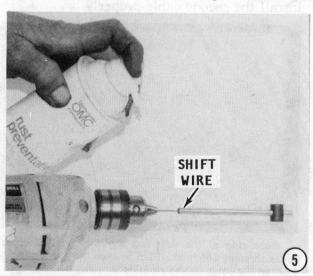

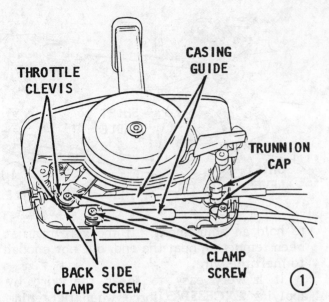

Figure 1

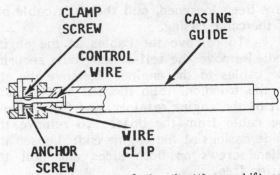

Cut-a-way section of the throttle or shift cable connection at the shift box. Note the crimp in the wire made by the clamp screw. The anchor screw is brought up just tight as explained in the text.

cable wire in the casing guide. The control wire is held in place with two screws -- the clamp screw already in place and the anchor screw. The wire **MUST** be crimped as described in "Critical Words" at the beginning of these Assembling procedures to hold the adjustment. This is accomplished by reaching through the access hole in the rear of the control box with a 3/32" Allen wrench and tightening the screw until it is up just snug against the wire. Now, tighten the screw **ONLY** one complete turn more to make the crimp. Install the second screw, from the front and bring it up just tight against the wire. The wire will now be held securely in the casing guide.

SPECIAL NOTE

The previous procedure is complete to install either the throttle or the shift cable. If both cables are to be replaced, the complete procedure must be followed again to install the second cable properly.

2- Snap the nylon trunnion caps onto the cable trunnion, and then position the trunnion in the remote control. If the cable has a spherical trunnion, use the anchor blocks included with the new cable instead of the nylon trunnion caps furnished with the remote control. The nylon caps may be discarded.

3- Insert the electric cable grommet into the remote control.

4- Mount the shift control box in the boat and secure it in place with the attaching screws. Install the shift control handle.

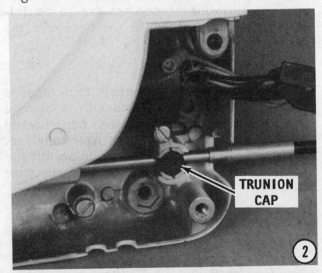

Figure 2

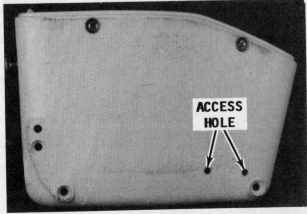

Back side of the shift box showing the two access holes through which an Allen wrench may be passed to tighten the inner cable retaining screws.

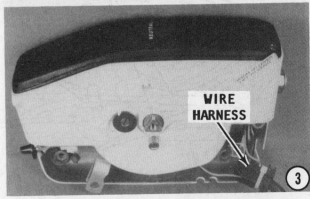

Figure 3

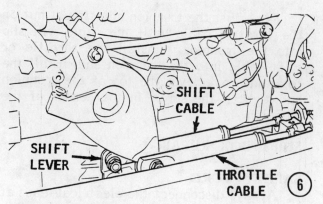

Install the access cover with the throttle cable positioned in the machined recess in the cover and the nylon trunnion caps in place. Clamp the control cables to the boat along the run to the engine.

Shift or Throttle Cable at the Engine

5- Check to be sure the shift control lever is in the **NEUTRAL** position and the warmup lever is in the **RUN** position. Insert the clamp in the casing guide. Start the Allen screws into the clamp. Work the casing guide down over the cable (shift or throttle cable), until the wire protrudes out into the inspection hole of the guide.

CRITICAL WORD

The flat side of the casing guide **MUST** face toward the engine. This flat side is necessary to allow the guide to move as the throttle or shift lever is operated.

Tighten one of the Allen screws until it makes contact with the wire, and then give it **ONLY** one more complete turn to make the crimp in the wire. Tighten the other Allen screw just snug against the wire.

SPECIAL NOTE

The previous procedure is complete to install either the throttle or the shift cable at the engine. If both cables are to be replaced, the complete procedure must be followed again to install the second cable properly.

6- Place the shift or throttle cable onto the shift or throttle lever studs. Secure the cables in place with the washers and locknuts.

7- Move the throttle lever on the engine until the idle stop screw makes contact with the stop. Pull firmly on the throttle casing guide and trunnion nut to remove any backlash in the cable run and at the remote control at the shift box. If this backlash is not removed, the engine may not return to a consistent idle speed.

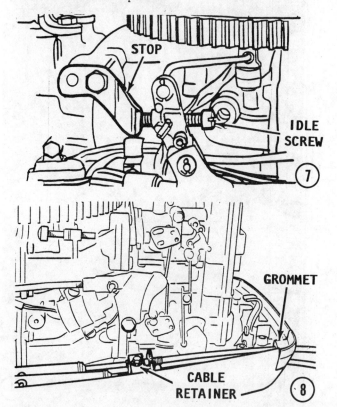

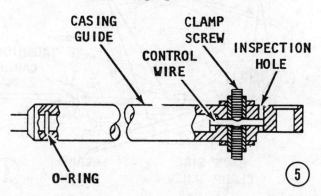

8- Adjust the trunnion adjustment nut on the throttle cable until the cable will slip into the trunnion. Install the throttle or shift cable into the trunnion and install the retaining cover over the top of the trunnion and tighten the screw in the center of the trunnion.

SHIFT BOX REPAIR

SAFETY WORD

Always disconnect the electrical leads at the battery terminals to prevent possible personal injury during the work.

1- Remove the throttle and shift handle by first removing the Allen screw in the center of the handle.

2- Remove the three Phillips screws on the lower side of the shift box, and then remove the cover.

3- Remove the screws from the top of the shift box, and then the arm rest. Remove the two screws from the back side of the shift box securing the upper panel to the outside of the shift box, and then lift off the panel.

4- Remove the screws from the inside of the shift box securing the shift box to the boat. Use an Allen wrench and working from the back side and another Allen wrench from the front side, remove the

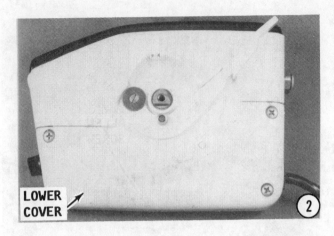

Allen screws in the control wire on the end of the casing guide. Remove these screws from the shift and the throttle wires.

GOOD WORDS

As the panel is lifted, notice how the arm and the mechanism for the throttle are mounted on one side of the panel. Notice the key switch, overheat horn, cam, and start-in-neutral switch installed on the other side of the panel. Also observe the spring and ball bearing under the plate installed under the shift cam.

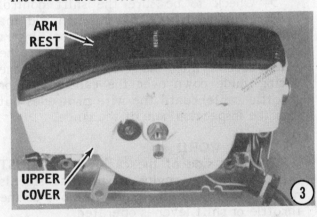

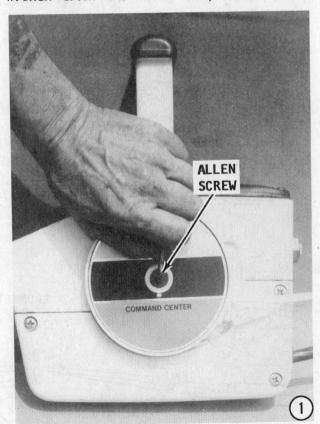

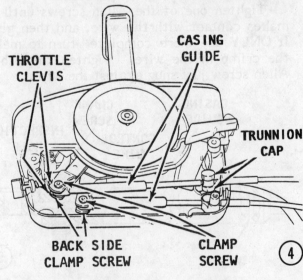

5- Lift the cam assembly from the panel. **TAKE CARE** not to lose the spring and ball bearing installed under the plate.

6- Remove the countersunk screw, flat washer, shift lever, and bushing from the housing. Remove the screw and cover plate containing the spring and ball bearing on the right side of the shift box.

CLEANING AND INSPECTING

Disassemble and clean mechanical parts in solvent, and then blow them dry with compressed air. **NEVER** dip electrical parts in solvent. Check wiring and electrical parts for continuity wtih a test light or an ohmmeter. Faulty electrical parts **MUST** be replaced.

Inspect mechanical parts for wear, cracks, or other damage. Questionable parts should be replaced to ensure satisfactory service.

Worn (left) and new (right) shift levers. Note the worn teeth indicating this lever is no longer fit for service.

Pay special attention to the shift lever teeth. The teeth are made of a hard plastic material. Worn teeth will result in hard shifting.

ASSEMBLING

GOOD WORDS

During the assembling work, take time to coat the friction areas of mechanical moving parts with OMC Multi-purpose grease.

1- Slide the bushing onto the shift lever post in the housing, and then install the lever with the countersunk side of the washer facing **UP**. Tighten the screw to the specifications given in the Appendix. Before installing the shift lever, the detent spring and ball must be removed from the retainer.

2- Lower the shift lever down over the shift cam with the center tooth of the shift lever indexes with the center tooth on the cam.

3- Install the ball and detent spring into the recess, and then install the cover over the spring.

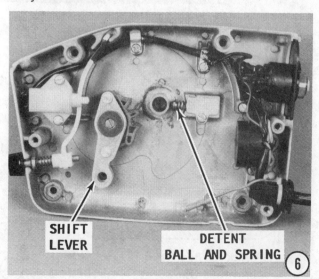

4- Place the top cover on the inner shift box with the lever cam follower seating in the shift lever cam channel. Tighten the screws on the back side of the shift box.

5- Install the arm rest to the top of the shift box and secure it in place with the retaining screws.

6- Slide the throttle and shift cables into the shift box and attach the casing guides in the shift and throttle levers. Tighten the Allen screws from the rear and side of the box. If difficulty is encountered during installation of the cables, see the more detailed instructions under Shift Cable Installation earlier in this section.

Install the shift box to the side of the boat. Check to be sure the cables are in their trunnions and the wiring harness is in the recess. Install the lower cover, and then install the shift handle and secure it with the Allen screw.

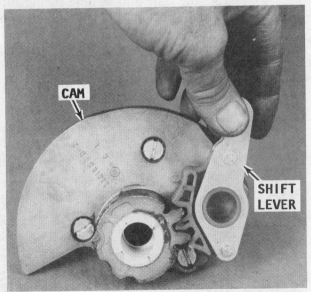

Shift lever and shift cam illustrating how the teeth of each indexes with the teeth of the other.

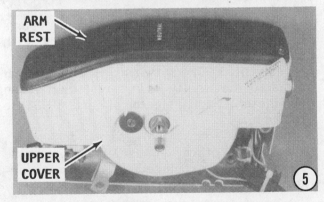

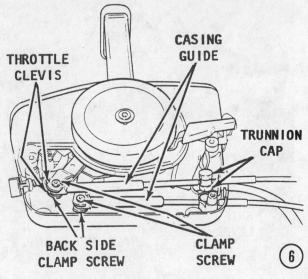

7-8 CABLE END FITTING INSTALLATION AT THE ENGINE END FOR ALL SHIFT BOXES EXCEPT SINGLE LEVER REMOTE CONTROL

GOOD WORDS

The procedures outlined in this section apply to the installation of the cable end fitting at the engine for all shift box installations.

MORE GOOD WORDS

The anchor on the engine, to which the trunnion is attached, has a "P" and an "S" stamped on the inside diameter or inside edge of the trunnion retainers. These letters identified **PORT** and **STARBOARD**.

INSTALLATION

Shift Cable End

1– Move the control lever at the shift control box to the **NEUTRAL** position. Slide the gear shift fitting onto the control wire. Check to be sure the inner wire passes completely through the small holes in the cable clamp. Clamp the anchor screws to prevent twisting the cable. The clamp and the anchor screws **MUST** be parallel to the trunnion on the gear shift cable.

2– Notice the flat and rounded areas of the casing guide. The flat edge **MUST** face **TOWARD** the engine. In this position, there is a flat area for the lever to ride during the shifting action. After the cable is in place in the casing guide, tighten the top screw until a definite crimp is made in the cable. If the screw is not tightened enough, the inner wire will slip during operation and the adjustment will be lost.

CRITICAL WORDS

Check the end of the cable to determine

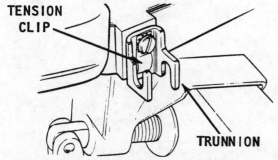

Trunion cap and tension clip. The tension clip must be installed to hold the proper adjustment.

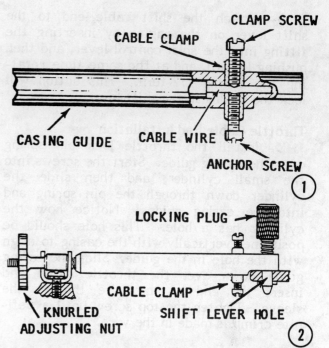

if the temper has been removed. If the end has a bluish appearance, it has been heated at an earlier date and the temper removed. The temper **MUST** be removed to permit the holding screw to make a crimp in the wire to hold an adjustment. If the wire has not been tempered, heat the end, but not enough to melt the wire. Bring the second screw up tight against the wire.

3– Insert the shift cable control vertically into the trunnion bracket and turn the cable to a horizontal position, as indicated by the arrows in the accompanying illustration.

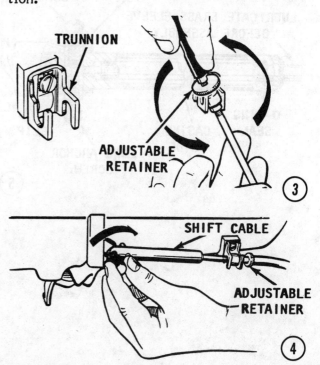

4- Attach the shift cable end to the shift lever on the engine by inserting the fitting into the shift control lever, and then pushing inward, and at the same time rotating the fitting 1/2-turn. This action will lock the fitting in the shift lever.

Throttle Cable End Installation

5- Install the throttle lock pin spring over the casing guide. Start the screws into the small cylinder, and then slide the cylinder down through the pin spring and into the casing guide. Notice how the cylinder has a hole. This hole should be positioned vertically with the casing to align with the hole in the guide. Slide the casing guide down over the throttle cable and insert the end of the wire through the sleeve. Tighten the top screw until a definite crimp is made in the wire.

CRITICAL WORDS

Check the end of the cable to determine if the temper has been removed. If the end has a bluish appearance, it has been heated at an earlier date and the temper removed. The temper **MUST** be removed to permit the holding screw to make a crimp in the wire to hold an adjustment. If the screw is not tightened to this degree, the wire will slip during operation and the adjustment will be

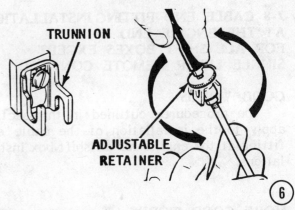

lost. If the wire has not been tempered, heat the end, but not enough to melt the wire. Bring the bottom screw up tight against the wire.

6- Install the trunnion retainers to the engine, if necessary. Check to be sure the retainer with **"P"** stamped on the inside is installed on the **PORT** side of the engine and the retainer with the **"S"** installed on the **STARBOARD** side. Connect the trunnion cap to the trunnion retainer. This is accomplished by holding the trunnion in a vertical position; inserting it into the retainer; and then turning it to the horizontal position, as shown.

7- Slide the guide over the pin onto the engine, and then snap the retainer clip over the end of the guide to lock it in place.

Cable Adjustments

See Chapter 8, Lower Unit, to properly adjust the shift cable and to adjust the throttle cable.

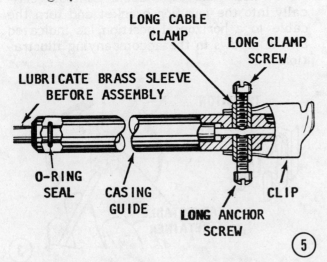

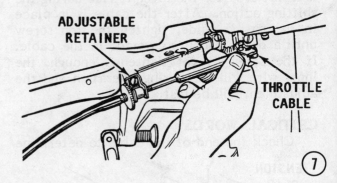

8
LOWER UNIT

8-1 DESCRIPTION

The lower unit is considered as that part of the outboard below the exhaust housing. The unit contains the propeller shaft, the driven and pinion gears, the driveshaft from the powerhead and the water pump. On models equipped with shifting capabilities, the forward and reverse gears, together with the clutch, shift assembly, and related linkage, are all housed within the lower unit.

The lower unit is removed by one of six methods depending on the model year and the engine horsepower.

1- The lower unit does not have shifting capabilities, therefore, removal of the lower unit is not an involved procedure.

2- The lower unit has shifting capabilities. The upper end of the shift rod indexes into the shift handle gear and the lower end of the rod indexes into the gear in the lower unit.

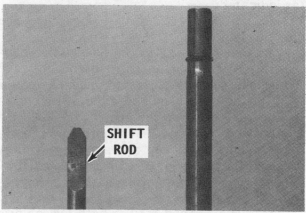

Shift rod (left) and driveshaft (right) after the lower unit has been separated from the exhaust housing. The flat on the shift rod must face in the direction shown toward the starboard side when the unit is in forward gear.

3- The lower unit is lowered a couple inches and the shift connector removed.

4- A window in the exhaust housing is opened and the shift connector disconnected.

5- A window in the lower unit is opened to disconnect the shift rod.

6- Green and blue shift wires are disconnected on the port side of the engine.

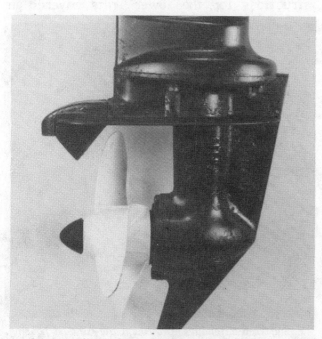

Lower unit used on the 2 hp to 4 hp engines. The shear pin is installed after the propeller is in place.

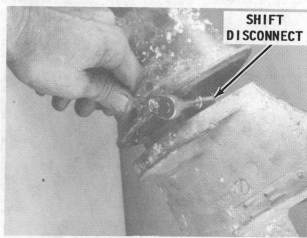

Disconnecting the shift connector after the lower unit has been separated slightly from the exhaust housing.

7- The shift rod is disconnected at the linkage under and to the rear of the bottom carburetor.

The engine and model year is given in each section heading. Therefore, the Table of Contents may be used to determine which set of procedures to follow for the engine being serviced.

CHAPTER COVERAGE

Nine different lower units are covered in this chapter with separate sections for each, as indicated:

Section 8-4 -- lower unit does not have shift capabilities.

Section 8-5 -- lower unit has shift capability but there is no shift disconnect, the upper end of the shift rod indexes into a part of the shift handle and the other end is splined to index into the gear.

Sections 8-7 and 8-8 -- both lower units have a split lower cap and are very similar, except the type of bearings used differ. Therefore, separate procedures are required.

Sections 8-9 and 8-10 -- both lower units have propeller exhaust and use a sliding clutch dog. A detent ball and spring are used to hold the unit in the neutral position. The shift mechanism on the 9.9 hp and 15 hp models is forward of the forward gear. On the 35 hp model, the shift mechanism is just aft of the reverse gear.

Section 8-11 -- lower unit has propeller exhaust with electric shift. Two solenoids

The two wires on the port side of the engine which must be disconnected before the lower unit is removed.

are used to affect the shift. One solenoid is used for neutral, and both solenoids are activated for the shift into reverse gear.

Section 8-12 -- lower unit has propeller exhaust and mechanical shift with a hydraulic assist pump.

Section 8-13 -- lower unit has propeller exhaust and mechanical shift with a sliding clutch dog. Movement of the clutch dog is mechanical utilizing a shift cradle and shift lever.

Each section is complete with detailed procedures. No troubleshooting directions are given for Sections 8-4 and 8-5. Section 8-6 contains detailed troubleshooting instructions for the lower units covered in

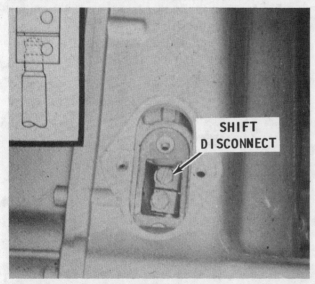

Window removed from a lower unit to gain access to the shift connector, as explained in the text. The detailed drawing, upper left, illustrates the relationship of the bolt to the shift rod.

Shift rod extending up through the exhaust housing. The disconnection is made under the carburetor.

Sections 8-7 and 8-8. The lower units covered in Sections 8-9, 8-10, 8-12, and 8-13, each contain their own troubleshooting procedures.

Check the Table of Contents and follow the procedures in the given section for the unit being serviced.

Water Pump

Water pump service work is by far the most common reason for removal of the lower unit. Each lower unit service section contains complete detailed procedures to rebuild the water pump. The instructions given to prepare for the water pump work must be performed as listed. However, once the pump is ready for installation, if no other work is to be performed on the lower unit, the reader may jump to the pump assembling procedures and proceed with installation of the water pump.

Each section is presented with complete detailed instructions for removal, disassembly, cleaning and inspecting, assembling, adjusting, and installation of only one type unit.

ILLUSTRATIONS

Because this chapter covers such a wide range of models over an extended period of time, the illustrations included with the

Typical water pump installation on the 4.5 hp and 7.5 hp engine.

procedural steps are those of the most popular lower units. In some cases, the unit being serviced may not appear to be identical with the unit illustrated. However, the step-by-step work sequence will be valid in all cases. If there is a special procedure for a unique lower unit, the differences will be clearly indicated in the step.

SPECIAL WORDS

All threaded parts are right-hand unless otherwise indicated.

If there is any water in the lower unit or metal particles are discovered in the gear lubricant, the lower unit should be completely disassembled, cleaned, and inspected.

Actually, problems in the lower unit can be classified into three broad areas:

1- Lack of proper lubrication in the lower unit. Most often this is caused by failure of the operator to check the gear oil level frequently and to add lubricant when required.

2- A faulty seal allowing water to enter the lower unit. Water allowed to remain in the lower unit over a period of non-use time will separate from the oil and can be destructive.

3- Excessive clutch dog and clutch ear wear on the forward and reverse gears. This condition is caused by excessive wear in the bellcrank under the powerhead. A worn bellcrank will result in sloppy shifting of the lower unit and cause the clutch components to wear and develop shifting problems. Improper shifting techniques at the shift box will also result in excessive wear to the clutch dog and clutch ears of the forward and reverse gears.

Time will also take its toll. Continued service over a long period of time will cause parts to wear and require replacement.

8-2 PROPELLER SERVICE

PROPELLER WITH SHEAR PIN REMOVAL

If the unit being serviced has the shear pin located between the propeller nut and the propeller, the propeller nut should be removed and the shear pin checked.

To remove the propeller, first pull the cotter key, and then remove the propeller nut, shear pin, and washer. Because the shear pin is not a tight fit, the propeller is

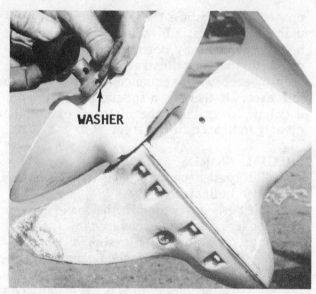

Arrangement of the shear pin, washer, and propeller nut on a typical lower unit. In this case, the shear pin is installed behind the propeller nut.

able to move on the pin and cause burrs on the hole. The propeller may be difficult to remove because of these burrs. To overcome this problem, the propeller hub has two grooves running the full length of the hub. Hold the shaft from turning, and then rotate the propeller 1/4 turn to position the grooves over the drive pin holes. The propeller can then be pulled straight off the shaft. After the propeller has been removed, file the drive pin holes on both sides of the shaft to remove the burrs.

If the propeller is the type with the shear pin installed next to the gearcase head, first remove the cotter key, then the propeller nut. Next, slide the propeller free of the shaft.

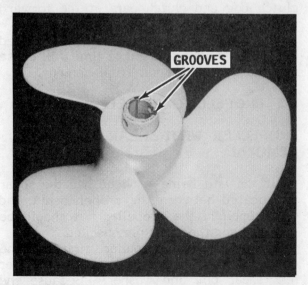

Propeller with the two grooves through the center to assist in removal from the shaft.

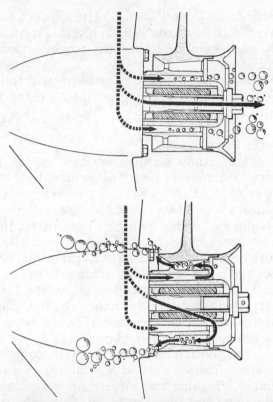

Cross-section drawing of the lower unit showing route of the exhaust gases with the unit in forward gear (top), and in reverse gear (bottom).

EXHAUST PROPELLER

Propellers with the exhaust passing through the hub **MUST** be removed more frequently than the standard propeller. Removal after each weekend use or outing is not considered excessive. These propellers do not have a shear pin. The shaft and propeller have splines which **MUST** be coated with an anti-corrosion lubricant prior to installation as an aid to removal the next time the propeller is pulled. Even with the lubricant applied to the shaft splines, the propeller may be difficult to remove.

Propeller with exhaust hub. The defuser ring is clearly visible.

The propeller with the exhaust hub is more expensive than the standard propeller and therefore, the cost of rebuilding the unit, if the hub is damaged, is justified.

A replaceable diffuser ring on the backside of the propeller disperses the exhaust away from the propeller blades as the boat moves through the water. If the ring becomes broken or damaged "ventilation" would be created pulling the exhaust gases back into the negative pressure area behind the propeller. This condition would create considerable air bubbles and reduce the effectiveness of the propeller.

PROPELLER WITH EXHAUST — REMOVAL

First, disconnect the high tension leads to the spark plugs to prevent accidental engine start. Next, pull the cotter pin from the propeller nut. Wedge a piece of wood between one of the propeller blades and the cavitation plate to prevent the propeller from rotating. Back off the castellated propeller nut and remove the splined washer. Pull the propeller straight off the shaft. It may be necessary to carefully tap on the front side of the propeller with a soft headed mallet to jar it loose. Remove the thrust washer from the propeller shaft.

"Frozen" Propeller

If the propeller appears to be "frozen" to the shaft, see Section 8-14 for special removal instructions. The thrust washer does not have to be removed unless it appears damaged.

PROPELLER INSTALLATION WITH SHEAR PIN

A FEW GOOD WORDS

The propeller washer, if used, and shear pin, play an extremely important role. When shifting gears during normal operation, or if the propeller should hit an underwater obstacle, the propeller is subjected to considerable shock. A washer is installed between the propeller and drive pin. This washer **MUST** always be in place for proper operation. If the hub should slip, the propeller will move back towards the propeller nut and lock against the drive pin. The washer is designed to stop propeller movement so the drive pin can be easily removed for service. Now, on with the installation.

Install the propeller. Coat the propeller shaft with an anti-corrosion grease. Install the propeller with the drive pin holes aligned. Install the washer and drive pin. Slide the propeller cap into place and secure it with the cotter pin.

If the unit being serviced uses the shear pin between the propeller and the bearing carrier, proceed as follows: Install the shear pin and then coat the propeller shaft with anti-corrosion grease; install the propeller; propeller nut; and then the cotter pin.

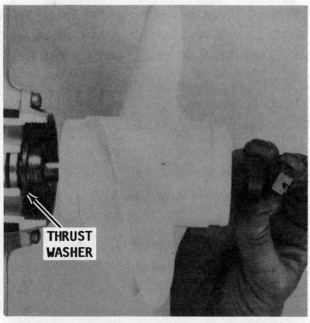

Propeller exhaust arrangement showing the thrust washer, propeller, splined washer, and propeller nut.

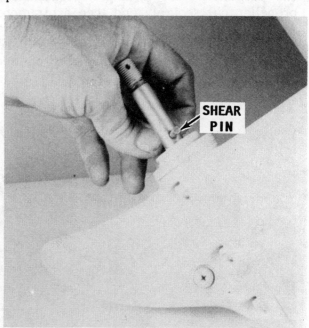

Removing the propeller shear pin from the propeller shaft. In this case, the shear pin is installed between the propeller and the gearcase head.

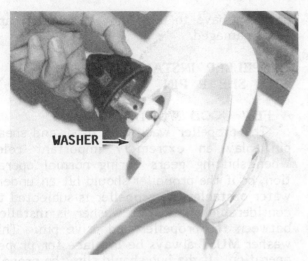

Arrangement of the shear pin, washer, and propeller nut on a typical lower unit. In this case, the shear pin is installed behind the propeller nut.

EXHAUST PROPELLER INSTALLATION

Slide the thrust washer onto the propeller shaft. Coat the propeller shaft with Perfect Seal No. 4, Triple Guard Grease, or similar good grade of lubricant to prevent the propeller from becoming "frozen" to the shaft. Slide the propeller onto the shaft with the splines in the propeller indexing with the splines on the shaft. Slide the splined washer onto the shaft. Thread the castellated nut onto the shaft. Jamb a piece of board between one of the propeller blades and the cavitation plate to prevent the propeller from turning. Tighten the propeller nut securely and then a bit more to align the hole through the nut with the hole through the propeller shaft. Install the cotter pin through the nut and propeller shaft.

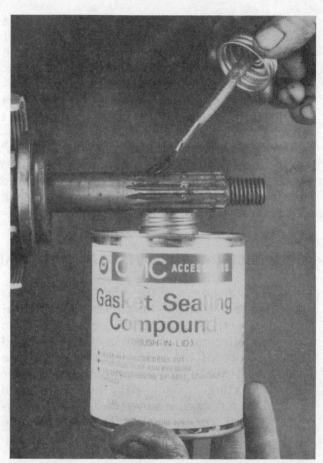

Applying gasket sealer to the propeller shaft splines to prevent the propeller from becoming "frozen" to the shaft.

Installing the thrust washer onto the propeller shaft of a propeller exhaust unit.

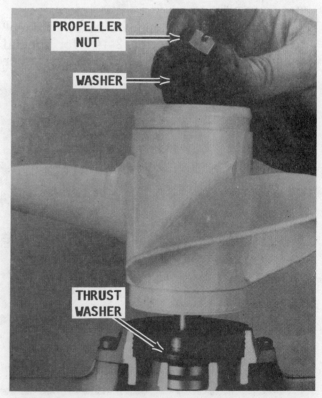

Installation of a propeller exhaust propeller with principle parts identified.

Tightening the nut on a propeller exhaust unit. The cotter pin is installed after the nut is secure.

8-3 LOWER UNIT LUBRICATION

DRAINING LOWER UNIT

Position a suitable container under the lower unit, and then remove the **FILL** screw and the **VENT** screw.

CRITICAL WORD

On many lower units, the Phillips screw securing the shift fork in place is located very close to the vent screw. On some units

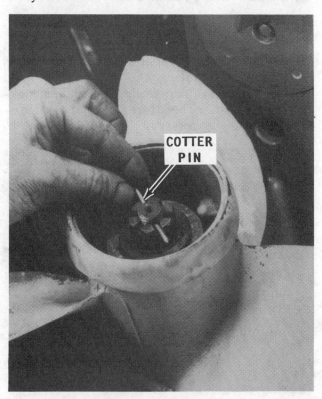

Installing the cotter pin through the castellated nut on a propeller exhaust unit.

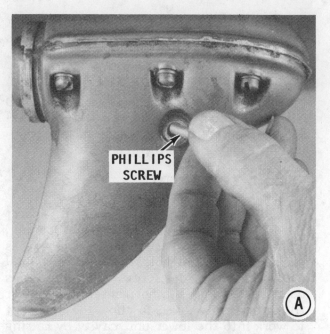

the Phillips screw is located on the other side. If the wrong screw is removed, **BAD NEWS, VERY BAD NEWS.** The lower unit will have to be disassembled in order to return the shift fork to its proper location, illustration **A**.

Allow the gear lubricant to drain into the container. As the lubricant drains, catch some with your fingers, from time-to-time, and rub it between your thumb and finger to determine if any metal particles are present. If metal is detected in the lubricant, the unit must be completely disassembled, inspected, and the damaged parts replaced, illustration **B**.

Check the color of the lubricant as it drains. A whitish or creamy color indicates the presence of water in the lubricant. Check the drain pan for signs of water separation from the lubricant. The presence of any water in the gear lubricant is **BAD NEWS**. The unit must be completely disassembled, inspected, the cause of the problem determined, and then corrected.

FILLING LOWER UNIT

Fill the lower unit with lubricant. Insert the lubricant tube into the bottom opening, and then fill the unit until lubricant is visible at the vent hole. The 2 hp model does not have a vent screw. Therefore, this unit must be laid in a horizontal position for filling and time taken to allow the lubricant to work into the lower unit cavity by raising the skeg slightly from time-to-time. Install the vent plug. Remove the gear lubricant tube and install the drain/fill plug.

After the lower plug has been installed, remove the vent plug again and using a squirt-type oil can, add lubricant through this vent hole. A squirt-type oil can must be used to allow the trapped air in the lower unit to escape at the same time the final lubricant is added. Once the unit is completely full, install and tighten the vent plug.

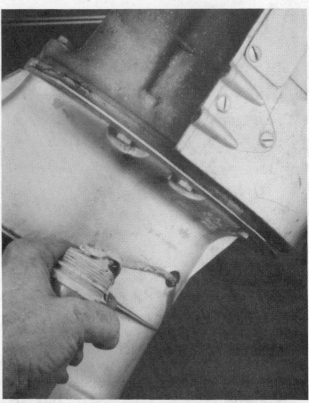

"Topping off" the lower unit using a squirt-type oil can through the vent hole, as described in the text.

8-4 NO SHIFT LOWER UNIT SERVICE ALL 1.25 HP TO 4.0 HP EXCEPT 4 DELUX (4 Delux — See Section 8-5)

Description

This is a very simple direct drive unit without any shift capabilities. Reverse is obtained by rotating the engine 180° and holding that position while the boat is moved sternward. Therefore, no shift rod disconnects are necessary.

Filling a lower unit with OMC Gearcase Lubricant. Notice the vent plug has been removed to allow air to escape, as the unit fills with lubricant.

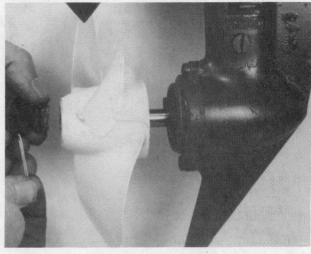

Installing a propeller onto a 2 hp to 4 hp unit. The propeller, propeller nut, and cotter pin, are shown.

TROUBLESHOOTING

The first item to check whenever loss of boat movement is encountered is the shear pin. The next area to check is the rubber hub in the propeller, if one is installed. A worn hub will give an indication the unit is not in gear.

The splines in the crankshaft or on the driveshaft may be damaged or worn and thus prevent rotation from the crankshaft to reach the propeller shaft. If the splines in the crankshaft are destroyed, the crankshaft will have to be replaced. See Chapter 3, Powerhead. If the splines on the driveshaft have been destroyed, the driveshaft must be replaced. Procedure to replace the driveshaft are included in each section of this chapter.

Frozen Powerhead

This condition is suggested when the operator unsuccessfully attempts to crank the engine with a hand starter. The flywheel will not rotate. Do not assume the engine is "frozen" until the lower unit has been removed and thoroughly checked. If the lower unit is "locked" (the driveshaft or propeller shaft will not rotate), the powerhead will have the indication of being "frozen" (failure to rotate the flywheel).

The first step to perform under these conditions is to "pull" the lower unit, and then again attempt to crank the engine. If the attempt is successful with the lower unit disconnected, the problem is in the

lower unit. If the attempt to crank the engine is still unsuccessful, the problem is in the powerhead.

Propeller Removal

Remove the propeller according to the procedures outlined in Section 8-2.

Draining the Lower Unit

Drain the lower unit according to the procedures outlined in Section 8-3.

GOOD WORDS

If water is discovered in the lower unit and the propeller shaft seal is damaged and requires replacement, the lower unit does **NOT** have to be removed in order to accomplish the work.

The seal may be replaced by first removing the two screws securing the cap in place and then tapping on the cap with a soft-headed mallet to jar it loose. The cap is then removed, the seal removed and replaced, and the cap installed and secured.

LOWER UNIT REMOVAL

ADVICE

If the only work to be performed is service of the water pump, be extremely **CAREFUL** to prevent the driveshaft from being pulled up and free of the pinion gear in the lower unit. **NEVER** carry the lower unit by the driveshaft. If the shaft should be released from the pinion, the lower unit **MUST** be disassembled to align the pinion gear and driveshaft, then the driveshaft installed.

1- Disconnect the spark plug wire from the plug. Remove the retaining bolts securing the lower unit to the exhaust housing. **CAREFULLY** pull directly downward, to prevent damage to the water tube, and remove the lower unit.

WATER PUMP REMOVAL

2- Remove the screws securing the water pump to the lower unit housing. It is very possible corrosion will cause the screw heads to break-off when an attempt to remove them is made. If this should happen, use a chisel and break away the water pump housing from the lower unit. **EXERCISE CARE** not to damage the lower unit housing.

3- After the screws have been removed, slide the water pump and impeller upward and free of the driveshaft. Remove the Woodruff key, and then the lower water pump plate,

GOOD WORDS

The 2.5hp, 4hp, Ultra, Junior, and Excel units since 1984 have a water intake screen beneath the lower water pump plate. Remove the screen and clean away any debris which might clog the water passageway and cause an overheating condition.

NOTE

If the only work to be performed is service of the water pump, proceed directly to Water Pump Installation, Step 5 on Page 8-14.

LOWER UNIT DISASSEMBLING

4- Remove the gearcase head and the two screws. Pull on the propeller shaft or tap on the gearcase head to separate the gearcase head from the lower unit housing.

On all units except the 2.5hp, 4hp, Ultra, Junior, and Excel prior to 1984: The driven

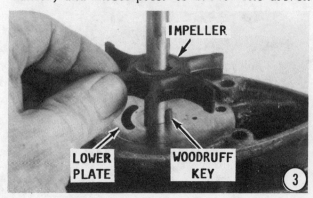

IMPELLER

LOWER PLATE   WOODRUFF KEY

③

④

gear is pressed onto the propeller shaft. Therefore, the propeller shaft and gear are considered as a complete assembly. If either is damaged and requires replacement, the two are purchased as an assembly.

On the 2.5hp, 4hp, Ultra, Junior, and Excel since 1984: Hold the forward gear still while rotating the propeller shaft **COUNTER-CLOCKWISE** and pull the gear and shaft apart. Remove the spring.

5- Pull upward on the driveshaft, and at the same time, reach inside the lower unit and remove the pinion gear.

Models 2.5hp, 4hp, Ultra, Junior, and Excel since 1984 have a thrust washer and bearing set infront of the forward gear and another thrust washer and bearing set behind the pinion gear. remove and identify both sets to make sure they are installed in their original locations.

SPECIAL WORDS

On the Weedless type lower unit, a thrust bearing is installed under the pinion

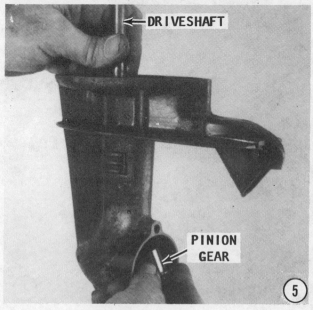

←DRIVESHAFT

PINION GEAR

⑤

gear. This thrust bearing can only be removed by tapping it out in the following manner: Turn the lower unit so the propeller shaft opening is facing downward. Now, gently rap the unit on a work bench or block of wood. The thrust bearing and pinion gear will be dislodged and fall free.

6- If the seal/s at the top of the lower unit housing under the water pump is to be replaced, remove the seal using any type seal remover. To remove the seal/s in the gearcase head, work the seal free by using a punch and mallet from the back side. Remove the O-ring.

CLEANING AND INSPECTING

Clean all water pump parts with solvent, and then dry them with compressed air. Inspect the water pump cover and base for cracks and distortion, possibly caused from overheating. Inspect the face plate and water pump insert for grooves and/or rough surfaces. If possible, **ALWAYS** install a complete new water pump while the lower unit is disassembled. A new impeller will ensure extended satisfactory service and give "peace of mind" to the owner. If the

old impeller must be returned to service, **NEVER** install it in reverse to the original direction of rotation. Installation in reverse will cause premature impeller failure.

Inspect the impeller side seal surfaces and the ends of the impeller blades for cracks, tears, and wear. Check for a glazed or melted appearance, caused from operating without sufficient water. If any question exists, and as previously stated, install a new impeller if at all possible.

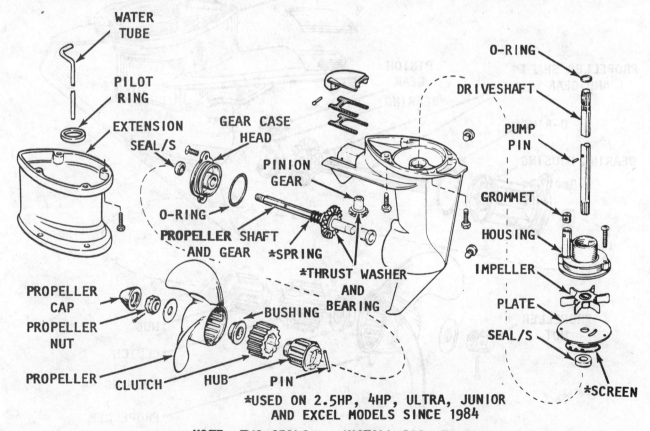

*USED ON 2.5HP, 4HP, ULTRA, JUNIOR
AND EXCEL MODELS SINCE 1984

NOTE: TWO SEALS -- INSTALL BACK-TO-BACK

Exploded drawing of a non-shifting lower unit, with principle parts identified.

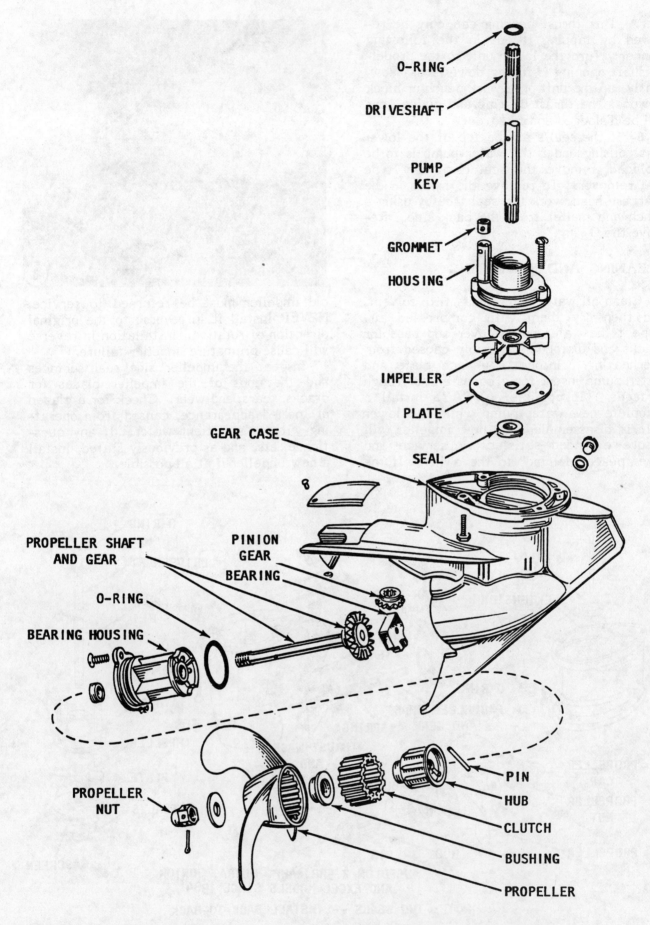

O-RING

DRIVESHAFT

PUMP KEY

GROMMET

HOUSING

IMPELLER

PLATE

SEAL

GEAR CASE

PINION GEAR

BEARING

PROPELLER SHAFT AND GEAR

O-RING

BEARING HOUSING

PIN

HUB

CLUTCH

BUSHING

PROPELLER

PROPELLER NUT

Exploded drawing of the Weedless lower unit gearcase, with major parts identified.

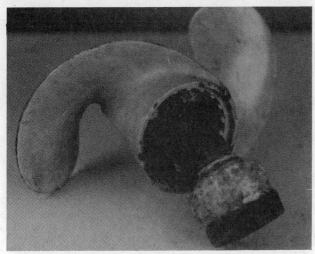

Propeller used on the 2 hp to 4 hp unit with remove-able hub. This hub is badly corroded. The hub and propeller should be replaced.

Clean all parts with solvent and dry them with compressed air. **DISCARD** all O-rings and gaskets. Inspect and replace the driveshaft if the splines are worn. Inspect the gearcase and exhaust housing for damage to the machined surfaces. Remove any nicks and refurbish the surfaces on a surface plate. Start with a No. 120 Emery paper and finish with No. 180.

Check the water intake screen and passages. Inspect the drive gear, pinion gear, and thrust washers. Replace these items if they appear worn.

LOWER UNIT ASSEMBLING

SPECIAL WORDS

The 2.5hp, 4hp, Ultra, Junior, and Excel since 1984, use a double seal arrangement on the propeller shaft and driveshaft. The two seals are installed **BACK-TO-BACK** (flat side to flat side). The outer seal will prevent the water from entering the lower unit and the inner seal will prevent lubricant in the lower unit from escaping.

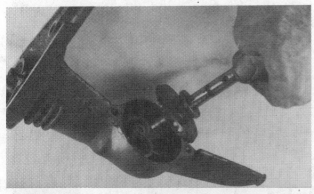

Checking the drive gear and propeller shaft on a 2 hp lower unit.

1- Tap a **NEW** seal/s into place on top of the lower unit housing.

2- Tap a **NEW** seal/s into place in the gearcase head. Install a **NEW** O-ring into the groove in the gearcase.

3- On the 2.5hp, 4hp, Ultra, Junior, and Excel units since 1984: Place the thrust bearing followed by the thrust washer down over the flat surface of the pinion gear.

All Models: Install the pinion gear into the recess in the lower unit housing. If the unit being serviced is the "Weedless" type gearcase, install the thrust bearing with the bosses on the bearing indexed between the two bosses in the gearcase. Hold the pinion

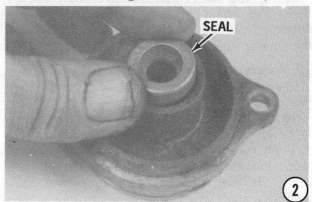

gear in place with one hand and with the other hand install the driveshaft down into the lower unit. Continue to hold the pinion gear, and at the same time, rotate the driveshaft slightly after it makes contact with the pinion gear to allow the splines on the shaft to index with the splines in the gear.

SPECIAL WORDS

After the driveshaft is installed, care must be exercised **NOT** to allow the drive- shaft to slip out of position in the pinion gear. This is especially important during water pump installation work. If the drive- shaft should come free, the lower unit must be disassembled in order to install the drive- shaft back into the pinion gear.

MORE SPECIAL WORDS

If servicing a 2.5hp, 4hp, Ultra, Junior, or Excel manufactured since 1984: Slide the spring over the forward gear shank with the spring tang facing the gear. Push the propeller shaft into the gear shank while rotating the shaft **COUNTERCLOCKWISE**, until the shaft seats. Slide the thrust bearing followed by the thrust washer over the forward end of the shaft.

All Models

4- Coat the propeller shaft and the gearcase O-ring with oil as an aid to instal- lation. Install the gearcase head over the propeller shaft. Slide the propeller shaft through the lower unit with the driven gear teeth indexed with the teeth of the pinion gear. It may be necessary to rotate the propeller shaft slightly in order to index the driven and pinion gear teeth. The teeth **MUST** engage fully and properly or the gear- case head will be damaged when the attach- ing screws are installed. Coat the screws securing the head to the lower unit with sealer, and then install the screws. **CARE-**

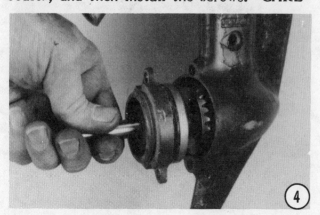

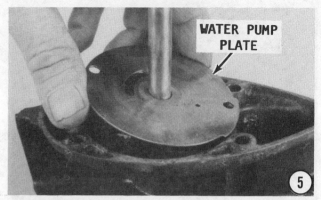

FULLY tap on the gearcase head with a soft-headed mallet and tighten the screws **EVENLY** and **ALTERNATELY.**

CRITICAL WORDS

If the screws are not tightened evenly, or the driven gear and pinion gear teeth are not fully and properly engaged, the gearcase head will be thrown out of line just a whisker, and the ears through which the bolts pass may snap off. **BAD NEWS!** A new gearcase head would have to be pur- chased.

If the unit being serviced is the "Weed- less" type, two sets of matching marks on the gearcase head and the lower unit **MUST** be aligned when the head is installed.

If the unit being serviced is the "High- Thrust" type, the gearcase has a hole which **MUST** face upward when the head is install- ed.

WATER PUMP INSTALLATION

5- Lay down a bead of sealer No. 1000 onto the lower unit surface. If servicing a 2.5hp, 4hp, Ultra, Junior, or Excel unit manufactured since 1984: Install the water intake screen over the driveshaft with the three prongs facing **UPWARD.**

All models: Slide the water pump plate down the driveshaft and onto the lower unit surface, or screen if used.

6- Insert the Woodruff key into the driveshaft groove.

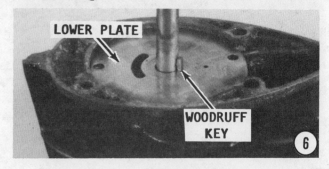

7- Slide the water pump impeller down the driveshaft and into place on top of the water pump base plate with the pump pin indexed in the impeller. Lubricate the inside surface of the water pump with light-weight oil.

8- Lower the water pump housing down the driveshaft and over the impeller. Rotate the driveshaft **CLOCKWISE** as the water pump housing is lowered to allow the impeller blades to assume their natural and proper position inside the housing. Continue to rotate the driveshaft and work the water pump housing downward until it is seated on the lower unit upper housing surface.

9- Rotate the driveshaft **CLOCKWISE** while the screws are tightened to prevent damaging the impeller vanes. If the impeller is not rotated, the housing could damage or cut the end of the vanes as the screws are brought up tight. The rotation allows them to spring back into a natural position. Place a **NEW** grommet into the water pump housing for the water pickup tube. If a new water pump was installed, this seal will already be in place. Install a **NEW** O-ring on the top of the driveshaft.

LOWER UNIT INSTALLATION

10- Clean and shine the water pump tube with lightweight sandpaper, and then

coat it with oil as an aid to installation. Apply oil to the grommet in the water pump housing as a further aid to installation. This tube is very small in size and will bend easily during installation if it has even a little difficulty passing through the rubber grommet in the water pump housing.

Bring the lower unit together to mate with the exhaust housing. Guide the water tube into the water pump housing grommet, and at the same time rotate the propeller shaft **CLOCKWISE**. Rotating the propeller

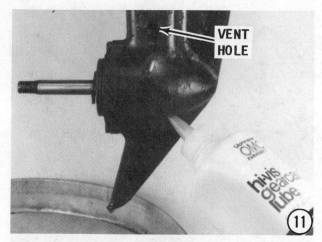

shaft will also rotate the driveshaft and allow the splines on the driveshaft to index with the splines of the engine crankshaft. Continue to work the lower unit closer to the exhaust housing until the mating surfaces make contact. Coat the retaining screws with sealer to prevent corrosion, and then start them in place. Tighten the retaining screws **EVENLY** and **ALTERNATELY**.

FILLING THE LOWER UNIT

11- Fill the lower unit with lubricant. Insert the lubricant tube into the bottom opening, and then fill the unit until lubricant is visible at the vent hole. Install the vent plug. Remove the gear lubricant tube, and install the drain/fill plug.

12- After the lower plug has been installed, remove the vent plug again and using a squirt-type oil can, add lubricant through this vent hole. A squirt-type oil can must be used to allow the trapped air in the lower unit to escape at the same time the final lubricant is added. Once the unit is completely full, install and tighten the vent plug.

PROPELLER INSTALLATION

FIRST, THESE GOOD WORDS

The propellers used on the outboards covered in this section have a removable clutch ring and a clutch hub, and bushing. Under normal conditions, these items are **NOT** removed from the propeller. However, if they have been removed for any number of reasons, they should be coated with OMC Type "A" lubricant prior to installation. The bushing is installed first, then the clutch hub, and finally the clutch ring, illustration **A**.

13- Install the shear pin. Apply a light coating of anti-corrosive lubricant onto the

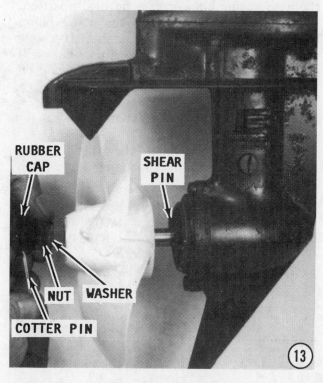

propeller shaft. Slide the propeller onto the shaft, then the washer, and finally the propeller nut, with the flange on the nut **TO-WARDS** the propeller. Tighten the nut securely. Install a cotter pin to prevent the nut from backing out. Slip the rubber cap over the propeller nut.

14- Perform a functional check of the completed work by mounting the engine in a test tank, in a body of water, or with a flush attachment connected to the lower unit. If the flush attachment is used, **NEVER** operate the engine above an idle speed, because the no-load condition on the propeller would allow the engine to **RUNAWAY** resulting in serious damage or destruction of the engine.

CAUTION: Water must circulate through the lower unit to the engine any time the engine is run to prevent damage to the water pump in the lower unit. Just five seconds without water will damage the water pump.

Start the engine and observe the tattle-tale flow of water from idle relief in the exhaust housing. The water pump installation work is verified. If a "Flushette" is connected to the lower unit, **VERY LITTLE** water will be visible from the idle relief port.

8-5 LOWER UNIT SERVICE SHIFT WITH NO DISCONNECT 4 DELUX, 4.5HP, 5HP, 6HP, 7.5HP, AND 8HP 1980 AND ON

Description

This unit has a shift mechanism that is controlled through a shift handle on the starboard side of the exhaust housing. A shaft extends through the housing and is connected to a gear on the inside. Another gear transfers the motion downward by means of a shaft indexed into the gear. Therefore, the shift rod is not disconnected in order to remove the lower unit.

The unit is shifted through a cam and plunger on the end of the shift rod. This cam and plunger arrangement forces the clutch dog forward and aft for **NEUTRAL** gear and **REVERSE** gear operation. The cam and plunger on the end of the shift rod are at rest when the unit is in **FORWARD** gear. Action takes place and movement is necessary to depress the pin in the propeller shaft to move the clutch dog for neutral and reverse gear.

Propeller Removal

Remove the propeller according to the detailed procedures outlined in Section 8-2.

Draining the Lower Unit

Drain the lower unit according to the procedures outlined in Section 8-3.

LOWER UNIT REMOVAL

ADVICE

If the only work to be performed is service of the water pump, be extremely **CAREFUL** to prevent the driveshaft from being pulled up and free of the pinion gear in the lower unit. **NEVER** carry the lower unit by the driveshaft or by the shift rod. If the shaft should be released from the pinion, the lower unit **MUST** be disassembled to align the pinion gear and driveshaft, then the driveshaft installed.

1- Disconnect the spark plug wire from the plug. Turn the propeller shaft **CLOCKWISE**, and at the same time shift the unit into **FORWARD** gear using the shift handle. Remove the retaining bolts securing the lower unit to the exhaust housing. Two bolts are located under the caviatation plate on each side of the zinc. A third bolt is located just above the cavitation plate at the forward leading edge of the exhaust housing. **CAREFULLY** pull directly downward on the lower unit, to prevent damage to the water tube, driveshaft, or shift rod, and remove the lower unit. The shift rod and the driveshaft will come free of the powerhead with the lower unit.

GOOD WORDS

If this shift rod is frozen in the gear at the shift handle, the rod will not come free with the lower unit. In this case, the shift handle mechanism must be disassembled in order to remove the gear on the end of the shift rod. The powerhead would have to be removed to gain access to the gears.

MORE GOOD WORDS

In **MOST** cases, if any unit being serviced has the 6-inch extension, it is **NOT** necessary to remove the extension in order to "drop" the lower unit.

2- Position the lower unit in a vertical position on the edge of the work bench resting on the cavitation plate. Secure the lower unit in this position with a C-clamp. The lower unit will then be held firmly in a favorable position for further service work. An alternate method is to cut a groove in a short piece of 2" x 6" wood to accommodate the lower unit with the cavitation plate resting on top of the wood. Clamp the wood in a vise and service work may then be performed with the lower unit erect (in its normal position), or inverted (upside down). In both positions, the cavitation plate is the supporting surface.

WATER PUMP DISASSEMBLING

TAKE TIME

Take time to notice the four bolts passing through the water pump. If a water pump replacement is the only work to be performed, **DO NOT** remove the three bolts in the gearcase cover. If these bolts should mistakenly be removed, the shift rod will be dislodged -- **BAD NEWS**. It is then necessary to rebuild the lower unit.

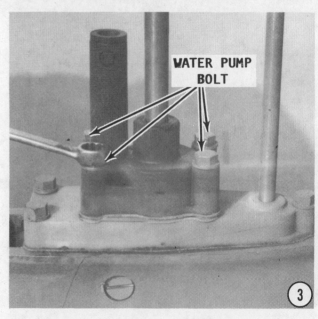

WATER PUMP BOLT

3- With the lower unit in position as described in Step 2 above, remove the O-ring from the top of the driveshaft. Remove the four bolts through the water pump. Work the water pump upward and free of the driveshaft.

4- Work the impeller up the driveshaft. Remove the impeller key.

5- Remove the water pump plate and gasket. Notice the plate and two gaskets under the water pump impeller. Notice how the plate is installed with the beveled, or rounded side of the plate facing UP. Both gaskets are identical and either one may be installed on top of the plate.

If the only work to be performed is service of the water pump, proceed directly to Page 8-33, Water Pump Installation.

LOWER UNIT DISASSEMBLING

6- Remove the two screws installed in the gearcase head. Tap the gearcase head in either direction with a soft-headed mallet, to turn the head in the housing. The

ears of the gearcase head will then protrude out the side of the lower unit. After the ears are exposed, tap on the bottom side of the ears and remove the gearcase head.

7- After the head is removed, the reverse gear, propeller shaft, and clutch dog can then be removed as an assembly. As the shaft is removed, TAKE CARE not to lose the plunger from the center of the shaft.

Slide the gearcase head and reverse gear from the propeller shaft. If working on a newer model, 1984 and on, also slide the thrust washer from the propeller shaft.

8- Remove the plunger pin from inside the propeller shaft end. Use a small screwdriver and work the coiled spring off the

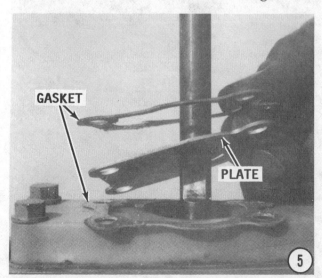

COILED SPRING

PIN   PUNCH

inside diameter of the clutch dog. Be very careful not to stretch the spring out of shape. If the spring is distorted, it is unfit for further service, because it will not hold the pin properly in the clutch dog. Therefore, if the spring is stretched out of shape, it must be replaced.

9-- Use special tool OMC No. 390766 to remove the pin from the clutch dog. Insert the tool into the forward gear end of the propeller shaft, push downward, and at the same time push the pin through the clutch dog. Ease up on the tool and remove the clutch dog from the propeller shaft. This special tool will depress the spring and allow the pin to be removed.

10-- If the tool is not available, hold the end of the propeller shaft on the bench or piece of wood, press down on the shaft, remove the plunger pin.

11-- Remove the spring from the end of the propeller shaft.

12-- Insert the propeller shaft down into the forward gear. Use a clenched fist and rap sharply on the end of the propeller shaft towards the cavitation plate. This action will disengage the forward gear bearing out of the cup and from the pinion gear. Remove the propeller shaft. Reach into the lower unit and remove the forward gear.

13-- Pull upward on the driveshaft, and at the same reach inside the lower unit and remove the pinion gear, the two thrust washers, and the thrust bearing. Notice how the washers are installed on the pinion gear.

SPRING

PIN

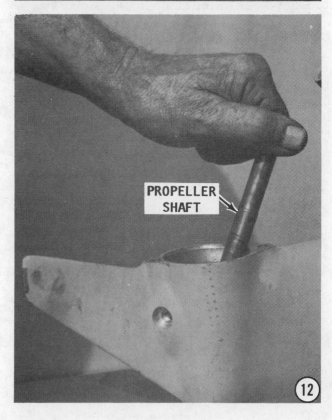

PROPELLER SHAFT

They **MUST** be installed in the same order and position from which they are removed.

14- Clean the shift rod and then coat it with oil, as an aid to removing the gearcase cover. Remove the three retaining bolts, and then slide the gearcase cover up and free of the shift rod.

15- Lift the shift rod free of the gearcase, and at the same time reach into lower unit cavity and remove the shift cam.

Removing the Forward Bearing Race

16- Use a slide hammer with fingers to remove the forward bearing race from the lower unit housing.

GOOD WORDS

A driveshaft bearing is installed at the top of the lower unit. A pinion gear bearing is installed at the lower end of the driveshaft. A bearing is attached to the forward gear. Under normal overhaul conditions, it is not necessary to remove these bearings unless they are unfit for further service. To determine if replacement is necessary, insert a finger into the bearing and feel for

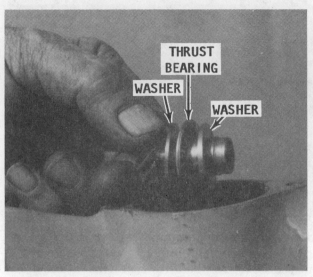

Pinion gear with the beveled washers and thrust washer in the correct order on the shank.

roughness. Check them with a flashlight. However, if they are to be removed, perform the following three steps.

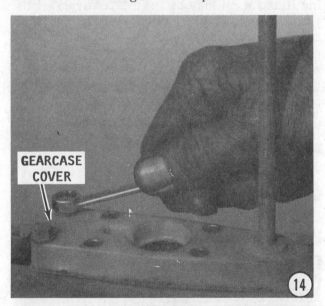

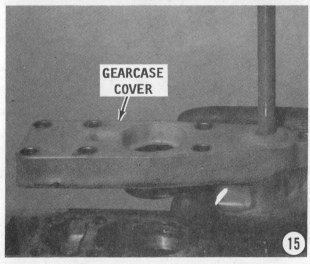

Upper driveshaft bearing and bearing housing. The housing is removed first, and then the bearing, but only if the bearing is unfit for further service.

Upper Gearcase Bearing Removal

17- A special puller is required to remove the upper bearing installed under the gearcase cover. The complete puller consists of a slide hammer, OMC No. 380658, and a bearing remover, OMC No. 380657. If the special puller is not available, a puller similar to the one shown in the accompanying illustration may be used effectively. Fit the fingers of the puller underneath the bearing and hold them in place with a heavy rubberband. Use the slide hammer to pull the bearing free of the recess. The slide hammer in this case, is used in the opposite application for which it was designed.

Pinion Gear Bearing Removal

18- If the bearing has been damaged to the point where it must be replaced, it may be driven out with a punch, or other suitable tool, because further damage is no loss. Drive the bearing downward into the bore of the lower unit.

Forward Gear Bearing Removal

This bearing should only be removed if the bearing or gear is damaged and is to be replaced.

19- A beveled clamp-type arrangement, as shown in the accompanying illustration, is required. Such a tool will exert a force between the bearing and the gear in order to remove the bearing without harm. Install the clamp and tighten it around the bearing. Now, push in the center of the bearing. The gear and bearing will separate.

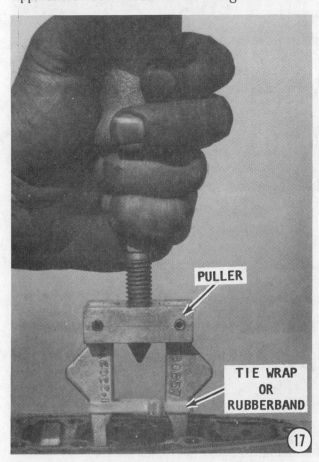

Seal Removal from the Gearcase Head

20- Remove the O-ring and then reinstall the head back onto the lower unit. Tighten the two retaining bolts. Use a slide hammer to remove the two seals from the gearcase head. Remove the two retaining bolts and again remove the head of the lower unit.

Seal Removal from the Gearcase Cover

21- Use special tool OMC No. 319880 or any type of punch to remove the seals. If the seal has been damaged to the point where it must be replaced, it may be driven out with a punch, or other suitable tool, because further damage is no loss.

Shift Rod Seal Removal

22- Use the shift rod to remove the seal. Work the shift rod in the seal, and then push outward and the seal will pop out of the gearcase cover. After the seal has been removed, remove the O-ring.

CLEANING AND INSPECTING

Clean all water pump parts with solvent, and then dry them with compressed air. Inspect the water pump cover and base for cracks and distortion, possibly caused from overheating. Inspect the face plate and water pump insert for grooves and/or rough surfaces. If possible, **ALWAYS** install a complete new water pump while the lower unit is disassembled. A new impeller will ensure extended satisfactory service and give "peace of mind" to the owner. If the old impeller must be returned to service,

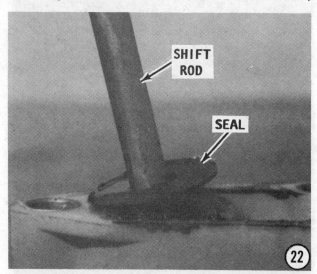

The side of the clutch dog shown must face toward the propeller.

NEVER install it in reverse to the original direction of rotation. Installation in reverse will cause premature impeller failure.

Inspect the impeller side seal surfaces and the ends of the impeller blades for cracks, tears, and wear. Check for a glazed or melted appearance, caused from operating without sufficient water. If any question exists, and as previously stated, install a new impeller if at all possible.

Clean all parts with solvent and dry them with compressed air. **DISCARD** all O-rings and gaskets. Inspect and replace the driveshaft if the splines are worn. Inspect the gearcase and exhaust housing for damage to the machined surfaces. Remove any nicks and refurbish the surfaces on a surface plate. Start with a No. 120 Emery paper and finish with No. 180.

Check the water intake screen and passages by removing the bypass cover, if one is used. Inspect the clutch dog, drive gears, pinion gear, and thrust washers. Replace these items if they appear worn. If the clutch dog and drive gear arrangement surfaces are nicked, chipped, or the edges

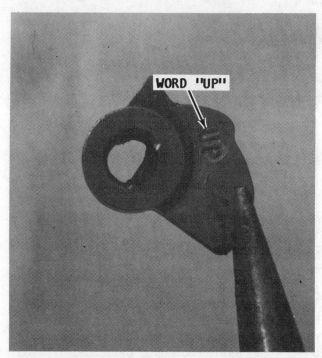

Shift cam with the word UP embossed on the arm to ensure proper installation.

rounded, the operator may be performing the shift operation improperly or the controls may not be adjusted correctly. These items **MUST** be replaced if they are damaged.

Inspect the dog ears on the inside of the forward and reverse gears. The gears must be replaced if they are damaged. Check the clutch dog retaining spring to be sure it is not distorted.

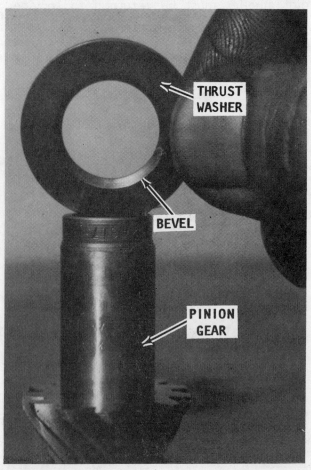

Pinion gear with the first thrust washer to be installed. The inside diameter of the washer is beveled.

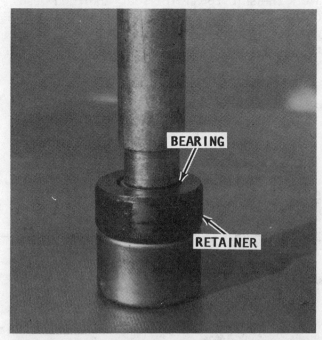

Removing the needle bearings from the upper drive-shaft bearing retainer.

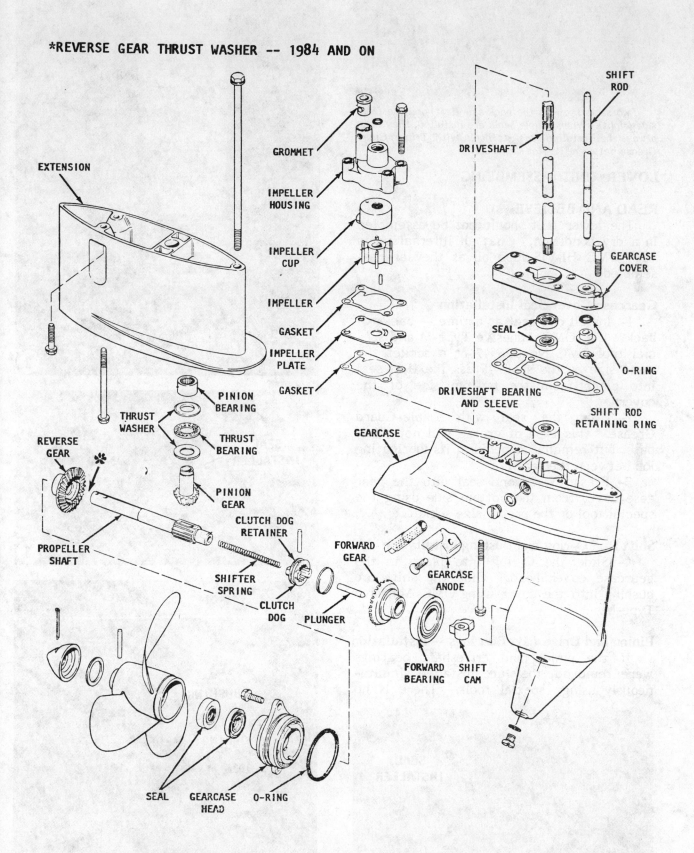

*REVERSE GEAR THRUST WASHER -- 1984 AND ON

GROMMET

IMPELLER HOUSING

IMPELLER CUP

IMPELLER

GASKET

IMPELLER PLATE

GASKET

EXTENSION

PINION BEARING

THRUST WASHER

THRUST BEARING

PINION GEAR

REVERSE GEAR

PROPELLER SHAFT

CLUTCH DOG RETAINER

SHIFTER SPRING

CLUTCH DOG

PLUNGER

FORWARD GEAR

FORWARD BEARING

SHIFT CAM

GEARCASE ANODE

GEARCASE

DRIVESHAFT BEARING AND SLEEVE

DRIVESHAFT

SHIFT ROD

GEARCASE COVER

SEAL

O-RING

SHIFT ROD RETAINING RING

SEAL

GEARCASE HEAD

O-RING

Exploded drawing of a lower unit for the 4.5 hp and 7.5 hp units, with major parts identified.

Two seals showing the back side (left) and the front side (right). When double seals are installed, they must always be installed back-to-back with Triple Guard Grease between the flat surfaces.

LOWER UNIT ASSEMBLING

READ AND BELIEVE

The lower unit should not be assembled in a dry condition. Coat all internal parts with OMC HI-VIS lube oil as they are assembled.

Gearcase Cover Seal Installation

1- Install one seal at a time -- back-to-back. Use OMC Adhesive **Type-M** and special tool OMC No. 326547 or a socket the same size as the seal. Press the first seal into place from the bottom side of the cover.

2- Coat the seal with Triple-Guard Grease. This type of grease will not dissipate, but remains to perform its lubricating job between the seals.

3- Press the second seal into the gear case cover from the bottom side using the special tool or the proper size socket.

Shift Rod O-ring and Bushing Installation

4- Slide the O-ring into place in the gearcase cover head. Glue the shift rod bushing into the cover using OMC Adhesive **Type-M**.

Pinion and Driveshaft Bearings -- Installation

If the pinion and driveshaft bearings were removed, they are installed simultaneously using special tools. There is no

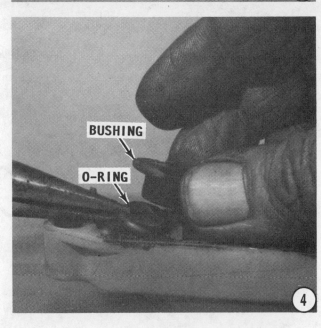

other way to install these bearings properly without the special OMC tools. Sorry about that!

5- Obtain special tool OMC No. 383173 and Spacer No. 383174. The pinion bearing is installed with the lettered side facing **DOWN**. The driveshaft bearing is installed with the lettered side facing **UP**. The tool will be pressing against the lettered side of each bearing. Work the bearings into place by tightening on the tool nut.

Forward Gear Bearing Race Installation

If the forward gear bearing race was removed, a new bearing can only be properly installed using a special OMC tool. As with the pinion and driveshaft bearings, there is no other way. Again, sorry about that!

6- Obtain special tool OMC No. 326025. Set the bearing race in the housing, and then use the tool and tap the race into position.

Forward Gear Bearing Installation

This bearing can be installed **WITHOUT** the use of a special tool.

7- Place the bearing in position on the forward gear with the taper facing **AWAY** from the gear. Press the bearing into position using a proper size socket. A short length of 1-1/4" O.D. pipe may also be used to install the bearing.

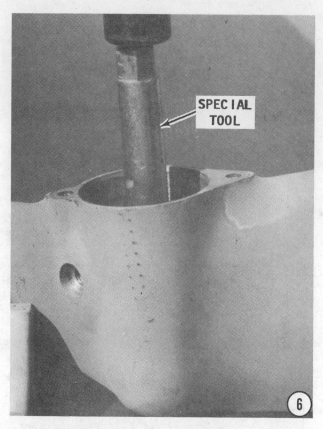

Gearcase Head Seal Installation

These two seals are of different sizes -- one small -- the other larger. The smaller (narrow) seal is installed on the inside of the gearcase head with the lip facing **INWARD**. The larger (wide) seal is installed on the outside of the head with the lip facing **OUTWARD**.

8- Install each seal from the proper side of the gearcase head. Use Triple Guard Grease between the seals. This type of grease will not dissipate, but remain to perform its lubricating job. Install the **O**-ring into the groove of the gearcase head.

Set the gearcase cover assembly aside for installation later.

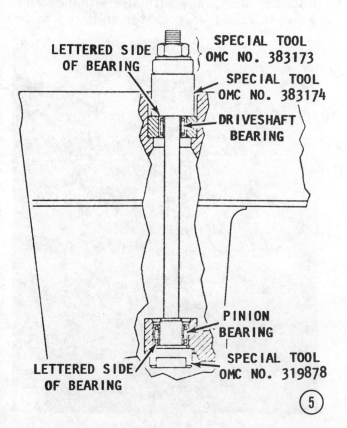

Figure 8 — SEAL, DRIVER

Shift Rod and Shift Cam Installation

9- Pickup the cam with a pair of needle-nose pliers and observe the word **UP** embossed on the arm of the cam. Take special notice of the irregular shape of the cam arm. The long flat surface must butt up against the back side of the lower unit housing. Also observe the inside diameter of the cam with the flat area on the side next to the cam arm.

10- Using the needle-nose pliers, lower the shift cam down into the lower unit with

Figure 10 — SHIFT CAM

the flat edge of the cam arm facing the **PORT** side and towards the **BACK** of the housing, as shown. Push the cam all the way in with the flat side of the cam against the back of the housing. In this position, when the shift rod is installed, the correct half-moon area on the inside diameter of the cam will index in the cam and the shift rod will be the proper position.

11- Snap the E-clip into the recess of the shift rod, as shown. Insert the shift rod down through the lower unit housing and into the shift cam with the retaining ring going in **FIRST**. Check the shift rod to be

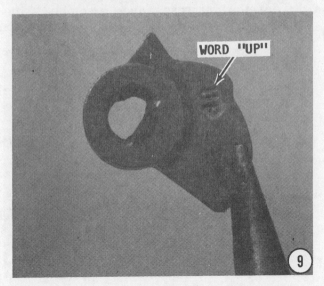

Figure 9 — WORD "UP"

Figure 11 — E-CLIP

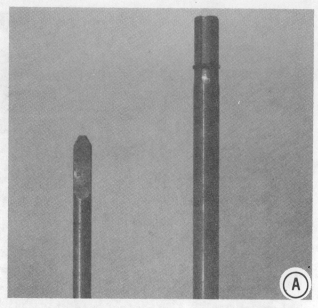

sure the flat area on the end of the rod is in the same plane (fore-and-aft) as shown in the accompanying illustration, **"A"**, when viewed from the **PORT** side.

SPECIAL NOTE

When the shift rod is rotated to the position shown in illustration **"B"**, as viewed from the **PORT** side, the unit will be in **NEUTRAL**. When the shift rod is rotated further to the position shown in illustration **"C"**, still viewed from the **PORT** side, the unit is in **REVERSE**.

Gear Case Cover Installation

12- Place a **NEW** gasket on the bottom of the gearcase cover. Coat the bottom side of the gearcase cover and the mating surface of the lower unit with OMC Adhesive **Type-M.** Coat the shift rod with oil as

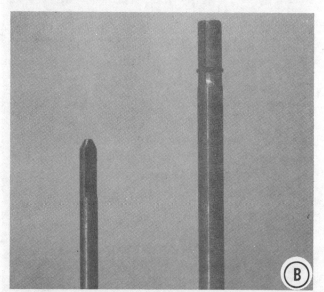

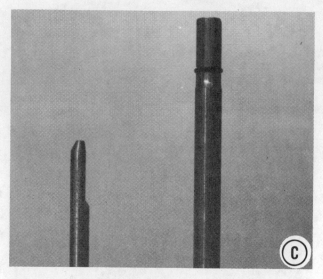

an aid to installing the gearcase cover. Slide the gearcase cover down over the shift rod. Secure the gearcase cover to the lower unit with the retaining bolts. Tighten the bolts evenly and alternately to the torque value given in the Appendix.

Pinion Gear Bearings
Installation Onto Pinion Gear

A bearing, two thrust washers, and a thrust bearing, are used on the pinion gear. The method and order of installation is most important. Notice how the inside diameter of one thrust washer is beveled. This washer is installed first with the beveled side facing **DOWN**. The thrust bearing is installed next, and finally the second thrust washer, with the outside beveled side facing **UP**.

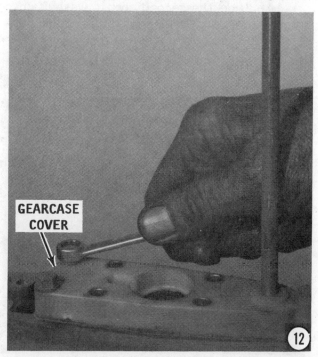

GEARCASE COVER

INSIDE BEVELED EDGE

⑬

OUTSIDE BEVELED EDGE

⑮

Lower Unit Assembling

16- Work the pinion gear assembly up into the lower unit housing through the pinion gear bearing previously installed into the housing.

17- Secure the lower unit in a horizontal position (90° from the normal position). Lower the forward gear into the lower unit housing with the gear tilted toward the cavitation plate and the gear teeth facing backward toward the opening.

18- Insert the propeller shaft through the opening and through the forward gear.

13- Coat the thrust washers, the thrust bearing, and the shank of the pinion gear with OMC HI-VIS Gearcase Lubricant. Slide the thrust washer with the beveled side facing the pinion gear shank.

14- Slide the thrust bearing onto the shank with the flat side facing **DOWN.**

15- Slide the second thrust washer onto the pinion gear shank with the outside beveled side facing **UP.**

THRUST BEARING

⑭

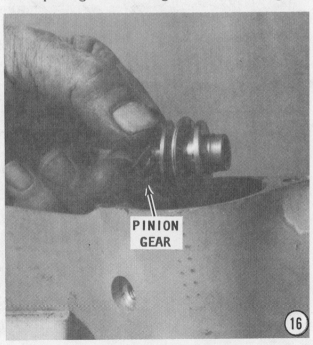

PINION GEAR

⑯

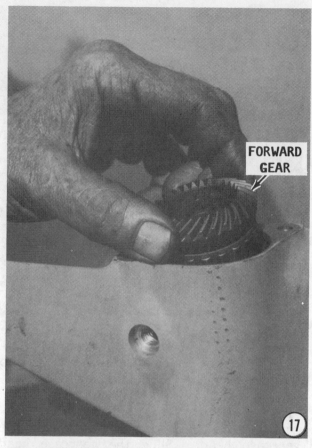

FORWARD GEAR

PROPELLER SHAFT

Forcefully pull the propeller shaft toward the skeg to snap the teeth of the forward into mesh with the teeth of the pinion gear. When the two gears have meshed, the propeller shaft will be in a vertical position. Withdraw the propeller shaft.

Propeller Shaft Assembling

19- Observe the clutch dog. Notice the mark on the clutch dog indicating the propeller end. This end **MUST** be installed toward the propeller.

20- Coat the spring with lubricant, and then slide it into the center of the propeller shaft at the forward gear end. Lubricate, and then slide the clutch dog onto the propeller shaft with the mark on the clutch dog facing **TOWARD** the propeller end of the shaft.

21- Insert special tool OMC No. 390766 into the end of the propeller shaft and depress the spring just installed. If the special tool is not available, a small screwdriver may be used. The job is not as easy

SPRING

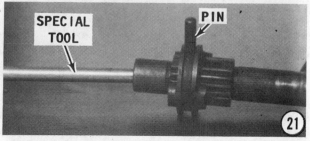

SPECIAL TOOL   PIN

COILED SPRING

22

as with the tool, but with a little patience and time, the spring can be depressed and the pin inserted. With the spring depressed, insert the pin through the clutch dog and the propeller shaft. Center the pin through the clutch dog, and then remove the special tool.

22- Install the clutch dog spring retainer into the groove in the clutch dog. **TAKE CARE** not to stretch the spring as it is stretched over the clutch dog. If the spring is stretched, it will allow the pin to come free of the clutch dog.

Propeller Shaft Assembly Installation

23- Apply some OMC HI-VIS lubricant to the plunger. Insert the square end of the plunger into the hole in the end of the propeller shaft. This plunger will make contact with the clutch dog pin through the propeller shaft. Lower the assembled propeller shaft into the lower unit, with care to prevent the plunger pin from being dislodged from the shaft. Work the end of the propeller shaft all the way into the forward gear.

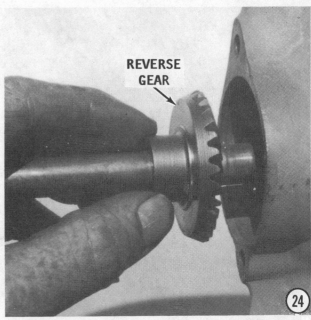

REVERSE GEAR

24

24- Models 1984 and later: Slide the thrust washer onto the propeller shaft.

All Models: Slide the reverse gear onto the propeller shaft and seat it in the housing.

25- Coat the propeller shaft with lubricant. Place the **O**-ring in position on the gearcase head. Slide the gearcase head down over the propeller shaft. Just before it reaches the lower unit surface, coat the **O**-ring with gasket sealer compound, and then secure the head to the lower unit with the attaching bolts.

26- Tighten the bolts **ALTERNATELY** and **EVENLY** to the torque value given in the Appendix.

PLUNGER

23

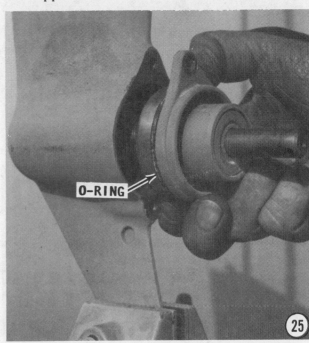

O-RING

25

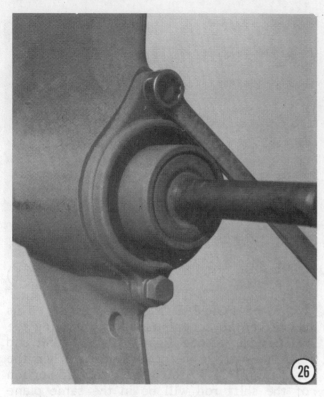

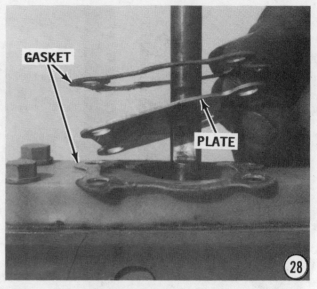

WATER PUMP INSTALLATION

27- Secure the lower unit in the vertical (normal) position. Slide the driveshaft down through the lower unit until it indexes with the pinion gear. If necessary, rotate the shaft slightly to allow the splines on the shaft to index with the splines in the pinion gear.

28- Apply OMC Sealer to both surfaces of both water pump gaskets. Place one gasket in position on the surface of the

gearcase cover. Place the water pump plate into position on the gasket with the rounded edge facing **UPWARD**. Place the second gasket in position on the top of the water pump plate.

29- Slide the impeller down the driveshaft. Before it reaches the keyway in the shaft, insert the key in the keyway, and then slide the impeller down onto the surface of the water pump plate.

30- Coat the inside surface of the water pump housing with light-weight oil, and then slide the housing down the driveshaft and over the impeller. Rotate the driveshaft as the water pump housing is moved into place to allow the impeller blades to enter and lay back properly inside the housing.

31- Apply a coating of Loctite to the water pump bolt threads. Secure the water pump to the lower unit with the bolts. Tighten the bolts **EVENLY** and **ALTERNATELY** to the torque value given in the

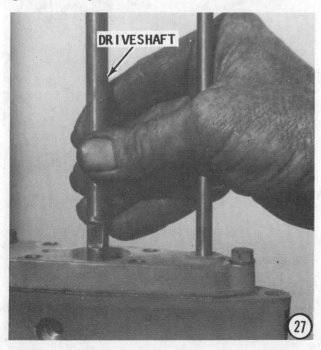

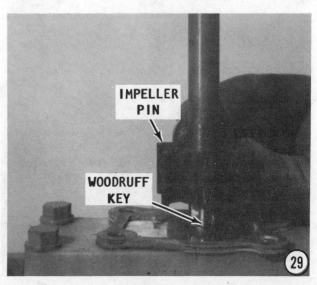

WATER PUMP HOUSING

Appendix. Slide the **O**-ring down the drive-shaft and into place on top of the water pump.

LOWER UNIT INSTALLATION

SPECIAL INSTRUCTIONS

Fill the lower unit with lubricant. Follow the procedures listed in Section 8-3.

Install the propeller, see Section 8-2.

Check to be sure the spark plug wires are disconnected from the plugs. Verify the shift handle is in the **FORWARD** gear position. Check to be sure the lower unit shift

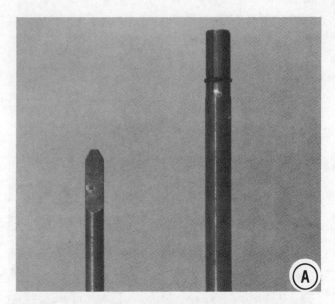

is in the **FORWARD** gear position. This check can be accomplished by verifying the shift rod is turned tightly **CLOCKWISE** and that the unit is in forward gear. When the unit is in **FORWARD** gear, the flat portion of the shift rod will be in the same plane (fore-and-aft) as shown in the accompanying illustration, **"A"**, when viewed from the **PORT** side.
copy .

Check to be sure the water pump tubes are clean, and then coat them with lubricant. Coat the **O**-rings with oil. Coat the

splines of the driveshaft with OMC Moly-lube.

32- Bring the lower unit together with the exhaust housing, and at the same time guide the water tubes and the driveshaft through the openings in the exhaust housing. Bring the assembled lower unit and the exhaust housing together with the powerhead, and at the same time guide the water tubes and the driveshaft into the place. Have an assistant slowly turn the propeller **CLOCKWISE** as the units are brought together to allow the splines on the end of the driveshaft to index with the splines in the crankshaft. Check to be sure the end of the shift rod indexes into the shift handle gear.

33- Install the retaining bolts securing the lower unit to the exhaust housing. Two bolts are installed up through the lower unit cavitation plate on each side of the zinc. A third bolt is installed down through the forward leading edge of the exhaust housing. Tighten the bolts **ALTERNATELY** and **EVENLY** to the torque value given in the Appendix. Install a new zinc and secure it in place with the Phillips screw.

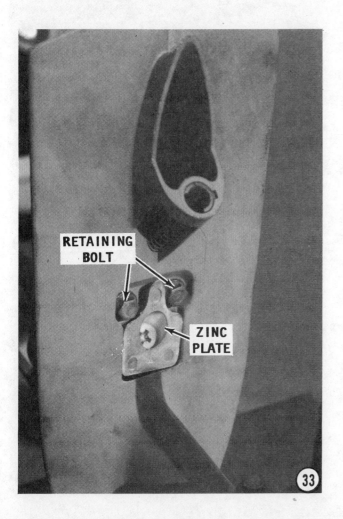

RETAINING BOLT

ZINC PLATE

33

FUNCTIONAL CHECK

Perform a functional check of the completed work by mounting the engine in a test tank, in a body of water, or with a flush attachment connected to the lower unit. If the flush attachment is used, **NEVER** operate the engine above an idle speed, because the no-load condition on the propeller would allow the engine to **RUNAWAY** resulting in serious damage or destruction of the engine.

CAUTION: Water must circulate through the lower unit to the engine any time the engine is run to prevent damage to the water pump in the lower unit. Just five seconds without water will damage the water pump.

Start the engine and observe the tattle-tale flow of water from idle relief in the exhaust housing. The water pump installation work is verified. If a "Flushette" is connected to the lower unit, **VERY LITTLE** water will be visible from the idle relief port. Shift the engine into the three gears and check for smoothness of operation and satisfactory performance.

8-6 TROUBLESHOOTING MANUAL SHIFT 6.0 HP TO 40 HP ALL MODELS EXCEPT 35 HP WITH PROPELLER EXHAUST

SPECIAL WORDS

The troubleshooting procedures in this section may be performed for any of the lower units covered in Sections 8-7 or 8-8.

Troubleshooting **MUST** be done **BEFORE** the unit is removed from the powerhead to permit isolating the problem to one area. Always attempt to proceed with troubleshooting in an orderly manner. The shotgun approach will only result in wasted time, incorrect diagnosis, replacement of unnecessary parts, and frustration.

The following procedures are presented in a logical sequence with the most prevalent, easiest, and less costly items to be checked listed first.

Unable to Shift into Forward or Reverse

Remove the propeller and check to determine if the shear pin has been broken, illustration **A**. If the unit being serviced has the shear pin at the rear of the propeller, the propeller should be removed and the shear pin checked at the rear of the propeller shaft, illustration **B**.

Check the Table of Contents for the horsepower and model year of the unit being serviced. Each section clearly outlines exactly how the shift disconnect is to be made. If the section indicates the powerhead must be removed in order to check the shift mechanism, then the powerhead must be removed. If the unit has a window in the exhaust housing, then the window must be removed. Hold the shift rod with a pair of pliers and at the same time attempt to move the shift lever on the starboard side of the engine. If it is possible to move the shift lever, the bellcrank is worn, illustration **C**.

If the engine is the type requiring the lower unit to be lowered slightly to gain access to the shift rod, proceed as follows: Lower the lower unit slightly, and then hold the shift rod with a pair of pliers and attempt to move the shift lever on the starboard side of the engine. If the lever can move, the bellcrank is worn and must be repaired.

Water in the Lower Unit

Water in the lower unit is usually caused by fish line becoming entangled around the propeller shaft behind the propeller and damaging the propeller seal, illustration **D**. If the line is not removed, it will cut the

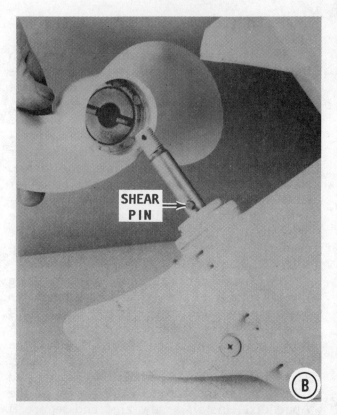

SHEAR PIN

B

BELLCRANK

C

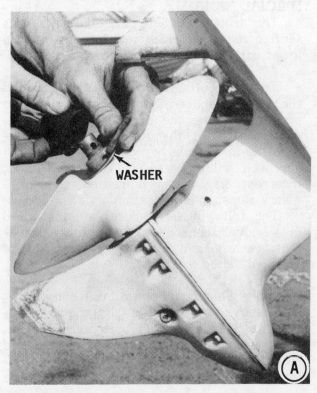

WASHER

A

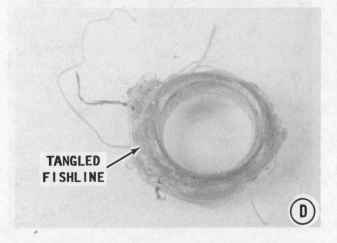

TANGLED FISHLINE

D

E

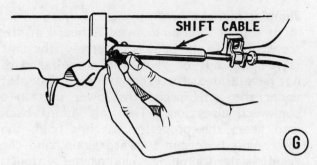

SHIFT CABLE

G

propeller shaft seal and allow water to enter the lower unit. Fish line has also been known to cut a groove in the propeller shaft.

The propeller should be removed each time the boat is hauled from the water at the end of an outing and any material entangled behind the propeller removed before it can cause expensive damage. The small amount of time and effort involved in pulling the propeller is repaid many times by reduced maintenance and service work, including the replacement of expensive parts, illustration **E**.

Slippage in the Lower Unit

If the shift seems to be slipping as the boat moves through the water: Check the propeller and the rubber hub, illustration **F**. If the propeller has been subjected to many strikes against underwater objects, it could slip on its hub. If the hub is damaged or excessively worn on the small propellers, it is not economical to have the hub or propeller rebuilt. A new propeller may be purchased for considerably less than meeting the expense of rebuilding an old worn propeller.

Difficult Shifting

Verify that the ignition switch is **OFF**, or better still, disconnect the spark plug wires

from the plugs, to prevent possible personal injury, should the engine start. Shift the unit into **REVERSE** gear at the shift control box, and at the same time have an assistant turn the propeller shaft to ensure the clutch is fully engaged. If the shift handle is hard to move, the trouble may be in the lower unit, with the shift cable, or in the shift box, if used.

Isolate the Problem: Disconnect the shift cable, if used, at the engine, illustration **G**. Operate the shift lever. If shifting is still hard, the problem is in the shift cable or control box, illustration **H**, see Chapter 7. If the shifting feels normal with the shift cable disconnected, the problem must be in the lower unit. To verify the problem is in the lower unit, have an assistant turn the propeller and at the same time move the shift cable back-and-forth. Determine if the clutch engages properly.

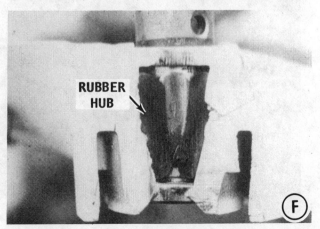

RUBBER HUB

F

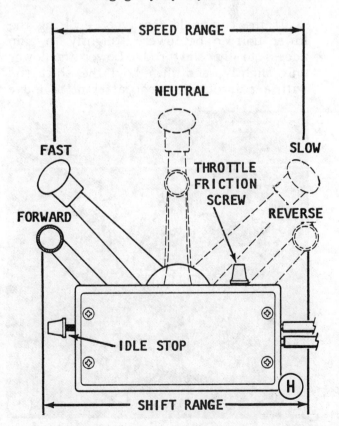

H

Jumping out of Gear

If a loud thumping sound is heard at the transom while the boat is underway, the unit is jumping out of gear, the propeller does not have a load, therefore the rushing water under the hull forces the lower unit in a backward direction. The unit jumps back into gear; the propeller catches hold; the lower unit is forced forward again, and the result is the thumping sound as the action is repeated. Normally this type of action occurs perhaps once a day, then more frequently each time the clutch is operated, until finally the unit will not stay in gear for even a short time.

The following areas must be checked to locate the cause:

1- Check the Table of Contents for the horsepower and model year of the unit being serviced. Each section clearly outlines exactly how the shift disconnect is to be made. If the section indicates the powerhead must be removed in order to check the shift mechanism, then the powerhead must be removed. If the unit has a window in the exhaust housing, then the window must be removed. Hold the shift rod with a pair of pliers and at the same time attempt to move the shift lever on the starboard side of the engine. If it is possible to move the shift lever, the bellcrank is worn.

If the engine is the type requiring the lower unit to be lowered slightly to gain access to the shift rod: Lower the lower unit slightly, and then hold the shift rod with a pair of pliers and attempt to move

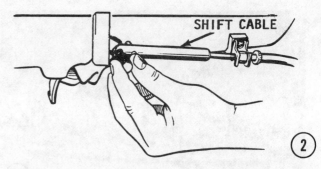

the shift lever on the starboard side of the engine. If the lever can move, the bellcrank is damaged and must be repaired.

2- Disconnect the shift cable at the engine. Attempt to shift the unit into forward gear with the shift lever on the starboard side of the engine and at the same time rotate the propeller in an effort to shift into gear. Shift the control lever at the control box into forward gear. Move the shift cable at the engine up to the shift handle and determine if the cable is properly aligned. The control lever may have jumped a tooth on the slider or on the shift lever arc. If a tooth has been jumped, the cable would lose its adjustment and the unit would fail to shift properly. If the inner cable should slip on the end cable guide, the adjustment would be lost.

3- Move the shift lever at the engine into the neutral position and the shift lever at the control box to the neutral position. Now, move the shift cable up to the shift lever and see if it is aligned. Shift the unit into reverse at the engine and shift the control lever at the control box into reverse. Move the cable up and see if it is aligned. If the cable is properly aligned, but the unit still jumps out of gear when the cable is connected, one of three conditions may exist.

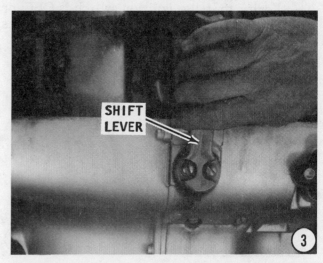

a- The bellcrank is worn excessively or damaged.

b- The shift rod connector is misaligned. This connector is used to link the upper shift rod with the lower rod. If the connector has not been installed properly, any shifting will be difficult.

c- Parts in the lower unit are worn from extended use.

Frozen Powerhead

This condition is suggested when the operator unsuccessfully attempts to crank the engine, either with a hand starter or with a starter motor. The flywheel will not rotate. Do not assume the engine is "frozen" until the lower unit has been removed and thoroughly checked. If the lower unit is "locked" (the driveshaft or propeller shaft will not rotate), the powerhead will have the indication of being "frozen" (failure to rotate the flywheel).

The first step to perform under these conditions is to "pull" the lower unit, and then again attempt to crank the engine. If the attempt is successful with the lower unit disconnected, the problem is in the lower unit. If the attempt to crank the engine is still unsuccessful, the problem is in the powerhead, illustration J.

FROZEN PISTON

J

8-7 LOWER UNIT SERVICE MANUAL SHIFT
6.0 HP — 1971-79
9.5 HP TO 20 HP — 1971-73
25 HP — 1971-85

Propeller Removal

Remove the propeller according to the detailed procedures outlined in Section 8-2.

Draining Lower Unit

Drain the lower unit according to the detailed procedures outlined in Section 8-3.

Shift Rod Disconnect

Three different shift rod connection arrangements may be encountered on the engines covered in this manual. The following three paragrpahs describe the connections and how they are to be handled for removal of the lower unit. The horsepower and model years are also given for each.

Pin in Upper Driveshaft
Used on All 6 hp Engines — 1971-79

"Pin in the upper driveshaft", means that the pin holds and pushes the seal and spring assembly against the powerhead and thus provides a bottom seal for the powerhead. After the lower unit attaching bolts have been removed, the flywheel must be rotated (to rotate the driveshaft) until the pin is aligned with two slots in the upper portion of the exhaust housing, illustration A. The lower housing can then be separated from

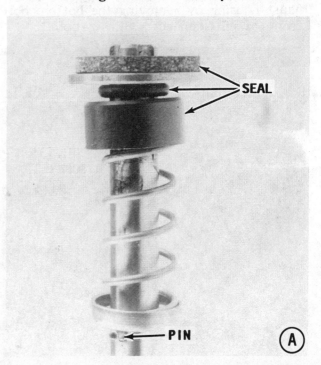

SEAL

PIN

A

the exhaust housing. If an attempt is made to force the lower unit from the exhaust housing without aligning the driveshaft pin, as just described, the pin may be broken and other items damaged. Loosen the attaching screws securing the lower unit to the exhaust housing. Allow the lower unit to drop approximately one inch, and then remove the bottom bolt in the shift connector illustration **B**. The lower unit may then be completely separated from the exhaust housing, illustration **C**.

Shift Disconnect Connector
Used on All 9.5 hp Engine — 1971-73

These units do not have the pin in the driveshaft. Loosen the attaching screws securing the lower unit to the exhaust housing. Allow the lower unit to drop approximately one inch, and then remove the bottom bolt in the shift connector, illustation **B**. The lower unit may then be completely separated from the exhaust housing, illustration **C**.

Window Removal To Gain Access
18 hp and 20 hp Engines — 1971-73
25 hp Engines 1971-85

Remove the metal plate from the port side of the engine. Access to the shift connector is gained through the opening.
Disconnect the shift rod from the exhaust housing by removing the bottom bolt from the shift connector, illustration **D**.

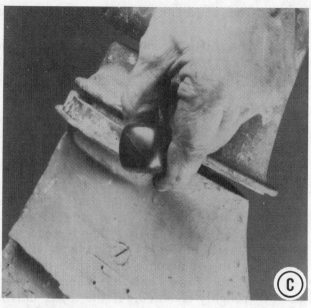

GOOD WORDS

In **MOST** cases, if any unit being serviced has the 6-inch extension, it is **NOT** necessary to remove the extension in order to "drop" the lower unit. However, as in most things in life, there are rare exceptions and here is one. If the lower unit is separated from the extension and the driveshaft connection is not accessible, then the extension will have to be removed to gain access to the coupler.

LOWER UNIT REMOVAL

1- After the shift rod has been disconnected, as described in the previous paragraphs, remove the bolts securing the lower unit to the housing. Some units may have an

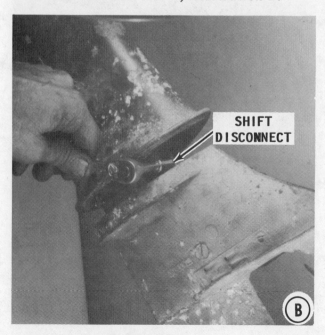

SHIFT DISCONNECT

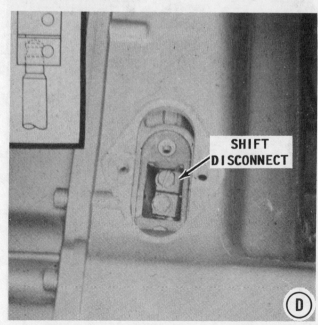

SHIFT DISCONNECT

additional bolt on each side and one at the rear of the engine. Work the lower unit loose from the exhaust housing. It is not uncommon for the water tube to be stuck in the water pump making separation of the lower unit from the exhaust housing difficult. However, with patience and persistence, the tube will come free of the pump and the lower unit separated from the exhaust housing.

MORE GOOD WORDS

Position the lower unit in a vertical position on the edge of the work bench resting on the cavitation plate. Secure the lower unit in this position with a C-clamp. The lower unit will then be held firmly in a favorable position for further service work. An alternate method is to cut a groove in a short piece of 2" x 6" wood to accommodate the lower unit with the cavitation plate resting on top of the wood. Clamp the wood in a vise and service work may then be performed with the lower unit erect (in its normal position), or inverted (upside down). In both positions, the cavitation plate is the supporting surface.

2- Remove the **O**-ring from the top of the driveshaft. Some units may have a pin installed in this location, instead of an **O**-ring. In this case, remove the pin from the driveshaft. The washer, springs, and other parts will have remained in the exhaust housing.

WATER PUMP REMOVAL

3- Remove the screws securing the water pump to the lower unit housing. It is very possible corrosion will cause the screw heads to break-off when an attempt to remove them is made. If this should happen,

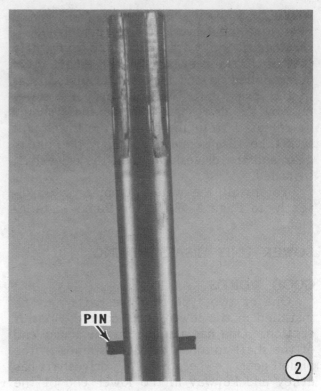

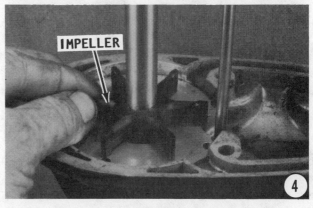
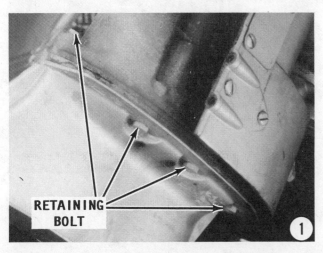

use a chisel and breakaway the water pump housing from the lower unit. **EXERCISE CARE** not to damage the lower unit housing.

4- After the screws have been removed, slide the water pump, impeller, the impeller key, and the lower water pump plate, upward and free of the driveshaft.

ADVICE

If the only work to be performed is service of the water pump, be extremely **CAREFUL** to prevent the driveshaft from being pulled up and free of the pinion gear in the lower unit. **NEVER** carry the lower unit by the driveshaft. If the shaft should be released from the pinion, the lower unit **MUST** be disassembled to align the pinion gear and the driveshaft, then the driveshaft installed.

To install the water pump, proceed directly to Page 8-52, Water Pump Installation.

LOWER UNIT DISASSEMBLING

GOOD WORDS

One of two type of driveshafts may be installed in the lower units covered in this section. One has a spline on the lower end of the shaft to index with the splines in the pinion gear. The other type driveshaft has a key and keyway in the lower end of the driveshaft. The key indexes with a matching keyway in the pinion gear.

5- **CAREFULLY** pull upward on the driveshaft. If the driveshaft comes free easily, the unit is the type with the splines on the end of the shaft. If the shaft will not come free, it is the type with the key and keyway. Therefore, the driveshaft will be removed later when the lower unit is disassembled. If the driveshaft comes free, remove it at this time.

6- Turn the lower unit upside down and again clamp it in the vise or slide it into the wooden block, if one is used. Carefully

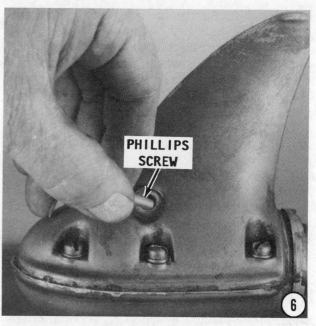

examine the lower portion of the unit. The cap is considered that part below the split with the skeg attached. The cap has a Phillips screw installed. Remove the screw.

7- Remove the attaching screws around the cap. These screws may be slotted-type or Phillips screws. **CAREFULLY** tap the cap to jar it loose, and then separate it from the lower unit housing. If the cap did not have a Phillips screw on the outside, observe the two slots inside the cap.

TAKE TIME

Before proceeding with the disassembly work, take time to study the arrangement of parts in the lower unit. You may elect to follow the practice of many professional mechanics and take a polaroid picture of the unit as an aid during the assembly work. Several engineering and production changes

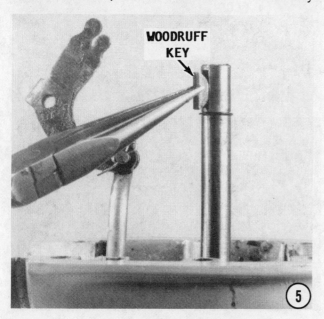

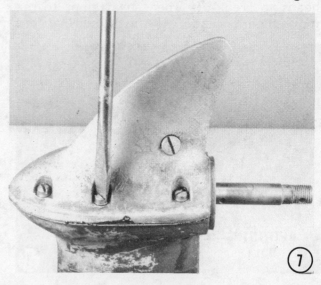

have been made to the lower unit over the years. Therefore, the positioning of the gears, shims, bearings, and other parts may vary slightly from one unit to the next.

To show each and every arrangement with a picture in this manual would not be practical. Even if it were done, the ability to associate the unit being serviced with the illustration would be almost impossible. Therefore, take time to make notes, scribble out a sketch, or take a couple photographs, illustration E.

8- Lift the shift lever out of the cradle, and then remove the cradle from the shift dog. Raise the propeller shaft and at the same time tap with a soft-headed mallet on the bottom side to jar it loose, then remove the shaft assembly from the lower unit. The forward and reverse gear including the bearings will all come out with the propeller shaft. The forward gear is the gear at the opposite end of the shaft from the propeller.

GOOD WORDS

Notice the bearing split on the back side of the forward gear. Also observe the pin in the housing and a matching slot in the bearing. The pin must index in the slot during installation. The reverse gear has a

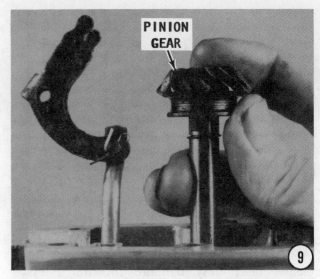

tab protruding from the bearing head. This tab indexes in a slot in the housing during installation. By taking note at this time of the particular type of installation for the unit being serviced, the task of installation will progress more smoothly. If the installation work is not performed properly, the lower unit housing will quickly be damaged requiring the purchase of a new unit.

9- If the unit being serviced is the type with a keyway in the driveshaft, remove the pinion gear from the shaft, then the key, and snap ring, before attempting to remove the driveshaft.

10- Slide the forward gear, babbitt bearing, and washer, free of the propeller shaft. Remove the clutch dog. Remove the reverse gearcase head, reverse gear, and washer, from the shaft.

11- Turn the lower unit housing right side up and again clamp it in the vise. Remove the bearing carrier and bearing assembly. This is accomplished by using a bearing carrier puller, as shown. An alternate method is to use two screwdrivers to remove the carrier from the lower unit. Sometimes the bearing carrier is difficult to remove. One effective method to release a stubborn bearing carrier is to heat the lower unit housing while attempting to remove the

Bearing carrier prior to removal. The two screws shown are used to align the carrier with the gasket during installation.

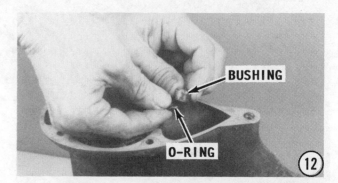

carrier. If this method is employed, **TAKE CARE** not to overheat the lower unit. Excessive heat may damage internal parts. Remove the gasket from underneath the bearing carrier housing.

12- Clean the upper part of the shift rod as an aid to pulling it through the bushing and O-ring. Pull the shift rod from the lower unit housing. The shift rod passes through an O-ring and bushing in the lower unit housing. These two items prevent water from entering the lower unit. A special tapered punch is required to remove the bushing from the lower unit housing. Obtain the special punch, and then remove the bushing, and the O-ring.

CLEANING AND INSPECTING

Clean all water pump parts with solvent, and then dry them with compressed air. Inspect the water pump cover and base for cracks and distortion, possibly caused from

overheating. Inspect the face plate and water pump insert for grooves and/or rough surfaces. If possible, **ALWAYS** install a complete new water pump while the lower unit is disassembled. A new impeller will ensure extended satisfactory service and give "peace of mind" to the owner. If the old impeller must be returned to service, **NEVER** install it in reverse to the original direction of rotation. Installation in reverse will cause premature impeller failure.

Inspect the impeller side seal surfaces and the ends of the impeller blades for cracks, tears, and wear. Check for a glazed or melted appearance, caused from operating without sufficient water. If any question exists, and as previously stated, install a new impeller if at all possible.

Clean all parts with solvent and dry them with compressed air. **DISCARD** all O-rings and gaskets. Inspect and replace the driveshaft if the splines are worn. Inspect the gearcase and exhaust housing for damage to the machined surfaces. Remove any nicks and refurbish the surfaces on a surface plate. Start with a No. 120 Emery paper and finish with No. 180.

Check the water intake screen and passages by removing the bypass cover, if one is used. Inspect the clutch dog, drive gears, pinion gear, and thrust washers. Replace these items if they appear worn. If the

Using a slide hammer to "pull" the bearing carrier from the upper portion of the lower unit.

clutch dog and drive gear arrangement surfaces are nicked, chipped, or the edges rounded, the operator may be performing the shift operation improperly or the controls may not be adjusted correctly. These items **MUST** be replaced if they are damaged.

Inspect the dog ears on the inside of the forward and reverse gears. The gears must be replaced if they are damaged.

A new shift cradle (right) compared with one badly worn and unfit for further service (left).

Worn clutch dog ears. This clutch dog must be replaced.

*Bearing head with a new seal and **O**-ring installed.*

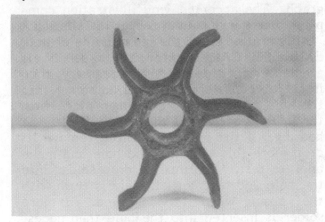

Badly worn water pump impeller. The best practice is to replace the impeller each time the lower unit is disassembled for service.

Badly worn propeller shaft. This damage may have started from an entangled fish line that was not removed, and then aggravated when water entered the lower unit.

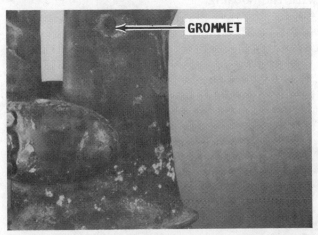

Grommet installed about half-way up the exhaust housing where the water tube passes through. This grommet should be removed and the area cleaned to ensure satisfactory water flow through the unit.

A rusted and corroded gear. Water was allowed to enter the lower unit through a badly worn seal and cause this damage to the gear and other expensive parts.

Badly worn pinion gear from a lower unit. The teeth of the gears must be carefully inspected for wear and damage.

Check the cradle that rides on the inside diameter of the clutch dog. The sides of the cradle must be in good condition, free of any damage or signs of wear. If damage or wear has occurred, the cradle must be replaced.

Check the shift lever and the two prongs that fit inside the cradle. Check to be sure the prongs are not worn or rounded. Damage or wear to the prongs indicates the lever must be replaced.

READ AND BELIEVE

The three accompanying lettered illustrations clearly show a lower unit that has been assembled **INCORRECTLY..**

Illustration F: Inspect very closely, the bushing bearing on the forward gear for any kind of indication the pin was missed when the housing was installed during the last repair work. Check the reverse gearcase head and if there is any indication of a pin mark, then check the housing for evidence the pin has been driven into the housing.

Some bearing carriers have a lip that indexes with a slot in the lower unit housing. Check for evidence the lip did not index properly.

Illustration G: If the pin has been driven down into the lower unit housing, it must be drilled out as described in the next paragraph. The accompanying illustration compares a proper and an improper installation.

Illustration H: If the pin must be drilled, **EXERCISE CARE** not too drill to deeply. If a hole is drilled deeper than necessary, then insert a couple drops of melted solder into the hole, and set the new pin in place with the **LARGE** portion of the pin flush with the housing. If the pin is not flush, remove it, drop more soldered into the hole and make another test. Continue to drop solder into the hole and test until the pin is flush with the housing when it is installed. If more solder is inserted into the hole than necessary, the pin may be tapped into the solder while it is still warm and the pin made flush with the housing.

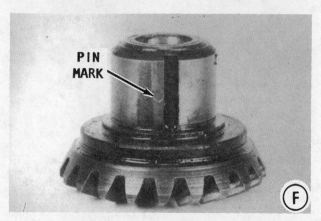

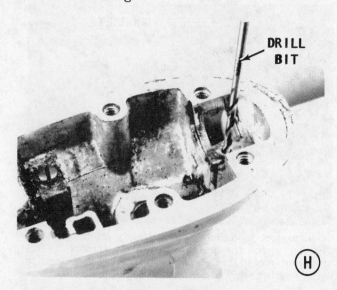

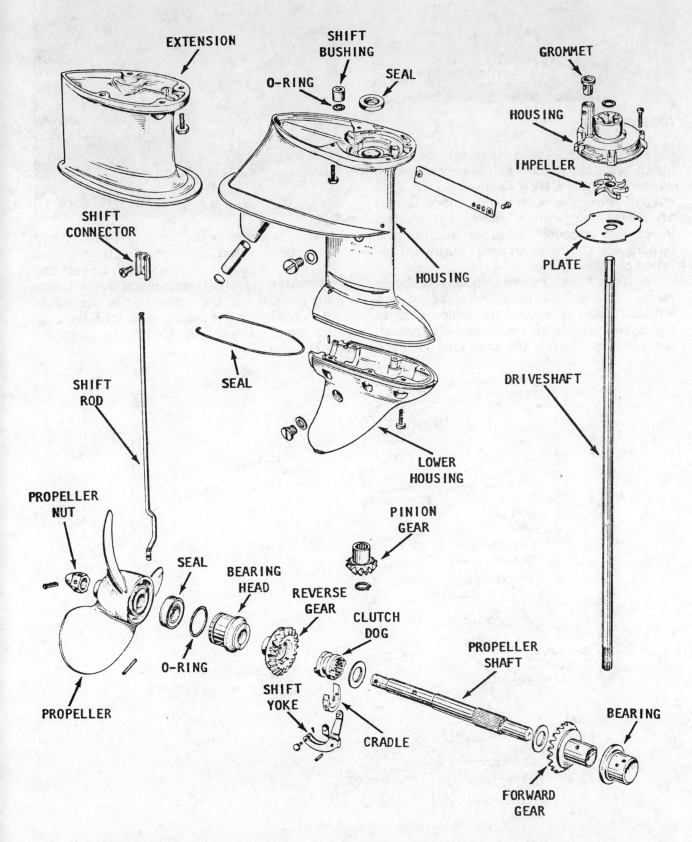

EXTENSION

SHIFT
BUSHING

O-RING

SEAL

GROMMET

HOUSING

IMPELLER

SHIFT
CONNECTOR

HOUSING

PLATE

SHIFT
ROD

SEAL

DRIVESHAFT

PROPELLER
NUT

SEAL

BEARING
HEAD

LOWER
HOUSING

PINION
GEAR

REVERSE
GEAR

CLUTCH
DOG

O-RING

PROPELLER
SHAFT

SHIFT
YOKE

PROPELLER

CRADLE

BEARING

FORWARD
GEAR

Exploded drawing of the lower unit for all the units covered in this section except the 6 hp units.

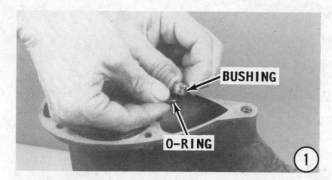

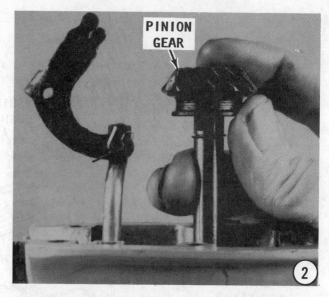

ASSEMBLING

1- Place the lower unit on the work bench with the water pump recess facing upward. Install a **NEW** O-ring into the shift cavity. Work the bushing into place on top of the O-ring with a punch and mallet. Inject just a couple drops of oil into the bushing and O-ring as an assist during installation of the shift rod.

2- Turn the exhaust housing upside down. If the unit being serviced uses a Woodruff key, to secure the pinion gear to the driveshaft, install the driveshaft through the housing. Install the snap ring into the

groove near the end of the driveshaft, and then the Woodruff key, and finally the pinion gear onto the driveshaft. Lower the assembled driveshaft into place in the lower unit housing. If the driveshaft is the splined-type, lower the pinion gear into place at this time. The driveshaft will be installed later.

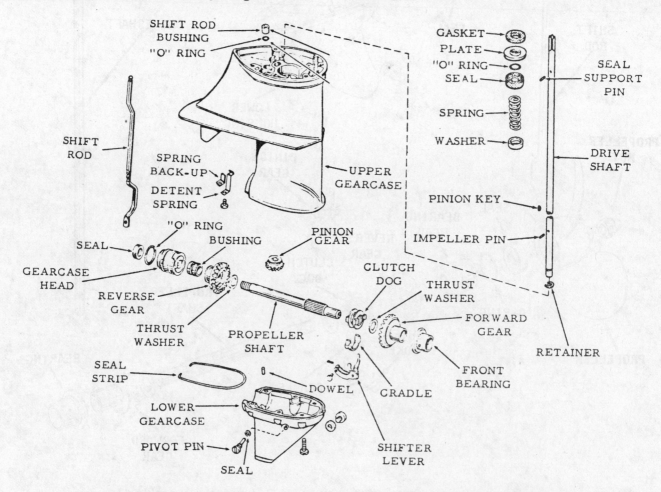

Exploded drawing of the lower unit for the 6 hp unit, with major parts identified.

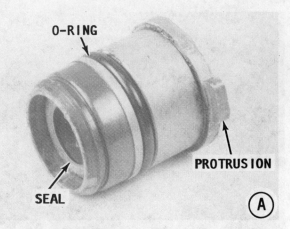

Assembling the Propeller Shaft

3- Slide the clutch dog onto the propeller shaft splines. Apply a light coating of lubricant to the washer and then insert it into the center of the forward gear. Slide the forward gear onto the end of the propeller shaft. Slide the forward gear bearing onto the shaft and into the forward gear.

4- Apply a light coating of lubricant to the washer, and then insert it into the center of the reverse gear. Slide the reverse gear onto the propeller shaft from the propeller end. Check to be sure a new O-ring and bearing seal has been installed into the gear case head, and then install the gearcase head assembly onto the propeller shaft.

CRITICAL WORDS

Look into the front part of the lower unit housing. Notice the pin protruding up from the housing. Now, observe the slot in the forward gear bearing. When the propeller shaft assembly is installed into the lower unit, this pin **MUST** index into the hole in the forward gear bearing. Also notice the protrusion on the end of the gearcase head, illustration **A**. This protrusion **MUST** index with the slot in the housing when the propeller shaft assembly is installed.

5- Check to be sure the pinion gear is properly located. Check to be sure the shift rod is clean and smooth (free of any burrs or corrosion). Coat the shift rod and the O-ring with oil as an aid to installation. Slide

the shift rod down through the O-ring and bushing into the gear case.

6- Slide the propeller shaft assembly into the lower unit housing. Check to be sure the slot in the forward gear bearing indexes with the pin in the lower unit, and the protrusion on the end of the gear case head indexes in the slot in the housing. On some models, a pin in the lower unit housing **MUST** index with a hole in the gearcase head.

7- Lubricate the cradle, and then slip it into the clutch dog groove.

8- Bring the shift lever down over the cradle and snap the fingers of the lever into the cradle. Check to be sure the clutch dog

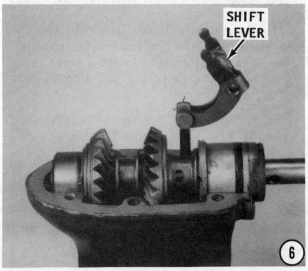

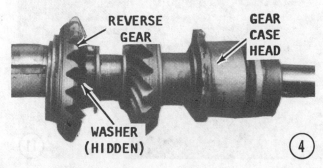

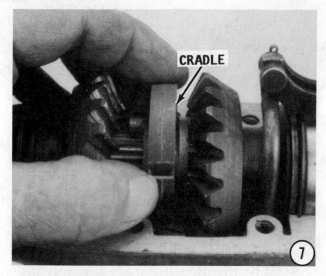

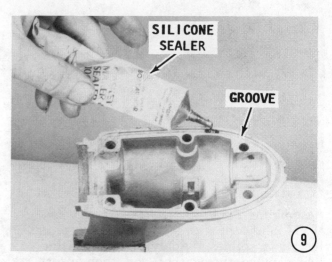

is in the **NEUTRAL** position. Push or pull on the shift rod to move it up or down until the clutch dog is in the center between the forward and reverse gears.

9– Lay down a bead of No. 1000 Sealer into the groove of the cap in preparation to installing the seal.

10– Place a **NEW** seal in the lower cap and hold the seal in the groove with sealer. Apply a small amount of silicone sealer on each side of the bearing gear case head. This sealer will form a complete seal when the lower unit cap is installed.

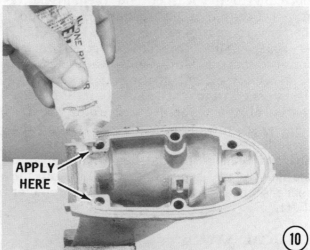

GOOD WORD

If time is taken to grind the end of the screw to a **SHORT** point, it will make the task of installation much easier, illustration **B.**   If the cap and shift lever are not aligned exactly, the screw will "seek" and make the alignment as it passes through. However, do not make a long point or the screw will not have enough support and would bend during operation of the shift lever.

11– Position the lower unit cap over the gear assembly onto the lower unit housing. If the unit being serviced uses a shift lever pin, work the cap until the pin indexes into the recess of the cap.

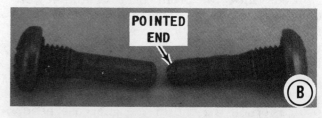

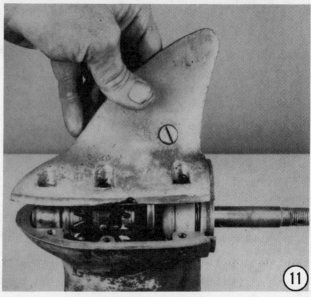

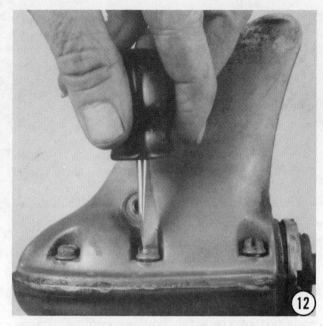

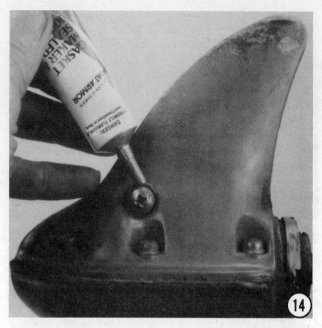

12- Apply a drop of sealer into the opening for each cap retaining screw to ensure a complete seal between the cap and the lower unit housing. Install the screws securing the cap to the lower unit housing. Tighten the screws **ALTERNATELY** and **EVENLY.**

13- A Phillips screw is used in the side of the cap. Use a flashlight and align the hole in the cap with the hole in the shift lever. Install the tapered Phillips screw into the housing and through the lever.

14- Apply a drop of good grade sealer to the threads, and then tighten the screw securely.

15- Turn the lower unit rightside up. Install a **NEW** gasket onto the upper surface of the lower unit.

16- Install the bearing housing and bearing assembly by sliding a couple of bolts through the housing to align the base gasket.

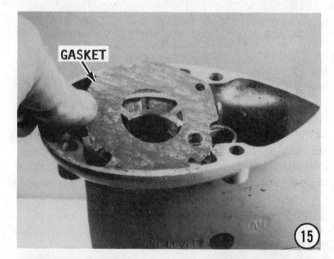

GASKET

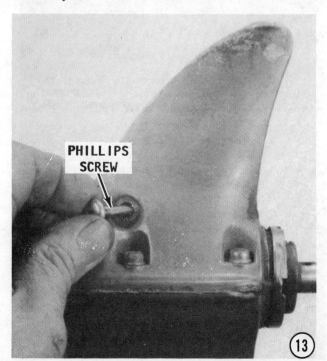

PHILLIPS SCREW

BASE PLATE

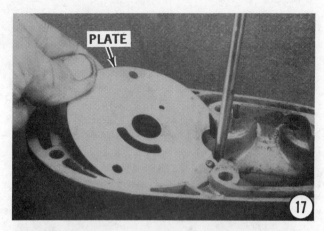

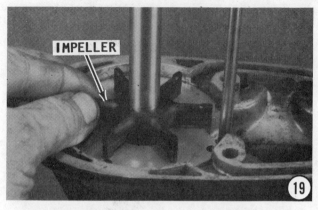

WATER PUMP INSTALLATION

17- Apply a coating of sealer to the upper surface of the lower unit. Install the water pump base plate.

18- If the unit being serviced, uses the splined-type driveshaft, slide the driveshaft into the lower unit, and then rotate the shaft very slowly. When the splines of the driveshaft index with the pinion gear, the shaft will drop slightly. Install the water pump pin or key.

19- Slide the water pump impeller down the driveshaft and into place on top of the water pump base plate with the pump pin or key indexed in the impeller. Lubricate the inside surface of the water pump with light-weight oil.

20- Lower the water pump housing down the driveshaft and over the impeller. Rotate the driveshaft CLOCKWISE as the water pump housing is lowered to allow the impeller blades to assume their natural and proper position inside the housing. Continue

to rotate the driveshaft and work the water pump housing downward until it is seated on the lower unit upper housing .

21- Rotate the driveshaft CLOCKWISE while the screws are tightened to prevent damaging the impeller vanes. If the impeller is not rotated, the housing could damage or cut the end of the vanes as the screws are brought up tight. The rotation allows them to spring back in a natural position. Place a NEW grommet into the water pump housing for the water pickup. If a new water pump was installed, this seal will already be in place.

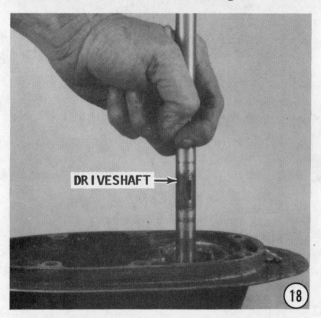

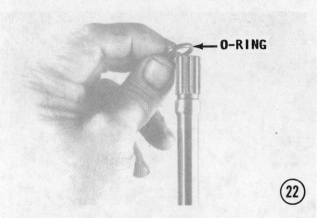

← O-RING

22

22- Install a **NEW O**-ring on the top of the driveshaft.

23- If the lower unit being serviced uses a pin on the top of the driveshaft, install the pin at this time. Shift the lower unit into **FORWARD** gear and at the same time rotate the propeller shaft **CLOCKWISE**. The lower unit assembling is now complete and ready to mate with the exhaust housing.

LOWER UNIT INSTALLATION

GOOD WORDS

If the unit being serviced uses the shift rod connector arrangement either through the window or before the lower unit and exhaust housings are fully mated, these words are extremely critical. Connecting the shift rod with the connector is not an easy task but can be accomplished as follows: First, notice the cutout area on the

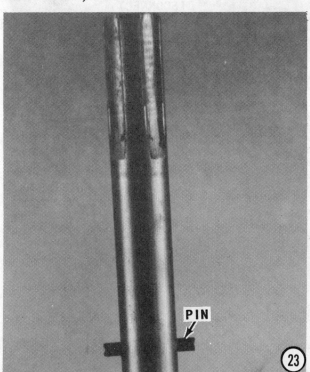

PIN

23

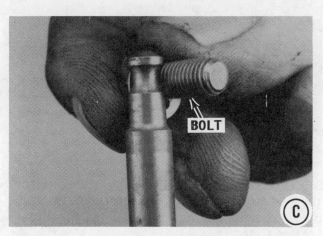

BOLT

C

end of the shift rod. This area permits the bolt to pass through the connector past the shift rod, and into the other side of the connector. It is this bolt that holds the shift rod in the connector. Now, in order for the bolt to be properly installed, the cutout area on the shift rod **MUST** be aligned in such a manner to allow the bolt to be properly installed. Therefore, as the lower unit is mated with the exhaust housing, exercise patience as the two units come together, to enable the bolt to be installed at the proper time. If the rod is allowed to move too far into the connector before the bolt is installed, it may be possible to force the bolt into place, past the shift rod. The threads on the bolt will be stripped, and the shift rod will eventually come out of the connector, illustration **C** and **D**.

24- Install the connector onto the lower unit shift rod, with the **NO THREAD** section facing towards the window. With the connector in this position, the bolt may be inserted through the connector and "catch" the threads on the far side. Install the connector bolt in the manner described in the previous paragraph.

SPECIAL WORDS

If the unit being serviced uses the shift rod connector arrangement, then the connector must be connected to the shift rod

SHIFT ROD
CONNECTOR →

D

BEFORE the lower unit is fully mated with the exhaust housing, as described in the previous step and the "Good Words", illustration C and D.

If the unit being serviced uses the driveshaft with the pin, extreme care must be exercised as the shaft is guided into the exhaust housing to allow the pin to index with the groove in the housing, illustration E.

If the unit being serviced uses the bolt through the window arrangement, insert the bolt into the connector. TAKE TIME to read and understand the "Good Words" just before Step 24, before making this connection. After the bolt is in place, install and secure the window with the attaching hardware, illustration F.

25- Check to be sure the water pickup tube is clean, smooth, and free of any corrosion. Coat the water pickup tube and grommet with lubricant as an aid to installation. Guide the lower unit up into the

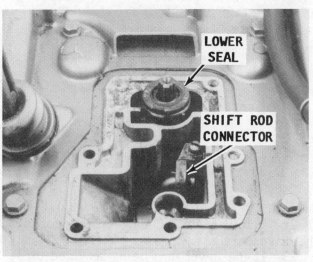

Shift rod connector on a lower unit requiring the powerhead to be removed to gain access to the connector.

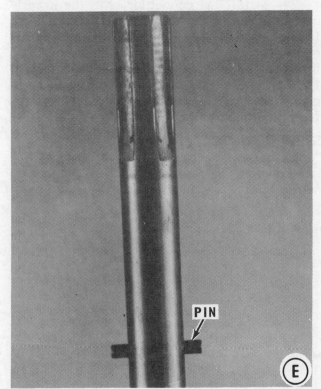

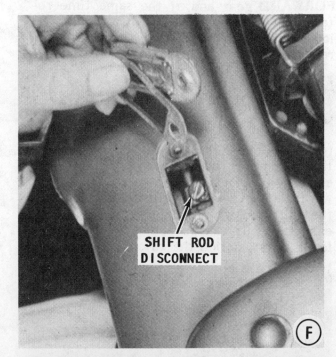

exhaust housing with the water tube sliding into the rubber grommet of the water pump. If the shift rod must be connected before the lower unit makes contact with the exhaust housing, make the connection at this time. Continue to work the lower unit towards the exhaust housing, and at the same time rotate the propeller shaft as an aid to indexing the driveshaft splines with the crankshaft. Start the bolts securing the lower unit to the exhaust housings together. Tighten the bolts **ALTERNATELY** and **EVENLY** to the torque value given in the Appendix.

SPECIAL PROCEDURES

Fill the lower unit with lubricant according to the procedures in Section 8-3.

Install the propeller, see Section 8-2.

26- Final adjustment for remote control units: Shift the lower unit into **NEUTRAL** gear. At the shift box, move the shift lever to the **NEUTRAL** position. If the pin on the end of the shift cable, does not align with the shift handle, move the adjusting knob until the pin aligns and will move into the shift handle. With the shift cable removed, move the lower unit into **FORWARD** gear and at the same time rotate the propeller **CLOCKWISE** to ensure the gears are fully indexed. At the control box, move the shift

lever into the **FORWARD** position. Again check to be sure the pin on the end of the shift cable aligns with the hole in the shift lever. Adjust the knob on the shift cable until the pin does align with the hole in the shift lever.

FUNCTIONAL CHECK

Perform a functional check of the completed work by mounting the engine in a test tank, in a body of water, or with a flush attachment connected to the lower unit. If the flush attachment is used, **NEVER** operate the engine above an idle speed, because the no-load condition on the propeller would allow the engine to **RUNAWAY** resulting in serious damage or destruction of the engine.

CAUTION: Water must circulate through the lower unit to the engine any time the engine is run to prevent damage to the water pump in the lower unit. Just five seconds without water will damage the water pump.

Start the engine and observe the tattletale flow of water from idle relief in the exhaust housing. The water pump installation work is verified. If a "Flushette" is connected to the lower unit, **VERY LITTLE** water will be visible from the idle relief port. Shift the engine into the three gears and check for smoothness of operation and satisfactory performance.

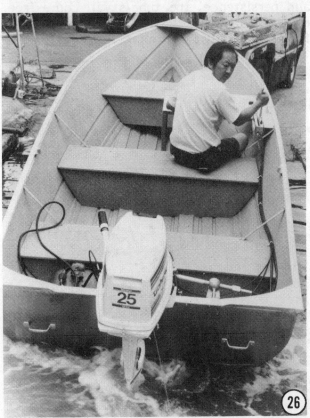

Small horsepower unit mounted in a test tank in preparation to making final adjustments and return to service.

8-8 LOWER UNIT SERVICE
MANUAL SHIFT — EARLY 40 HP TO 1976
(40 HP SINCE 1977 — SEE SEC. 8-13)

Propeller Removal

Remove the propeller according to the detailed procedures outlined in Section 8-2.

Draining Lower Unit

Drain the lower unit according to the detailed procedures outlined in Section 8-3.

GOOD WORDS

Only one type of shift mechanism and removal procedures are used on the engines covered in this section. Access to the shift disconnect is through a window on the starboard side of the engine.

LOWER UNIT REMOVAL

1- Remove the metal plate from the port side of the engine. Access to the shift connector is gained through the opening. Disconnect the shift rod from the exhaust housing by removing the bottom bolt from the shift connector.

GOOD WORDS

In **MOST** cases, if any unit being serviced has the 6-inch extension, it is **NOT** necessary to remove the extension in order to "drop" the lower unit.

2- Remove the attaching hardware securing the rear exhaust housing cover, installed on all 40 hp models. This cover **MUST** be removed to gain access to one of

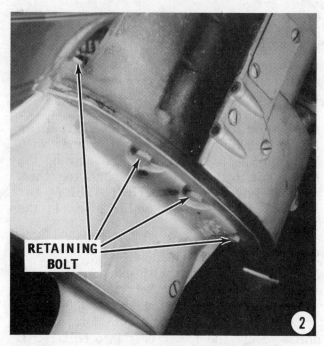

RETAINING BOLT

the bolts securing the lower unit to the exhaust housing. Remove the bolts securing the lower unit to the housing. Some units may have an additional bolt on each side and one at the rear of the engine. Work the lower unit loose from the exhaust housing. It is not uncommon for the water tube/s to be stuck in the water pump making separation of the lower unit from the exhaust housing difficult. However, with patience and persistence, the tube/s will come free of the pump and the lower unit separated from the exhaust housing.

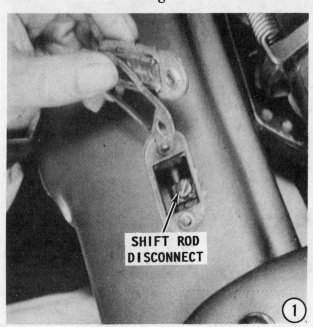

SHIFT ROD DISCONNECT

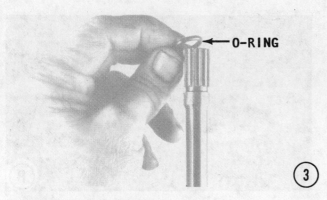

O-RING

③

WATER PUMP HOUSING

⑤

MORE GOOD WORDS

Position the lower unit in a vertical position on the edge of the work bench resting on the cavitation plate. Secure the lower unit in this position with a C-clamp. The lower unit will then be held firmly in a favorable position for further service work. An alternate method is to cut a groove in a short piece of 2" x 6" wood to accommodate the lower unit with the cavitation plate resting on top of the wood. Clamp the wood in a vise and service work may then be performed with the lower unit erect (in its normal position), or inverted (upside down). In both positions, the cavitation plate is the supporting surface, illustration **A**.

3- Remove and **DISCARD** the O-ring from the top of the driveshaft.

WATER PUMP REMOVAL

AUTHOR'S APOLOGY

The photographs taken for this section involved a water pump from an electric shift lower unit. However, the pump and the service procedures are identical for the manual shift. Therefore, disregard any wiring shown in the photographs.

4- Remove the O-ring from the top of the water pump. Remove the screws securing the water pump to the lower unit housing. It is very possible corrosion will cause

the screw heads to break-off when an attempt to remove them is made. If this should happen, use a chisel and breakaway the water pump housing from the lower unit. **EXERCISE CARE** not to damage the lower unit housing.

5- After the screws have been removed, slide the water pump, impeller, the impeller key, and the lower water pump plate upward and free of the driveshaft.

GOOD NEWS

If the only work to be performed is service of the water pump, disregard the following steps, and proceed directly to the Water Pump Assembling portion of this section on Page 8-67, Water Pump Installation.

LOWER UNIT DISASSEMBLING

6- CAREFULLY pull upward on the driveshaft and remove it from the lower unit.

7- Turn the lower unit upside down and again clamp it in the vise or slide it into the wooden block, if one is used. Carefully examine the lower portion of the unit. The cap is considered that part below the split with the skeg attached. Remove the Phillips screw from the starboard side of the

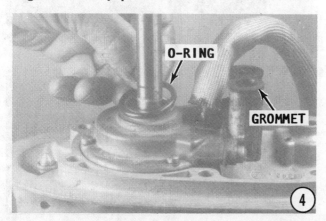

O-RING

GROMMET

④

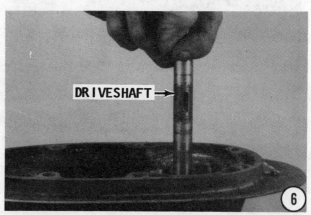

DRIVESHAFT

⑥

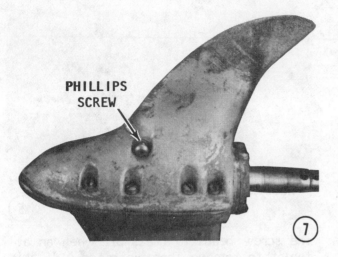

PHILLIPS
SCREW

7

lower housing. This screw passes through the shift yoke and threads into the other side of the housing.

8- Remove the attaching screws around the cap. These screws may be slotted-type or Phillips screws. **CAREFULLY** tap the cap to jar it loose, and then separate it from the lower unit housing.

TAKE TIME

Before proceeding with the disassembly work, take time to study the arrangement of parts in the lower unit. You may elect to follow the practice of many professional mechanics and take a polaroid picture of the unit as an aid during the assembly work. Several engineering and production changes have been made to the lower unit over the years. Therefore, the positioning of the gears, shims, bearings, and other parts may vary slightly from one unit to the next.

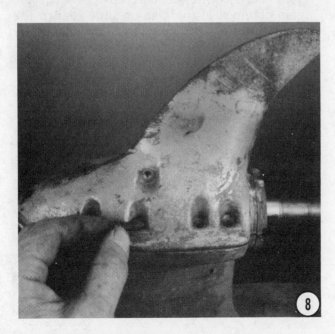

8

B

To show each and every arrangement with a picture in this manual would not be practical. Even if it were done, the ability to associate the unit being serviced with the illustration would be almost impossible. Therefore, take time to make notes, scribble out a sketch, or take a couple photographs, illustration **B**.

9- Lift the shift lever out of the cradle, and then remove the cradle from the shift dog. Raise the propeller shaft and at the same time tap with a soft-headed mallet on the bottom side to jar it loose, then remove the shaft assembly from the lower unit. The forward and reverse gear including the bearings will all come out with the propeller shaft. The forward gear is the gear at the opposite end of the shaft from the propeller.

GOOD WORDS

Notice that the forward gear bearing is a tapered bearing with a race and that the taper faces outward, **AWAY** from the gear. Also observe the seal retainer on the propeller end of the propeller shaft. Now, notice the matching pin in the lower unit housing. During the installation work, the retainer

PIN

9

MUST be installed with the pin indexed in the hole. Take note of the snap ring installed between the thrust washer and the reverse gear bearing. One more item of particular interest. Notice the two sides of the thrust washer. One side is as a normal washer, but the other side is a babbitt. The babbitt side MUST face toward the reverse gear during installation. The washer also has two dog ears, one facing upward and the other downward, illustration C.

By taking note at this time of these items and exactly how they are installed, the task of assembling and installation will progress more smoothly.

10- Remove the attaching screws, and then the U-shaped bracket from the top of the pinion gear

11- Reach into the lower housing and remove the pinion gear.

12- Slide the tapered bearing, forward gear, washer, and clutch dog, off the propeller shaft.

13- Remove the seal retainer, reverse bearing, snap ring, washer, reverse gear and bearing, and the washer, from the propeller end of the shaft.

14- Turn the lower unit housing right side up and again clamp it in the vise. Remove the upper seal using a seal puller. An alternate method, if the puller is not available, is to use two screwdrivers and

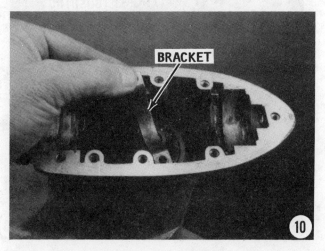

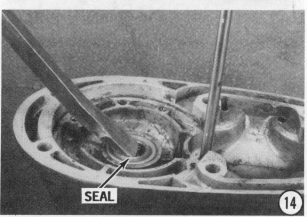

Using a slide hammer to remove the seals or bearing from underneath the water pump base plate.

work the seal out of the housing. **TAKE CARE** not to damage the seal recess as the seal is being removed. A babbitt bearing is installed under the seal. Late model units may have caged needle bearings installed. Normally, it is **NOT** necessary to remove this bearing. However, check the bearing surface with a finger and if any roughness is felt, the bearing **MUST** be replaced.

15- Clean the upper portion of the shift rod as an aid to pulling it through the bushing and O-ring. Pull the shift rod from the lower unit housing.

16- The shift rod passes through an O-ring and bushing in the lower unit housing. These two items prevent water from entering the lower unit. A special tapered punch

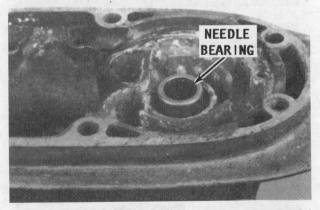

Caged needle bearing set installed on newer model lower units instead of a babbitt bearing.

is required to remove the bushing from the lower unit housing. Obtain the special punch, and then remove the bushing, and the O-ring.

CLEANING AND INSPECTING

Clean all water pump parts with solvent, and then dry them with compressed air. Inspect the water pump cover and base for cracks and distortion, possibly caused from overheating. Inspect the face plate and water pump insert for grooves and/or rough surfaces. If possible, **ALWAYS** install a complete new water pump while the lower unit is disassembled. A new impeller will ensure extended satisfactory service and give "peace of mind" to the owner. If the old impeller must be returned to service, **NEVER** install it in reverse to the original direction of rotation. Installation in reverse will cause premature impeller failure.

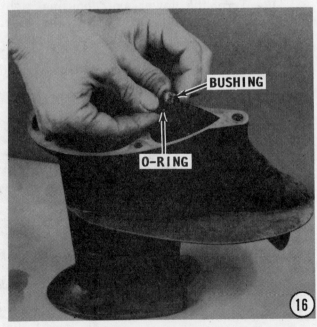

Babbitt bearing unfit for further service. This type bearing must not show any sign of water damage.

Inspect the impeller side seal surfaces and the ends of the impeller blades for cracks, tears, and wear. Check for a glazed or melted appearance, caused from operating without sufficient water. If any question exists, and as previously stated, install a new impeller if at all possible.

Clean all parts with solvent and dry them with compressed air. **DISCARD** all O-rings and gaskets. Inspect and replace the driveshaft if the splines are worn. Inspect

Caged ball bearing set destroyed due to lack of lubrication, vibration, corrosion, metal particles, or all the above.

View of a badly corroded lower unit. Water entered and was allowed to remain over an extended period of time causing extensive damage.

Badly worn pinion gear from a lower unit. The teeth of the gears must be carefully inspected for wear and damage.

A rusted and corroded gear. Water was allowed to enter the lower unit through a badly worn seal and cause this damage to the gear and other expensive parts.

A new shift cradle (right) compared with one badly worn and unfit for further service (left).

the gearcase and exhaust housing for damage to the machined surfaces. Remove any nicks and refurbish the surfaces on a surface plate. Start with a No. 120 Emery paper and finish with No. 180.

Check the water intake screen and passages by removing the bypass cover, if one is used. Inspect the clutch dog, drive gears, pinion gear, and thrust washers. Replace these items if they appear worn. If the clutch dog and drive gear arrangement surfaces are nicked, chipped, or the edges rounded, the operator may be performing the shift operation improperly or the controls may not be adjusted correctly. These items **MUST** be replaced if they are damaged.

Inspect the dog ears on the inside of the forward and reverse gears. The gears must be replaced if they are damaged.

Check the cradle that rides on the inside diameter of the clutch dog. The sides of the

Distorted tapered bearing unfit for further service.

cradle must be in good condition, free of any damage or signs of wear. If damage or wear has occurred, the cradle must be replaced.

Check the shift lever and the two prongs that fit inside the cradle. Check to be sure the prongs are not worn or rounded. Damage or wear to the prongs indicates the lever must be replaced.

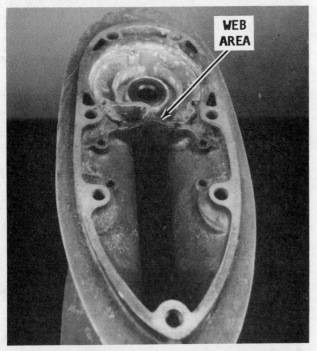

The web area indicated should be inspected carefully for the slightest sign of a hairline crack. The smallest evidence of a crack is cause to replace the housing because exhaust gases will find their way into the water pump area when the engine is operated and and the boat planing. These exhaust gases will cause the engine to run hot.

Using a punch to remove the seal from the bearing retainer.

Installing a seal into the bearing retainer. Notice the O-ring has been installed.

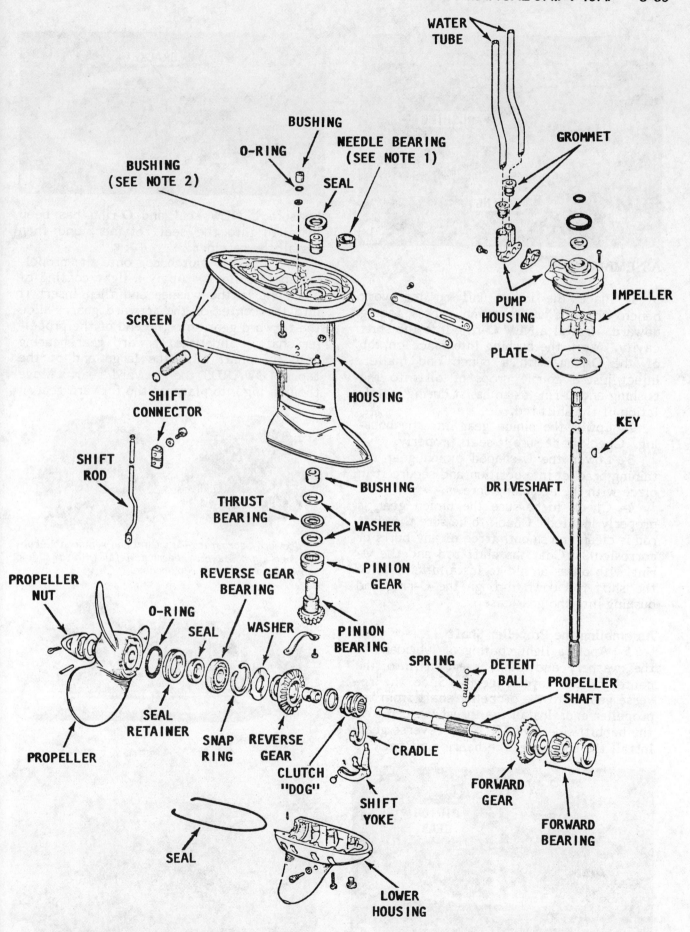

WATER TUBE

GROMMET

BUSHING

O-RING

NEEDLE BEARING (SEE NOTE 1)

BUSHING (SEE NOTE 2)

SEAL

IMPELLER

PUMP HOUSING

PLATE

SCREEN

HOUSING

KEY

SHIFT CONNECTOR

SHIFT ROD

BUSHING

DRIVESHAFT

THRUST BEARING

WASHER

PINION GEAR

PROPELLER NUT

REVERSE GEAR BEARING

PINION BEARING

O-RING

SEAL

WASHER

SPRING

DETENT BALL

PROPELLER SHAFT

SEAL RETAINER

SNAP RING

REVERSE GEAR

CRADLE

CLUTCH "DOG"

FORWARD GEAR

PROPELLER

SHIFT YOKE

FORWARD BEARING

SEAL

LOWER HOUSING

Exploded drawing of the lower unit for the 40 hp units, with major parts identified. Notice the double water tubes out of the water pump and the detent spring and two detent balls in the propeller shaft.

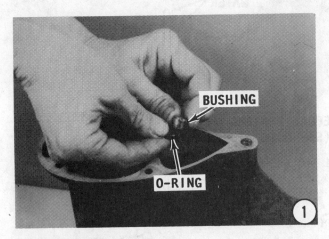

ASSEMBLING

1- Place the lower unit on the work bench with the water pump recess facing upward. Install a **NEW** O-ring into the shift cavity. Work the bushing into place on top of the O-ring with a punch and mallet. Inject just a couple drops of oil into the bushing and O-ring as an assist during installation of the shift rod.

2- Lower the pinion gear into the housing. Check to be sure it seats properly.

3- Lower the U-shaped pinion gear retaining bracket into position and secure it in place with the attaching screws.

4- Check to be sure the pinion gear is properly located. Check to be sure the shift rod is clean and smooth (free of any burrs or corrosion). Coat the shift rod and the O-ring with oil as an aid to installation. Slide the shift rod down through the O-ring and bushing into the gear case.

Assembling the Propeller Shaft

5- Apply a light coating of lubricant to the washer, and then insert it into the center of the reverse gear. Slide the reverse gear onto the propeller shaft from the propeller end. Install the thrust washer with the babbitt side **TOWARDS** the reverse gear. Install the snap ring, the bearing. Check to

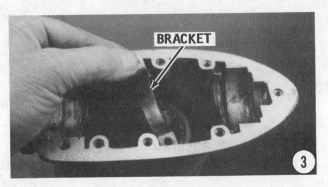

be sure a **NEW** seal and O-ring has been installed into the seal retainer, and then install the retainer.

6- Slide the clutch dog onto the propeller shaft splines. Apply a light coating of lubricant to the washer and then insert it into the center of the forward gear. Slide the forward gear onto the end of the propeller shaft. Slide the forward gear bearing onto the shaft with the large end of the taper **TOWARDS** the forward gear. Move the bearing into place on the forward gear.

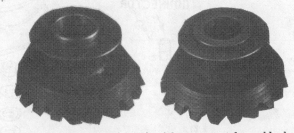

Forward bearing (right) with a non-replaceable babbitt bearing. Reverse bearing (left) with a sliding babbitt bearing that is replaceable.

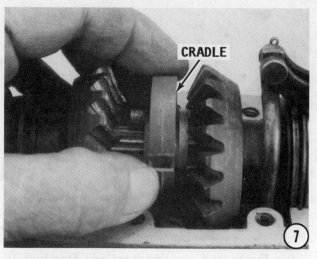

Check to be sure a new O-ring and bearing seal has been installed into the gear case head, and then install the gearcase head assembly onto the propeller shaft.

CRITICAL WORDS

The seal retainer has a hole and the lower housing of the lower unit has a pin. This pin **MUST** index into the hole in the retainer when the propeller shaft is installed. If the pin is not seated properly in the hole, the seal retainer will work part way out of the housing and the lubricant in the lower unit will be lost, illustration **A**.

7- Slide the propeller shaft assembly into the lower unit housing. Check to be sure the forward and reverse gear index with the pinion gear and the hole in the seal retainer indexes with the pin in the lower unit housing. Lubricate the cradle, and then slip it into the clutch dog groove.

8- Bring the shift lever down over the cradle and snap the fingers of the lever into the cradle. Check to be sure the clutch dog is in the **NEUTRAL** position. Push or pull on the shift rod to move it up or down until the clutch dog is in the center between the forward and reverse gears.

9- Lay down a bead of No. 1000 Sealer into the groove of the cap in preparation to installing the seal.

10- Place a **NEW** seal in the lower cap and hold the seal in the groove with sealer. Apply a small amount of silicone sealer on each side of the bearing gear case head. This sealer will form a complete seal when

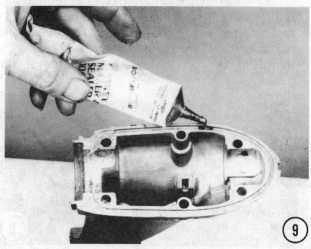

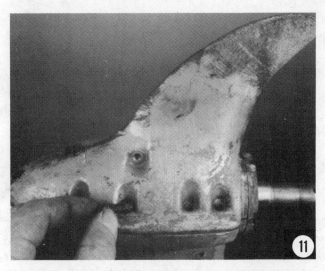

the lower unit cap is installed. Position the lower unit cap over the gear assembly onto the lower unit housing.

11- Apply a drop of sealer into the opening for each cap retaining screw to ensure a complete seal between the cap and the lower unit housing. Install the screws securing the cap to the lower unit housing. Tighten the screws **ALTERNATELY** and **EVENLY**.

GOOD WORD

If time is taken to grind the end of the screw to a **SHORT** point, it will make the task of installation much easier. If the cap and shift lever are not aligned exactly, the screw will "seek" and make the alignment as it passes through. However, do not make a long point or the screw will not have enough support and would bend during operation of the shift lever, illustration **B.**

12- Use a flashlight and align the hole in the cap with the hole in the shift lever. Install the tapered Phillips screw into the housing and through the lever. Tighten the screw securely.

13- Install the babbitt or needle bearing, if it was removed. The babbitt bearing may be installed using the proper size socket and hammer. If the caged needle bearing is installed tap on the numbered side of the bearing.

14- Coat the outside edge of a **NEW** seal with No. 1000 sealer, and then tap the seal into place in the top of the upper lower unit housing.

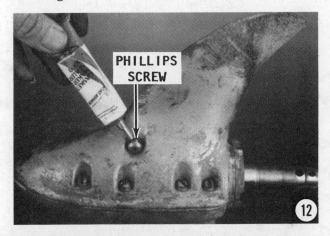

PHILLIPS SCREW

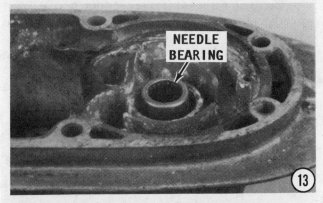

NEEDLE BEARING

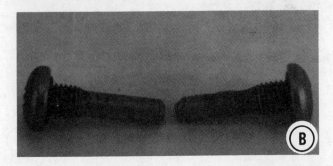

SEAL

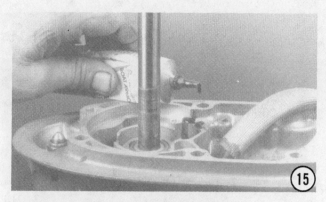

Old type standard water pump (left) with new type (right). The new type must be used if the lower unit has been updated (replaced) with a new unit.

WATER PUMP INSTALLATION

15- Apply a coating of sealer to the upper surface of the lower unit.

16- Install the water pump base plate. Slide the driveshaft into the lower unit, and then rotate the shaft very slowly. When the splines of the driveshaft index with the pinion gear, the shaft will drop slightly. Install the water pump pin or key.

17- Slide the water pump impeller down the driveshaft and into place on top of the water pump base plate with the pump pin or key indexed in the impeller. Lubricate the inside surface of the water pump with lightweight oil.

18- Lower the water pump housing down the driveshaft and over the impeller. Rotate the driveshaft **CLOCKWISE** as the water pump housing is lowered to allow the impeller blades to assume their natural and proper position inside the housing. Continue to rotate the driveshaft and work the water pump housing downward until it is seated on the water pump plate.

ALWAYS rotate the driveshaft **CLOCKWISE** while the screws are tightened to prevent damaging the impeller vanes. If the impeller is not rotated, the housing could damage or cut the end of the vanes as the screws are brought up tight. The rotation allows them to spring back in a natural position.

Place **NEW** grommets into the water pump housing for the water pickup. If a new water pump was installed, this seal will already be in place.

19- Install a **NEW** O-ring on the top of the driveshaft.

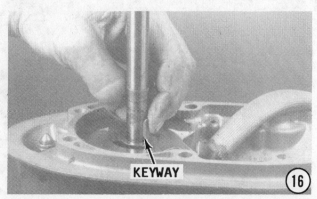

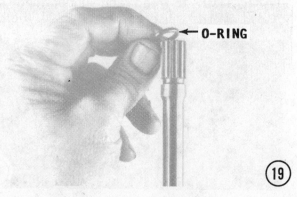

LOWER UNIT INSTALLATION

GOOD WORDS

Connecting the shift rod with the coupler is not an easy task but can be accomplished as follows: First, notice the cutout area on the end of the shift rod. This area permits the bolt to pass through the connector, past the shift rod, and into the other side of the connector. It is this bolt that holds the shift rod in the connector. Now, in order for the bolt to be properly installed, the cutout area on the shift rod **MUST** be aligned in such a manner to allow the bolt to be properly installed. Therefore, as the lower unit is mated with the exhaust housing, exercise patience as the two units come together, to enable the bolt to be installed at the proper time. If the rod is allowed to move too far into the connector before the bolt is installed, it may be possible to force the bolt into place, past the shift rod. The threads on the bolt will be stripped, and the shift rod will eventually come out of the connector, illustration **C** and **D**.

20– Install the connector, onto the lower unit shift rod, with the **NO THREAD** section facing towards the window. With the connector in this position, the bolt may be inserted through the connector and "catch" the threads on the far side. Install the connector bolt in the manner described in the previous paragraph.

21– Check to be sure the water pickup tubes are clean, smooth, and free of any corrosion. Coat the water pickup tubes and grommets with lubricant as an aid to installation. Guide the lower unit up into the exhaust housing with the water tube sliding into the rubber grommet of the water pump. Continue to work the lower unit towards the exhaust housing, and at the same time rotate the propeller shaft as an aid to indexing the driveshaft splines with the crankshaft.

22– Insert the bolt into the connector. **TAKE TIME** to read and understand the "Good Words" just before Step 20, before

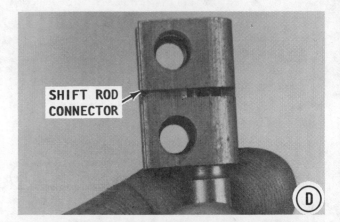

SHIFT ROD CONNECTOR

D

BOLT

20

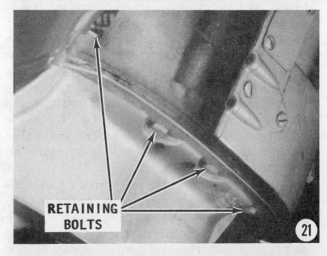

RETAINING BOLTS

21

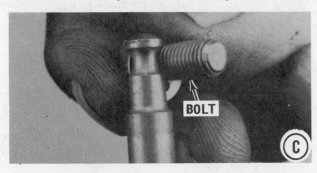

BOLT

C

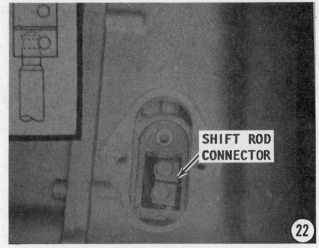

SHIFT ROD CONNECTOR

22

making this connection. After the bolt is in place, install and secure the window with the attaching hardware.

Start the bolts securing the lower unit to the exhaust housing. Tighten the bolts **EVENLY** and **ALTERNATELY** to the torque value given in the Appendix.

23- Install the rear cover over the exhaust housing. This cover is installed only on the 40 hp engines. When the cover is installed, check to be sure the idle relief rubber tube on the upper side underneath the powerhead fits into the recess of the cover. Secure the cover in place with the attaching hardware.

Filling the Lower Unit

Fill the lower unit with lubricant according to the procedures outlined in Seciton 8-3.

Propeller Installation

Install the propeller, see Section 8-2.

Final Adjustment

24- Final adjustment for remote control units: Shift the lower unit into **NEUTRAL** gear. At the shift box, move the shift lever to the **NEUTRAL** position. If the pin on the

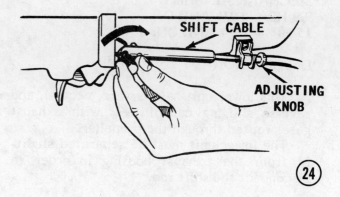

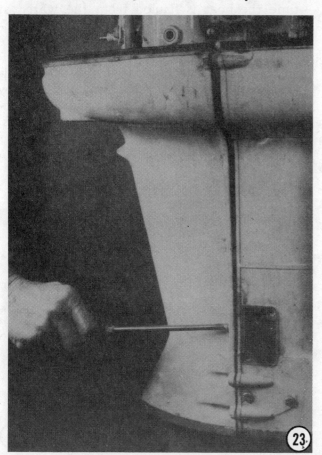

end of the shift cable, does not align with the shift handle, move the adjusting knob until the pin aligns and will move into the shift handle. With the shift cable removed, move the lower unit into **FORWARD** gear and at the same time rotate the propeller **CLOCKWISE** to ensure the gears are fully indexed. At the control box, move the shift lever into the **FORWARD** position. Again check to be sure the pin on the end of the shift cable aligns with the hole in the shift lever. Adjust the knob on the shift cable until the pin does align with the hole in the shift lever.

FUNCTIONAL CHECK

Perform a functional check of the completed work by mounting the engine in a test tank, in a body of water, or with a flush attachment connected to the lower unit. If the flush attachment is used, **NEVER** operate the engine above an idle speed, because the no-load condition on the propeller would allow the engine to **RUNAWAY** resulting in serious damage or destruction of the engine.

CAUTION: Water must circulate through the lower unit to the engine any time the engine is run to prevent damage to the water pump in the lower unit. Just five seconds without water will damage the water pump.

Start the engine and observe the tattle-tale flow of water from idle relief in the exhaust housing. The water pump installation work is verified. If a "Flushette" is connected to the lower unit, **VERY LITTLE** water will be visible from the idle relief port. Shift the engine into the three gears and check for smoothness of operation and satisfactory performance.

8-9 PROPELLER EXHAUST
MECHANICAL SHIFT
9.9 HP — 1974 AND ON
15 HP — 1974 AND ON

DESCRIPTION

This lower unit has forward, neutral, and reverse shifting capabilities, with exhaust gases routed through the propeller.

The lower unit must be separated slightly from the exhaust housing in order to disconnect the shift rod.

GOOD WORDS

Propellers with the exhaust passing through the hub **MUST** be removed more frequently than the standard propeller. Removal after each weekend use or outing is not considered excessive. These propellers do not have a shear pin. The shaft and propeller have splines which **MUST** be coated with an anti-corrosion lubricant prior to installation as an aid to removal the next time the propeller is pulled. Even with the lubricant applied to the shaft splines, the propeller may be difficult to remove.

The propeller with the exhaust hub is more expensive than the standard propeller, therefore, the cost of rebuilding the unit, if the hub is damaged, is justified, illustration **"A"**.

A replaceable diffuser ring on the aft side of the propeller disperses the exhaust gases away from the propeller blades. If the ring becomes broken or damaged "ventillation" would be created pulling the exhaust gases back into the negative pressure area behind the propeller. This condition would create considerable air bubbles and reduce the effectiveness of the propeller.

If the propeller is "frozen" onto the propeller shaft, see the special instructions for removal outlined in Section 8-14, Frozen Propeller Removal.

TROUBLESHOOTING

Troubleshooting **MUST** be done **BEFORE** the unit is removed from the powerhead to permit isolating the problem to one area. Always attempt to proceed with troubleshooting in an orderly manner. The shotgun approach will only result in wasted time, incorrect diagnosis, replacement of unnecessary parts, and frustration.

The following procedures are presented in a logical sequence with the most prevalent, easiest, and less costly items to be checked, listed first.

Water in the Lower Unit

Water in the lower unit is usually caused by fish line becoming entangled around the propeller shaft behind the propeller and damaging the propeller seal. If the line is not removed, it will cut the propeller shaft seal and allow water to enter the lower unit. Fish line has also been known to cut a groove in the propeller shaft, illustration **"B"**.

The propeller should be removed each time the boat is hauled from the water at the end of an outing and any material entangled behind the propeller removed before it can cause extensive damage. The small amount of time and effort involved in pulling the propeller is repaid many times by reduced maintenance and service work, including the replacement of expensive parts, illustration **"C"**.

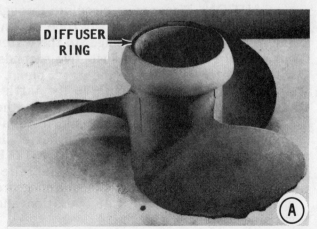

DIFFUSER RING

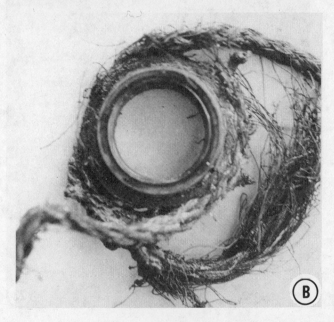

Slippage in the Lower Unit

If the shift seems to be slipping as the boat moves through the water: Check the propeller and the rubber hub. If the propeller has been subjected to many strikes against underwater objects, it could slip on its hub. If the hub is damaged or excessively worn on the small propellers, it is not economical to have the hub or propeller rebuilt. A new propeller may be purchased for considerably less than meeting the expense of rebuilding an old worn propeller, illustration **"D"**.

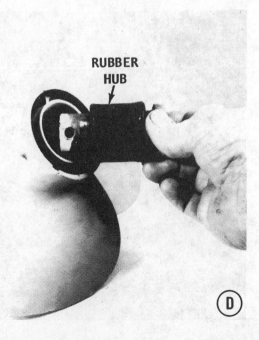

Jumping Out of Gear

If a loud thumping sound is heard at the transom while the boat is underway, the unit is jumping out of gear, resulting in a no-load condition on the propeller. When this happens, the rushing water under the hull forces the lower unit in a backward direction. The unit jumps back into gear; the propeller catches hold; the lower unit is forced forward again; and the result is the thumping sound as the action is repeated. Normally this type of action occurs perhaps once a day, then more frequently each time the clutch is operated, until finally the unit will not stay in gear for even a short time.

The following areas must be checked to locate the cause:

1- Check the upper shift rod at the connection underneath the carburetor.

2- Attempt to correct the problem with better shift habits by the operator.

3- Check for water in the lower unit as described earlier in this section.

4- Normal wear at the upper and lower shift rod connection may cause the unit to jump out of gear.

Frozen Powerhead

This condition is suggested when the operator unsuccessfully attempts to crank the engine, either with a hand starter or with a starter motor. The flywheel will not rotate. Do not assume the engine is "frozen" until the lower unit has been removed

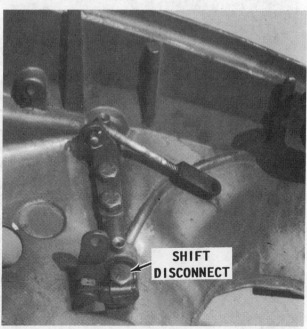

Bellcrank located under the carburetor. The shift linkage is disconnected at this point, as described in the text.

and thoroughly checked. If the lower unit is "locked" (the driveshaft or propeller shaft will not rotate), the powerhead will have the indication of being "frozen" (failure to rotate the flywheel), illustration **"E"**.

The first step to perform under these conditions is to "pull" the lower unit, and then again attempt to crank the engine. If the attempt is successful with the lower unit disconnected, the problem is in the lower unit. If the attempt to crank the engine is still unsuccessful, the problem is in the powerhead.

LOWER UNIT REMOVAL

SPECIAL WORDS

The lower unit must be separated slightly from the exhaust housing in order to disconnect the shift rod.

The bearing carrier in the lower unit may be removed and the seals replaced without removing the lower unit. **HOWEVER,** this is not considered good practice because there is no way to check the bearings for damage if water has been allowed to enter the lower unit. Therefore, if the bearing carrier seals must be replaced, the lower unit should be removed and a complete inspection made of all other bearings and seals for water damage, illustration **"F"**.

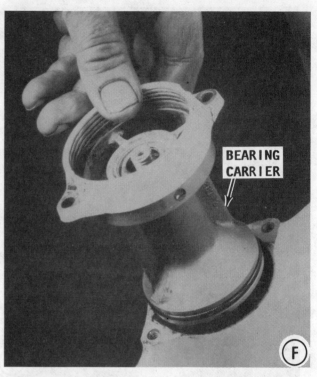

Preliminary Tasks

Drain the lower unit according to the procedures outlined in Section 8-3.

Remove the propeller, see Section 8-2. If the propeller is "frozen" to the shaft, see special instructions outlined in Section 8-14.

1- Remove the engine hood. Disconnect the spark plug leads at the spark plugs. Remove the bolts securing the lower unit to the exhaust housing or to the 6-inch extension. In **MOST** cases, it is not necessary to remove the 6-inch extension in order to "drop" the lower unit. Separate the lower

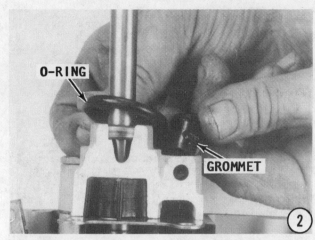

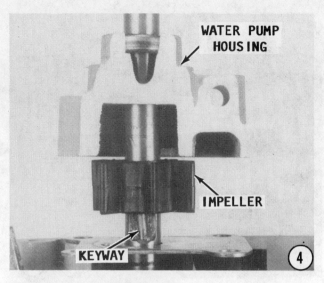

unit slightly from the exhaust housing or the 6-inch extension. Reach in and remove the bottom bolt through the shift rod connector. The shift rod is now free to separate from the connector. **CAREFULLY** remove the lower unit straight away from the exhaust housing or the 6-inch extension. **TAKE CARE** not to pull backward, sideways, etc., when removing the lower unit, because the driveshaft may be bent.

WATER PUMP REMOVAL

2- Remove the O-ring from the top of the driveshaft. Remove the O-ring and grommet from the top of the water pump.

3- Remove the bolts securing the water pump housing to the lower unit. Notice how the bolts from the aft holes are longer than the two bolts forward. Slide the impeller housing up and free of the driveshaft.

4- Remove the impeller and driveshaft key. Remove the water pump plate and gaskets from the lower unit.

If the only work to be performed is service of the water pump, proceed directly to Page 8-86, Water Pump Installation.

Bearing Carrier Removal
FIRST, THESE WORDS

A hole is drilled through the propeller shaft just forward of the reverse gear. A detent spring is installed in the hole along with a detent ball on each side of the propeller shaft. The clutch dog is installed over the top of the two detent balls. This arrangement assists in holding the unit in the specific gear desired. When the bearing carrier, propeller shaft, and associated parts are removed, the clutch dog will remain in the lower unit. As soon as the clutch dog slides free of the propeller shaft, the detent balls and spring will fly free of the shaft, but be contained within the lower unit housing. Therefore, after the propeller shaft, bearing carrier, etc., are removed, take time to retrieve the two detent balls and the spring from inside the lower unit, illustration **G**.

5- Remove the two screws securing the bearing carrier in the lower unit.

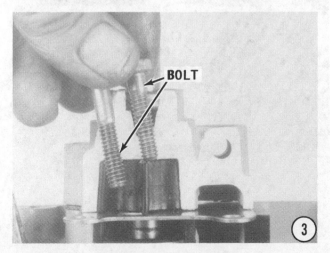

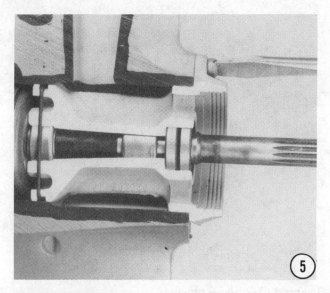

5

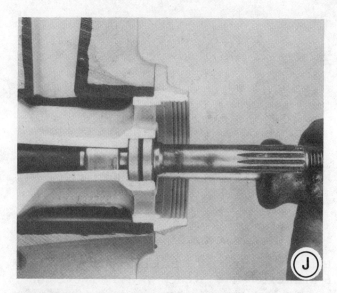

J

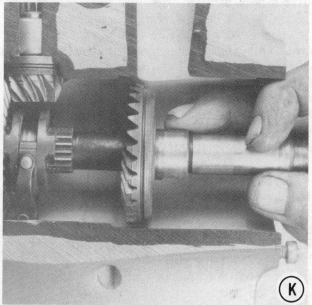

K

GOOD WORDS

Three methods are available to free the carrier from the lower unit.

The first method involves the use of a special tool, OMC No. 386631. This tool is installed over the propeller shaft; the propeller nut is threaded onto the shaft behind the tool; and then the tool extended with a wrench on both sides of the shaft; "pulling" the propeller shaft, bearing carrier, and reverse gear from the lower unit.

On some bearing carriers, two threaded holes are provided in the carrier. The second method involves the use of a fly-wheel puller. Two long bolts are installed into the threaded holes of the bearing carrier, while the center bolt is tightened against the end of the propeller shaft to pull the carrier free, illustration #H.

The third method of removing the bearing carrier also utilizes the two threaded holes in the carrier. A slide hammer with a long rod is attached to the carrier and the carrier removed in that manner, illustration J. The propeller shaft and reverse gear are then removed, illustration K.

6- Reach in with a pair of needle-nose pliers and remove the clutch dog from the cradle, as shown.

H

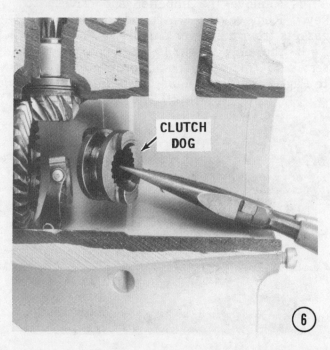

CLUTCH DOG

6

7- Remove the driveshaft from the lower unit and at the same time, reach in and remove the pinion gear, two thrust washers, and the thrust bearing. Pay particular attention to the washers and how they are positioned on top of the pinion gear. One washer is noticeably thicker than the other. Also, one washer is beveled on the inside diameter and the other washer is beveled on the outside diameter. Take time to identify one of the washers with a dab of paint or other mark to ensure they will be installed in the same location from which they were removed.

8- Back out the shift rod from the yoke by turning it COUNTERCLOCKWISE until it

is free. Withdraw the shift rod from the lower unit.

9- Remove the Phillips screw from outside of the lower unit. This is the screw very close to the lubricant drain plug. The screw secures the shift lever and yoke assembly in position. After the screw is removed, the shift lever, yoke assembly, and forward gear may be removed from the lower unit.

10- Remove the forward gear tapered bearing from the lower unit.

11- Remove the forward gear bearing race. This race need NOT be removed unless the forward gear tapered bearing is to be replaced. The bearing and the race are sold as a matched pair.

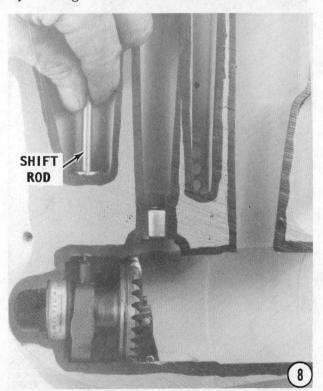

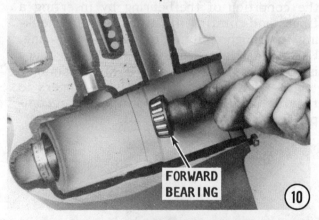

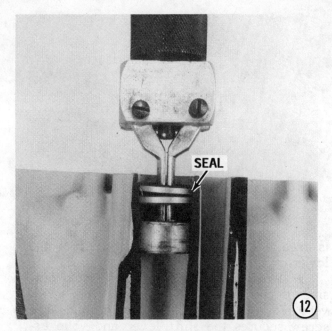

(12)

Upper Driveshaft Seals -- Removal

12- Use a slide hammer with finger-type pullers and remove the two seals. Notice how the two seals are installed back-to-back. It is most important that they be installed in the same position from which they were removed.

Upper Driveshaft Bearing Removal

This bearing need **NOT** be removed, unless it is unfit for further service. Check the condition of the bearing by inserting a finger and rotating the bearing while checking for rough spots or evidence of binding. Use a flashlight and check to be sure there is no evidence of corrosion or other damage.

13- Use a slide hammer with fingers to remove the bearing, as shown in the accompanying illustration.

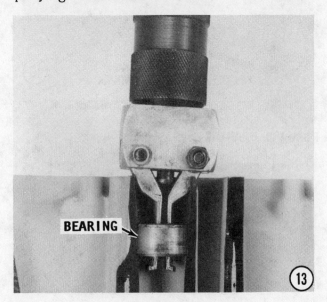

(13)

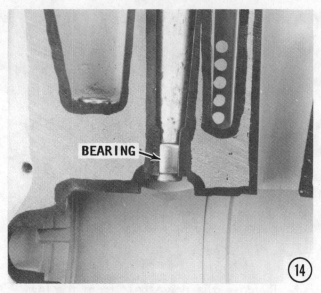

(14)

Pinion Gear Bearing -- Removal

As with the upper driveshaft bearing, the lower bearing need **NOT** be removed unless it is unfit for further service. Check its condition in the same manner as described for the upper bearing.

14- Use a drift punch or other suitable tool and drive the bearing out of position into the lower unit. A special tool is not required because if the bearing is to be removed it is unfit for further service and additional damage will be of no consequence.

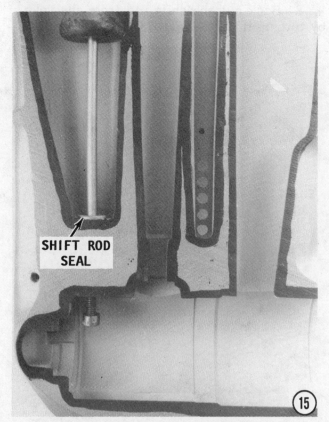

(15)

Shift Rod O-ring and Bushing Removal

15- Insert a length of 1/4" rod down through the O-ring and bushing. Thread a nut onto the lower end of the rod. (The edges of the nut must be first rounded to permit the nut to pass through the opening in the housing). Attach a slide hammer to the rod and "pull" the O-ring, bushing and washer.

Bearing Carrier Seal Removal

16- Use a puller with two fingers and work the puller down into the seals. Take up on the puller and remove the seals. Notice how the seals are installed back-to-back. They must be installed in this manner to provide a proper seal to prevent the lubricant in the lower unit from escaping and water from entering.

Bearing Carrier Bearing Removal

The bearings in the bearing carrier need **NOT** be removed unless they are unfit for further service. Check their condition for damage and corrosion. Insert a finger inside and rotate the bearing while checking for roughness or any sign of binding. Use a flashlight and check for evidence of corrosion.

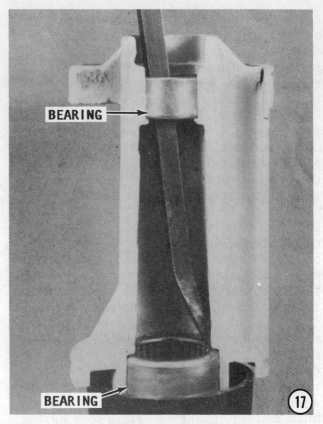

17- Use a punch and drive the bearing out. The bearing is being removed because it is unfit for service, therefore further damage is of no consequence.

CLEANING AND INSPECTING

Clean all water pump parts with **solvent**, and then dry them with compressed air. Inspect the water pump cover and base for

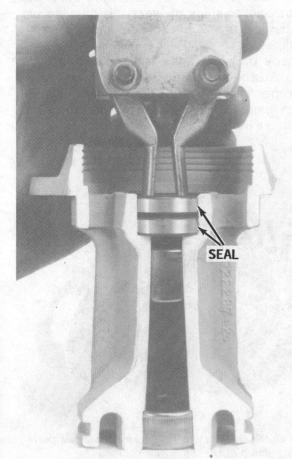

Badly worn pinion gear from a lower unit. The teeth of the gears must be carefully inspected for wear and damage.

cracks and distortion, possibly caused from overheating. Inspect the face plate and water pump insert for grooves and/or rough surfaces. If possible, **ALWAYS** install a complete new water pump while the lower unit is disassembled. A new impeller will ensure extended satisfactory service and give "peace of mind" to the owner. If the old impeller must be returned to service, **NEVER** install it in reverse to the original direction of rotation. Installation in reverse will cause premature impeller failure.

Inspect the impeller side seal surfaces and the ends of the impeller blades for cracks, tears, and wear. Check for a glazed or melted appearance, caused from operating without sufficient water. If any question exists, and as previously stated, install a new impeller if at all possible.

Clean all parts with solvent and dry them with compressed air. **DISCARD** all O-rings and gaskets. Inspect and replace the driveshaft if the splines are worn. Inspect the gearcase and exhaust housing for damage to the machined surfaces. Remove any nicks and refurbish the surfaces on a surface plate. Start with a No. 120 Emery paper and finish with No. 180.

Check the water intake screen and passages by removing the bypass cover, if one is used. Inspect the clutch dog, drive gears, pinion gear, and thrust washers. Replace these items if they appear worn. If the clutch dog and drive gear arrangement surfaces are nicked, chipped, or the edges

The ears on this clutch dog and the teeth on the gear are badly worn. Both items are unfit for further service.

rounded, the operator may be performing the shift operation improperly or the controls may not be adjusted correctly. These items **MUST** be replaced if they are damaged.

Inspect the dog ears on the inside of the forward and reverse gears. The gears must be replaced if they are damaged.

Check the cradle that rides on the inside diameter of the clutch dog. The sides of the cradle must be in good condition, free of any damage or signs of wear. If damage or wear has occurred, the cradle must be replaced.

Check the shift lever and the two prongs that fit inside the cradle. Check to be sure the prongs are not worn or rounded. Damage or wear to the prongs indicates the lever must be replaced.

A rusted and corroded gear. Water was allowed to enter the lower unit through a badly worn seal and cause this damage to the gear and other expensive parts.

New clutch dog and gear. Compare these two parts with those shown at the top of this column.

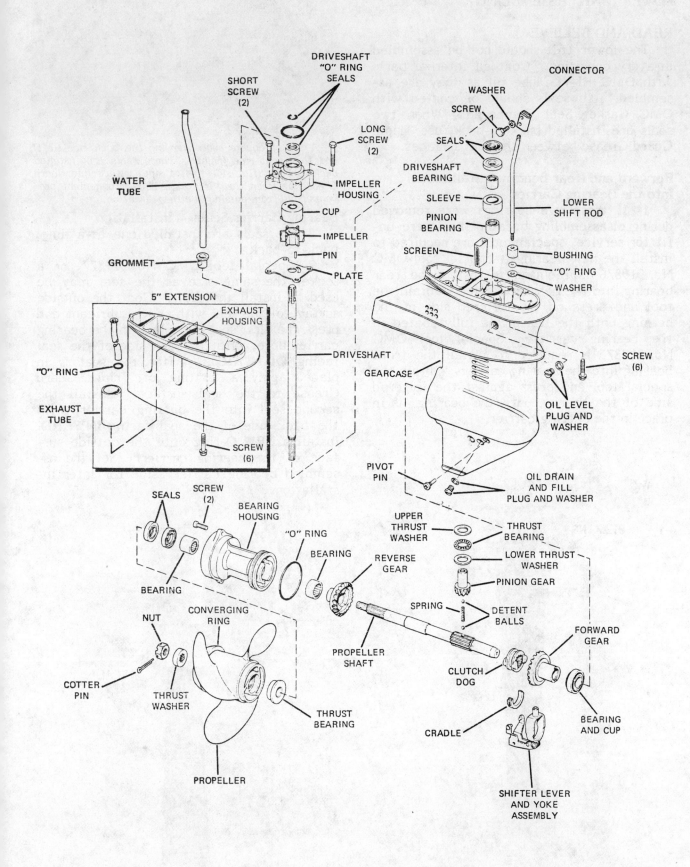

Exploded drawing of the lower unit for the 9.9 hp and 15 hp units with major parts identified. Notice the detent spring and two detent balls in the propeller shaft.

LOWER UNIT ASSEMBLING

READ AND BELIEVE

The lower unit should not be assembled in a dry condition. Coat all internal parts with OMC HI-VIS lube oil as they are assembled. All seals should be coated with OMC Gasket Seal Compound. When two seals are installed back-to-back, use Triple Guard Grease between the seal surfaces.

Forward and Rear Bearing Installation Into the Bearing Carrier

1- If these two bearings were removed during disassembling because they were unfit for service, special tools are required to install the new bearings. Special tool OMC No. 319876 is required to install the rear bearing into the bearing carrier. Obtain the tool and press on the lettered side of the bearing until the bearing is fully seated in the bearing carrier. Special tool OMC No. 319875 is required to install the front bearing into the bearing carrier. Obtain the special tool and press against the lettered side of the bearing until the bearing is in place in the bearing carrier.

A set of double seals showing the back side (left) and the front side (right). These seals are installed back-to-back (flat side-to-flat side) with Triple Guard Grease between the surfaces. This arrangement prevents fluid from passing in either direction.

Bearing Carrier Seals — Installation

These seals are installed one at-a-time, back-to-back.

2- Special tool OMC No. 319877, or a socket the same size as the seal may be used to install the seals. Coat the outside surface of the seal with OMC Lubricant and press the first seal into place in the bearing carrier bore, with the back side of the seal facing **OUTWARD**. After the seal is in place, apply a coating of Triple Guard Grease to the seal surface. Install the second seal into the bearing carrier, with the back side of the seal facing **INWARD**. Install a **NEW O**-ring onto the outside surface of the bearing carrier. Set the assembled bearing carrier aside for later installation.

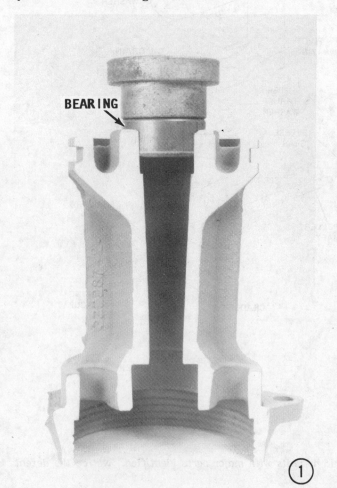

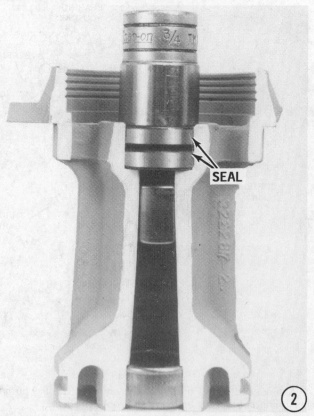

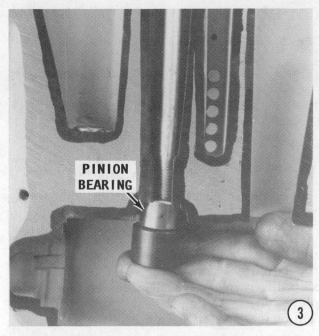

PINION BEARING

③

SEAL

⑤

Pinion Gear Bearings -- Installation
SPECIAL WORDS

The lower pinion gear bearing **MUST** be installed from the bottom cavity of the lower unit. The upper pinion gear bearing is pressed into place from the top.

3- If the lower pinion gear bearing was removed during disassembling because it was unfit for further service, obtain special tools OMC No. 319878 and No. 383173. Using these special tools is the only way the bearing may be installed properly. Actually, the bearing is "pulled" up into place with the special tools. Place the bearing in position for installation with the lettered side facing **DOWN.** Insert the special tool through the opening in the lower unit and through the bearing. Thread the bolt into the special tool and "pull" the bearing up into place in the lower unit.

4- To install the upper pinion gear bearing, obtain special tool OMC No. 319931 or No. 326566. Place the bearing in position with the lettered side facing **UP.** Use the special tool and press the bearing into place.

5- Install the upper driveshaft seals back-to-back. Coat the surfaces between the two seals with Triple Guard Grease.

Shift Rod O-ring and Bushing -- Installation

6- Lower the washer, O-ring, and bushing for the shift rod into place in the lower unit housing. Obtain special tool OMC No. 304515. Tap the bushing into place with a hammer and the special tool until the bushing is fully seated in the housing. If the special tool is not available, a socket or other similar tool may be used to install the bushing **PROVIDED** the tool will not damage the bushing during the installation process.

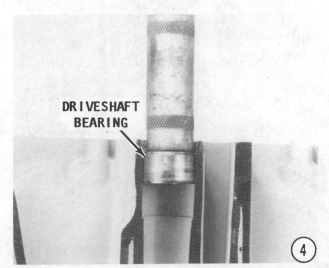

DRIVESHAFT BEARING

④

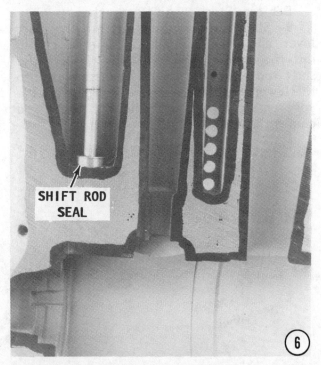

SHIFT ROD SEAL

⑥

Forward Gear Bearing Race — Installation

7- Obtain special tool OMC No. 319929 and drive handle OMC No. 311880. Seat the tool into the bearing race. Drive the race into place in the lower unit. If the leading edge of the lower unit is placed on a block of wood, or other solid non-mar surface, when the race is driven into place, each blow will be firm without "bounce" and the lower unit housing will not be damaged or scarred.

8- Insert the forward gear bearing into the race installed in the previous step.

WORDS OF ADVICE

The following operation, installation of the forward gear shift lever and cradle, requires time and patience. Make an effort to keep a cool tool without becoming frustrated, and the installation will be accomplished in a reasonable time.

A special flexible tool is listed, OMC No. 319991, to install these parts. The cable is threaded into the yoke and fed up through the lower unit. When the assembled propeller shaft is inserted into the lower unit, the cable is pulled through the lower unit and thus the short extension of the yoke will be guided into the shift rod hole. However, if the tool is not available, time and patience will result in victory, and the unit will function properly. Without the special tool, proceed as outlined in the next step.

ADVICE

Perform Steps 9 and 10 together.

9- Slide the forward gear into the shift lever and yoke assembly. Slip the cradle into the fingers of the shift lever. Coat the shift rod with oil. Insert the assembly into the lower unit and work the top part of the yoke up into the recess where the shift rod comes through.

FORWARD BEARING

⑧

10- At the same time, insert the shift rod down through the lower unit and shift rod bushing. Thread the shift rod into the shift lever yoke assembly four or five complete turns. An adjustment for the shift rod will be made later in Step 21.

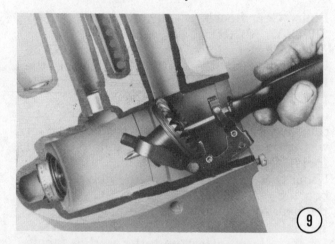

⑨

FORWARD BEARING RACE

⑦

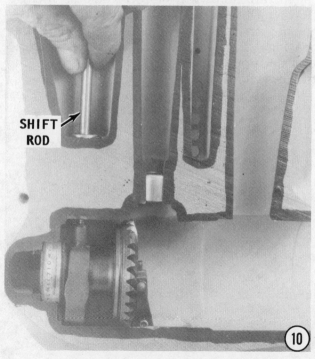

SHIFT ROD

⑩

14- Slide the pinion gear shank up into the pinion gear bearing, and at the same time, lower the driveshaft down through the lower unit and into the pinion gear. Rotate the driveshaft very slightly after it makes contact with the pinion gear to allow the splines on the shaft to index with the splines of the gear.

Pinion Gear Installation

11- Observe the two pinion gear washers. One washer is beveled on an inner edge and the other washer is beveled on an outer egde. Slide the washer with the inner edge bevel onto the shank of the pinion gear with the bevel facing **DOWN.**

12- Slide the thrust bearing onto the shank of the pinion gear.

13- Slide the remaining beveled washer onto the shank with the bevel on the outer egde facing **UP.**

Clutch Dog Installation

15- Pick up the clutch dog and notice the grooves on one side of the outside surface. When the clutch dog is installed,

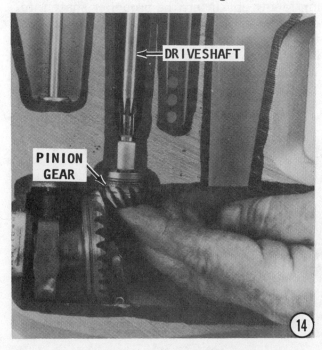

GROOVE   Ⓐ

these grooves **MUST** face the forward gear, illustration **A.** Grasp the clutch dog with a pair of needle-nose pliers and insert it into the cradle of the yoke assembly with the grooved side going in **FIRST.** Life is not a bowl of cherries, and this is not the easiest task, but if the cradle is tilted back slightly it will help with the installation. Work slowly and with patience, and the clutch dog will be properly seated in the yoke assembly.

Propeller Shaft Installation

16- Coat the propeller shaft, the detent spring, and two detent balls, with needle bearing grease or similar lubricant to hold them in place during installation of the shaft. The detent balls do not seat all the way into the shaft holes. Therefore, if grease is not used, they will not remain in place while the propeller shaft is installed into the lower unit.

Slide the spring into the shaft and place the two balls in position. Place the lower unit in a horizontal position with the opening facing up. Insert the propeller shaft into the lower unit with the shaft in the position that places the balls on each side, as shown. In this position the balls should be aligned

CLUTCH DOG   ⑮

with the ramps in the clutch dog. The detent balls must ride in a groove inside the clutch dog.

Continue moving the propeller shaft into the lower unit housing until the shaft indexes into the clutch dog. Align the detent balls with the ramps on the clutch dog by turning the shaft very slowly. If the detent balls will not align in the center of the ramp, withdraw the shaft slightly to clear the splines in the clutch dog, and then rotate the shaft 180° and the balls should align with the ramp when the splines index again. The spring will allow the detent balls to be depressed slightly and then to be held in place in the clutch dog groove. As the shaft moves into its proper place, a definite "click" sound will be heard, indicating the detent balls have "popped" into the clutch dog groove and the unit is in **NEUTRAL.**

17- Coat the threads, and then install the Phillips screw through the outside of the lower unit housing to secure the shift lever and yoke assembly in place. This is accomplished by reaching inside the lower unit and moving the shift lever and yoke assembly to align the hole in the housing with the hole in the assembly. When the holes are aligned, install the screw and tighten it securely.

DETENT BALL

CLUTCH DOG   ⑯

Reverse Gear Installation

18- Insert the reverse gear into the lower unit over the propeller shaft.

Bearing Carrier Installation

19- Thoroughly lubricate the bearings in the bearing carrier with HI-VIS grease. Check to be sure the O-ring is in place. Insert the bearing carrier into the lower unit housing.

20- Coat the threads of the attaching screws with OMC Sealant. Secure the bearing carrier in place with the screws. Tighten the screws **EVENLY** and **ALTERNATELY** to the torque value given in the Appendix.

Shift Rod Adjustment

21- After installation, the bend in the shift rod must be toward the forward (leading) edge of the lower unit. Lay a straightedge over the top of the lower unit housing. Measure the distance from the straightedge to the top corner edge of the shift connector. This dimension must be between 1/16"-13/32" (1.6-10.3mm). To adjust, thread the shift rod into, or out of, the yoke assembly until the required dimension is obtained.

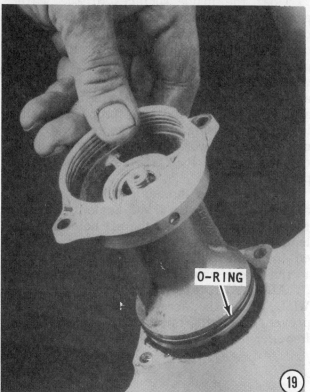

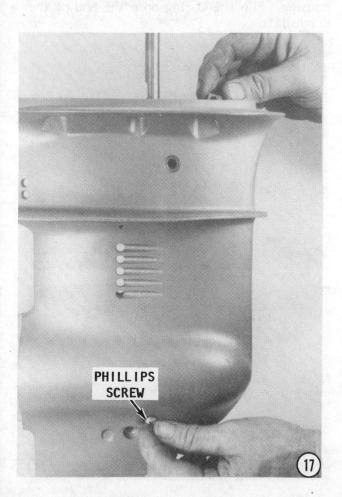

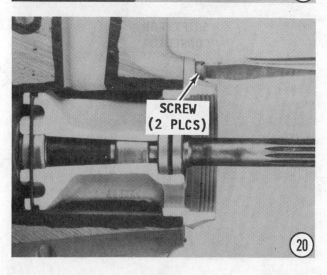

WATER PUMP INSTALLATION

22- Apply a thin bead of OMC Sealant Type-**M** onto the bottom side of the water pump plate where it will contact the lower unit. Slide the plate down over the driveshaft and into place on the lower unit surface. Apply a coating of needle bearing grease in the keyway of the driveshaft. Slide the impeller down the driveshaft. Just before it covers the driveshaft keyway, insert the key into the keyway. Slide the impeller the remaining way down the driveshaft until it seats on the water pump plate with the key indexed into the slot of the impeller.

Coat the inside surfaces of the water pump housing with light-weight oil. Lower the water pump housing down the driveshaft over the impeller and onto the water pump plate. Rotate the driveshaft **CLOCKWISE** while lowering the water pump housing to allow the impeller blades to enter and to assume their natural and proper position inside the housing. Continue to rotate the driveshaft and work the water pump housing downward until it is seated on the water pump plate.

23- Coat the threads of the water pump attaching bolts with OMC Sealant. Install the short bolts through the aft holes and the long bolts through the forward holes. Tighten the bolts **EVENLY** and **ALTER-NATELY** to the torque value given in the Appendix.

ALWAYS rotate the driveshaft **CLOCK-WISE** while the screws are tightened to prevent damaging the impeller vanes. If the impeller is not rotated, the housing could

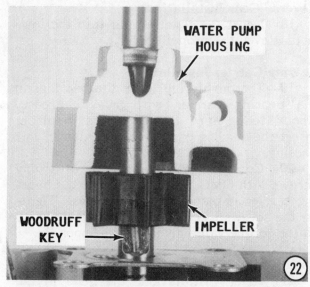

damage or cut the end of the vanes as the screws are brought up tight. The rotation allows them to spring back in a natural position.

24- Place **NEW** grommets into the water pump housing for the water pickup. If a new water pump was installed, these grommets will already be in place. Check to be sure the seal is in place on top of the water pump housing. Slip the **O**-ring onto the end of the driveshaft.

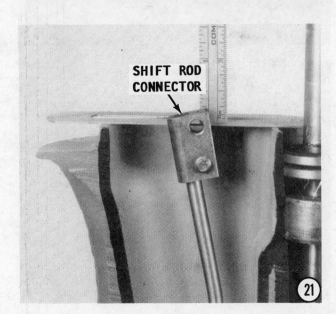

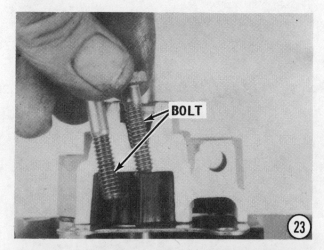

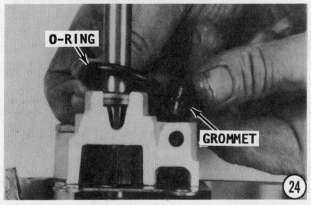

Filling the Lower Unit

Fill the lower unit with lubricant according to the procedures in Section 8-3.

Propeller Installation

Install the propeller, see Section 8-2.

LOWER UNIT INSTALLATION

GOOD WORDS

Connecting the shift rod with the connector is not an easy task but can be accomplished as follows: First, notice the cutout area on the end of the shift rod. This area permits the bolt to pass through the connector, past the shift rod, and into the other side of the connector. It is this bolt that holds the shift rod in the connector. Now, in order for the bolt to be properly installed, the cutout area on the shift rod **MUST** be aligned in such a manner to allow the bolt to be properly installed. Therefore, as the lower unit is mated with the exhaust housing, exercise patience as the two units come together, to enable the bolt to be installed at the proper time. If the rod is allowed to move too far into the connector before the bolt is installed, it may be possible to force the bolt into place, past the shift rod. The threads on the bolt will be stripped, and the shift rod will eventually come out of the connector, illustration **B** and **C.**

25- Install the connector onto the lower unit shift rod, with the **NO THREAD** section facing towards the starboard side of the lower unit. With the connector in this position, the bolt may be inserted through the connector and "catch" the threads on the far side. Install the connector bolt in the manner described in the previous paragraph.

26- Check to be sure the water pickup tubes are clean, smooth, and free of any corrosion. Coat the water pickup tubes and

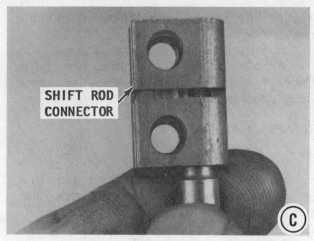

grommets with lubricant as an aid to installation. Guide the lower unit up into the exhaust housing with the water tube sliding into the rubber grommet of the water pump. Continue to work the lower unit towards the exhaust housing, and at the same time rotate the propeller shaft as an aid to indexing the driveshaft splines with the crankshaft. Insert the bolt into the connector. **TAKE TIME** to read and understand the "Good

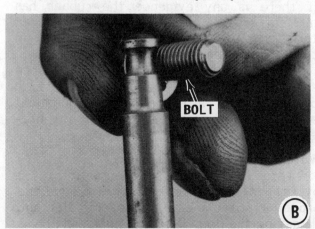

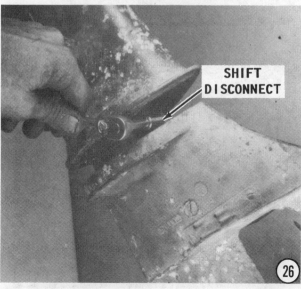

Words" just before Step 25, before making this connection.

Start the bolts securing the lower unit to the exhaust housing. Tighten the bolts **EVENLY** and **ALTERNATELY** to the torque value given in the Appendix.

FUNCTIONAL CHECK

27- Perform a functional check of the completed work by mounting the engine in a test tank, in a body of water, or with a flush attachment connected to the lower unit. If the flush attachment is used, **NEVER** operate the engine above an idle speed, because the no-load condition on the propeller would allow the engine to **RUNAWAY** resulting in serious damage or destruction of the engine.

CAUTION: Water must circulate through the lower unit to the engine any time the engine is run to prevent damage to the water pump in the lower unit. Just five seconds without water will damage the water pump.

Start the engine and observe the tattletale flow of water from idle relief in the exhaust housing. The water pump installation work is verified. If a "Flushette" is connected to the lower unit, **VERY LITTLE** water will be visible from the idle relief port. Shift the engine into the three gears and check for smoothness of operation and satisfactory performance.

8-10 PROPELLER EXHAUST MECHANICAL SHIFT
20 HP AND 30 HP — 1981 & ON
25 HP — 1986 & ON
35 HP — 1976-84

DESCRIPTION

As the name implies, the unit covered in this section is a mechanical shift, propeller exhaust lower unit. Forward, neutral, and reverse shift capabilities are incorporated. A pinion gear is splined onto the lower end of the driveshaft. This pinion gear rotates constantly while the engine is operating and drives the forward and reverse gears. A clutch dog splined to the propeller shaft is centered between the two gears when the unit is in neutral gear. A shift lever causes the clutch dog to engage either the forward or reverse gear. Power is then transferred from the direction gear through the clutch dog to the propeller shaft and propeller.

TROUBLESHOOTING

Troubleshooting **MUST** be done **BEFORE** the unit is removed from the exhaust housing, to permit isolating the problem to one area. Always attempt to proceed with troubleshooting in an orderly manner. The shotgun approach will only result in wasted time, incorrect diagnosis, replacement of unnecessary parts, and frustration.

The following procedures are presented in a logical sequence with the most prevalent, easiest, and less costly items to be checked, listed first.

Unable to Shift into Forward or Reverse

Remove the propeller according to the procedures outlined in Section 8-2. Make a careful check of the rubber hub to determine if it has been slipping in the propeller. If there is any evidence the rubber has melted, or if pieces of rubber have been torn from the hub, it is a clear indication the hub has been slipping.

If the check reveals the hub has been slipping, the propeller must be sent to a propeller shop with the proper equipment and trained personnel to perform the necessary service work.

Water in the Lower Unit

Water in the lower unit is usually caused by fish line becoming entangled around the propeller shaft ahead of the propeller and damaging the propeller seal. If the line is not removed, it will cut the propeller shaft seal and allow water to enter the lower unit. Fish line has also been known to cut a groove in the propeller shaft.

The shift rod seal may be damaged and require replacement. The seal under the water pump may be damaged and allowing water to enter the lower unit.

The propeller should be removed each time the boat is hauled from the water at the end of an outing and any material entangled behind the propeller removed before it can cause extensive damage. The small amount of time and effort involved in pulling the propeller is repaid many times by reduced maintenance and service work, including the replacement of expensive parts.

Slippage in the Lower Unit

If the shift seems to be slipping as the boat moves through the water: First, check the propeller and the rubber hub. If the propeller has been subjected to many strikes against underwater objects, it could slip on its hub. If the hub is damaged or excessively worn on the small propellers, it is questionable whether it economical to have the hub or propeller rebuilt. Sometimes a new propeller may be purchased for less than meeting the expense of rebuilding an old worn propeller. It will pay to check it out.

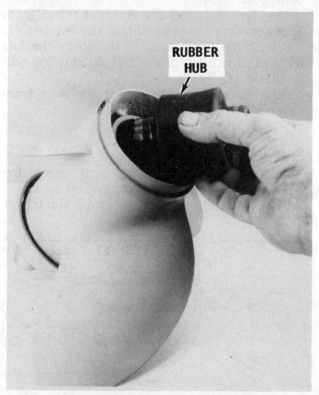

The rubber hub on the propeller exhaust unit was found to be slipping. A new hub is being installed.

Difficult Shifting

Verify that the ignition switch is **OFF**, or better still, disconnect the spark plug wires from the plugs, to prevent possible personal injury, should the engine start. Shift the unit into **REVERSE** gear at the shift control box, and at the same time have an assistant turn the propeller shaft to ensure the clutch is fully engaged. If the shift handle is hard to move, the trouble may be in the lower unit, the shift cable, the handle passing through the exhaust housing, or in the shift box if one is used.

To Isolate the Problem:

Disconnect the shift cable, if used, at the engine. Operate the shift lever at the shift box. If shifting is still hard, the problem is in the shift cable or control box, see Chapter 7. If the shifting feels normal with the shift cable disconnected, the problem must be in the lower unit or in the area where the shift lever passes through the cowling to the bellcrank. Lack of lubrication is usually the cause of problems with the shift lever and bellcrank. To verify the problem is in the lower unit, have an assistant turn the propeller and at the same time move the shift cable back-and-forth. Determine if the clutch engages properly.

Jumping Out of Gear

If a loud thumping sound is heard at the transom while the boat is underway, the unit is jumping out of gear, resulting in a no-load condition on the propeller. When this happens, the rushing water under the hull forces the lower unit in a backward direction. The unit jumps back into gear; the propeller catches hold; the lower unit is forced forward again; and the result is the thumping sound as the action is repeated. Normally this type of action occurs perhaps once a day, then more frequently each time the clutch is operated, until finally the unit will not stay in gear for even a short time.

The following areas must be checked to locate the cause:

1- Check the bellcrank under the powerhead. Remove the window on the port and starboard side of the lower unit. If working on a 1976-79 model, remove the window in the exhaust housing. Hold the shift rod with a pair of pliers and at the same time attempt to move the shift lever on the starboard side of the engine. If it is possible to move the shift lever, the bellcrank is damaged.

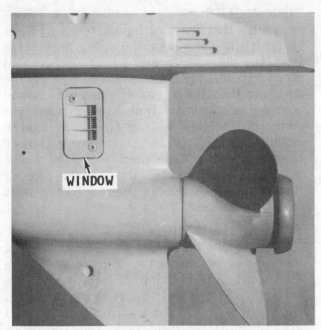

Window in a lower unit to permit access to the shift rod disconnect. Notice how the ramps face forward to allow water to enter the cooling system. If the window is not installed properly, the engine will quickly overheat from lack of cooling water, causing damage to the pump impeller and the powerhead.

2- Disconnect the shift cable at the engine. Attempt to shift the unit into forward gear with the shift lever on the starboard side of the engine and at the same time rotate the propeller in an effort to shift into gear. Shift the control lever at the control box into forward gear. Move the shift cable at the engine up to the shift handle and determine if the cable is properly aligned. If the inner cable should slip on the end cable guide, the adjustment would be lost.

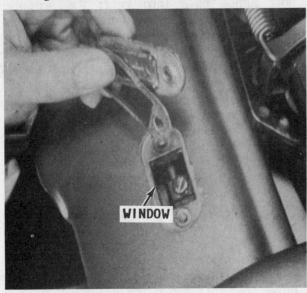

Window and gasket removed from the exhaust housing with the shift rod disconnect bolt shown.

View into the exhaust housing after the power head has been removed. The shift rod bellcrank is visible.

3- Move the shift lever at the engine into the neutral position and the shift lever at the control box to the neutral position. Now, move the shift cable up to the shift lever and see if it is aligned. Shift the unit into reverse at the engine and shift the control lever at the control box into reverse. Move the cable up and see if it is aligned. If the cable is properly aligned, but the unit still jumps out of gear when the cable is connected, one of three conditions may exist.

a- The bellcrank is worn excessively or damaged.

b- The shift rod connector is misaligned. This connector is used to link the upper shift rod with the lower rod. If the connector has not been installed properly, any shifting will be difficult.

c- Parts in the lower unit are worn from extended use.

Frozen Powerhead

This condition is suggested when the operator unsuccessfully attempts to crank the engine, either with a hand starter or

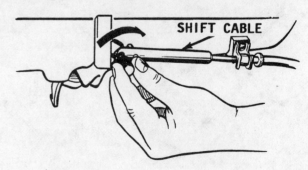

Detailed drawing to depict removal of the cable from the shift handle.

with a starter motor. The flywheel will not rotate. Do not assume the engine is "frozen" until the lower unit has been removed and thoroughly checked. If the lower unit is "locked" (the driveshaft or propeller shaft will not rotate), the powerhead will have the indication of being "frozen" (failure to rotate the flywheel).

The first step to perform under these conditions is to "pull" the lower unit, and then again attempt to crank the engine. If the attempt is successful with the lower unit disconnected, the problem is in the lower unit. If the attempt to crank the engine is still unsuccessful, the problem is in the powerhead.

LOWER UNIT SERVICE

Access to the shift connector for this unit is gained by removing the window in the lower unit (models 1980 and on), or by removing the window in the exhaust housing (models 1976-79).

Propeller Removal

Remove the propeller according to the procedures outlined in Section 8-2.

Damaged pistons in a "frozen" powerhead. If the shift problem is isolated to the powerhead, the lower unit need not be disassembled.

Draining Lower Unit

Drain the lubricant in the lower unit, see Section 8-3.

GOOD WORDS

If water is discovered in the lower unit and the determination is made the propeller shaft seal is damaged and requires replacement, the lower unit does **NOT** have to be removed in order to accomplish the work.

The bearing carrier can be removed and the seal replaced without disassembling the lower unit. **HOWEVER**, and this is a big **HOWEVER**, such a procedure is not considered good shop practice, but merely a quick-fix. If water has entered the lower unit, the unit should be disassembled and a detailed check made to determine if any other seals, bearings, bearing races, O-rings or other parts have been rendered unfit for further service by the water.

LOWER UNIT REMOVAL

1- Disconnect the spark plug wires from the spark plugs. Remove the port and starboard water inlet screens, below the anti-cavitation plate. After the screens have been removed, the shift rod connection will be visible inside the lower unit. Notice the two nuts on the shift rod. Use two wrenches and loosen the top nut. Back off the upper nut onto the upper portion of the shift rod.

2- Work a knife or similar tool into the split of the black plastic keeper and remove the keeper from the upper shift rod. Shift the unit into forward gear, and then remove the nut from the upper shift rod.

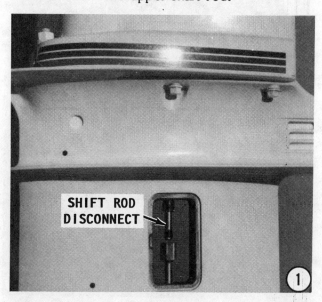

SHIFT ROD DISCONNECT

①

3- If servicing a 1976-79 model, remove the window in the exhaust housing and then remove the bottom screw from the shift connector. The shift rod will then be disconnected when the lower unit is separated from the exhaust housing.

4- Remove the forward nut and two bolts, one on each side of the lower unit, securing the unit to the exhaust housing.

5- Separate the lower unit straight away from the exhaust housing. **TAKE CARE** not to twist or pull backward, sideways, etc., when removing the lower unit, because the driveshaft may be bent. If there is restricted clearance between the bottom of the lower unit and the floor, tilt the complete unit forward in order to gain the necessary room for the lower unit to clear.

6- Position the lower unit in a vertical position on the edge of the work bench resting on the cavitation plate. Secure the lower unit in this position with a C-clamp. The lower unit will then be held firmly in a favorable position for further service work. An alternate method is to cut a groove in a short piece of 2" x 6" wood to accommodate the lower unit with the cavitation plate resting on top of the wood. Clamp the wood in a vise and service work may then be performed with the lower unit erect (in its

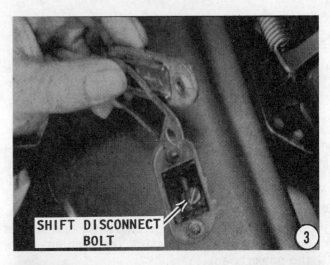

SHIFT DISCONNECT BOLT

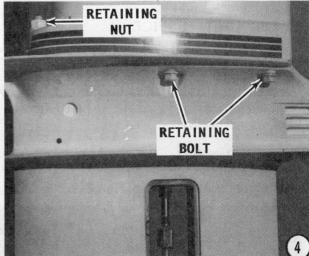

RETAINING NUT

RETAINING BOLT

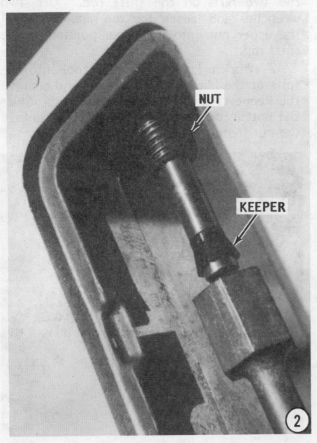

NUT

KEEPER

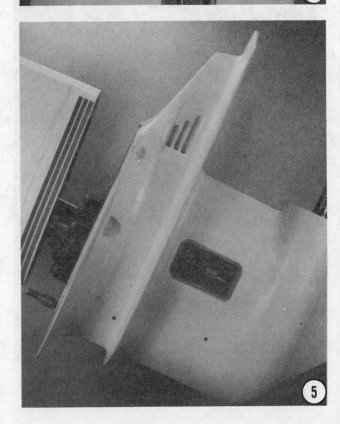

normal position), or inverted (upside down). In both positions, the cavitation plate is the supporting surface.

WATER PUMP REMOVAL

ADVICE

If the only work to be performed is service of the water pump, be extremely **CAREFUL** to prevent the driveshaft from being pulled up and free of the pinion gear in the lower unit. **NEVER** carry the lower unit by the driveshaft or by the shift rod. If the shaft should be released from the pinion gear, the lower unit **MUST** be disassembled to align the pinion gear and driveshaft, then the driveshaft installed.

7- Remove the bolts securing the water pump housing to the lower unit. Slide the water pump housing up and free of the driveshaft. Remove the water pump impeller and key from the driveshaft. Remove the long spacer and bushing from the pump housing. Slide the gasket, pump plate, and second gasket, up and free of the driveshaft. Discard the gaskets. New gaskets are included in a water pump repair kit.

GOOD WORDS

If the only work to be performed is service of the water pump, proceed directly to Page 8-110, Water Pump Installation.

LOWER UNIT DISASSEMBLING

8- Grasp the driveshaft firmly and withdraw it from the lower unit.

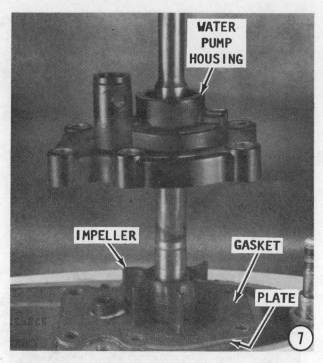

Bearing Carrier Removal

9- Use a thin-wall socket and remove the bolts securing the bearing carrier in the lower unit.

SPECIAL WORDS

Two methods are available to pull the bearing carrier from the lower unit. One method involves the use of a flywheel puller and a couple of bolts. The second method requires the use of a special OMC puller and

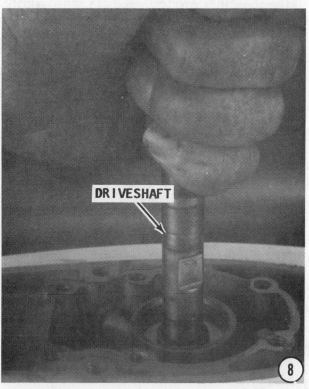

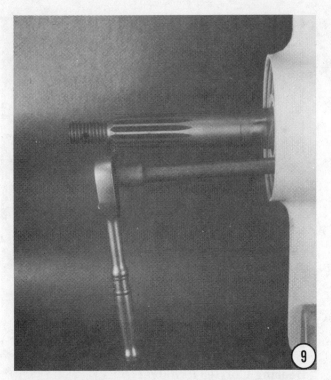

9

a couple bolts. Both methods are described in the following step. If the bearing carrier is stubborn and refuses to budge, apply heat to the outside surface of the lower unit while taking up on either type of puller. **TAKE CARE** not to overheat the lower unit. Aluminum will start to bubble at a relatively low temperature.

10- Obtain an OMC flywheel puller and two 1/4" x 20 bolts. Thread the bolts into the bearing carrier opposite one another. Take up on the center nut of the puller and pull the bearing carrier free of the lower unit.

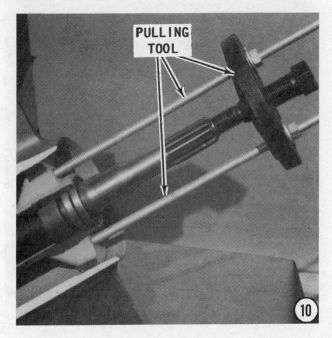

PULLING TOOL

10

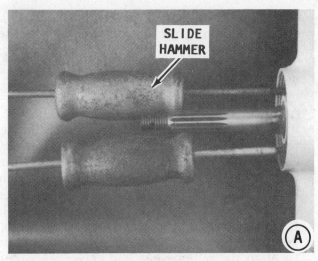

SLIDE HAMMER

A

The second method involves the use of a special puller tool, OMC No. 378103 and two long 1/4" x 20 bolts. Thread the bolts into the bearing carrier. Use the special puller to remove the bearing carrier, as shown in the accompanying illustration **"A"**.

WARNING

The next step involves a dangerous procedure and should be executed with care while wearing **SAFETY GLASSES**. The retaining ring is under tremendous tension in the groove and while it is being removed. If it should slip off the Truarc pliers, it will travel with incredible speed causing personal injury if it should strike a person. Therefore, continue to hold the ring and pliers firm after the ring is out of the groove and clear of the lower unit. Place the ring on the floor and hold it securely with one foot before releasing the grip on the pliers. An alternate method is to hold the ring inside a trash barrel, or other suitable container, before releasing the pliers.

11- Obtain a pair of Truarc pliers. Insert the tips of the pliers into the holes of

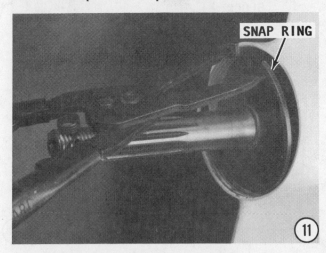

SNAP RING

11

the retaining ring. Now, **CAREFULLY** re-move the retaining ring from the groove and gear case without allowing the pliers to slip. Release the grip on the pliers in the manner described in the above **WARNING.** Remove the retainer plate. As the plate is removed, notice which surface is facing into the hous-ing, as an aid during installation.

12- If servicing a 1980 or later model, use the proper size wrench extended through the window of the lower unit, and back the shift rod out of the shift yoke by rotating it **COUNTERCLOCKWISE** until it is free.

13- If servicing a 1976-79 model, simply back the shift rod out **COUNTERCLOCK-WISE** until it is free of the housing.

14- Using a pair of needle-nose pliers, reach in, grasp the shift yoke, and slide it back off the propeller shaft.

NOW THESE WORDS

A hole is drilled through the propeller shaft just forward of the reverse gear. A detent spring is installed in the hole along with a detent ball on each side of the propeller shaft. The clutch dog is installed over the top of the two detent balls. This arrangement assists in holding the unit in the specific gear desired.

When the bearing carrier, propeller shaft, and associated parts are removed, the clutch dog will remain in the lower unit. As soon as the clutch dog slides free of the propeller shaft, the detent balls and spring

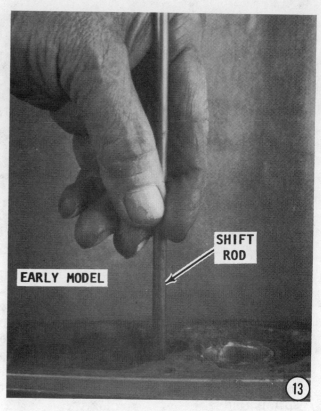

will fly free of the shaft, but will be con-tained within the lower unit housing. There-fore, after the propeller shaft, bearing car-rier, etc., are removed (next step) take time to retrieve the two detent balls and the spring from inside the lower unit.

Propeller Shaft -- Removal

15- Remove the Phillips screw from the outside of the lower unit. This is the screw securing the shift lever in place and is located close to the lubricant drain plug.

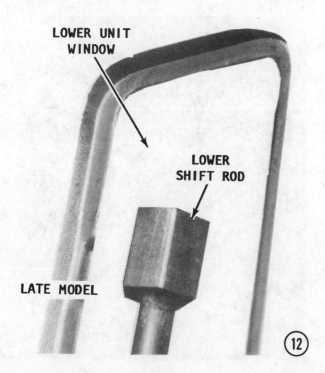

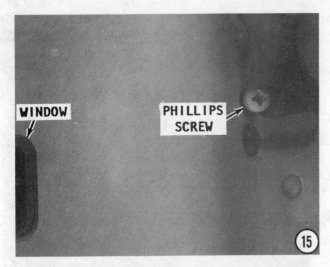

Figure 15 — WINDOW, PHILLIPS SCREW

16- After the screw has been removed, grasp the propeller shaft firmly and withdraw it from the lower unit. The reverse gear, clutch dog, cradle, and associated parts will come out with the shaft. If these parts fail to come out with the shaft, reach in and remove them one at-a-time. Remember the two detent balls and the detent spring.

17- Remove the pinion gear, two thrust washers, and thrust bearing. Take time to notice the arrangement of the two thrust washers and the bearing. Notice how one thrust washer is beveled on the inside diameter and the other is beveled on the outside diameter. It is extremely important that these washers and the thrust bearing are installed properly.

18- Reach in and remove the forward gear. TAKE CARE not to lose the thrust washer installed in the inside diameter of the forward gear.

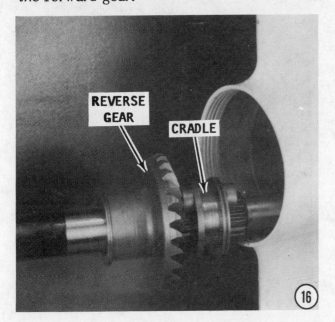

Figure 16 — REVERSE GEAR, CRADLE

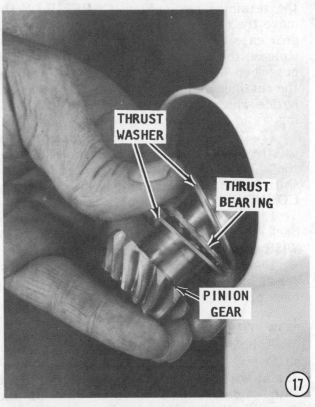

Figure 17 — THRUST WASHER, THRUST BEARING, PINION GEAR

19- Remove the forward gear tapered bearing from the lower unit.

Upper Driveshaft Seals — Removal
20- Use a slide hammer with external jaws. Fit the jaws down inside the seals, and then operate the slide hammer to remove the seals. Notice how the seals were installed back-to-back (flat side against flat side).

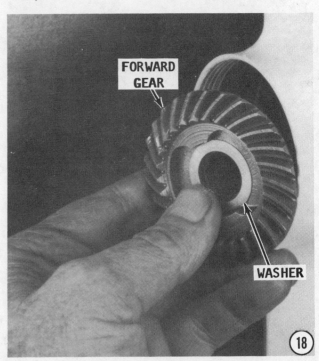

Figure 18 — FORWARD GEAR, WASHER

VERY CRITICAL WORDS

The upper driveshaft bearing, the pinion gear bearing, and the forward gear bearing race **NEED NOT** be removed unless they are unfit for further service. These bearings, especially the pinion gear bearing, can only be installed using special OMC tools. Even with the tools, the task is not an easy one. Therefore, determine their condition by first checking with a flashlight for signs of corrosion or damage, and then by inserting a finger into the bearing and rotating it while checking for "rough" spots or binding. If they appear to be in satisfactory condition, "let a sleeping dog lie." Continue with the other work.

Upper Driveshaft Bearing — Removal

21— The upper driveshaft bearing is housed in a sleeve. If the bearing is to be removed, first punch out the bearing downward into the housing. Use special tool OMC No. 391010, and remove the bearing sleeve from the housing. Turn the lower unit upside down and the upper driveshaft bearing will fall free.

Pinion Gear Bearing — Removal

22- Use any type of punch to drive the bearing free. Further damage to the bearing is of no concern because it is being removed due to its unfitness for further service.

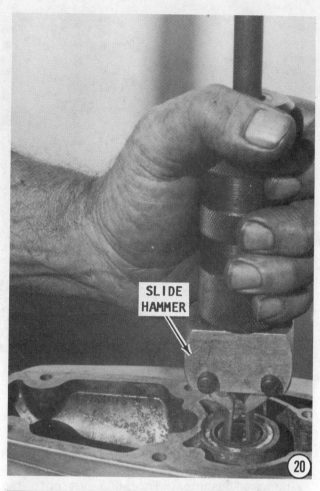

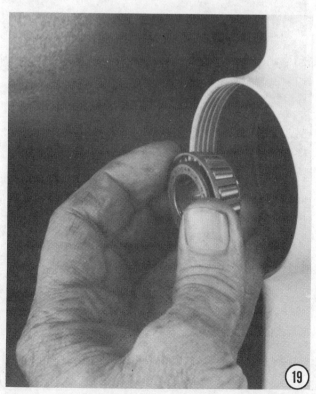

PUNCH

Forward Gear Bearing Race — Removal

23- Use a slide hammer equipped with fingers and "pull" the race from the housing.

Shift Rod O-ring and Bushing — Removal

24- Insert a length of 1/4" rod, with threads on both ends, down through the O-ring and bushing. Thread a nut onto the lower end of the rod. (The edges of the nut must be first rounded to permit the nut to pass through the opening in the housing.) Attach a slide hammer to the rod and "pull" the O-ring, bushing and washer.

SLIDE HAMMER

SHIFT BUSHING

Bearing Carrier Seals — Removal

25- Use a puller with two fingers and work the puller down into the seals. Take up on the puller and remove the seals. Notice how the seals are installed back-to-back. They must be installed in this manner to provide a proper seal to prevent the lubricant in the lower unit from escaping and water from entering.

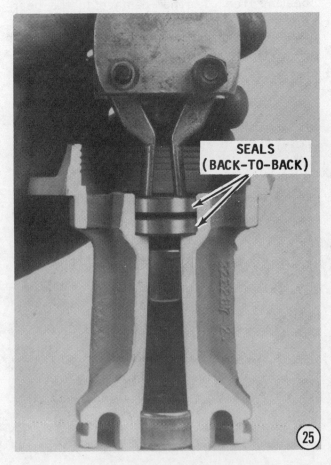

SEALS (BACK-TO-BACK)

An alternate method to remove the seal is to wedge a heavy duty screwdriver underneath the seal, and then to pull back on the bearing carrier, as shown, illustration "B". The seal will pop out. Repeat the procedure for the second seal.

Bearing Carrier Bearing — Removal

The bearings in the bearing carrier need **NOT** be removed unless they are unfit for further service. Check their condition for damage and corrosion. Insert a finger inside and rotate the bearing while checking for roughness or any sign of binding. Check for evidence of corrosion.

26- Use a slide hammer with fingers and remove the bearings outward from the carrier. An alternate method is to use a punch and drive the bearing free of the carrier, illustration "C". If the bearing is damaged and no longer fit for service, further damage will be of no consequence.

CLEANING AND INSPECTING

Clean all water pump parts with solvent, and then dry them with compressed air. Inspect the water pump cover and base for cracks and distortion, possibly caused from overheating. Inspect the face plate and

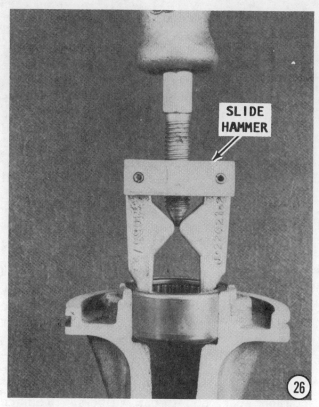

water pump insert for grooves and/or rough surfaces. If possible, **ALWAYS** install a complete new water pump while the lower unit is disassembled. A new impeller will ensure extended satisfactory service and

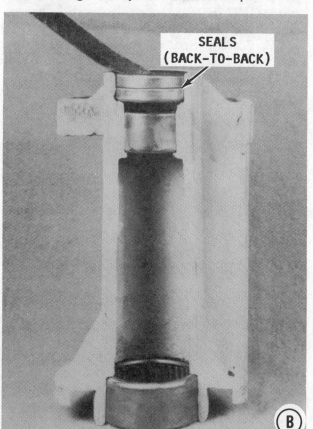

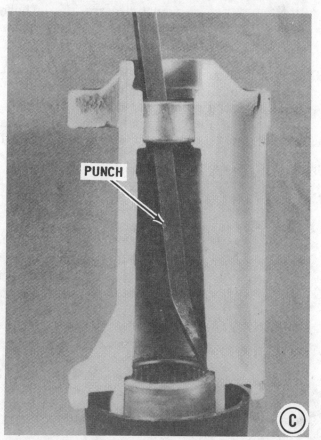

give "peace of mind" to the owner. If the old impeller must be returned to service, **NEVER** install it in reverse to the original direction of rotation. Installation in reverse will cause premature impeller failure.

Inspect the impeller side seal surfaces and the ends of the impeller blades for cracks, tears, and wear. Check for a glazed or melted appearance, caused from operating without sufficient water. If any question exists, and as previously stated, install a new impeller if at all possible.

Clean all parts with solvent and dry them with compressed air. **DISCARD** all O-rings and gaskets. Inspect and replace the driveshaft if the splines are worn. Inspect the gearcase and exhaust housing for damage to the machined surfaces. Remove any nicks and refurbish the surfaces on a surface plate. Start with a No. 120 Emery paper and finish with No. 180.

Check the water intake screen and passages by removing the bypass cover, if one is used. Inspect the clutch dog, drive gears, pinion gear, and thrust washers. Replace these items if they appear worn. If the clutch dog and drive gear arrangement surfaces are nicked, chipped, or the edges rounded, the operator may be performing the shift operation improperly or the controls may not be adjusted correctly. These items **MUST** be replaced if they are damaged.

Inspect the dog ears on the inside of the forward and reverse gears. The gears must be replaced if they are damaged.

Check the cradle that rides on the inside diameter of the clutch dog. The sides of the cradle must be in good condition, free of

New clutch dog and gear. Compare these two parts with those shown at the bottom of the previous column.

any damage or signs of wear. If damage or wear has occurred, the cradle must be replaced.

Check the shift lever and the two prongs that fit inside the cradle. Check to be sure the prongs are not worn or rounded. Damage or wear to the prongs indicates the lever must be replaced.

LOWER UNIT ASSEMBLING

READ AND BELIEVE

The lower unit should **NOT** be assembled in a dry condition. Coat all internal parts with OMC HI-VIS lube oil as they are assembled. All seals should be coated with OMC Gasket Seal Compound. When two seals are installed back-to-back, use Triple Guard Grease between the seal surfaces.

A rusted and corroded gear. Water was allowed to enter the lower unit through a badly worn seal and cause this damage to the gear and other expensive parts.

The ears on this clutch dog and the teeth on the gear are badly worn. Both items are unfit for further service.

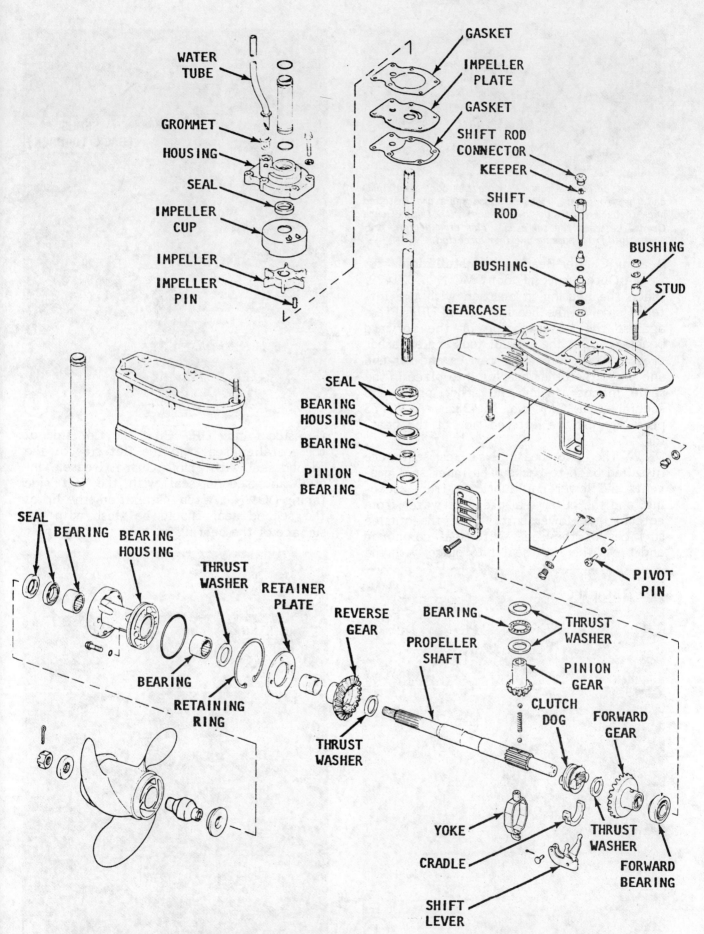

WATER TUBE

GROMMET

HOUSING

SEAL

IMPELLER CUP

IMPELLER

IMPELLER PIN

GASKET

IMPELLER PLATE

GASKET

SHIFT ROD CONNECTOR

KEEPER

SHIFT ROD

BUSHING

BUSHING

STUD

BUSHING

GEARCASE

SEAL

BEARING HOUSING

BEARING

PINION BEARING

SEAL

BEARING

BEARING HOUSING

THRUST WASHER

RETAINER PLATE

REVERSE GEAR

BEARING

PROPELLER SHAFT

THRUST WASHER

PINION GEAR

CLUTCH DOG

FORWARD GEAR

BEARING

RETAINING RING

THRUST WASHER

YOKE

CRADLE

THRUST WASHER

FORWARD BEARING

PIVOT PIN

SHIFT LEVER

Exploded view of a 20hp, 25hp, 30hp, and 35hp lower unit with major parts identified.

A set of double seals showing the back side (left) and the front side (right). These seals are installed back-to-back (flat side-to-flat side) with Triple Guard Grease between the surfaces. This arrangement prevents fluid from passing in either direction.

Bearing Carrier Bearings — Installation

1- Obtain special tool OMC NO. 321429. Place the bearing in position with the lettered side of the bearing facing **UP**. Press against the lettered side of the forward bearing, using the special tool and an arbor press. Turn the bearing carrier end-for-end and press the other bearing into place in the same manner, using the arbor press and special tool OMC No. 321428. **ALWAYS** press against the lettered side of the bearing.

2- The two bearing carrier seals are installed back-to-back. The inner seal prevents the lower unit lubricant from escaping, and the outer seal prevents water from entering the lower unit. Coat the outside surfaces of the seals with seal compound and press the first seal into place with the

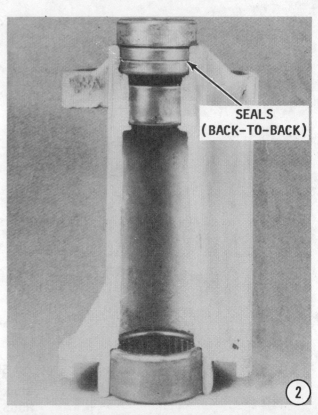

flat side facing **UP**. Coat the flat side of the installed seal and the flat side of the second seal with Triple Guard Grease. Install the second seal with the flat side facing **DOWN**. After installation, the lip of the second seal should be flush with the surface of the bearing carrier.

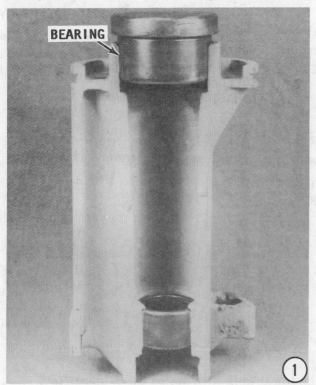

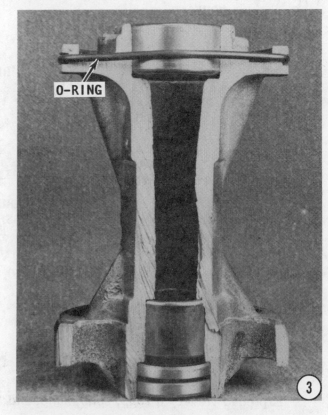

3- Check to be sure the O-ring groove of the bearing carrier is clean. Coat the O-ring with OMC HI-VIS lube oil and then install it into the groove. Set the bearing carrier aside for later installation.

Forward Gear Bearing Race -- Installation

4- If the forward gear bearing race was removed, install a new race by first coating the race with OMC HI-VIS lube oil. Obtain special tool, OMC No. 319929 and driver handle, OMC No. 311880. Drive the race into place squarely.

Shift Rod Bushing and O-ring — Installation

5- For the 1980 and later models: Obtain special tool, OMC No. 304515. Install the washer and then the O-ring. Install the bushing using the special tool. If the special tool is not available, a tool the same size as the outside diameter as the bushing may be used to drive the bushing into place. The bushing does not go in hard. **TAKE CARE** not to distort the inside diameter of the bushing.

GOOD WORDS

Two different lower units have been used with the late model 35 hp lower units from 1976. Changes in the assembling procedures for these two units is clearly indicated in the following steps.

Pinion Gear Bearing — Installation

6- Special tool OMC No. 391257 is required to install this bearing. Assemble the

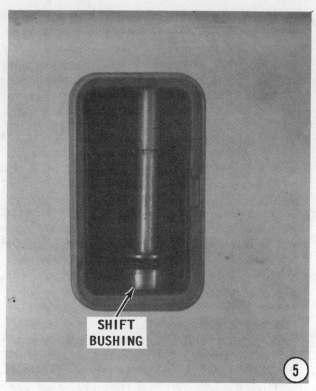

SHIFT BUSHING

5

tool, as shown in the accompanying illustration. Drive the bearing into place from the top until the plate of the tool makes contact with the lower unit. The bearing will then be installed to the proper depth in the lower unit.

For 1976 models **ONLY**: Special tool OMC No. 321516 and No. 318122 are required to install the bearing. Use a flat washer and a 1/2" x 13 x 1" screw from tool, OMC No. 316910. Drive the bearing into place with the lettered side facing the tool.

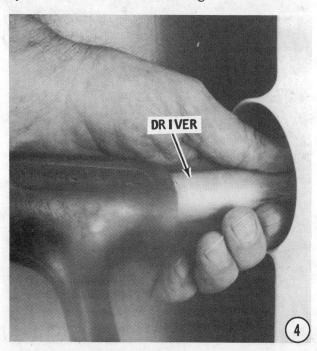

DRIVER

4

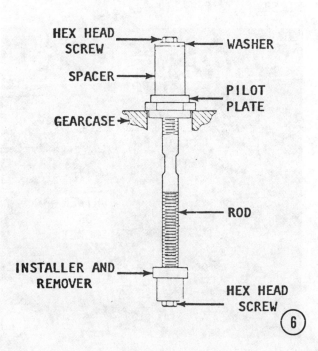

HEX HEAD SCREW — WASHER

SPACER

PILOT PLATE

GEARCASE

ROD

INSTALLER AND REMOVER

HEX HEAD SCREW

6

Upper Driveshaft Bearing — Installation

7- Obtain special tool, OMC No. 322923. Press the bearing into the bearing retainer with the lettered side of the bearing facing the tool. Press the assembled bearing and retainer into the lower unit housing until it seats. A socket the same size as the outside diameter of the retainer may be used. **TAKE CARE** not to damage the bearing during installation.

For 1976 model **ONLY:** When the bearing is purchased, the bearing is installed in the retainer. If the bearing retainer was not removed, press the bearing into the retainer in the housing. Press the bearing with the lettered side facing the tool, until the bottom of the bearing is flush with the bottom of the retainer, illustration **"A"**.

Driveshaft Seals — Installation

8- These seals are installed back-to-back, flat side-to-flat side, illustration **"B"**. The inner seal prevents the lower unit lubricant from escaping, and the outer seal prevents water from entering the lower unit. Coat the outside surface of the first seal with Seal Compound. Install the seal with the flat side facing **UP.** Coat the flat side of the installed seal and the flat side of the second seal with Triple Guard Grease. Install the second seal with the flat side facing **DOWN.**

Forward Gear Bearing — Installation

9- Insert the bearing into the race with the tapered side of the bearing going in **FIRST.**

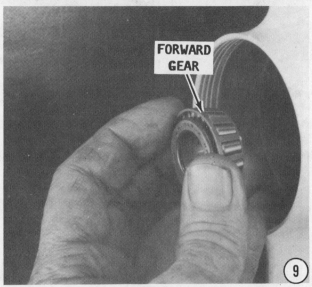

Pinion Gear -- Installation

10- Install the thin thrust washer onto the pinion gear shank with the bevel on the inside diameter facing **DOWN**.

11- Slide the thrust bearing onto the shank.

12- Slide the thick second washer onto the pinion gear shank with the bevel facing **UP**.

13- Insert the assembled pinion gear into the opening in the lower unit housing.

Forward Gear -- Installation

14- Insert the thrust washer into the forward gear. Lower the forward gear down into the lower unit housing past the pinion gear. By tilting the gear slightly it will pass the pinion gear. The teeth of the forward gear must index with the teeth of the pinion gear.

Shift Mechanism — Installation

15- Lower the shift lever into the lower unit housing and into place in the recess.

Propeller Shaft — Installation

VERY VERY GOOD WORDS

Two illustrations accompany most of the next 10 steps. One set is numbered and is

THRUST BEARING

11

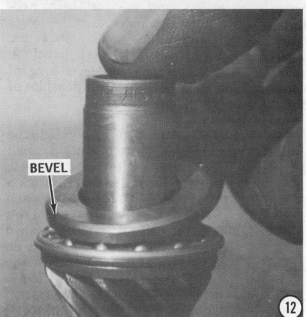

BEVEL

12

BEVEL

PINION GEAR SHANK

10

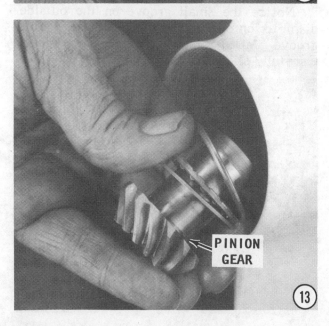

PINION GEAR

13

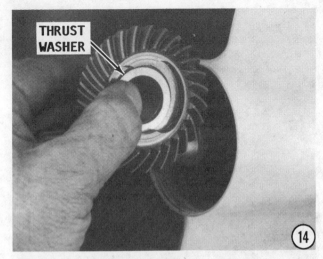

the same as the step number. The other set has a letter identification. **NOW**, the numbered illustrations show the work being performed in the normal manner with the lower unit. The lettered illustrations show the work being performed on a work bench in order to give you a clear understanding of what is happening and the relationship of the parts inside the lower unit.

16- Insert the detent spring through the hole in the propeller shaft. Apply a small amount of needle bearing grease to the hole on both sides of the propeller shaft. Stick a detent ball into place in the grease on each side of the shaft. Slide the clutch dog down the propeller shaft with the cutouts aligned with the two detent balls. If the cutouts do not align with the detent balls, withdraw the clutch dog, rotate it 180°, and then slide it back onto the shaft, reference illustration "C".

Notice the small groove on the outside diameter on one side of the clutch dog. This groove **MUST** face the forward gear. Carefully slide the clutch dog over the two

detent balls and into the neutral position on the shaft.

17- Lubricate the large groove in the clutch dog with needle bearing grease, reference illustration "D". This illustration will be helpful in understanding the relationship of the forward gear, forward gear bearing, and the shift lever. Notice how the shaft passes over the top of the shift lever. Slide

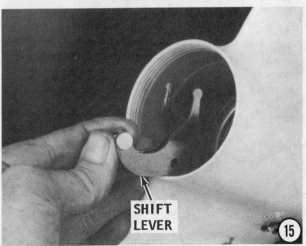

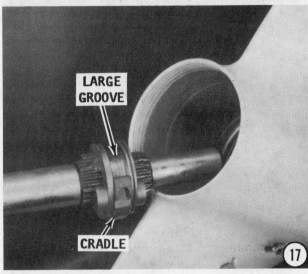

SHIFT
LEVER

D

CRADLE

SHIFT
LEVER

E

the cradle into the large groove of the clutch dog. Insert the assembled propeller shaft into the lower unit, with the forward end of the shaft indexed into the forward gear and forward gear bearing. At the same time, the shaft is worked over the top of the shift lever.

18- Slide a screwdriver blade under the shift lever and work the fork fingers up into the cradle, reference illustration **"E"**.

19- After the fingers are in place in the cradle, move the shift lever until the hole in the lever is aligned with the hole in the lower unit housing, reference illustration **"F"**. When the holes align, start the Phillips screw through the housing and into the lever. Apply a coating of OMC 1000 sealer onto the threads of the screw. Tighten the screw securely.

Reverse Gear — Installation

20- Insert the thrust washer into the reverse gear. Slide the reverse gear down the propeller shaft into the lower unit housing.

PHILLIPS
SCREW

19

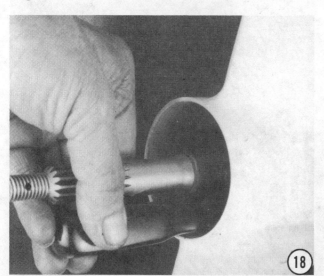

18

PHILLIPS
SCREW

F

REVERSE GEAR

SHIFT YOKE

SHIFT LEVER

21- Slide the shift yoke down the propeller shaft and hook it into the shift lever, reference illustation "G".

22- On the 1980 and later models: Coat the shift rod with light-weight oil and then insert the shift rod down through the O-ring and thread it into the shift yoke. Tighten the shift rod securely in the yoke, with a wrench, reference illustation "H".

On the 1976-79 models: Thread the shift rod into the shift yoke with the bend in the rod facing the driveshaft, reference illustration "J".

Bearing Carrier — Installation

23- Slide the retainer plate onto the propeller shaft and against the lower unit housing, with the lip of the plate indexed into the short slot on the bottom side of the lower unit, reference illustration "K".

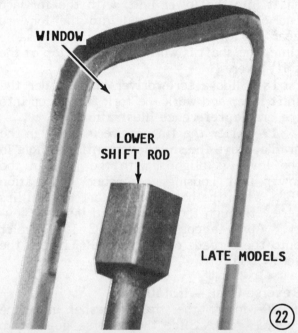

WINDOW

LOWER SHIFT ROD

LATE MODELS

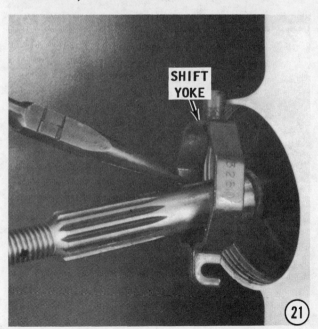

SHIFT YOKE

LOWER SHIFT ROD

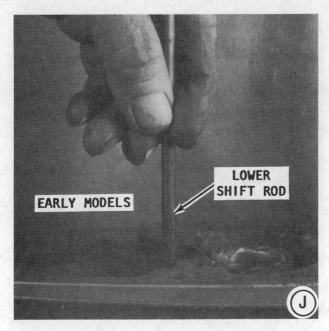

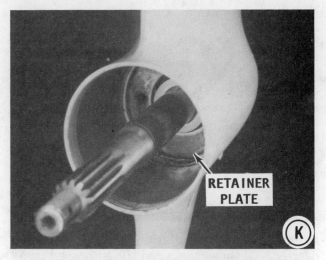

WARNING

This next step can be dangerous. The snap ring is placed under tremendous tension with the Truarc pliers while it is being placed into the groove. Therefore, wear **SAFETY GLASSES** and exercise care to prevent the snap ring from slipping out of the pliers. If the snap ring should slip out, it would travel with incredible speed and cause personal injury if it struck a person.

24- Use a pair of Truarc pliers and install the Truarc snap ring into the groove in the lower unit, next to the retainer plate. Check to be sure the ring is properly seated all the way around in the groove.

25- Obtain two 1/4" rods about 12" long with 1/4 x 28 threads on one end. Thread these two rods into the retainer plate to act

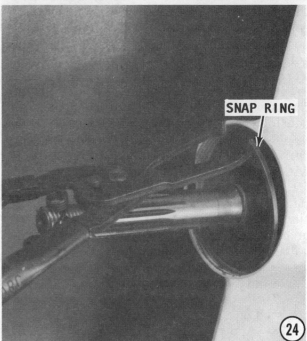

as guides for installation of the bearing carrier. Observe the word **UP** embossed into the metal of the bearing carrier rim, reference illustration **"L"**. Slide the bearing carrier onto the propeller shaft over the guide rods and into the lower unit, with the word **UP** facing **UPWARD** in relation to the lower unit housing. Check to be sure the thrust washer is seated in the recess of the bearing carrier towards the reverse gear.

26- Slide new **O**-rings onto the bearing carrier bolts. Coat the threads of the bolts with OMC Sealer. Install two bolts through the bearing carrier and into the retainer. Back out the two guide rods used to install the bearing carrier. Install and tighten the bearing carrier bolts securely to the torque value given in the Appendix.

Driveshaft — Installation

27- Install the guide bushing down over the stud on top of the housing, as shown.

28- Apply sealer to the upper housing surface.

29- Coat the driveshaft with lightweight oil as an aid to installation. Slide the driveshaft down into the lower unit. As the driveshaft is lowered, rotate the driveshaft slightly to permit the splines on the shaft to index with the splines of the pinion gear.

WATER PUMP INSTALLATION

30- Slide a **NEW** water pump gasket down the driveshaft and into place on the housing. Coat the upper surface of the gasket with sealer. Slide the water pump plate down the driveshaft and into place on top of the gasket. Check to be sure the small driveshaft grommet that seats in the plate and through the gaskets is installed with the small side facing **UP**. Coat both sides of a **NEW** second water pump gasket with sealer, and then slide it down the driveshaft and into place on top of the water pump plate. Check to be sure the

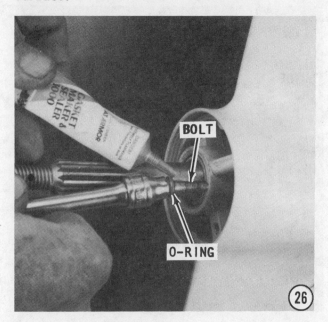

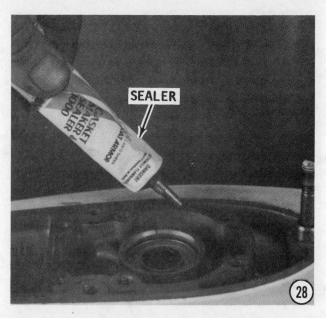

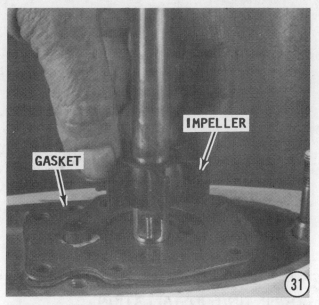

holes in both gaskets, the plate, and in the housing are aligned. If the holes do not align, one of the gaskets or the plate is upside down. Correct the error.

31- Slide the water pump impeller down the driveshaft. Just before the impeller covers the cutout for the impeller pin, install the pin. Align the slot in the impeller with the impeller pin, and then continue to work the impeller down the driveshaft until it is firmly in place on the surface of the upper water pump gasket.

32- Check to be sure **NEW** seals and O-rings have been installed in the water pump. Lubricate the inside surface of the water pump with light-weight oil. Lower the water pump housing down the driveshaft and over the impeller. **ALWAYS** rotate the

driveshaft slowly **CLOCKWISE** as the housing is lowered over the impeller to allow the impeller blades to assume their natural and proper position inside the housing. Continue to rotate the driveshaft and work the water pump housing downward until it is seated on the gasket and plate.

33- Coat the threads of the water pump attaching bolts with sealer, and then secure the pump in place with the bolts. Tighten the bolts **ALTERNATELY** and **EVENLY**. Check to be sure a **NEW** grommet has been installed in the top of the water pump. Install a **NEW** O-ring onto the top of the driveshaft.

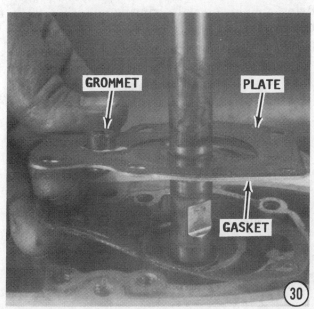

(33)

LOWER UNIT INSTALLATION

GOOD WORDS
1976 thru 1979 Models ONLY

Connecting the shift rod with the connector is not an easy task but can be accomplished as follows: First, notice the cutout area on the end of the shift rod. This area permits the bolt to pass through the connector, past the shift rod, and into the other side of the connector. It is this bolt that holds the shift rod in the connector. Now, in order for the bolt to be properly installed, the cutout area on the shift rod **MUST** be aligned in such a manner to allow the bolt to be properly installed. Therefore, as the lower unit is mated with the exhaust housing, exercise patience as the two units come together, to enable the bolt to be installed at the proper time. If the rod is allowed to

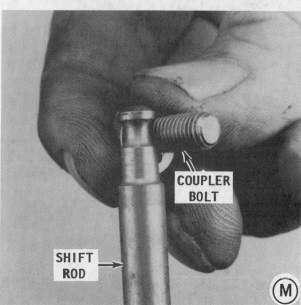

COUPLER BOLT

SHIFT ROD

(M)

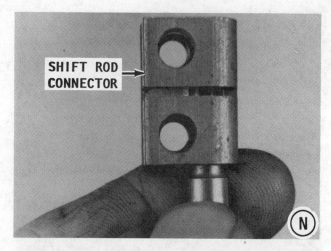

SHIFT ROD CONNECTOR

(N)

move too far into the connector before the bolt is installed, it may be possible to force the bolt into place, past the shift rod. The threads on the bolt will be stripped, and the shift rod will eventually come out of the connector, illustration "M" and "N".

34— Install the connector onto the lower unit shift rod, with the **NO THREAD** section facing towards the window. With the connector in this position, the bolt may be inserted through the connector and "catch" the threads on the far side. Install the connector bolt in the manner described in the previous paragraph.

For All Models

35— Check to be sure the water tubes are clean, smooth, and free of any corrosion. Coat the water pickup tubes and grommets with lubricant as an aid to installation. Check to be sure the spark plug wires are disconnected from the spark plugs. Bring the lower unit housing together with the exhaust housing, and at the same time, guide the water tube into the rubber grommet of the water pump. As the two units

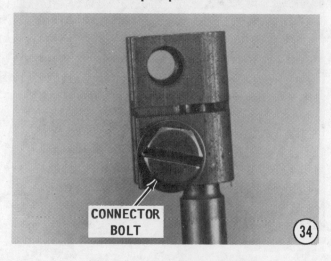

CONNECTOR BOLT

(34)

come together, rotate the flywheel slowly to permit the splines of the driveshaft to index with the splines of the crankshaft.

36- After the surfaces of the lower unit and exhaust housing make contact, start the nut on the stud on the leading edge of the lower unit. Start the four bolts on the bottom side of the lower unit. **DO NOT** tighten this hardware at this time.

For the 1976 thru 1979 Models ONLY
The shift connection is made through the window in the exhaust housing.

37- Insert the bolt into the connector. **TAKE TIME** to read and understand the "Good Words" just before Step 34, before making this connection. After the bolt is in place, install and secure the window with the attaching hardware. Tighten the bolts and the nut securing the lower unit to the exhaust housing **ALTERNATELY** and **EVENLY** to the torque value given in the Appendix. Install the outer plate and gasket to the exhaust housing.

For Models since 1980
The shift connection is made through the water pickup openings in the lower unit.

38- At the powerhead, move the shift lever to the **FORWARD** gear position. Slip the upper shift lever nut upward, and then snap the keeper onto the end of the shift

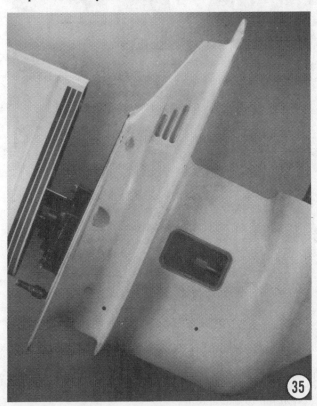

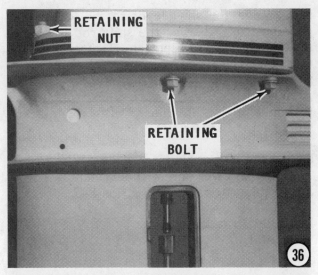

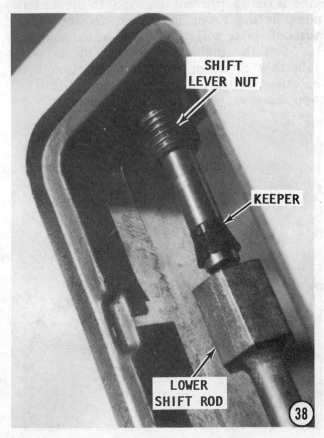

rod. Move the shift lever to the **REVERSE** gear position. Lower the shift rod into the lower unit shift rod section. Tighten the nut to secure the lower portion of the shift rod to the upper portion.

39- Install and secure the two water pickup windows in place, with the cutout slots facing **FORWARD**.

Filling the Lower Unit
Fill the lower unit with lubricant according to the procedures in Section 8-3.

Propeller Installation
Install the propeller, see Section 8-2.

FUNCTIONAL CHECK

Perform a functional check of the completed work by mounting the engine in a test tank, in a body of water, or with a flush attachment connected to the lower unit. If the flush attachment is used, **NEVER** operate the engine above an idle speed, because the no-load condition on the propeller would allow the engine to **RUNAWAY** resulting in serious damage or destruction of the engine.

CAUTION: Water must circulate through the lower unit to the engine any time the engine is run to prevent damage to the water pump in the lower unit. Just five seconds without water will damage the water pump.

Start the engine and observe the tattle-tale flow of water from idle relief in the exhaust housing. The water pump installation work is verified. If a "Flushette" is

connected to the lower unit, **VERY LITTLE** water will be visible from the idle relief port. Shift the engine into the three gears and check for smoothness of operation and satisfactory performance.

8-11 ELECTRIC SHIFT
TWO SOLENOIDS
50 HP 1971-72

DESCRIPTION

The lower unit covered in this section is a three shift position, hydraulic activated, solenoid controlled, propeller exhaust unit. A hydraulic pump mounted in the forward portion of the lower unit provides the force required to shift the unit. Two solenoids installed in the lower unit, above the pump, control and operate the pump valve. The pump valve directs the hydraulic force to place the clutch dog in the desired position for neutral, forward, or reverse gear position. One solenoid controls the valve for the neutral position. Both solenoids control the valve for the reverse position. When

Cut-a-way view of a complete lower unit prior to disassembling. This type illustration is most helpful in gaining an appreciation of the internal parts and their relationship to one another.

neither solenoid is activated, the unit is at rest in the forward gear position.

In simple terms, something must be done (a solenoid activated and hydraulic pressure applied) to move the unit into neutral or reverse gear position. If no action is taken (shift mechanism at rest) the unit is in the forward gear position.

A full 12-volts is required to activate the solenoids. This means shifting is not possible if the battery should become low for any number of reasons. A potentially dangerous condition could exist because the unit could not be taken out of forward gear. Therefore, the only way to stop forward boat movement would be to shut the engine down. A low battery would also mean the electric starter motor would fail to crank the engine properly for engine start. Such a condition would require hand starting in an emergency, if a second battery were not available.

The lower unit houses the driveshaft and pinion gear, the forward and reverse driven gears, the propeller shaft, clutch dog, hydraulic pump, two solenoids, and the necessary shims, bearings, and associated parts to make it all work properly.

ONE MORE WORD

A useful piece of information to remember is that the green wire carries current for the neutral position; and the green and blue wires carry current for the reverse gear operation.

TROUBLESHOOTING

Preliminary Checks

Whenever the lower unit fails to shift properly the first place to check is the condition of the battery. Determine if the battery contains a full charge. Check the condition of the battery terminals, the battery leads to the engine, and the electrical connections.

The second area to check is the quantity and quality of the lubricant in the lower unit. If the lubricant level is low, contaminated with water, or is broken down because of overuse, the shift mechanism may be affected. Water in the lower unit is **VERY BAD NEWS** for a number of reasons, particularly when the lower unit contains electrical or hydraulic components. Electrical parts short out and hydraulic units will not function with water in the system.

BEFORE making any tests, remove the propeller, see Section 8-2. Check the propeller carefully to determine if the hub has been slipping and giving a false indication the unit is not in gear. If there is any doubt, the propeller should be taken to a shop properly equipped for testing, before the time and expense of disassembling the lower unit is undertaken. The expense of the propeller testing and possible rebuild is justified.

The following troubleshooting procedures are presented on the assumption the battery, including its connections, the lower unit lubricant, and the propeller have all been checked and found to be satisfactory.

Lower Unit Locked

Determine if the problem is in the powerhead or in the lower unit. Attempt to rotate the flywheel. If the flywheel can be moved even slightly in either direction, the problem is most likely in the lower unit. If it is not possible to rotate the flywheel, the problem is a "frozen" powerhead. To absolutely verify the powerhead is "frozen", separate the lower unit from the exhaust housing and then again attempt to rotate the flywheel. If the attempt is successful, the problem is definitely in the lower unit. If the attempt to rotate the flywheel, with the lower unit removed, still fails, a "frozen" powerhead is verified.

A "frozen" powerhead with burned pistons.

Unit Fails to Shift
Neutral, Forward, or Reverse

Disconnect the green and blue electrical wires from the lower unit at the engine.

Voltmeter Tests

Separate the green and blue wires at the engine by first sliding the sleeve back, and then making the disconnect. Connect one lead of the voltmeter to the green wire to the control panel and the other lead to a good ground on the engine. Turn the ignition key to the **ON** position. With the shift box handle in the forward position, the voltmeter should indicate **NO** voltage. Move the test lead from the green wire to the blue wire to the control panel. With the shift lever still in the forward position, the voltmeter should indicate **NO** voltage.

Move the shift lever to the **NEUTRAL** position. With the voltmeter still connected to the blue wire, **NO** voltage should be indicated. Move the test lead to the green wire. Voltage **SHOULD** be indicated.

Move the shift lever to the **REVERSE** position. Voltage **SHOULD** be indicated on the green wire **AND** on the blue wire, reference illustration **"A"**.

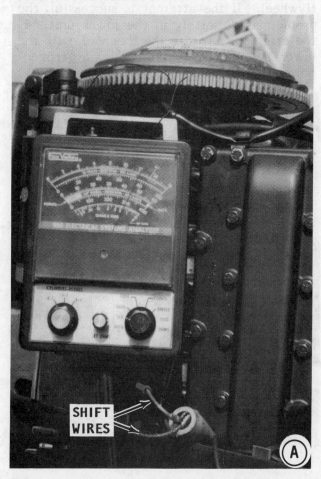

If the desired results are not obtained on any of these tests, the problem is in the shift box switch or the wiring under the control panel. See Chapter 7.

Ohmmeter Tests

Set the ohmmeter to the low scale. Connect one lead to the green wire to the lower unit, and the other lead to a good ground. The meter should indicate 5 to 7 ohms. Connect the meter to the blue wire to the lower unit and ground. The meter should again indicate from 5 to 7 ohms, reference illustration **"B"**.

BAD NEWS

If the unit fails the voltmeter and ohmmeter tests just outlined, the only course of action is to disassemble the lower unit to determine and correct the problem.

LOWER UNIT SERVICE

Propeller Removal

If the propeller was not removed, as directed for the troubleshooting, remove it now, according to the procedures outlined in Section 8-2.

Draining the Lower Unit

Drain the lower unit of lubricant, see Section 8-3.

GOOD WORDS

If water is discovered in the lower unit and the propeller shaft seal is damaged and requires replacement, the lower unit does **NOT** have to be removed in order to accomplish the work.

The bearing carrier can be removed and the seal replaced without disassembling the lower unit. **HOWEVER,** such a procedure is not considered good shop practice, but merely a quick-fix. If water has entered the lower unit, the unit should be disassembled and a detailed check made to determine if any other seals, bearings, bearing races, O-rings or other parts have been rendered unfit for further service by the water.

LOWER UNIT REMOVAL

1- Disconnect and ground the spark plug wires. Slide back the insulators on the shift wires at the engine. Disconnect the blue wire from the blue and the green wire from the green, at the engine.

HELPFUL WORD

Obtain a piece of electrical wire, about 5 ft. long. Connect one end to the green wire and the other end to the blue wire. Tape the connections. Now, when the lower unit is separated from the exhaust housing, the ends of the wire will feed down through the exhaust housing. When the lower unit is free, disconnect the wire ends from the blue and green wires and leave the wire loop in the exhaust housing. When it is time to

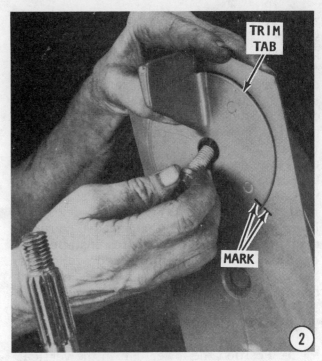

bring the lower unit together with the exhaust housing, the wire ends will be connected again and the blue and green wires easily pulled back up through the exhaust housing. No sweat! Alright, on with the work.

2- Scribe a mark on the trim tab and a matching mark on the lower unit to ensure the trim tab will installed in the same position from which it is removed. Remove the attaching hardware, and then remove the trim tab.

3- Use a 1/2" socket with a short extension and remove the bolt from inside the

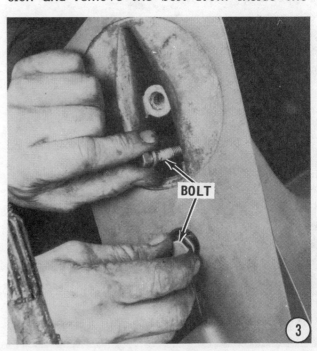

trim tab cavity. Remove the 5/8" counter-sunk bolt located just ahead of the trim tab position.

4- Remove the four 9/16" bolts, two on each side, securing the lower unit to the exhaust housing. Work the lower unit free of the exhaust housing. If the unit is still mounted on a boat, tilt the engine forward to gain clearance between the lower unit and the deck (floor, ground, whatever). **EXERCISE CARE** to withdraw the lower unit straight away from the exhaust housing to prevent bending the driveshaft. Once the lower unit is free of the exhaust housing, stop and disconnect the wires installed as described in the **"Helpful Word"** following Step 1. Leave the wire loop in the exhaust housing as an aid during installation.

WATER PUMP REMOVAL

5- Position the lower unit in the vertical position on the edge of the work bench resting on the cavitation plate. Secure the lower unit in this position with a C-clamp. The lower unit will then be held firmly in a favorable position during the service work. An alternate method is to cut a groove in a short piece of 2" x 6" wood to accommodate the lower unit with the cavitation plate resting on top of the wood. Clamp the wood in a vise and service work may then be performed with the lower unit erect (in its normal position), or inverted (upside down). In both positions, the cavitation plate is the supporting surface.

TAKE TIME

Take time to notice how the shift wires are routed and anchored in position with a clamp on top of the water pump. It is extremely important for the shift wires to be routed and secured in the same position during installation, reference illustration **"C"**.

6- Remove the O-ring from the top of the driveshaft. Remove the bolts securing the water pump to the lower unit housing and the clamp securing the shift wires in place. Leave the clamp on the shift cable as an aid during installation. Pull the water pump housing up and free of the driveshaft.

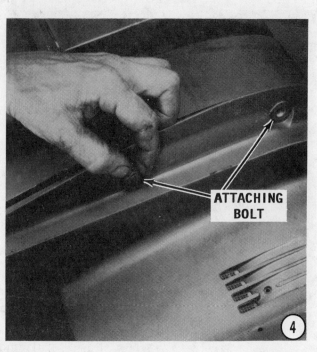

ATTACHING BOLT

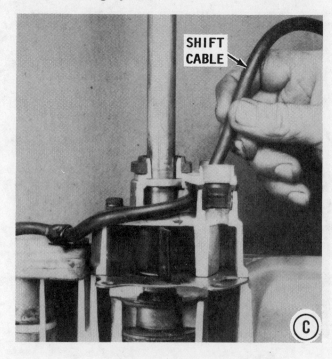

SHIFT CABLE

7- Slide the water pump impeller up and free of the driveshaft. Pop the impeller Woodruff key out of the driveshaft keyway. Slide the water pump base plate up and off of the driveshaft.

GOOD WORDS

If the only work to be performed is service of the water pump, proceed directly to Page 8-137, Water Pump Installation.

Shift Solenoid — Removal

8- Remove the shift solenoid cover located just aft of the water pump position. Take care not to lose the wavy washer installed under the cover. Grasp the upper (green) shift solenoid and withdraw the solenoids and shift rod from the lower unit cavity as an assembly.

Bearing Carrier — Removal

9- Remove the four 5/16" bolts from inside the bearing carrier. Notice how each bolt has an O-ring seal. These O-rings should be replaced each time the bolts are removed. Also observe the word UP embossed into the metal rim of some bearing carriers. This word must face UP in relation to the lower unit during installation. Clean the surface and if the word "UP" does not show, the position of the carrier during installation is not important.

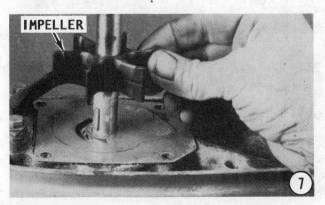

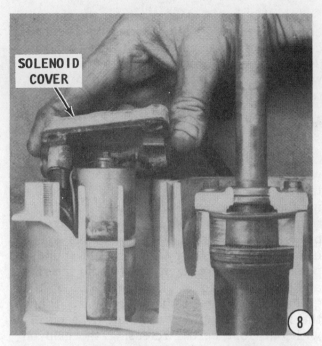

10- Remove the bearing carrier using one of the methods described in the following paragraphs, under Special Words.

SPECIAL WORDS

Several models of bearing carriers are used on the lower units covered in this section.

The bearing carriers are a very tight fit into the lower unit opening. Therefore, it is not uncommon to apply heat to the outside surface of the lower unit with a torch, at the same time the puller is being worked to remove the carrier. TAKE CARE not to overheat the lower unit.

One model carrier has two threaded holes on the end of the carrier. These threads permit the installation of two long bolts. These bolts will then allow the use of a flywheel puller to remove the bearing carrier, illustration 10.

Another model does not have the threaded screw holes. To remove this type bearing carrier, a special puller with arms must be used. The arms are hooked onto the carrier web area, and then the carrier removed, reference illustration **"D"**.

WARNING

The next step involves a dangerous procedure and should be executed with care while wearing **SAFETY GLASSES**. The retaining rings are under tremendous tension in the groove and while they are being removed. If a ring should slip off the Truarc pliers, it will travel with incredible speed causing personal injury if it should strike a person. Therefore, continue to hold the ring and pliers firm after the ring is out of the groove and clear of the lower unit. Place the ring on the floor and hold it securely with one foot before releasing the grip on the pliers. An alternate method is to hold the ring inside a trash barrel, or other suitable container, before releasing the pliers.

11- Obtain a pair of Truarc pliers. Insert the tips of the pliers into the holes of the first retaining ring. Now, **CAREFULLY** remove the retaining ring from the groove and gear case without allowing the pliers to slip. Release the grip on the pliers in the manner described in the above **WARNING**. Remove the second retaining ring in the same manner. The rings are identical and either one may be installed first.

12- Remove the retainer plate. As the plate is removed, notice which surface is facing into the housing, as an aid during installation.

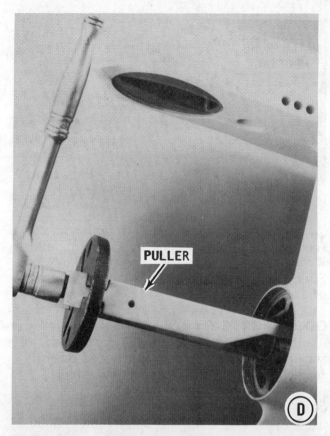

PULLER

D

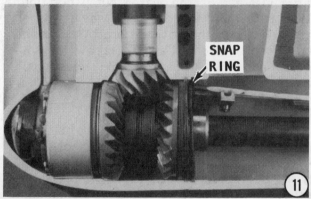

SNAP RING

11

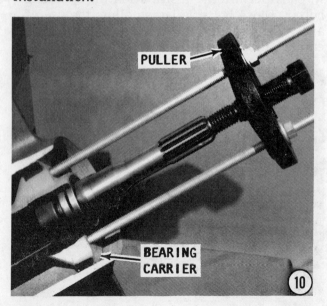

PULLER

BEARING CARRIER

10

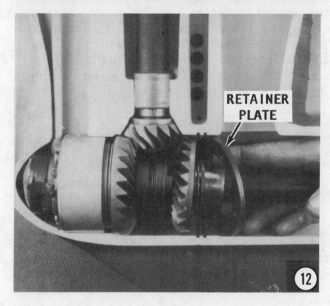

RETAINER PLATE

12

13- Reach inside the cavity and remove the thrust bearing, thrust washer and the reverse gear.

Propeller Shaft -- Removal

14- Grasp the propeller shaft firmly and withdraw it from the lower unit.

"FROZEN" PROPELLER SHAFT

On rare occasions, especially if water has been allowed to enter the lower unit, it may not be possible to withdraw the propeller shaft as described in Step 14. The shaft may be "frozen" in the hydraulic pump due to corrosion.

If efforts to remove the propeller shaft after the bearing carrier has been removed fail: Obtain a block of wood 2"x 4" approx. one foot in length. Drill a hole in the center of the flat side, large enough for the propeller shaft to pass through. Place the block over the shaft. Slide some thick large washers over the shaft and thread the propeller nut onto the shaft.

With the skeg clamped securely in a vise equipped with soft jaws, attempt to pull the shaft free. If necessary hammer on the wood, rotating the block at intervals to prevent wedging the shaft in any one direction.

Pinion Gear -- Removal

Special tool, OMC No. 316612 is required to turn the driveshaft in order to remove the pinion gear nut.

15- Obtain the special tool and slip it over the end of the driveshaft with the splines of the tool indexed with the splines on the driveshaft. Hold the pinion gear nut with the proper size wrench, and at the same time rotate the driveshaft, with the

special tool and wrench **COUNTERCLOCK-WISE** until the nut is free. If the special tool is not available, clamp the driveshaft in a vise equipped with soft jaws, in an area below the splines but not in the water pump impeller area. Now, with the proper size wrench on the pinion gear nut, rotate the complete lower unit **COUNTERCLOCKWISE** until the nut is free. This procedure will probably require the driveshaft to be loosened in the vise several times and reclamped in order to affect rotation of the lower unit and wrench. After the nut is free, proceed with the next step. The driveshaft will be withdrawn from the pinion gear.

Driveshaft -- Removal

16- Remove the four bolts from the top of the lower unit securing the bearing housing. **CAREFULLY** pry the bearing housing upward away from the lower unit, then slide it free of the driveshaft. An alternate method is to again clamp the driveshaft in a vise equipped with soft jaws. Use a soft-headed mallet and tap on the top side of the bearing housing. This action will jar the housing loose from the lower unit. Continue tapping with the mallet and the bearing housing, O-rings, shims, thrust washer, thrust bearing, and the driveshaft will all

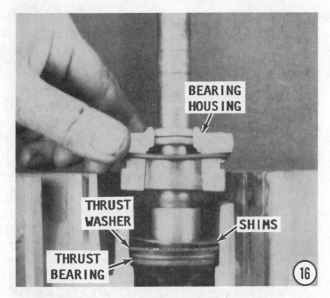

BEARING HOUSING

THRUST WASHER

SHIMS

THRUST BEARING

16

breakaway from the lower unit and may be removed as an assembly.

17- Remove the pinion gear from the lower unit cavity. Remove the forward gear from the hydraulic pump.

Hydraulic Pump — Removal

18- Obtain two long rods with 1/4" x 20 threads on both ends. Thread the two rods into the hydraulic pump housing. Attach a slide hammer to the rods and secure it with a nut on the end of each rod. Check to be sure the slide hammer is installed onto the rods **EVENLY** to allow an even pull on the pump. If the slide hammer is not installed to the rods properly, the pump may become tightly wedged in the lower unit. Operate the slide hammer and pull the hydraulic pump free. If the pump should happen to become lodged in the lower unit, stop operating the slide hammer **IMMEDIATELY.** Tap the hydraulic pump back into place in the lower unit and start the removal procedure over.

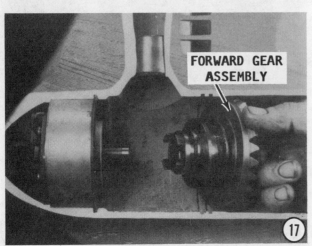

FORWARD GEAR ASSEMBLY

17

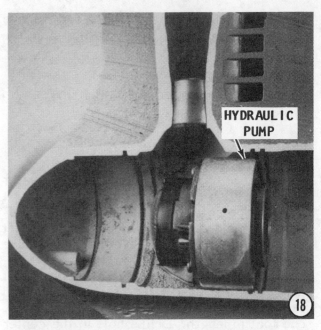

HYDRAULIC PUMP

18

Lower Driveshaft Bearing — Removal

This bearing cannot be removed without the aid of a special tool. Therefore, **DO NOT** attempt to remove this bearing unless it is unfit for further service. To check the bearing, first use a flashlight and inspect it for corrosion or other damage. Insert a finger into the bearing, and then check for "rough" spots or binding while rotating it.

19- For 1971 models **ONLY:** Use a punch, or similar tool and drive the lower driveshaft bearing out of position and into the lower unit cavity.

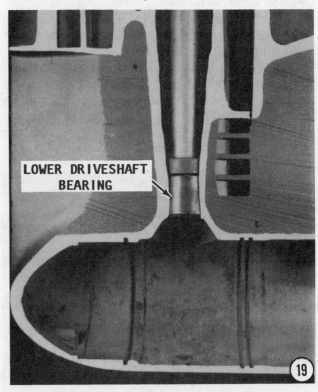

LOWER DRIVESHAFT BEARING

19

For 1972 models **ONLY**: Remove the **ALLEN** screw from the water pickup slots in the starboard side of the lower unit housing, reference illustration **"E"**. This screw secures the bearing in place and **MUST** be removed before an attempt is made to remove the bearing. The bearing must actually be **"PULLED"** upward to come free. **NEVER** make an attempt to "drive" it down and out or the lip in the lower unit holding the bearing will be broken off. **VERY BAD NEWS**. The lower unit housing would have to be replaced. Obtain special tool, OMC No. 385546. Use the special tool and "pull" the bearing from the lower unit, reference illustration **"F"**.

Propeller Shaft — Disassembling

20- Notice the spring retainer on the outside surface of the clutch dog. Use a small screwdriver and work one end of the spring up onto the shoulder of the clutch dog. Continue working the spring out of the groove until it is free. **TAKE CARE** not to distort the spring. Place one end of the propeller shaft on the bench and push the pin free of the clutch dog. Raise the propeller end of the shaft upward and the piston, retainer, and spring, will come free of the shaft.

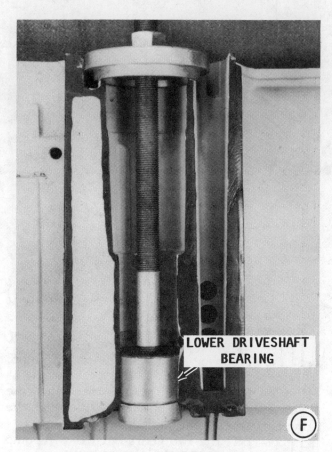

LOWER DRIVESHAFT BEARING

21- Notice how the small end of the retainer fits into the spring. Also notice the hole in the retainer. During installation, this hole must align with the hole in the propeller shaft and clutch dog to allow the pin to pass through. Slide the clutch dog free of the propeller shaft.

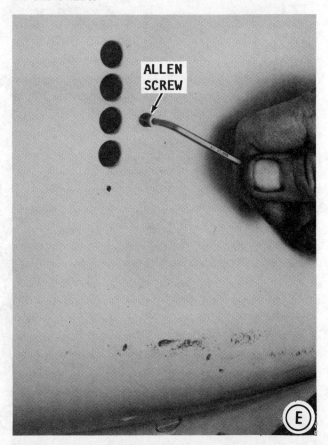

ALLEN SCREW

SPRING

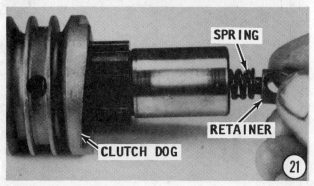

SPRING RETAINER CLUTCH DOG

(22)

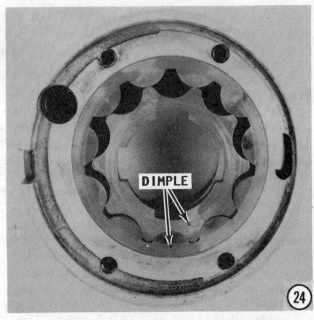

(24)

Hydraulic Pump — Disassembling

22- Remove the screw from the center of the screen on the back side of the pump. Remove the screen.

23- Remove the screws securing the valve housing to the pump, and then lift the housing free of the pump.

24- Lift the two gears out off the pump housing and **HOLD** them just as they were removed. Check the face of each gear for an indent mark (a dot, dimple, or similar identification). The identification mark will indicate how the gear **MUST** face in the housing. Make a note of how the mark faces, outward or inward, to **ENSURE** the gears will be installed properly in the same position from which they were removed.

Bearing Carrier — Disassembling

The bearings in the carrier need **NOT** be removed unless they are unfit for further service. Insert a finger and rotate the bearing. Check for "rough" spots or binding. Inspect the bearing for signs of corrosion or other types of damage. If the bearings must be replaced, proceed with the next step.

25- Use a seal remover to remove the two back-to-back seals or clamp the carrier in a vise and use a pry bar to pop each seal out.

26- Use a drift punch to drive the bearings free of the carrier. The bearings are being removed because they are unfit for service, therefore, additional damage is of no consequence.

(23)

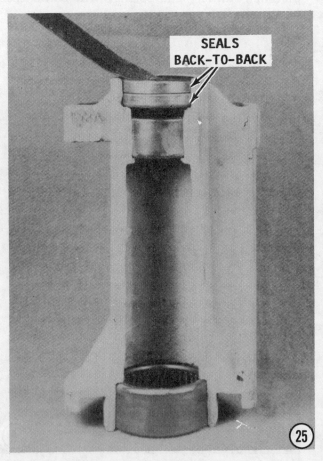

(25)

Solenoid and Shift Assembly

It is not recommended to attempt service of this assembly. If troubleshooting has been performed and the determination made the unit or any part is faulty, the **ONLY** satisfactory solution is to purchase and install a new assembly.

CLEANING AND INSPECTING

Wash all, except **ELECTRICAL**, parts in solvent and dry them with compressed air. Discard all **O**-rings and seals that have been removed. A new seal kit for this lower unit is available from the local dealer. The kit will contain the necessary seals and O-rings to restore the lower unit to service.

Inspect all splines on shafts and in gears for wear, rounded edges, corrosion, and damage.

Carefully check the driveshaft and the propeller shaft to verify they are straight and true without any sign of damage. A complete check must be performed by turning the shaft in a lathe. This is only necessary if there is evidence to suspect the shaft is not true.

Check the water pump housing for corrosion on the inside and verify the impeller

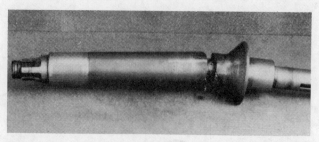

A two-section driveshaft with a weld section that has failed. This area of the driveshaft should be carefully checked anytime the lower unit is disassembled.

and base plate are in good condition. Actually, good shop practice dictates to rebuild or replace the water pump each time the lower unit is disassembled. The small cost is rewarded with "peace of mind" and satisfactory service.

Inspect the lower unit housing for nicks, dents, corrosion, or other signs of damage. Nicks may be removed with No. 120 and No. 180 emory cloth. Make a special effort to ensure all old gasket material has been removed and mating surfaces are clean and smooth.

Inspect the water passages in the lower unit to be sure they are clean. The screen may be removed and cleaned.

Check the gears and clutch dog to be sure the ears are not rounded. If doubt exists as to the part performing satisfactorily, it should be replaced.

Inspect the bearings for "rough" spots, binding, and signs of corrosion or damage.

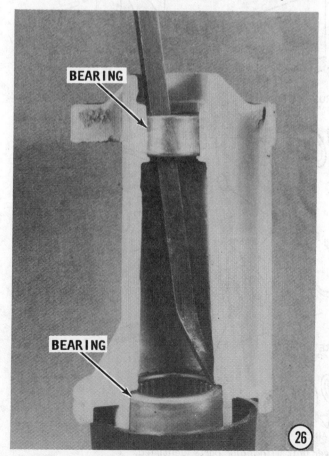

BEARING

BEARING

26

Damaged hydraulic pump. Water in the lower unit and a broken gear was the cause of this pump being destroyed.

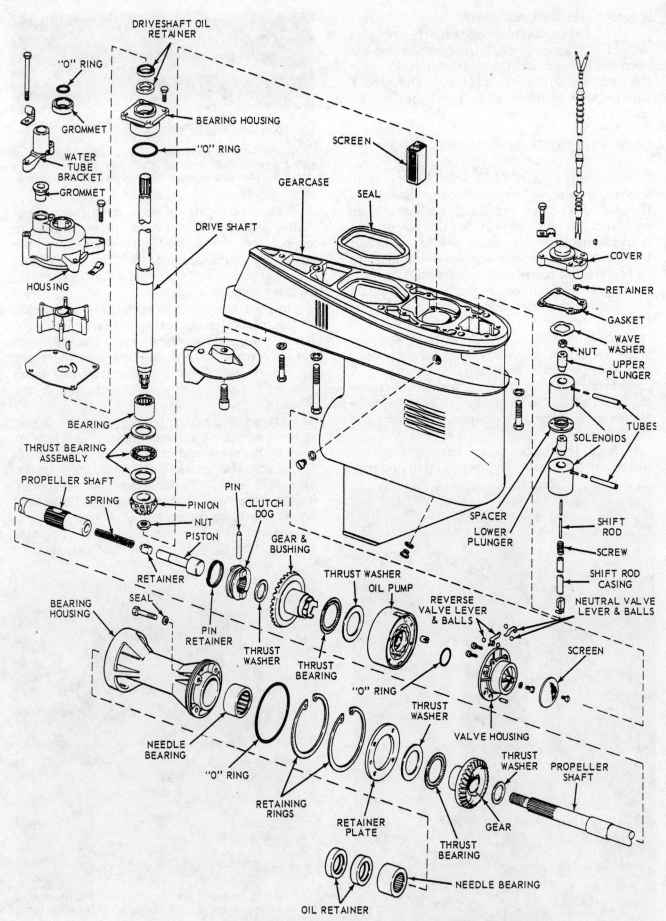

Exploded view of a 50 hp -- 1971-72 lower unit with major parts identified.

Damaged reverse gear (left) and forward gear (right). This damage was caused from water entering the lower unit.

Test the neutral and reverse solenoids with an ohmmeter. A reading of 5 to 7 ohms is normal and indicates the solenoid is in satisfactory condition.

ASSEMBLING

READ AND BELIEVE

The lower unit should **NOT** be assembled in a dry condition. Coat all internal parts with OMC HI-VIS lube oil as they are assembled. All seals should be coated with OMC Gasket Seal Compound. When two

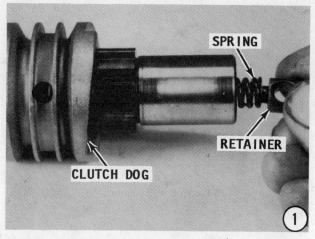

seals are installed back-to-back, use Triple Guard Grease between the seal surfaces.

AUTHORS APOLOGY

The accompanying illustrations show a rubber seal on the end of the hydraulic pump and a snap ring installed in front of the pump. During 1971 and 1972, these two items were not used. Therefore, disregard the seal and snap ring, for these two years.

Propeller Shaft -- Assembling

1- Slide the clutch dog onto the propeller shaft with the face of the dog marked **"PROP END"** facing toward the propeller end of the shaft, reference illustration **"A"**. Before the splines of the clutch dog engage the splines of the propeller shaft, rotate the dog until the hole for the pin appears to align with the hole through the propeller shaft. Slide the clutch dog onto the splines until the hole in the dog aligns with the hole in the shaft. If the hole is off just a bit, slide the clutch dog back off the splines, rotate it one spline in the required direction, and then slide it into place.

Insert the spring into the end of the propeller shaft. Secure the spring in place with the spring retainer. Install the retainer with the small end going into the propeller shaft **FIRST**.

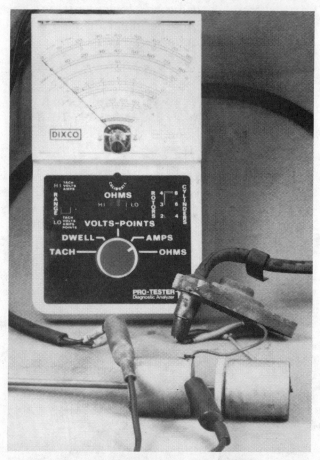

Using an ohmmeter to test the solenoids, as explained in the text.

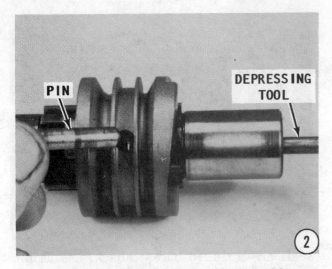

2- Depress the spring retainer and insert the pin through the clutch dog, the shaft, spring retainer, and out the other side of the shaft and clutch dog. Center the pin through the clutch dog.

3- Install the spring-type pin retainer around the clutch dog to secure the pin in place. **TAKE CARE** not to distort the pin retainer during the installation process.

Bearing Carrier Bearings & Seals Installation

4- Install the reverse gear bearing into the bearing carrier by pressing against the **LETTERED** side of the bearing with the proper size socket. Press the forward gear bearing into the bearing carrier in the same manner. Press against the **LETTERED** side of the bearing.

5- Coat the outside surfaces of the seals with HI-VIS oil. Install the first seal with the flat side facing **OUT**. Coat the flat surface of both seals with Triple Guard Grease, and then install the second seal with the flat side going in **FIRST**. The seals are

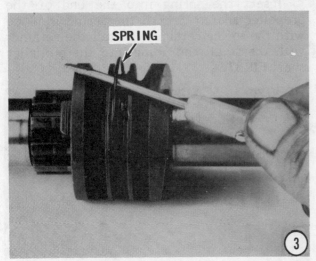

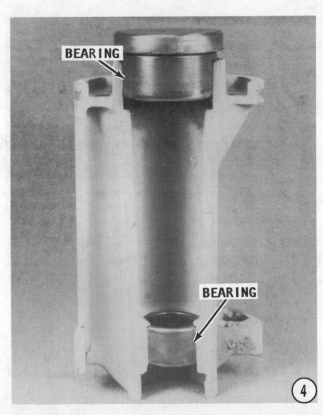

then back-to-back with the grease between the two surfaces. The outside seal prevents water from entering the lower unit and the inside seal prevents the lubricant in the lower unit from escaping.

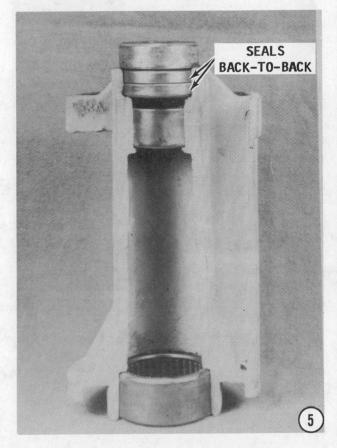

A set of double seals showing the back side (left) and the front side (right). These seals are installed back-to-back (flat side-to-flat side) with Triple Guard Grease between the surfaces. This arrangement prevents fluid from passing in either direction.

Hydraulic Pump — Assembling

6- Check the note made during disassembling, per Step 23, to determine how the identifying marks (dots, dimples, whatever) on the gears must face -- inward or outward. The gears **MUST** be installed in the same position from which they were removed.

7- Install the rear valve housing with the tang on the outside edge of the housing indexed with the small slot in the pump housing. Secure the valve housing in place with the attaching screws tightened securely.

8- Place the screen in position on the back side of the valve housing, and then secure it in place with the screw.

9- Install the forward gear into the pump housing. It may be necessary to work the gears around in the pump to permit the tangs on the forward gear shank to index in the slots in the housing. Set the assembly aside for later installation.

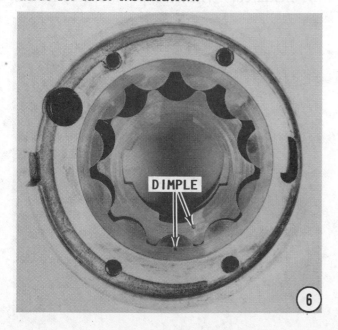

Two types of hydraulic pumps. The pump on the left is the most common with the shift rod passing through a hole in two levers on top of the pump. The pump on the right has the shift rod passing directly through a hole in the pump top.

Lower Driveshaft Bearing — Installation

10- Obtain tool, OMC No. 385546. Assemble the tool with the washer, guide sleeve, and remover portion of the tool, in the order given. The shoulder of the tool must face **DOWN**. Place the bearing onto the end of the tool, with the lettered side of the bearing facing the tool. Drive the bearing down until the large washer on the tool makes contact with the surface of the lower unit. The bearing is then seated to the proper depth.

11- If an Allen screw is used to secure the bearing in place, apply Loctite to the threads, amd then install the screw through the lower unit, as shown.

Hyraulic Pump — Installation
First, These Words

Observe the tang on the backside of the pump. This tang **MUST** face directly up in relation to the lower unit housing to permit installation of the shift rod into the pump. Also notice the pin on the backside of the pump. This pin **MUST** index into a matching hole in the housing to restrain the pump from rotating.

12- Secure the lower unit housing in the horizontal position with the bearing carrier opening facing up. Remove the forward gear from the pump. Obtain two long 1/4 x

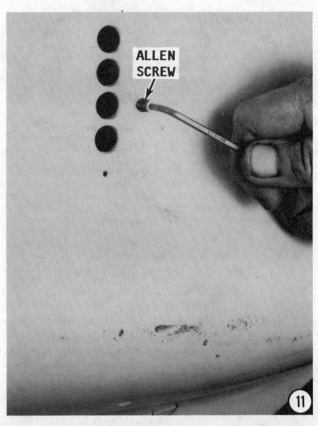

20 rods with threads on both ends. Thread the rods into the pump and then lower the pump into the lower unit housing. To index the pin on the back of the pump housing into the hole in the lower unit is not an easy task. However, exercise patience and rotate the pump ever so slowly. A helpful hint at this point: As the pump is being lowered into the cavity, align the opening and tang on top of the pump in the approximate position your eye indicates the shift rod may be installed. When the pin indexes, it will

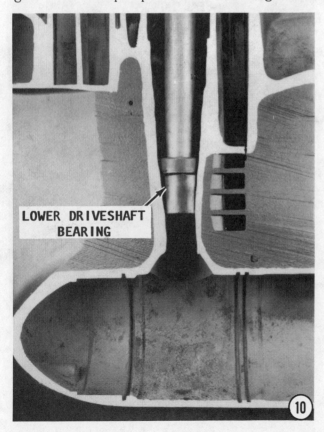

LOWER DRIVESHAFT BEARING

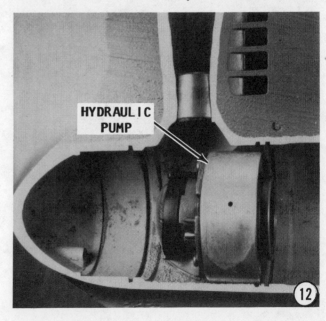

HYDRAULIC PUMP

not be possible to rotate the pump. The pump **MUST** be properly seated to permit installation of the shift rod and the pinion gear. After the pump is in place, remove the two rods used during installation.

13- Coat the plunger with oil, and then lower the large end of the plunger into the hydraulic pump. Check to be sure it seats all the way into place.

14- Install the thrust washer and thrust bearing into the pump with the flat side of the bearing facing **OUTWARD**. Lower the forward gear into the pump. Work the gear slowly until the teeth index with the teeth of the pump gear. This should not be too difficult because the forward gear was installed once, and then removed in the previous step. However, it is entirely possible the gears moved when the pump was installed. Therefore, use a flashlight and check the position of the gears. If necessary, use a long shank screwdriver and rotate the gears until they are close to center, then install the forward gear.

Driveshaft and Pinion Gear -- Installation

CRITICAL WORDS

The driveshaft and pinion gear must be assembled prior to installation, and then checked with a special shimming gauge. This shimming must be accomplished properly, the unit disassembled, and then installed into the lower unit. Use of the shimming gauge is the **ONLY** way to determine the proper amount of shimming required at the upper end of the driveshaft. The following detailed step outlines the procedure.

15- Clamp the driveshaft in a vise equipped with soft jaws and in such a manner that the splines, water pump area, or other critical portions of the shaft cannot

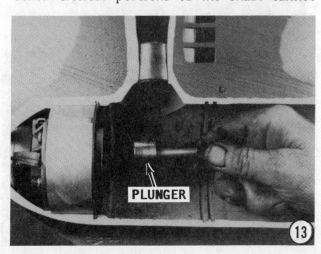

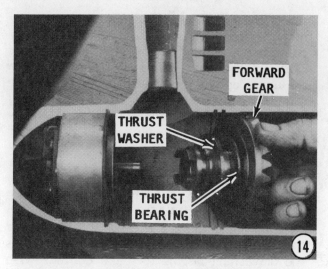

be damaged. Slide the pinion gear onto the driveshaft with the bevel of the gear teeth facing toward the lower end of the shaft. Install the pinion gear nut and tighten it to a torque value of 40 to 45 ft-lbs. No parts should be installed on the upper end of the driveshaft at this point. Slide the same amount of shim material removed during disassembling, onto the driveshaft and seat it against the driveshaft shoulder.

Obtain special shimming tool, OMC No. 315767. Slip the special tool down over the driveshaft and onto the upper surface of the top shim. Measure the distance between the

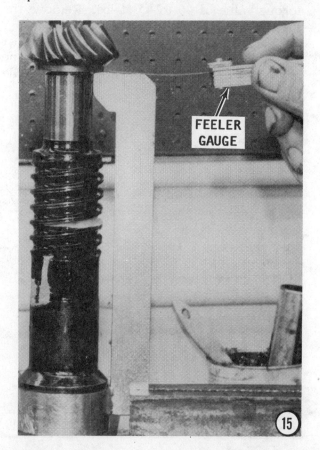

top of the pinion gear and the bottom of the tool. The tool should just barely make contact with the pinion gear surface for **ZERO** clearance. Add or remove shims from the upper end of the driveshaft to obtain the required **ZERO** clearance. Remove the tool and set the shims aside for installation later. Back off the pinion gear nut and remove the pinion gear. Remove the driveshaft from the vise.

16- Insert the pinion gear into the cavity in the lower unit with the flat side of the bearing facing **UPWARD**. Hold the pinion gear in place and at the same time lower the driveshaft down into the lower unit. As the driveshaft begins to make contact with the pinion gear, rotate the shaft slightly to permit the splines on the shaft to index with the splines of the pinion gear. After the shaft has indexed with the pinion gear, thread the pinion gear nut onto the end of the shaft. Obtain special tool, OMC No. -316612. Slide the special tool over the upper end of the driveshaft with the splines of the tool indexed with the splines on the shaft. Attach a torque wrench to the special tool. Now, hold the pinion gear nut with the proper size wrench and rotate the driveshaft **CLOCKWISE** with the special tool until the pinion gear nut is tightened to a torque value of 40 to 45 ft-lbs. Remove the special tool.

17- Slide the thrust bearing, thrust washer, and the shims, set aside after the shim gauge procedure in Step 15, onto the driveshaft.

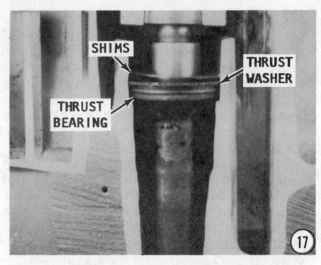

18- Install the two seals back-to-back into the opening on top of the bearing housing. Coat the outside surface of the **NEW** seals with OMC Lubricant. Press the first seal into the housing with the flat side of the seal facing **OUTWARD**. After the seal is in place, apply a coating of Triple Guard Grease to the flat side of the installed seal and the flat side of the second seal. Press the second seal into the bearing housing with the flat side of the seal facing **INWARD**. Insert a **NEW** O-ring into the bottom opening of the bearing housing.

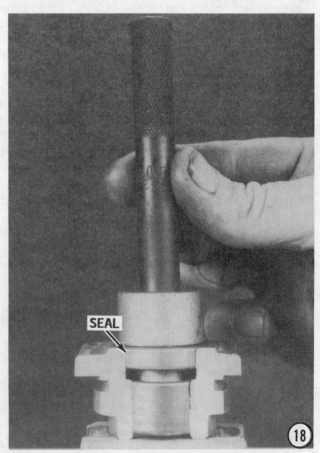

19- Wrap friction tape around the splines of the driveshaft to protect the seals in the bearing housing as the housing is installed. Now, slide the assembled bearing housing down the driveshaft and seat it in the lower unit housing. Secure the housing in place with the four bolts. Tighten the bolts **ALTERNATELY** and **EVENLY**.

Propeller Shaft — Installation

20- Secure the lower unit in the horizontal position with the bearing carrier opening facing **UPWARD**. Check the inside diameter of the forward gear to be sure the small thrust washer in the gear is still in place. Lower the propeller shaft assembly down into the lower unit with the inside diameter of the shaft indexing over the shaft of the piston.

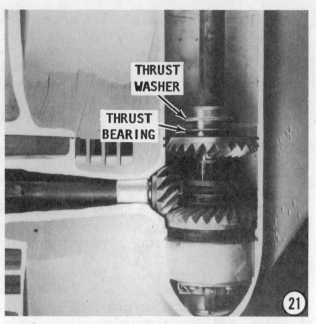

21- Apply a light coating of grease to the inside surface of the reverse gear to hold the thrust washer in place. Insert the thrust washer into the reverse gear. Lower the thrust bearing and thrust washer down onto the shank of the reverse gear. Slide the reverse gear down the propeller shaft with the splines of the reverse gear indexing with the splines of the shaft.

22- Insert the retainer plate into the lower unit against the reverse gear.

WARNING

This next step can be dangerous. Each snap ring is placed under tremendous tension with the Truarc pliers while it is being

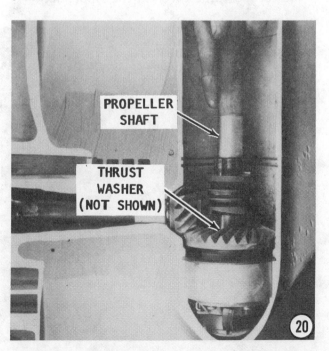

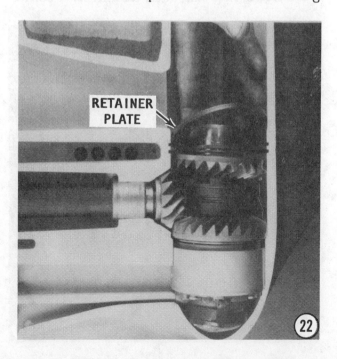

placed into the groove. Therefore, wear **SAFETY GLASSES** and exercise care to prevent the snap ring from slipping out of the pliers. If the snap ring should slip out, it would travel with incredible speed and cause personal injury if it struck a person.

23– Install the Truarc snap rings one at-a-time following the precautions given in the **WARNING** and the **ADVICE** given in the following paragraph.

WORDS OF ADVICE

The two snap rings index into separate grooves in the lower unit housing. As the first ring is being installed, depth perception may play a trick on yours eyes. It may appear that the first ring is properly indexed all the way around in the proper groove, when in reality, a portion may be in one groove and the remainder in the other groove. Should this happen, and the Truarc pliers be released from the ring, it is extremely difficult to get the pliers back into the ring to correct the condition. If necessary use a flashlight and carefully check to be sure the first ring is properly seated all the way around **BEFORE** releasing the grip on the pliers. Installation of the second ring is not so difficult because the one groove is filled with the first ring.

24– Obtain two long 1/4" rods with threads on one end. Thread the rods into the retainer plate opposite each other to act as guides for the bearing carrier.

25– Check the bearing carrier to be sure a **NEW O**-ring has been installed. Position

the carrier over the guide pins with the embossed word **UP** on the rim of the carrier facing **UP** in relation to the lower unit housing. Now, lower the bearing carrier down over the guide pins and into place in the lower unit housing.

26– Slide **NEW** little O-rings onto each bolt, and apply some OMC Sealer onto the threads. Install the bolts through the carrier and into the retaining plate. After a couple bolts are in place, remove the guide pins and install the remaining bolts. Tighten the bolts **EVENLY** and **ALTERNATELY** to the torque value given in the Appendix.

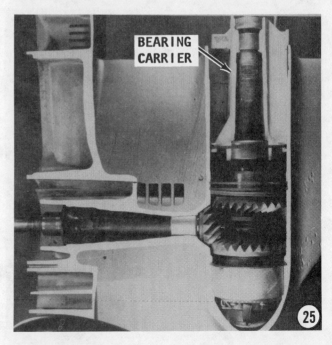

Caption: BOLT — 26

Shift Cable, Solenoids, and Rod — Assembling

CRITICAL WORDS

The following procedures **MUST** be performed exactly as given and in the order presented. Do not attempt any shortcuts or anticipate what will be done next. All parts must be installed and adjusted to the letter, for the unit to function properly. Separate the upper (green) solenoid from the lower (blue) solenoid, by pulling them apart. An inner shift rod will be released from the shift rod casing. Check to be sure the cap on the bottom of the shift rod casing is in place.

27- Lower the blue solenoid down into the lower unit housing. Continue lowering the solenoid until the cap on the end of the shift rod casing seats on top of the valve

Caption: PLUNGER — 28

lever and check ball assembly. Check to be sure the solenoid is fully seated in the housing. The lower plunger should be flush with the top of the solenoid.

28- If the plunger is not flush with the solenoid, remove the solenoid and screw the lower plunger up or down on the shift rod casing, then install the solenoid again and check the plunger.

29- Install the spacer with the lip on the spacer facing **UPWARD**.

30- Lower the green solenoid into the lower unit housing. Insert the shift rod into the shift rod casing.

31- The upper plunger must be flush with the top surface of the solenoid. If it is not flush with the solenoid, remove the solenoid, loosen the nut, and make an adjustment. Install the solenoid again and check the upper plunger.

Caption: LOWER SOLENOID — 27

Caption: SPACER — 29

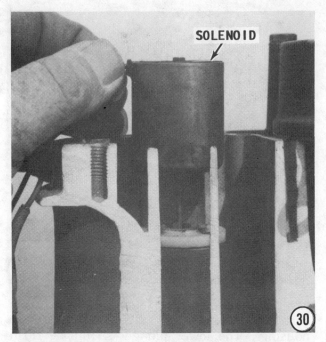

SOLENOID

30

32- Check to be sure the inside plunger indexes into the hole in the hydraulic pump.

33- Install the wavy washer and a **NEW** gasket into the lower unit housing. Work the shift wires down into the lower unit cavity. Lower the cover into place in the lower unit housing. The wavy washer will give you a false pressure against the cap. Therefore, it takes a bit of patience to be sure the wavy washer indexes into the recess of the cover. Start the bolts securing the cover. Tighten the bolts **ALTERNATELY** and **EVENLY**. As the bolts are tightened, make continuous checks to be sure the wavy washer and the green solenoid fit up into the cover as the bolts are

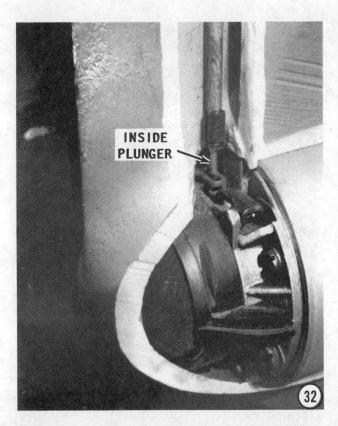

INSIDE PLUNGER

32

tightened. This may not be accomplished on the first attempt, but keeping cool and working slowly will be rewarded with success.

34- Obtain an ohmmeter. Ground one lead of the meter, and then check the blue and green wires for continuity. The ohmmeter should indicate 5 to 7 ohms.

ADJUSTMENT

31

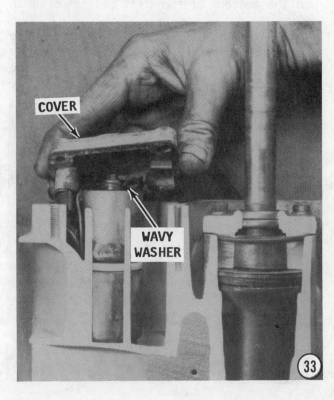

COVER

WAVY WASHER

33

WATER PUMP INSTALLATION

FIRST, THESE WORDS

An improved water pump is available as a replacement. If the old water pump housing is unfit for further service, only the new pump housing can be purchased. It is strongly recommended to replace the water pump with the improved model while the lower unit is disassembled. The accompanying illustration shows the original equipment (left) compared with the improved pump (right), reference illustration **"A"**.

The new pump must be assembled before it is installed. Therefore, the following steps outline procedures for both pumps.

To assemble and install a replacement pump, perform steps 35 thru 43, then jump to Step 46.

To install an original equipment pump, proceed direectly to Step 44.

Assembling an Improved Pump Housing

35- Remove the water pump parts from the container. Insert the plate into the housing, as shown. The tang on the bottom side of the plate **MUST** index into the short slot in the pump housing.

36- Slide the pump liner into the housing with the two small tabs on the bottom side indexed into the two cutouts in the plate.

37- Coat the inside diameter of the liner with light-weight oil. Work the impeller into the housing with all of the blades bent back to the right, as shown. In this position,

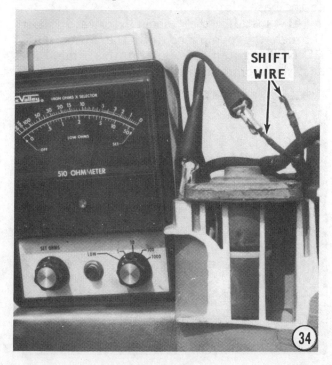

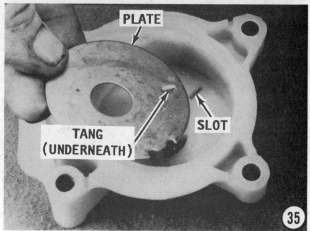

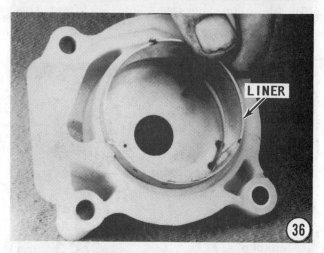

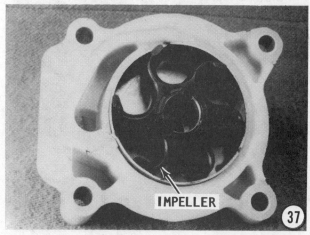

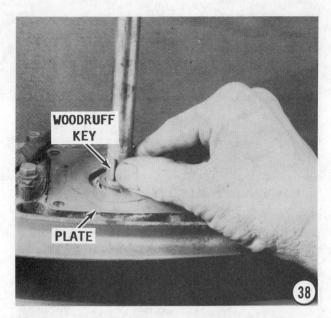

the blades will rotate properly when the pump housing is installed. Remember, the pump and the blades will be rotating **CLOCKWISE** when the housing is turned over and installed in place on the lower unit.

38- Coat the mating surface of the lower unit with 1000 Sealer. Slide the water pump base plate down the driveshaft and into place on the lower unit. Insert the Woodruff key into the key slot in the driveshaft.

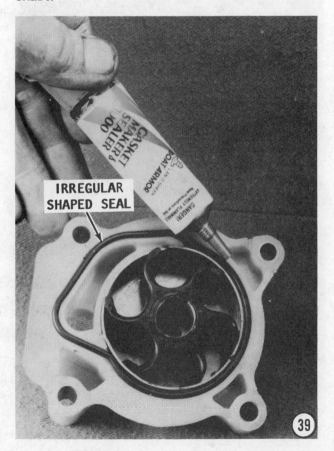

39- Lay down a very thin bead of 1000 sealer into the irregular shaped groove in the housing. Insert the seal into the groove, and then coat the seal with the 1000 Sealer.

Water Pump Installation

40- Begin to slide the water pump down the driveshaft, and at the same time observe the position of the slot in the impeller. Continue to work the pump down the driveshaft, with the slot in the impeller indexed over the Woodruff key. The pump must be fairly well aligned before the key is covered because the slot in the impeller is not visible as the pump begins to come close to the base plate.

41- Install the short forward bolt through the pump and into the lower unit. **DO NOT** tighten this bolt at this time. Insert the grommet into the pump housing.

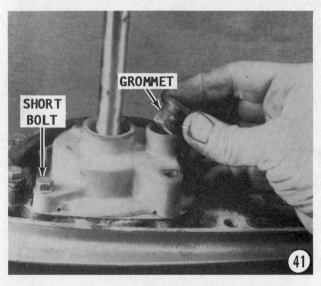

42

42- Install the grommet retainer and water tube guide onto the pump housing. Install the remaining pump attaching bolts. Tighten the bolts **ALTERNATELY** and **EVENLY,** and at the same time rotate the driveshaft **CLOCKWISE.** If the driveshaft is not rotated while the attaching bolts are being tightened, it is possible to pinch one of the impeller blades underneath the housing.

43- Slide the large grommet down the driveshaft and seat it over the pump collar. This grommet does not require sealer. Its function is to prevent exhaust gases from entering the water pump. Proceed directly to Lower Unit Installation, Step 47.

Water Pump Installation
Original Equipment

Perform the following two steps to install an original water pump.

44- Coat the water pump plate mating surface on the lower unit with 1000 Sealer. Slide the water pump plate down the driveshaft and **BEFORE** it makes contact with the sealer check to be sure the bolt holes in the plate will align with the holes in the housing. The plate will only fit one way. If the holes will not align, remove the plate, turn it over and again slide the plate down the driveshaft and into place on the housing.

43

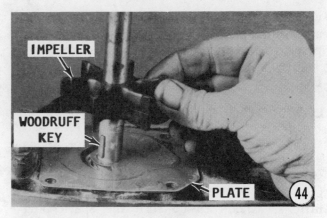

44

This checking will prevent accidently getting the sealer on both sides of the plate. If by chance sealer does get on the top surface it **MUST** be removed before the water pump impeller is installed.

Slide the water pump impeller down the driveshaft. Just before the impeller covers the cutout for the Woodruff key, install the key, and then work the impeller on down, with the slot in the impeller indexed over the Woodruff key. Continue working the impeller down until it is firmly in place on the surface of the pump plate.

45- Check to be sure **NEW** seals and O-rings have been installed in the water pump. Lubricate the inside surface of the water pump with light-weight oil. Lower the water pump housing down the driveshaft and over the impeller. **ALWAYS** rotate the driveshaft slowly **CLOCKWISE** as the housing is lowered over the impeller to allow the impeller blades to assume their natural and proper position inside the housing. Continue to rotate the driveshaft and work the water pump housing downward until it is seated on the plate. Coat the threads of the water pump attaching screws with sealer, and then secure the pump in place with the screws. Install the solenoid cable bracket with the same bolt and in the same position from

45

which it was removed. On some units the solenoid cable fits into a recess of the water pump and is held in place in that manner. Tighten the screws **ALTERNATELY** and **EVENLY.**

LOWER UNIT INSTALLATION

46- Check to be sure the water tubes are clean, smooth, and free of any corrosion. Coat the water pickup tubes and grommets with lubricant as an aid to installation. Check to be sure the spark plug wires are disconnected from the spark plugs. Bring the lower unit housing together with the exhaust housing, and at the same time, guide the water tube into the rubber grommet of the water pump. Connect the ends of the wire left in the exhaust housing during removal to the blue and green wires from the lower unit. Tape the connections, and then oil the shift cable as an aid to slipping the cable through the exhaust housing. Continue to bring the two units together and at the same time: pull the wires through the exhaust housing; guide the water pickup tubes into the rubber grommet of the water pump; and, rotate the flywheel slowly to permit the splines of the driveshaft to index with the splines of the crankshaft. This may sound like it is necessary to do four things at the same time, and

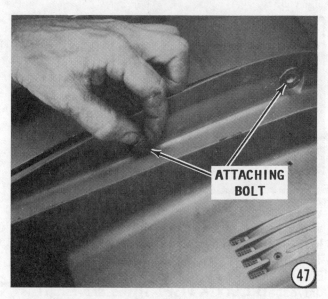

so it is. Therefore, make an earnest attempt to secure the services of an assistant for this task.

47- After the surfaces of the lower unit and exhaust housing are close, dip the attaching bolts in OMC Sealer and then start them in place. Two bolts are used on each side of the two housings.

48- Install the retaining bolt in the recess of the trim tab and another bolt in the cavitation plate. Tighten the bolts **ALTERNATELY** and **EVENLY** to the torque value given in the Appendix.

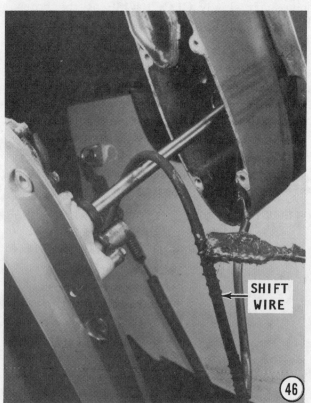

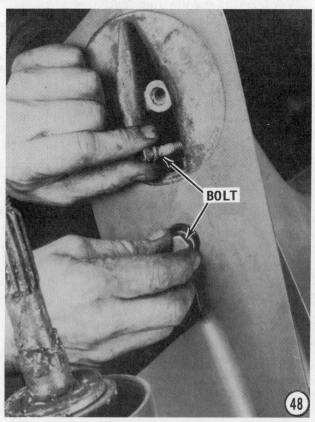

49- Install the trim tab with the mark made on the tab during disassembling aligned with the mark made on the lower unit housing.

50- Disconnect and remove the "pull" wire used to feed the electrical wires up through the exhaust housing.

51- Obtain an ohmmeter. Ground the meter and again check the green and blue wires for continuity. The meter should indicate 5 to 7 ohms. Connect the green wire to the green wire and the blue wire to the blue wire. After the connections have been made, slide the sleeve down over the connections. A bit of oil on the wires will allow the sleeves to slide more easily.

Filling the Lower Unit

Fill the lower unit with lubricant according to the procedures in Section 8-3.

Propeller Installation

Install the propeller, see Section 8-2.

FUNCTIONAL CHECK

Perform a functional check of the completed work by mounting the engine in a test tank, in a body of water, or with a flush attachment connected to the lower unit. If the flush attachment is used, **NEVER** operate the engine above an idle speed, because the no-load condition on the propeller would allow the engine to **RUNAWAY** resulting in serious damage or destruction of the engine.

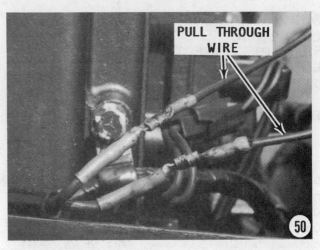

CAUTION: Water must circulate through the lower unit to the engine any time the engine is run to prevent damage to the water pump in the lower unit. Just five seconds without water will damage the water pump.

Start the engine and observe the tattle-tale flow of water from idle relief in the exhaust housing. The water pump installation work is verified. If a "Flushette" is connected to the lower unit, **VERY LITTLE** water will be visible from the idle relief port. Shift the engine into the three gears and check for smoothness of operation and satisfactory performance. Remember, when the unit is in forward gear, it is at rest with no current flow to either solenoid.

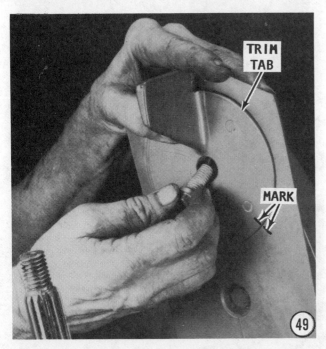

8-12 MECHANICAL SHIFT
HYDRAULIC ASSIST
SHIFT DISCONNECT UNDER
LOWER CARBURETOR
50 HP 1973-74

DESCRIPTION

The lower unit covered in this section is a three shift position, hydraulic activated, propeller exhaust unit. A hydraulic pump mounted in the forward portion of the lower unit provides the force required to shift the unit. A shift rod extends down through the exhaust housing and lower unit to an assist cylinder. A push rod connects the assist cylinder and the hydraulic pump. The pump valve directs the hydraulic force to place the clutch dog in the desired position for neutral, forward, or reverse gear position.

When the shift rod is in the upward position, the system is at rest and the lower unit is in forward gear.

In simple terms, something must be done to move the unit into neutral or reverse gear position. If no action is taken (shift mechanism at rest) the unit is in the forward gear position.

As the shift control lever is moved at the control box, the shift rod moves in the assist cylinder; the valve in the cylinder is activated; the push rod is moved, allowing oil from the pump to pass up through the hollow push rod into the assist cylinder; the shift mechanism is thereby "assisted" for easier and smoother shifting.

The lower unit houses the driveshaft and pinion gear, the forward and reverse driven gears, the propeller shaft, clutch dog, hydraulic pump, assist cylinder and the necessary shims, bearings, and associated parts to make it all work properly.

The water pump is considered a part of the lower unit.

TROUBLESHOOTING

Preliminary Checks

At rest, and without the engine running, the lower unit is in forward gear. Mount the engine in a test tank, in a body of water, or with a flush attachment connected to the lower unit. If the flush attachment is used, **NEVER** operate the engine above an idle speed, because the no-load condition on the propeller would allow the engine to **RUN-AWAY** resulting in serious damage or destruction of the engine.

CAUTION: Water must circulate through the lower unit to the engine any time the engine is run to prevent damage to the water pump in the lower unit. Just five seconds without water will damage the water pump.

Cut-a-way view of a 50 hp, 1973-74 lower unit with major parts exposed. This type of illustration will be most helpful during work on the lower unit.

Helm-type shift box and steering wheel.

Attempt to shift the unit into **NEU-TRAL** and **REVERSE**. It is possible the propeller may turn very slowly while the unit is in neutral, due to "drag" through the various gears and bearings. If difficult shifting is encountered, the problem is in the shift linkage or in the lower unit.

The second area to check is the quantity and quality of the lubricant in the lower unit. If the lubricant level is low, contaminated with water, or is broken down because of overuse, the shift mechanism may be affected. Water in the lower unit is **VERY BAD NEWS** for a number of reasons, particularly when the lower unit contains hydraulic components. Hydraulic units will not function with water in the system.

Remember, when the engine is not running, the unit should be in **FORWARD** gear.

BEFORE making any tests, remove the propeller, see Section 8-2. Check the propeller carefully to determine if the hub has been slipping and giving a false indication the unit is not in gear, reference illustration **"A"**. If there is any doubt, the propeller should be taken to a shop properly equipped for testing, before the time and expense of disassembling the lower unit is undertaken. The expense of the propeller testing and possible rebuild is justified.

The following troubleshooting procedures are presented on the assumption the lower unit lubricant, and the propeller have been checked and found to be satisfactory.

Lower Unit Locked

Determine if the problem is in the powerhead or in the lower unit. Attempt to rotate the flywheel. If the flywheel can be moved even slightly in either direction, the problem is most likely in the lower unit. If it is not possible to rotate the flywheel, the problem is a "frozen" powerhead. To absolutely verify the powerhead is "frozen", separate the lower unit from the exhaust housing and then again attempt to rotate the flywheel. If the attempt is successful, the problem is definitely in the lower unit. If the attempt to rotate the flywheel, with the lower unit removed, still fails, a "frozen" powerhead is verified, reference illustration **"B"**.

Unit Fails to Shift
Neutral, Forward, or Reverse

With the outboard mounted on a boat, in a test tank, or with a flush attachment connected, disconnect the shift lever at the engine. Attempt to manually shift the unit into **NEUTRAL, REVERSE, FORWARD**. At the same time move the shift handle at the shift box and determine that the linkage and shift lever are properly aligned for the shift positions. If the alignment is not correct, adjust the shift cable at the trunnion. It is

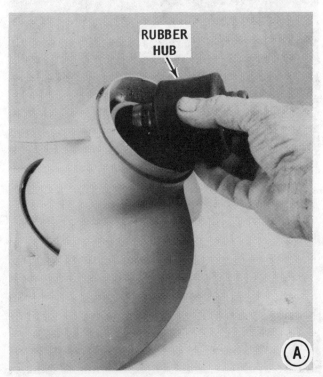

RUBBER HUB

(A)

(B)

also possible the inner wire may have slipped in the connector at the shift box. This condition would result in a lack of inner cable to make a complete "throw" on the shift handle. Check to be sure the shift handle is moved to the full shift position and the linkage to the lower unit is moved to the full shift position.

If an adjustment is required at the shift box, see Chapter 7.

If it is not possible to shift the unit into gear by manually operating the shift rod while the engine is running, the lower unit requires service as described in this section.

LOWER UNIT SERVICE

Propeller Removal

If the propeller was not removed, as directed for the troubleshooting, remove it now, according to the procedures outlined in Section 8-2.

Draining the Lower Unit

Drain the lower unit of lubricant, see Section 8-3.

GOOD WORDS

If water is discovered in the lower unit and the propeller shaft seal is damaged and

requires replacement, the lower unit does **NOT** have to be removed in order to accomplish the work.

The bearing carrier can be removed and the seal replaced without disassembling the lower unit. **HOWEVER** such a procedure is not considered good shop practice, but merely a quick-fix. If water has entered the lower unit, the unit should be disassembled and a detailed check made to determine if any other seals, bearings, bearing races, O-rings or other parts have been rendered unfit for further service by the water.

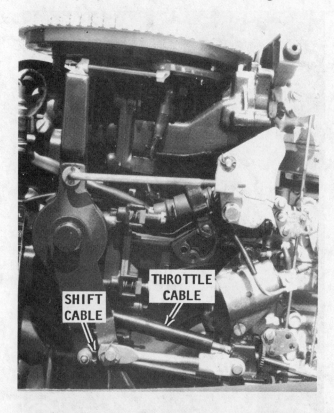

Shift and throttle linkage on the 50 hp, 1973-74 engines.

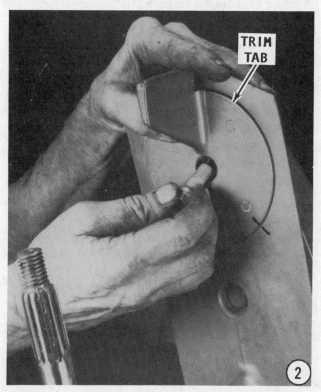

LOWER UNIT REMOVAL

1- Disconnect and ground the spark plug wires. Remove the bolt or bolts from the shift disconnect coupler under the lower carburetor.

2- Scribe a mark on the trim tab and a matching mark on the lower unit to ensure the trim tab will be installed in the same position from which it is removed. Use an Allen wrench and remove the trim tab.

3- Use a 1/2" socket with a short extension and remove the bolt from inside the trim tab cavity. Remove the 5/8" countersunk bolt located just ahead of the trim tab position.

4- Remove the four 9/16" bolts, two on each side, securing the lower unit to the exhaust housing.

5- Work the lower unit free of the exhaust housing. If the unit is still mounted on a boat, tilt the engine forward to gain clearance between the lower unit and the deck (floor, ground, whatever). **EXERCISE CARE** to withdraw the lower unit straight away from the exhaust housing to prevent bending the driveshaft.

WATER PUMP REMOVAL

6- Position the lower unit in the vertical position on the edge of the work bench resting on the cavitation plate. Secure the

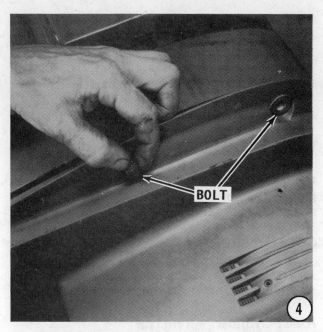

lower unit in this position with a C-clamp. The lower unit will then be held firmly in a favorable position during the service work. An alternate method is to cut a groove in a short piece of 2" x 6" wood to accommodate the lower unit with the cavitation plate resting on top of the wood. Clamp the wood in a vise and service work may then be performed with the lower unit erect (in its

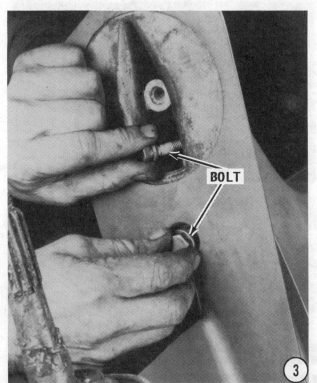

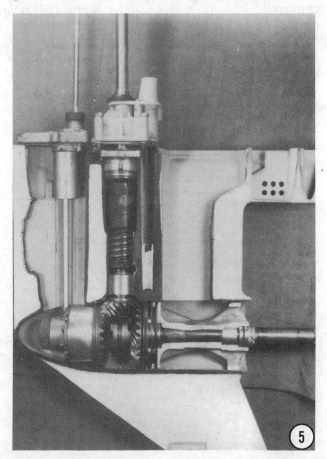

normal position), or inverted (upside down). In both positions, the cavitation plate is the supporting surface.

7- Remove the O-ring from the top of the driveshaft. Remove the bolts securing the water pump to the lower unit housing. Pull the water pump housing up and free of the driveshaft. Slide the water pump impeller up and free of the driveshaft.

8- Pop the impeller Woodruff key out of the driveshaft keyway. Slide the water pump base plate up and off of the driveshaft.

GOOD WORDS

If the only work to be performed is service of the water pump, proceed directly to Page 8-167, Water Pump Installation.

Assist Cylinder — Removal

9- Remove the four screws at the front edge of the lower unit. Lift and slide the cover up and free of the shift rod.

Bearing Carrier — Removal

10- Remove the four 5/16" bolts from inside the bearing carrier. Notice how each bolt has an O-ring seal. These O-rings should be replaced each time the bolts are removed. In most cases, the word **UP** is embossed into the metal of the bearing carrier rim. This word must face **UP** in relation to the lower unit during installation.

11- Remove the bearing carrier using one of the methods described in the following paragraphs, under Special Words.

SPECIAL WORDS

Several models of bearing carriers are used on the lower units covered in this section.

The bearing carriers are a very tight fit into the lower unit opening. Therefore, it is not uncommon to apply heat to the outside surface of the lower unit with a torch, at the same time the puller is being worked to remove the carrier. **TAKE CARE** not to overheat the lower unit.

One model carrier has two threaded holes on the end of the carrier. These threads permit the installation of two long bolts. These bolts will then allow the use of a flywheel puller to remove the bearing carrier, illustration 11.

Another model does not have the threaded screw holes. To remove this type bearing carrier, a special puller with arms must be used. The arms are hooked onto the carrier web area, and then the carrier removed, reference illustration "C".

WARNING

The next step involves a dangerous procedure and should be executed with care while wearing **SAFETY GLASSES**. The retaining rings are under tremendous tension in the groove and while they are being removed. If a ring should slip off the Truarc pliers, it will travel with incredible speed

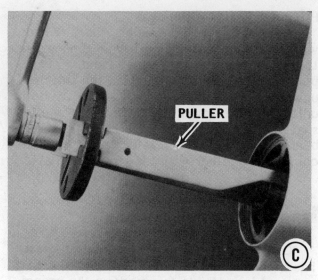

causing personal injury if it should strike a person. Therefore, continue to hold the ring and pliers firm after the ring is out of the groove and clear of the lower unit. Place the ring on the floor and hold it securely with one foot before releasing the grip on the pliers. An alternate method is to hold the ring inside a trash barrel, or other suitable container, before releasing the pliers.

12- Obtain a pair of Truarc pliers. Insert the tips of the pliers into the holes of the first retaining ring. Now, **CAREFULLY** remove the retaining ring from the groove and gear case without allowing the pliers to slip. Release the grip on the pliers in the manner described in the above **WARNING**. Remove the second retaining ring in the same manner. The rings are identical and either one may be installed first. Remove the retainer plate.

13- As the plate is removed, notice which surface is facing into the housing, as an aid during installation.

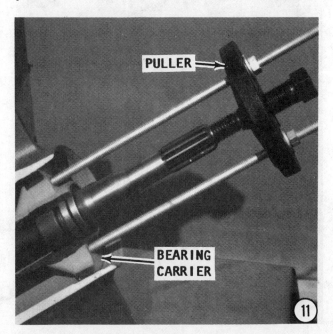

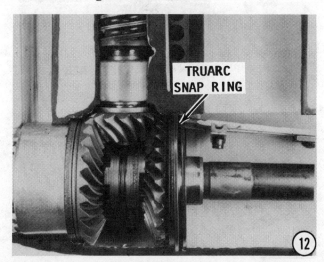

Propeller Shaft — Removal

14- Grasp the propeller shaft firmly with one hand and at the same time grasp the shift rod with your other hand. Now, pull the propeller shaft out and the shift rod upward at the same time. Remove both units from the lower unit. As the propeller shaft is withdrawn, the reverse gear, thrust bearing, washer, and retainer plate will all come out together with the propeller shaft.

"FROZEN" PROPELLER SHAFT

On rare occasions, especially if water has been allowed to enter the lower unit, it may not be possible to withdraw the propeller shaft as described in Step 14. The shaft may be "frozen" in the hydraulic pump due to corrosion.

If efforts to remove the propeller shaft after the bearing carrier has been removed fail: Obtain a block of wood 2"x 4" approx. one foot in length. Drill a hole in the center of the flat side, large enough for the propeller shaft to pass through. Place the block over the shaft. Slide some thick large washers over the shaft and thread the propeller nut onto the shaft.

With the skeg clamped securely in a vise equipped with soft jaws, attempt to pull the shaft free. If necessary hammer on the wood, rotating the block at intervals to prevent wedging the shaft in any one direction.

The second and less desirable method is as follows: Clamp the propeller shaft horizontally in a vise equipped with soft jaws. Two blocks of wood may be substituted for the soft jaws, but **NEVER** clamp the shaft in

the vise without protection, to prevent damage to the threads or splines. Use a soft head mallet and strike the lower unit with quick sharp blows midway between the anti-cavitation plate and the propeller shaft. This action will drive the lower unit from the propeller shaft. **TAKE CARE** to hold the lower unit to keep it in line with the propeller shaft and prevent wedging the shaft in any one direction. Holding the lower unit will also prevent the housing from falling to the floor when the housing comes free of the shaft.

Pinion Gear — Removal

Special tool, OMC No. 316612 is required to turn the driveshaft in order to remove the pinion gear nut.

15- Obtain the special tool and slip it over the end of the driveshaft with the splines of the tool indexed with the splines on the driveshaft. Hold the pinion gear nut with the proper size wrench, and at the same time rotate the driveshaft, with the special tool and wrench **COUNTERCLOCKWISE** until the nut is free. If the special tool is not available, clamp the driveshaft in a vise equipped with soft jaws, in an area below the splines but not in the water pump

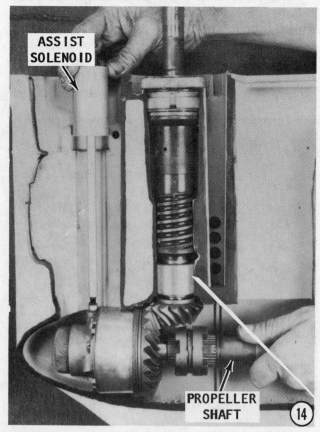

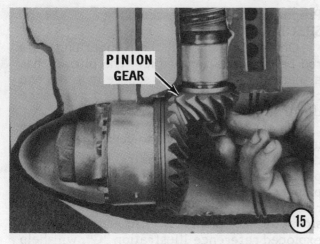

PINION GEAR

⑮

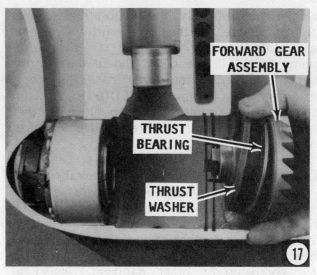

FORWARD GEAR ASSEMBLY

THRUST BEARING

THRUST WASHER

⑰

impeller area. Now, with the proper size wrench on the pinion gear nut, rotate the complete lower unit **COUNTERCLOCKWISE** until the nut is free. This procedure will require the lower unit to be rotated, then the wrench released, the lower unit turned back, the wrench again attached to the nut, and the unit rotated again. Continue with this little maneuver until the nut is free. After the nut is free, proceed with the next step. The driveshaft will be withdrawn from the pinion gear.

Driveshaft — Removal

16- Remove the four bolts from the top of the lower unit securing the bearing housing. **CAREFULLY** pry the bearing housing upward away from the lower unit, then slide it free of the driveshaft. An alternate method is to again clamp the driveshaft in a

vise equipped with soft jaws. Use a soft-headed mallet and tap on the top side of the bearing housing. This action will jar the housing loose from the lower unit. Continue tapping with the mallet and the bearing housing, **O**-rings, shims, thrust washer, thrust bearing, and the driveshaft will all breakaway from the lower unit and may be removed as an assembly.

17- Remove the pinion gear from the lower unit cavity. Remove the forward gear, thrust washer, and thrust bearing, from the hydraulic pump.

Hydraulic Pump — Removal

18- Obtain two long rods with 1/4" x 20 threads on both ends. Thread the two rods into the hydraulic pump housing. Attach a slide hammer to the rods and secure it with a nut on the end of each rod. Check to be sure the slide hammer is installed onto the rods **EVENLY** to allow an even pull on the pump. If the slide hammer is not installed

BEARING HOUSING

⑯

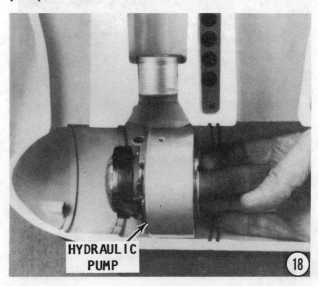

HYDRAULIC PUMP

⑱

to the rods properly, the pump may become tightly wedged in the lower unit. Operate the slide hammer and pull the hydraulic pump free. If the pump should happen to become lodged in the lower unit, stop operating the slide hammer **IMMEDIATELY.** Tap the hydraulic pump back into place in the lower unit and start the removal procedure over.

Lower Driveshaft Bearing — Removal

This bearing cannot be removed without the aid of a special tool. Therefore, **DO NOT** attempt to remove this bearing unless it is unfit for further service. To check the bearing, first use a flashlight and inspect it for corrosion or other damage. Insert a finger into the bearing, and then check for "rough" spots or binding while rotating it.

19- Remove the Allen screw, if used, from the water pickup slots in the starboard side of the lower unit housing. This screw secures the bearing in place and **MUST** be removed before an attempt is made to remove the bearing. The bearing must be actually **"PULLED"** upward to come free. **NEVER** make an attempt to "drive" it down and out.

20- Obtain special tool, OMC No. 385546. Use the special tool and "pull" the bearing from the lower unit.

Propeller Shaft — Disassembling
SPECIAL WORDS

On the **1973** model, the plunger and shift rod bearing are removable. The plunger has a snap ring on the end to secure it in the end of the propeller shaft, exploded reference illustration **"F"**.

On the **1974** model, a spring is installed into the end of the propeller shaft. Three detent balls are used on the shift rod and plunger assembling -- one ball between the plunger and the spring, and the other two ride partially indexed into holes on opposite sides of the plunger. The accompanying exploded reference illustration **"G"**, will help clarify the relationship of these parts.

21- Notice the spring retainer on the outside surface of the clutch dog. Use a small screwdriver and work one end of the spring up onto the shoulder of the clutch dog. Continue working the spring out of the groove until it is free. **TAKE CARE** not to distort the spring.

22- Place one end of the propeller shaft on the bench and push the pin free of the clutch dog. Raise the propeller end of the shaft upward and the piston, plunger, detent balls, and spring, will come free of the shaft, 1974 models only, reference illustration **"E"**.

23- Slide the clutch dog free of the propeller shaft.

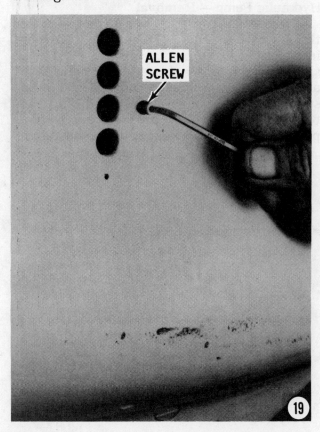

ALLEN SCREW

19

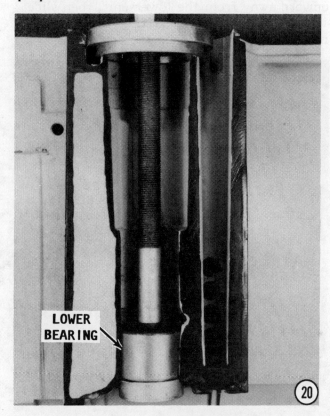

LOWER BEARING

20

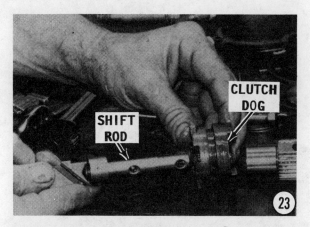

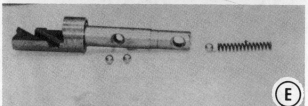

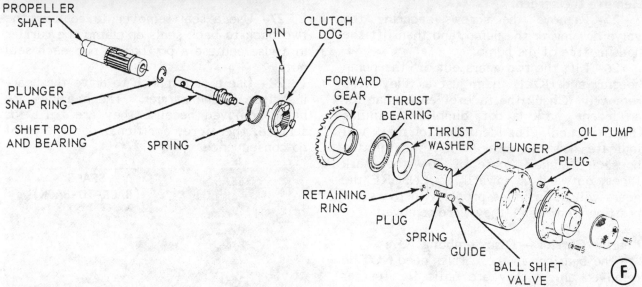

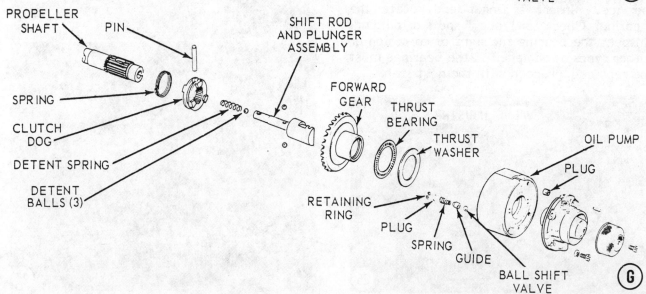

Hydraulic Pump — Disassembling

24- Remove the screw from the center of the screen on the back side of the pump. Remove the screen.

25- Remove the screws securing the valve housing to the pump, and then lift the housing free of the pump.

26- Lift the two gears out off the pump housing and **HOLD** them just as they were removed. Check the face of each gear for an indent mark (a dot, dimple, or similar identification). The identification mark will indicate how the gear **MUST** face in the housing. Make a note of how the mark faces, outward or inward, to **ENSURE** the gears will be installed properly in the same position from which they were removed.

Bearing Carrier — Disassembling

The bearings in the carrier need **NOT** be removed unless they are unfit for further service. Insert a finger and rotate the bearing. Check for "rough" spots or binding. Inspect the bearing for signs of corrosion or other types of damage. If the bearings must be replaced, proceed with the next step.

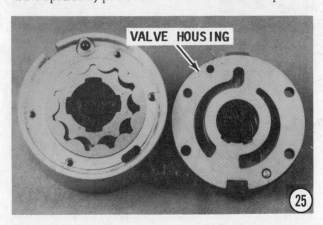

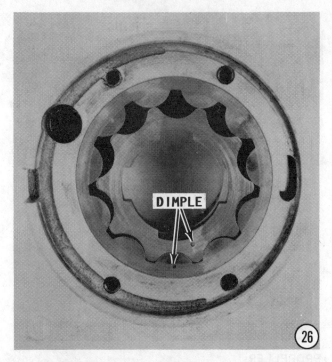

27- Use a seal remover to remove the two back-to-back seals or clamp the carrier in a vise and use a pry bar to pop each seal out.

28- Use a drift punch to drive the bearings free of the carrier. The bearings are being removed because they are unfit for service, therefore, additional damage is of no consequence.

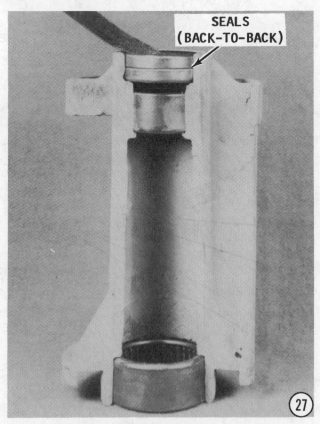

Assist Cylinder -- Disassembling

29- Back the shift rod out of the assist cylinder by rotating it **COUNTERCLOCk-WISE** until it is free. The threads of the shift rod provide the adjustment, which will be made at the time of installation. Therefore, it is not necessary to count the number of turns at the time of removal.

Two special tools are required to disassemble the assist cylinder further. The assist cylinder can be purchased as a complete assembly. Therefore, it may be less expensive to purchase a new unit rather than buying the special tools and attempting an overhaul.

30- The two special tools required, are both OMC No. 386112. One is used on top of the cylinder and the other on the bottom. Obtain the two tools. Notice the hex head on both ends of the tool. Secure one end in a vise. Slide the lower push rod of the assist cylinder through the special tool and seat it onto the two prongs of the tool. Position the other tool onto the top of the cylinder and back off the cylinder cap.

31- Remove the valve and O-ring from the cylinder cap. Remove the two O-rings from the lower end of the cylinder.

32- Remove the retaining pin from the piston and the lower push rod will come free.

ASSIST CYLINDER

29

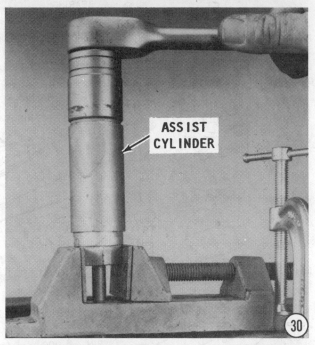

ASSIST CYLINDER

30

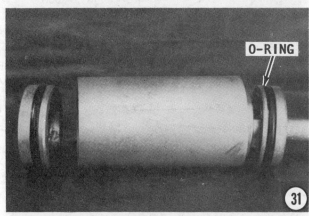

O-RING

31

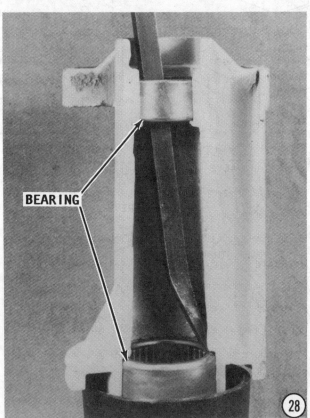

BEARING

28

PIN

32

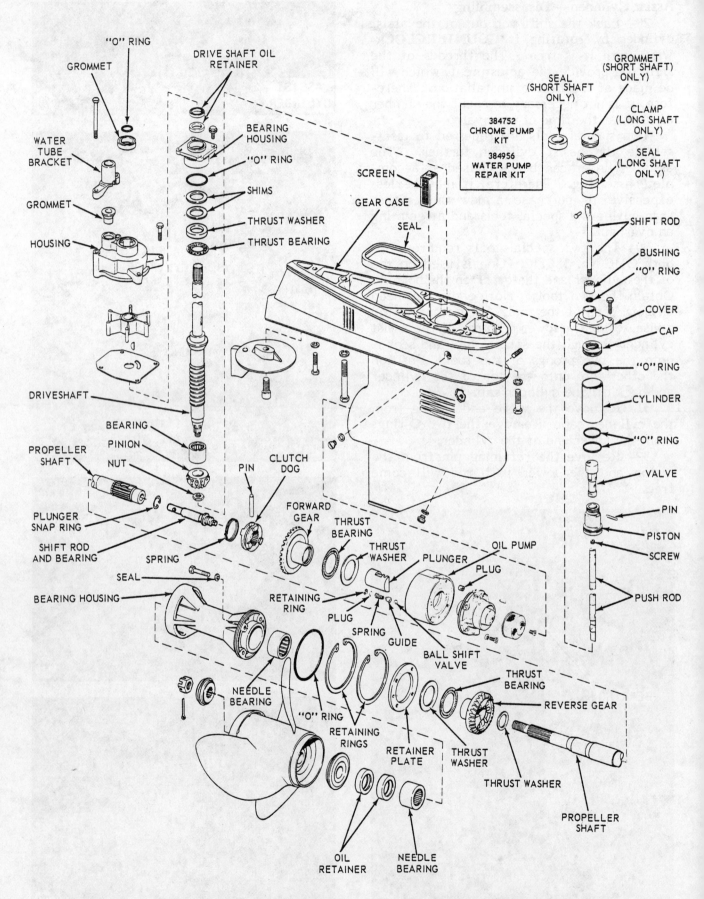

Exploded view of a 50 hp -- 1973 lower unit with major parts identified.

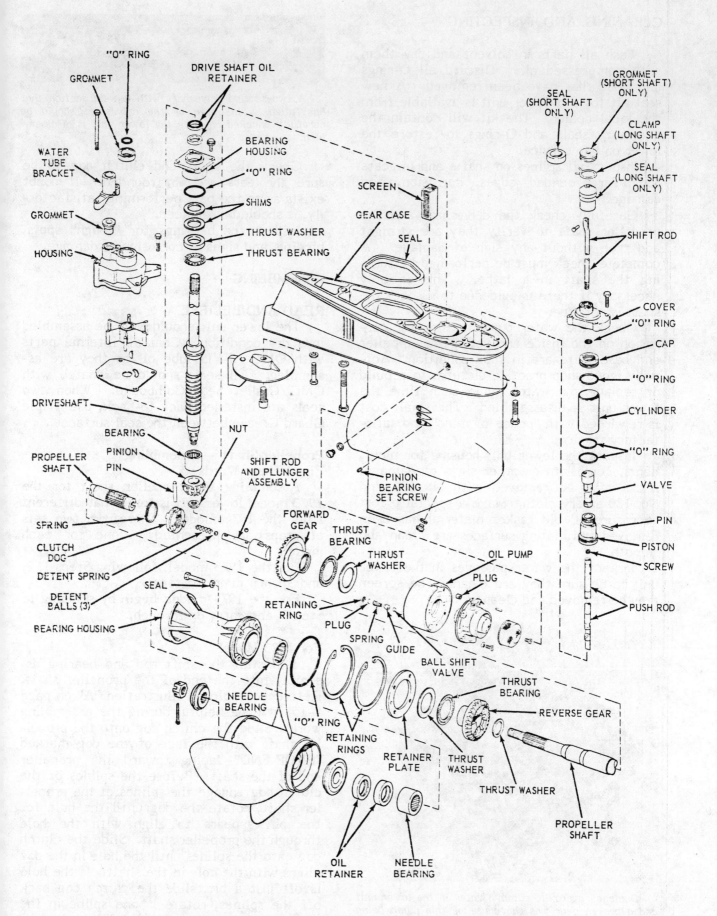

"O" RING
GROMMET
DRIVE SHAFT OIL RETAINER
SEAL (SHORT SHAFT ONLY)
GROMMET (SHORT SHAFT ONLY)
CLAMP (LONG SHAFT ONLY)
SEAL (LONG SHAFT ONLY)
WATER TUBE BRACKET
BEARING HOUSING
"O" RING
SHIMS
SHIFT ROD
GROMMET
THRUST WASHER
THRUST BEARING
SCREEN
GEAR CASE
SEAL
HOUSING
COVER
"O" RING
CAP
"O" RING
CYLINDER
DRIVESHAFT
"O" RING
BEARING
NUT
VALVE
PINION
SHIFT ROD AND PLUNGER ASSEMBLY
PROPELLER SHAFT
PIN
PIN
PISTON
PINION BEARING SET SCREW
SCREW
SPRING
FORWARD GEAR
THRUST BEARING
PUSH ROD
CLUTCH DOG
THRUST WASHER
OIL PUMP
DETENT SPRING
RETAINING RING
PLUG
DETENT BALLS (3)
SEAL
PLUG
SPRING
BEARING HOUSING
GUIDE
BALL SHIFT VALVE
NEEDLE BEARING
THRUST BEARING
REVERSE GEAR
"O" RING
RETAINING RINGS
RETAINER PLATE
THRUST WASHER
THRUST WASHER
PROPELLER SHAFT
OIL RETAINER
NEEDLE BEARING

Exploded view of a 50 hp -- 1974 lower unit with major parts identified.

CLEANING AND INSPECTING

Wash all parts in solvent and dry them with compressed air. Discard all O-rings and seals that have been removed. A new seal kit for this lower unit is available from the local dealer. The kit will contain the necessary seals and O-rings to restore the lower unit to service.

Inspect all splines on shafts and in gears for wear, rounded edges, corrosion, and damage.

Carefully check the driveshaft and the propeller shaft to verify they are straight and true without any sign of damage. A complete check must be performed by turning the shaft in a lathe. This is only necessary if there is evidence to suspect the shaft is not true.

Check the water pump housing for corrosion on the inside and verify the impeller and base plate are in good condition. Actually, good shop practice dictates to rebuild or replace the water pump each time the lower unit is disassembled. The small cost is rewarded with "peace of mind" and satisfactory service.

Inspect the lower unit housing for nicks, dents, corrosion, or other signs of damage. Nicks may be removed with No. 120 and No. 180 emery cloth. Make a special effort to ensure all old gasket material has been removed and mating surfaces are clean and smooth.

Inspect the water passages in the lower unit to be sure they are clean. The screen may be removed and cleaned.

Damaged hydraulic pump. Water in the lower unit and a broken gear was the cause of this pump being destroyed.

A two-section driveshaft with a weld section that has failed. This area of the driveshaft should be carefully checked anytime the lower unit is disassembled.

Check the gears and clutch dog to be sure the ears are not rounded. If doubt exists as to the part performing satisfactorily, it should be replaced.

Inspect the bearings for "rough" spots, binding, and signs of corrosion or damage.

ASSEMBLING

READ AND BELIEVE

The lower unit should **NOT** be assembled in a dry condition. Coat all internal parts with OMC HI-VIS lube oil as they are assembled. All seals should be coated with OMC Gasket Seal Compound. When two seals are installed back-to-back, use Triple Guard Grease between the seal surfaces.

Propeller Shaft -- Assembling
First, These Words

Assembling the propeller shaft for the 1973 model lower unit is somewhat different from the 1974 model. Therefore, two sets of steps are presented -- one for each model.

For the 1973 model, following Steps 1, 2, and 3, then jump to Step 7.

For the 1974 model, begin by skipping to Step 3, and carry on through.

For 1973 Model

1- Install the shift rod and bearing assembly into the end of the propeller shaft. Reference exploded illustration "A" on page 8-157 will be helpful during the assembling work. Slide the clutch dog onto the propeller shaft with the face of the dog marked **"PROP END"** facing toward the propeller end of the shaft. Before the splines of the clutch dog engage the splines of the propeller shaft, rotate the dog until the hole for the pin appears to align with the hole through the propeller shaft. Slide the clutch dog onto the splines until the hole in the dog aligns with the hole in the shaft. If the hole is off just a bit, slide the clutch dog back off the splines, rotate it one spline in the required direction, and then slide it into

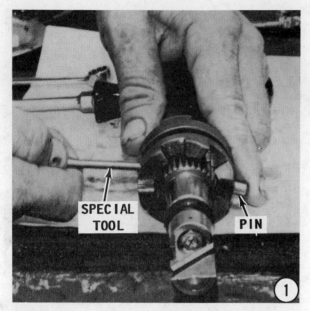

place. Insert the pin through the clutch dog, the shaft, and out the other side of the shaft and clutch dog. Center the pin through the clutch dog.

2- Install the spring-type pin retainer around the clutch dog to secure the pin in place. **TAKE CARE** not to distort the pin retainer during the installation process.

For 1974 Model

3- In order to assemble the shift rod and clutch dog onto the propeller shaft, it will be necessary to make a simple detent spring depressing tool. This can be accomplished by cutting off a piece of 9/32" diameter round bar stock approximately 6-inches long. Flatten one end of the bar stock, as shown in the accompanying illustration.

4- Insert two of the detent balls into the detent ball hole in the side of the shift rod. Reference exploded illustration "B" on the next page will be helpful during the assembling work. Insert the third detent ball into the end of the shift rod, and then place the spring through the hole in the end of the

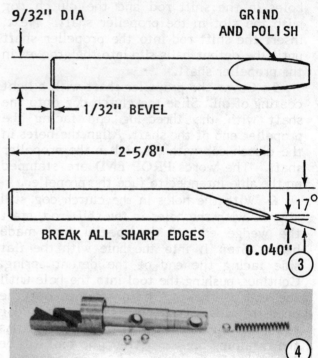

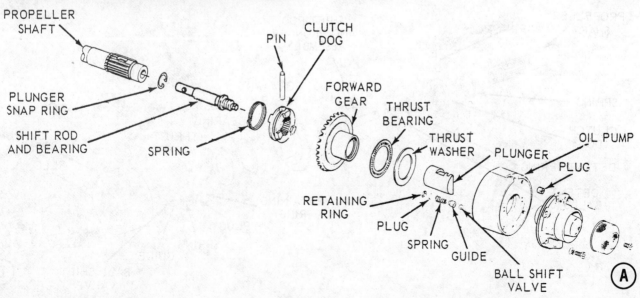

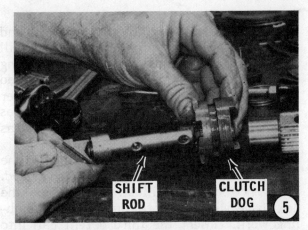

SHIFT ROD   CLUTCH DOG   5

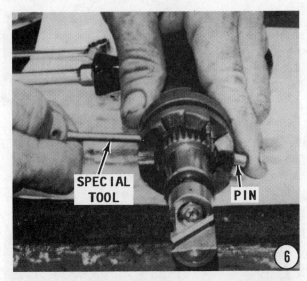

SPECIAL TOOL   PIN   6

SPRING   7

shift rod. Now, hold the balls in the shaft, and at the same time carefully align the holes in the shift rod and the clutch dog with the slot in the propeller shaft. Next, insert the shift rod into the propeller shaft until the detent balls slip into the grooves in the propeller shaft.

5- Cover the propeller shaft with a light coating of oil. Slide the clutch dog onto the shaft with the three-lug end facing the propeller end of the shaft. Align the holes in the clutch dog with the slot in the propeller shaft. The words **PROP END** are stamped on the side intended to face the propeller.

6- With the holes in the clutch dog still aligned with the holes in the shift rod, start the wedge end of the special tool made before Step 1, into the hole with the flat side facing the end of the detent spring. Continue pushing the tool into the hole until the end of the tool barely comes out the opposite side. Now, press the clutch dog retaining pin in through the clutch dog. As the retaining pin is inserted, the tool will be forced out.

7- Secure the pin in place with the retaining spring. **TAKE CARE** to be sure none of the spring coils overlap or the spring is not distorted in any way. Only through careful attention to installation of the retaining pin and spring can proper operation of the shift mechanism be expected. Set the completed assembly to one side until ready for installation into the lower unit.

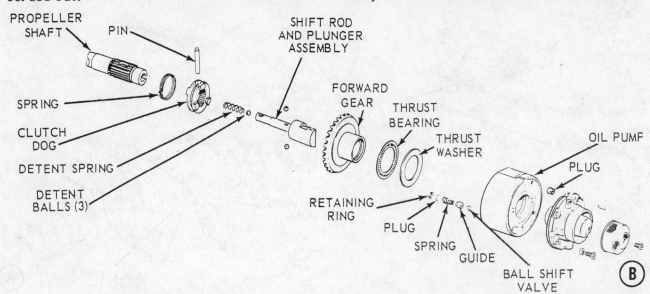

PROPELLER SHAFT   PIN   SHIFT ROD AND PLUNGER ASSEMBLY

SPRING

CLUTCH DOG

DETENT SPRING

DETENT BALLS (3)

FORWARD GEAR   THRUST BEARING   THRUST WASHER

RETAINING RING   PLUG   SPRING   GUIDE   BALL SHIFT VALVE

OIL PUMP   PLUG

A   B

Bearing Carrier Bearings & Seals Installation

8- Install the reverse gear bearing into the bearing carrier by pressing against the **LETTERED** side of the bearing with the proper size socket. Press the forward gear bearing into the bearing carrier in the same manner. Press against the **LETTERED** side of the bearing.

9- Coat the outside surfaces of the seals with HI-VIS oil. Install the first seal with the flat side facing **OUT**. Coat the flat surface of both seals with Triple Guard Grease, and then install the second seal with the flat side going in **FIRST**. The seals are then back-to-back with the grease between the two surfaces. The outside seal prevents water from entering the lower unit and the inside seal prevents the lubricant in the lower unit from escaping.

Hydraulic Pump — Assembling

10- Check the note made during disassembling, per Step 26, to determine how the identifying marks (dots, dimples, whatever) on the gears must face — inward or outward. The gears **MUST** be installed in the same position from which they were removed.

11- After the gears have been installed into the pump, use a straightedge and check to be sure the gears are level in the pump. If the gears cannot be made level, either the gears or the pump must be replaced.

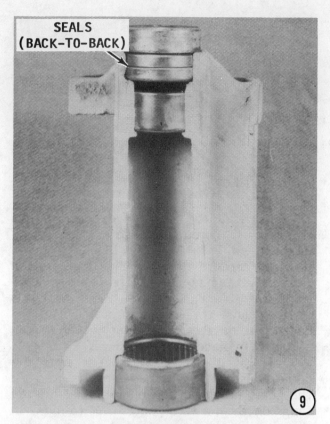

SEALS (BACK-TO-BACK)

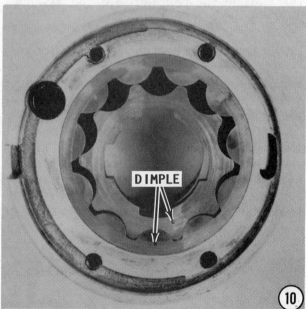

DIMPLE

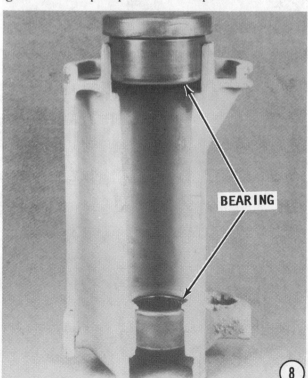

BEARING

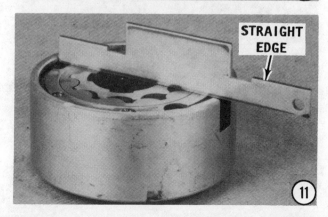

STRAIGHT EDGE

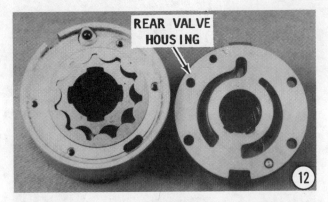

12- Install the rear valve housing with the tang on the outside edge of the housing indexed with the small slot in the pump housing. Secure the valve housing in place with the attaching screws tightened securely.

13- Place the screen in position on the back side of the valve housing, and then secure it in place with the screw.

14- Install the forward gear into the pump housing. It may be necessary to work the gears around in the pump to permit the tangs on the forward gear shank to index in the slots in the housing. Set the assembly aside for later installation.

Lower Driveshaft Bearing — Installation

15- Obtain tool, OMC No. 385546. Assemble the tool with the washer, guide sleeve, and remover portion of the tool, in the order given. The shoulder of the tool must face **DOWN**. Place the bearing onto the end of the tool, with the lettered side of the bearing facing the tool. Drive the bearing down until the large washer on the tool makes contact with the surface of the lower unit. The bearing is then seated to the proper depth.

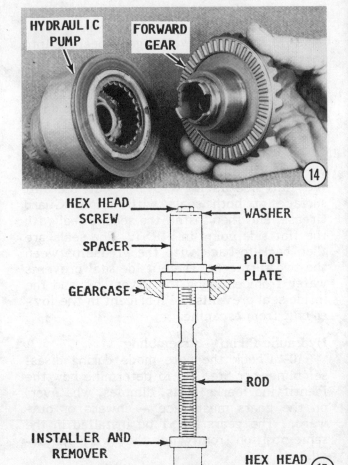

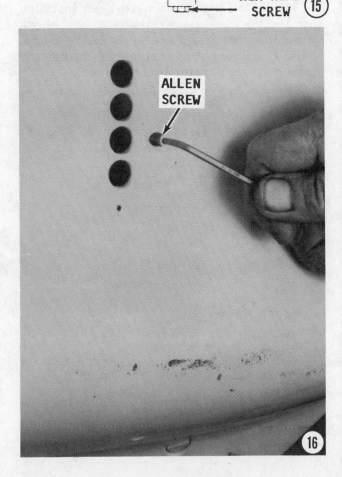

16- After the bearing has been installed, install the Allen screw through the water pickup vent ramps to secure the bearing in place. Use Loctite on the screw threads before installation.

Hyraulic Pump — Installation
First, These Words

Observe the hole on the topside of the pump. This tang **MUST** face directly up in relation to the lower unit housing to permit installation of the shift rod into the pump. Also notice the pin on the backside of the pump. This pin **MUST** index into a matching hole in the housing to restrain the pump from rotating.

17- Secure the lower unit housing in the horizontal position with the bearing carrier opening facing up. Remove the forward gear from the pump. Obtain two long 1/4 x 20 rods with threads on both ends. Thread the rods into the pump and then lower the pump into the lower unit housing. To index the pin on the back of the pump housing into the hole in the lower unit is not an easy task. However, exercise patience and rotate the pump ever so slowly. A helpful hint at this point: As the pump is being lowered into the cavity, align the opening and tang on top of the pump in the approximate position your eye indicates the shift rod may be installed. When the pin indexes, it will not be possible to rotate the pump. The pump **MUST** be properly seated to permit installation of the shift rod and the pinion gear. After the pump is in place, remove the two rods used during installation.

18- Install the thrust washer and thrust bearing into the pump with the flat side of the bearing facing **OUTWARD**. Lower the forward gear into the pump. Work the gear slowly until the teeth index with the teeth of the pump gear. This should not be too difficult because the forward gear was installed once, and then removed in the previous step. However, it is entirely possible

the gears moved when the pump was installed. Therefore, use a flashlight and check the position of the gears. If necessary, use a long shank screwdriver and rotate the gears until they are close to center, then install the forward gear.

Driveshaft and Pinion Gear — Installation
CRITICAL WORDS

The driveshaft and pinion gear must be assembled prior to installation, and then checked with a special shimming gauge. This shimming must be accomplished properly, the unit disassembled, and then installed into the lower unit. Use of the shimming gauge is the **ONLY** way to determine the proper amount of shimming required at the upper end of the driveshaft. The following detailed step outlines the procedure.

19- Clamp the driveshaft in a vise equipped with soft jaws and in such a manner that the splines, water pump area, or other critical portions of the shaft cannot be damaged. Slide the pinion gear onto the driveshaft with the bevel of the gear teeth facing toward the lower end of the shaft. Install the pinion gear nut and tighten it to a torque value of 40 to 45 ft-lbs. No parts should be installed on the upper end of the driveshaft at this point. Slide the shims removed during disassembly onto the driveshaft and seat them against the driveshaft shoulder. Obtain special shimming tool, OMC No. 315767. Slip the special tool down over the driveshaft and onto the upper surface of the top shim. Measure the distance between the top of the pinion gear and the bottom of the tool. The tool should just barely make contact with the pinion gear surface for **ZERO** clearance. Add or remove shims from the upper end of the driveshaft

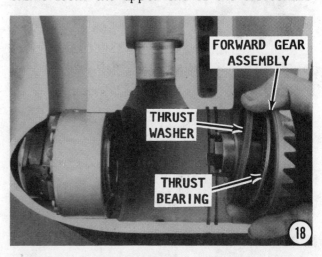

to obtain the required **ZERO** clearance. Remove the tool and set the shims aside for installation later. Back off the pinion gear nut and remove the pinion gear. Remove the driveshaft from the vise.

20- Insert the pinion gear into the cavity in the lower unit with the flat side of the bearing facing **UPWARD**. Hold the pinion gear in place and at the same time lower the driveshaft down into the lower unit. As the driveshaft begins to make contact with the pinion gear, rotate the shaft slightly to permit the splines on the shaft to index with the splines of the pinion gear. After the shaft has indexed with the pinion gear, thread the pinion gear nut onto the end of the shaft. Obtain special tool, OMC No. - 316612. Slide the special tool over the upper end of the driveshaft with the splines of the tool indexed with the splines on the shaft. Attach a torque wrench to the special tool. Now, hold the pinion gear nut with the proper size wrench and rotate the driveshaft **CLOCKWISE** with the special tool until the pinion gear nut is tightened to a torque value of 40 to 45 ft-lbs. Remove the special tool.

21- Slide the thrust bearing, thrust washer, and the shims, set aside after the shim gauge procedure in Step 19, onto the driveshaft.

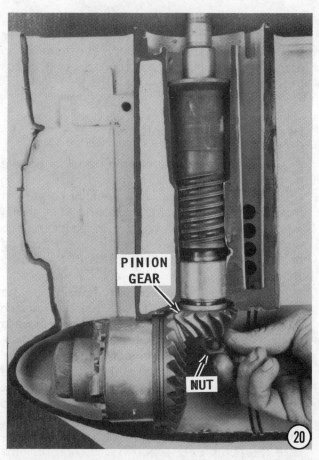

22- Install the two seals back-to-back into the opening on top of the bearing housing. Coat the outside surface of the **NEW** seals with OMC Lubricant. Press the first seal into the housing with the flat side of the seal facing **OUTWARD**. After the seal is in place, apply a coating of Triple

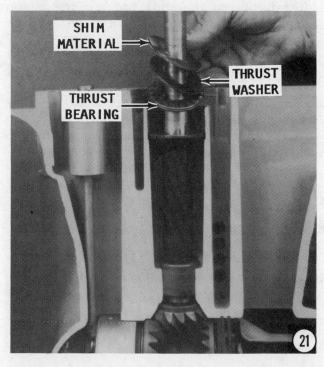

A set of double seals showing the back side (left) and the front side (right). These seals are installed back-to-back (flat side-to-flat side) with Triple Guard Grease between the surfaces. This arrangement prevents fluid from passing in either direction.

BEARING HOUSING

㉓

Guard Grease to the flat side of the installed seal and the flat side of the second seal. Press the second seal into the bearing housing with the flat side of the seal facing **INWARD**. Insert a **NEW** O-ring into the bottom opening of the bearing housing.

23- Wrap friction tape around the splines of the driveshaft to protect the seals in the bearing housing as the housing is installed. Now, slide the assembled bearing housing down the driveshaft and seat it in the lower unit housing. Secure the housing in place with the four bolts. Tighten the bolts **ALTERNATELY** and **EVENLY**.

Propeller Shaft -- Installation
First, Some Good Words

Installation of the propeller shaft, assist valve, and plunger is not a simple task. However, with patience, attention to detail, and an understanding of the installation sequence, the work can proceed smoothly without serious problems or the need to disassemble and repeat certain steps. The best word of advice is to read through the following procedure at least twice, or until the various parts and their relationship to each other including the installation sequence is thoroughly familiar. Before these

units are installed in the lower unit, some assembling on the bench can be performed.

Assemble the propeller shaft and shifter as outlined at the beginning of this section, Steps 1 and 2 for the 1973 model; Steps 3 thru 6 for the 1974 model. Observe how the boss on the plunger slides into the cut-a-way in the shift rod, reference illustration "C". This arrangement **MUST** be accomplished when these units are installed in the lower unit. Also observe that the shift rod on the end of the propeller shaft is heavy on the back side causing the rod to rotate. Therefore, when the propeller shaft is held horizontal, the heavy side will turn the shift rod downward.

24- Lay the lower unit on a flat surface with the port side facing up. Adjust the unit on the bench to offer a clear view into the assist cylinder bore and also into the propeller shaft opening. Now, hold the propeller shaft with the heavy side of the shift rod down and insert it through the oil pump and forward gear assembly.

25- Push the shaft in as far as possible. Use a flashlight to illuminate the interior of the shift assist cylinder bore. Observe the cut-a-way in the shifter rod. If you cannot see the cut-a-way, move the propeller shaft inward or outward until it is visible. With the propeller shaft in this position, **CARE-**

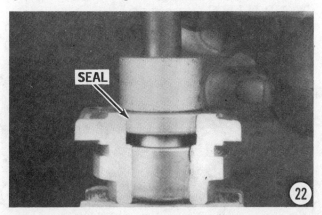

SEAL

㉒

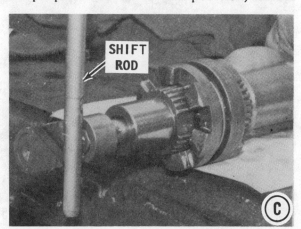

SHIFT ROD

Ⓒ

SHIFT ROD

(24)

FULLY insert the push rod and assist cylinder assembly into the gearcase with the boss on the cylinder facing down.

Continue to slowly move the assembly into the gearcase until it makes contact with the plunger. When the valve and plunger are properly engaged, flat-to-flat, the valve and plunger cannot be rotated. Exert a **GENTLE** downward pressure on the assist cylinder and valve assembly and at the same time, slowly **EASE** the propeller shaft backward to engage the plunger keyway with the push rod key. The gentle downward pressure on the shift assist assembly will help the push rod key to slip into the plunger keyway. After the proper alignment has been made, as just described, push the propeller shaft and shift assist assembly inward until full engagement is reached.

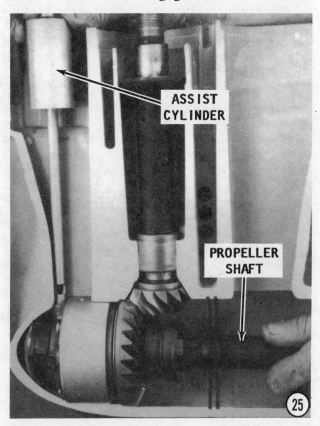

ASSIST CYLINDER

PROPELLER SHAFT

(25)

When the propeller shaft and the shift assist assembly are properly engaged, the assist valve and push rod **CANNOT** be rotated.

Solenoid Cover and Gasket — Installation

26- Slide the solenoid cover up onto the upper part of the shift rod, and then place a **NEW** gasket into place on the bottom side of the cover. Thread the shift rod into the valve assembly in the assist solenoid. Rotate the shift rod inward about three complete turns for a rough preliminary adjustment. Check to be sure the unit is in **NEUTRAL**. Now, measure the distance from the top of the housing to the center of the hole in the shift rod, reference illustration **"D"**. This distance must be within 1/32" (0.8mm) as follows for a 1973 or 1974 model.

Long shaft -- 21 7/32" 54.6cm
Short shaft -- 16 7/32" 41.9cm

Rotate the shift rod inward or outward until the required measurement is obtained. Lower the cover and gasket down onto the lower unit housing. Secure the cover in place with the attaching hardware. Tighten the bolts **ALTERNATELY** and **EVENLY**.

Reverse Gear — Installation

27- Apply a light coating of grease to the inside surface of the reverse gear to hold the thrust washer in place. Place the thrust washer onto the front side of the reverse gear. Install the thrust bearing and thrust washer onto the shank on the back side of the reverse gear. Slide the reverse gear down the propeller shaft and index the teeth of the gear with the teeth of the pinion gear.

28- Insert the retainer plate into the lower unit against the reverse gear.

W A R N I N G
This next step can be dangerous. Each snap ring is placed under tremendous tension

COVER

(26)

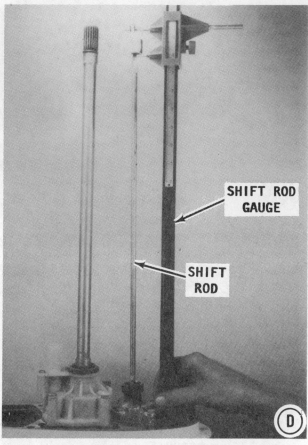

SHIFT ROD GAUGE

SHIFT ROD

D

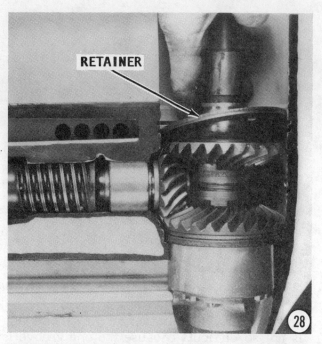

RETAINER

28

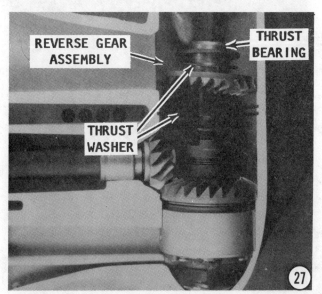

REVERSE GEAR ASSEMBLY

THRUST BEARING

THRUST WASHER

27

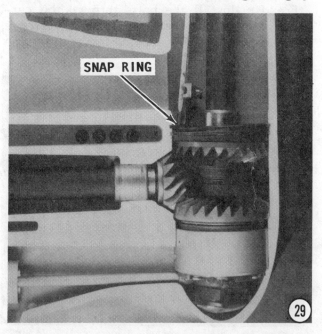

SNAP RING

29

with the Truarc pliers while it is being placed into the groove. Therefore, wear **SAFETY GLASSES** and exercise care to prevent the snap ring from slipping out of the pliers. If the snap ring should slip out, it would travel with incredible speed and cause personal injury if it struck a person.

29- Install the Truarc snap rings one at-a-time following the precautions given in the **WARNING** and the **ADVICE** given in the following paragraph.

WORDS OF ADVICE

The two snap rings index into separate grooves in the lower unit housing. As the first ring is being installed, depth perception may play a trick on yours eyes. It may appear that the first ring is properly indexed all the way around in the proper groove, when in reality, a portion may be in one groove and the remainder in the other groove. Should this happen, and the Truarc pliers be released from the ring, it is extremely difficult to get the pliers back into the ring to correct the condition. If necessary use a flashlight and carefully check to be sure the first ring is properly seated all the way around **BEFORE** releasing the grip

View into the lower unit showing the Truarc snap rings properly locked in the grooves.

on the pliers. Installation of the second ring is not so difficult because the one groove is filled with the first ring.

30- Obtain two long 1/4" rods with threads on one end. Thread the rods into the retainer plate opposite each other to act as guides for the bearing carrier.

31- Check the bearing carrier to be sure a **NEW O**-ring has been installed. Position the carrier over the guide pins with the embossed word **UP** on the rim of the carrier facing **UP** in relation to the lower unit housing, reference illustration **"E"**. Now, lower the bearing carrier down over the guide pins and into place in the lower unit housing.

32- Slide **NEW** little O-rings onto each bolt, and apply some OMC Sealer onto the threads. Install the bolts through the carrier and into the retaining plate. After a couple bolts are in place, remove the guide

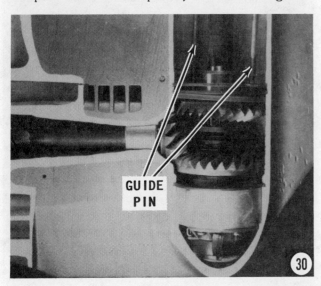

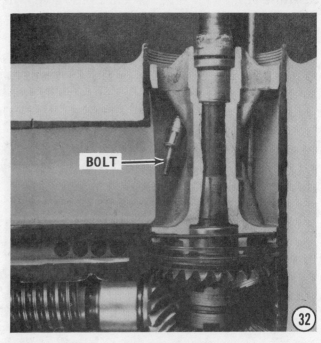

pins and install the remaining bolts. Tighten the bolts **EVENLY** and **ALTERNATELY** to the torque value given in the Appendix.

WATER PUMP INSTALLATION

FIRST, THESE WORDS

An improved water pump is available as a replacement. If the old water pump housing is unfit for further service, only the new pump housing can be purchased. It is strongly recommended to replace the water pump with the improved model while the lower unit is disassembled. The accompanying illustration shows the original equipment (left) compared with the improved pump (right), reference illustration **"F"**.

The new pump must be assembled before it is installed. Therefore, the following steps outline procedures for both pumps.

To assemble and install a replacement pump, perform steps 33 thru 41, then jump to Step 46.

To install an original equipment pump, proceed directly to Step 42.

Assembling an Improved Pump Housing

33- Remove the water pump parts from the container. Insert the plate into the housing, as shown. The tang on the bottom side of the plate **MUST** index into the short slot in the pump housing.

34- Slide the pump liner into the housing with the two small tabs on the bottom side indexed into the two cutouts in the plate.

35- Coat the inside diameter of the liner with light-weight oil. Work the impeller into the housing with all of the blades bent back to the right, as shown. In this position, the blades will rotate properly when the pump housing is installed. Remember, the pump and the blades will be rotating **CLOCKWISE** when the housing is turned over and installed in place on the lower unit.

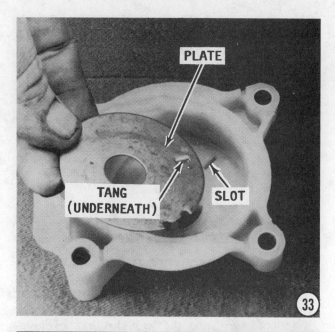

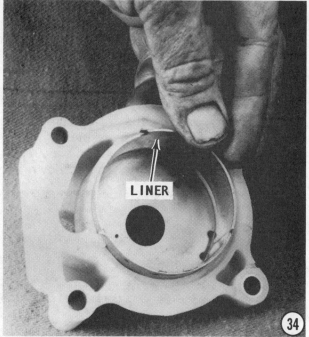

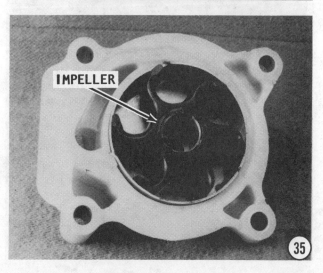

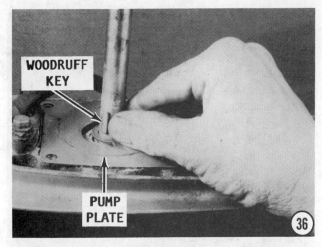

36– Coat the mating surface of the lower unit with 1000 Sealer. Slide the water pump base plate down the driveshaft and into place on the lower unit. Insert the Woodruff key into the key slot in the driveshaft.

37– Lay down a very thin bead of 1000 sealer into the irregular shaped groove in the housing. Insert the seal into the groove, and then coat the seal with the 1000 Sealer.

Water Pump Installation

38– Begin to slide the water pump down the driveshaft, and at the same time observe the position of the slot in the impeller. Continue to work the pump down the

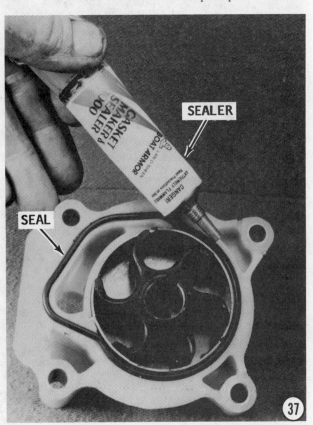

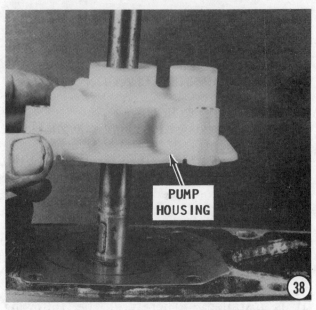

driveshaft, with the slot in the impeller indexed over the Woodruff key. The pump must be fairly well aligned before the key is covered because the slot in the impeller is not visible as the pump begins to come close to the base plate.

39– Install the short forward bolt through the pump and into the lower unit. **DO NOT** tighten this bolt at this time. Insert the grommet into the pump housing.

40– Install the grommet retainer and water tube guide onto the pump housing. Install the remaining pump attaching bolts. Tighten the bolts **ALTERNATELY** and **EVENLY,** and at the same time rotate the driveshaft **CLOCKWISE.** If the driveshaft is not rotated while the attaching bolts are being tightened, it is possible to pinch one of the impeller blades underneath the housing.

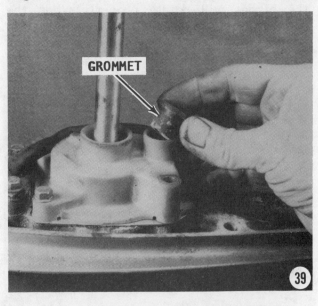

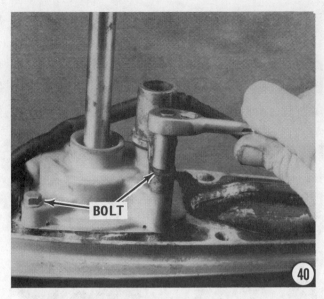

41- Slide the large grommet down the driveshaft and seat it over the pump collar. This grommet does not require sealer. Its function is to prevent exhaust gases from entering the water pump. Proceed directly to Lower Unit Installation, Step 44.

Water Pump Installation
Original Equipment

Perform the following two steps to install an original water pump.

42- Coat the water pump plate mating surface on the lower unit with 1000 Sealer. Slide the water pump plate down the driveshaft and **BEFORE** it makes contact with the sealer check to be sure the bolt holes in the plate will align with the holes in the housing. The plate will only fit one way. If the holes will not align, remove the plate, turn it over and again slide the plate down the driveshaft and into place on the housing. This checking will prevent accidently getting the sealer on both sides of the plate. If by chance sealer does get on the top surface it **MUST** be removed before the water pump impeller is installed.

Slide the water pump impeller down the driveshaft. Just before the impeller covers the cutout for the Woodruff key, install the key, and then work the impeller on down, with the slot in the impeller indexed over the Woodruff key. Continue working the impeller down until it is firmly in place on the surface of the pump plate.

43- Check to be sure **NEW** seals and O-rings have been installed in the water pump. Lubricate the inside surface of the water pump with light-weight oil. Lower the water pump housing down the driveshaft and over the impeller. **ALWAYS** rotate the driveshaft slowly **CLOCKWISE** as the housing is lowered over the impeller to allow the impeller blades to assume their natural and

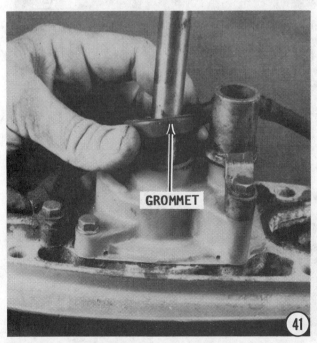

proper position inside the housing. Continue to rotate the driveshaft and work the water pump housing downward until it is seated on the plate. Coat the threads of the water pump attaching screws with sealer, and then secure the pump in place with the screws. Install the solenoid cable bracket with the same bolt and in the same position from which it was removed. On some units the solenoid cable fits into a recess of the water pump and is held in place in that manner. Tighten the screws **ALTERNATELY** and **EVENLY**.

LOWER UNIT INSTALLATION

44- Check to be sure the water tubes are clean, smooth, and free of any corrosion. Coat the water pickup tubes and grommets with lubricant as an aid to installation. Check to be sure the spark plug wires are disconnected from the spark plugs. Bring the lower unit housing together with the exhaust housing, and at the same time, guide the water tube into the rubber grommet of the water pump. Continue to bring the two units together, at the same time guiding the water pickup tubes into the rubber grommet of the water pump, and, simultaneously rotating the flywheel slowly to permit the splines of the driveshaft to index with the splines of the crankshaft. This may sound like it is necessary to do four things at the same time, and so it is. Therefore, make an earnest attempt to secure the services of an assistant for this task.

45- After the surfaces of the lower unit and exhaust housing are close, dip the attaching bolts in OMC Sealer and then start them in place. Install the two bolts on each side of the two housings. Install the bolt in the recess of the trim tab. Install the bolt

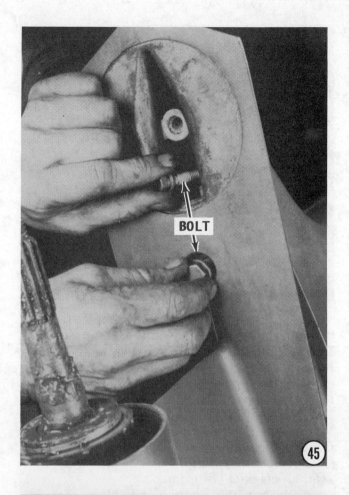

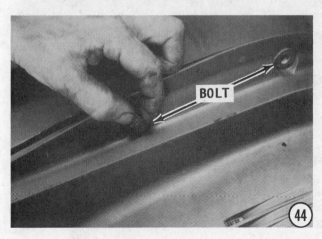

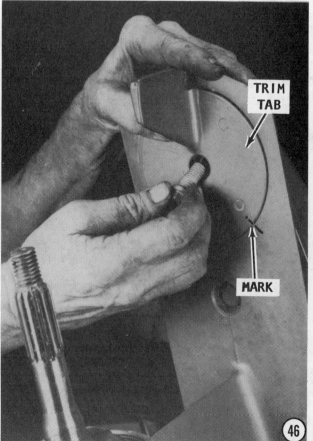

just forward of the trim tab that extends up through the cavitation plate. Tighten the bolts **ALTERNATELY** and **EVENLY** to the torque value given in the Appendix.

46- Install the trim tab with the mark made on the tab during disassembling aligned with the mark made on the lower unit housing.

47- The shift rod is visible beneath and to the rear of the lower carburetor. Install the shift connector with a bolt or pin, depending on the model being serviced.

Filling the Lower Unit

Fill the lower unit with lubricant according to the procedures in Section 8-3.

Propeller Installation

Install the propeller, see Section 8-2.

FUNCTIONAL CHECK

Perform a functional check of the completed work by mounting the engine in a test tank, in a body of water, or with a flush attachment connected to the lower unit. If the flush attachment is used, **NEVER** operate the engine above an idle speed, because the no-load condition on the propeller would allow the engine to **RUNAWAY** resulting in serious damage or destruction of the engine.

CAUTION: Water must circulate through the lower unit to the engine any time the engine is run to prevent damage to the water pump in the lower unit. Just five seconds without water will damage the water pump.

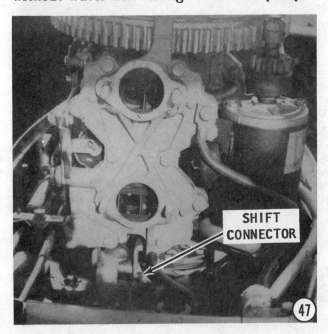

SHIFT CONNECTOR

47

Start the engine and observe the tattle-tale flow of water from idle relief in the exhaust housing. The water pump installation work is verified. If a "Flushette" is connected to the lower unit, **VERY LITTLE** water will be visible from the idle relief port. Shift ngine into the three gears and check for smoothness of operation and satisfactory performance. Remember, when the unit is in forward gear, it is at rest with no current flow to either solenoid.

**8-13 MECHANICAL SHIFT
SHIFT DISCONNECT UNDER
LOWER CARBURETOR
40 HP 1976 AND ON
50 HP 1972-73
50 HP 1975 and 1980 AND ON
55 HP 1975-83
60 HP 1980-85**

DESCRIPTION

The lower unit covered in this section is complete mechanical shift unit. A shift cable connects the shift box to the shift linkage at the engine. A shift rod extends from the engine down through the exhaust housing to the lower unit.

Shifting into forward, neutral, and reverse, is accomplished directly through mechanical means from the shift control handle through the cable and linkage to the clutch dog in the lower unit.

The lower unit houses the driveshaft and pinion gear, the forward and reverse driven gears, the propeller shaft, shift lever, cradle, shift shaft, clutch dog, shift rod, and the necessary shims, bearings, and associated parts to make it all work properly. A detent ball and spring is installed on some models.

The water pump is considered a part of the lower unit.

Two different lower units are covered in this section. One unit is used with the electric start model and the other with manual start. The model years actually overlap. The upper driveshaft bearing and the shift mechanism differ between the two units. These differences are clearly indicated in the procedural steps and illustrations.

TROUBLESHOOTING

Preliminary Checks

At rest, and without the engine running, the lower unit is in forward gear. Mount the engine in a test tank, in a body of water, or with a flush attachment connected to the lower unit. If the flush attachment is used, **NEVER** operate the engine above an idle speed, because the no-load condition on the propeller would allow the engine to **RUN-AWAY**, resulting in serious damage or destruction of the engine.

CAUTION: Water must circulate through the lower unit to the engine any time the engine is run to prevent damage to the water pump in the lower unit. Just five seconds without water will damage the water pump.

Attempt to shift the unit into **NEUTRAL** and **REVERSE**. It is possible the propeller may turn very slowly while the unit is in neutral, due to "drag" through the various gears and bearings. If difficult shifting is encountered, the problem is in the shift linkage or in the lower unit.

The second area to check is the quantity and quality of the lubricant in the lower unit. If the lubricant level is low, contaminated with water, or is broken down because of overuse, the shift mechanism may be affected. Water in the lower unit is **VERY BAD NEWS** for a number of reasons, particularly when the lower unit contains hydraulic components. Hydraulic units will not function with water in the system.

Remember, when the engine is not running, the unit should be in **FORWARD** gear.

BEFORE making any tests, remove the propeller, see Section 8-2. Check the propeller carefully to determine if the hub has

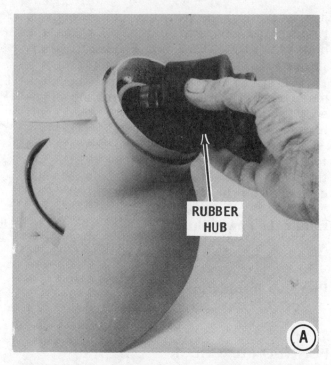

been slipping and giving a false indication the unit is not in gear, reference illustration "A". If there is any doubt, the propeller should be taken to a shop properly equipped for testing, before the time and expense of disassembling the lower unit is undertaken. The expense of the propeller testing and possible rebuild is justified.

The following troubleshooting procedures are presented on the assumption the lower unit lubricant, and the propeller have been checked and found to be satisfactory.

Helm-type shift box and steering wheel.

Lower Unit Locked

Determine if the problem is in the powerhead or in the lower unit. Attempt to rotate the flywheel. If the flywheel can be moved even slightly in either direction, the problem is most likely in the lower unit. If it is not possible to rotate the flywheel, the problem is a "frozen" powerhead. To absolutely verify the powerhead is "frozen", separate the lower unit from the exhaust housing and then again attempt to rotate the flywheel. If the attempt is successful, the problem is definitely in the lower unit. If the attempt to rotate the flywheel, with the lower unit removed, still fails, a "frozen" powerhead is verified, reference illustration "B".

Unit Fails to Shift
Neutral, Forward, or Reverse

With the outboard mounted on a boat, in a test tank, or with a flush attachment connected, disconnect the shift lever at the engine. Attempt to manually shift the unit into **NEUTRAL, REVERSE, FORWARD**. At the same time move the shift handle at the shift box and determine that the linkage and shift lever are properly aligned for the shift

Shift and throttle linkage on the 50 hp, 1973-74 engines.

positions. If the alignment is not correct, adjust the shift cable at the trunnion. It is also possible the inner wire may have slipped in the connector at the shift box. This condition would result in a lack of inner cable to make a complete "throw" on the shift handle. Check to be sure the shift handle is moved to the full shift position and the linkage to the lower unit is moved to the full shift position.

If an adjustment is required at the shift box, see Chapter 7.

If it is not possible to shift the unit into gear by manually operating the shift rod while the engine is running, the lower unit requires service as described in this section.

LOWER UNIT SERVICE

Propeller Removal

If the propeller was not removed, as directed for the troubleshooting, remove it now, according to the procedures outlined in Section 8-2.

Draining the Lower Unit

Drain the lower unit of lubricant, see Section 8-3.

GOOD WORDS

If water is discovered in the lower unit and the propeller shaft seal is damaged and requires replacement, the lower unit does **NOT** have to be removed in order to accomplish the work.

The bearing carrier can be removed and the seal replaced without disassembling the lower unit. **HOWEVER**, such a procedure is not considered good shop practice, but merely a quick-fix. If water has entered the lower unit, the unit should be disassembled and a detailed check made to determine if any other seals, bearings, bearing races, O-rings or other parts have been rendered unfit for further service by the water.

LOWER UNIT REMOVAL

1- Disconnect and ground the spark plug wires. Remove the starter motor. Remove the bolt or bolts from the shift connector under the lower carburetor.

2- Scribe a mark on the trim tab and a matching mark on the lower unit to ensure the trim tab will be installed in the same position from which it is removed. Use an Allen wrench and remove the trim tab.

SHIFT DISCONNECT

3- On some units, an attaching bolt is installed inside the trim tab cavity. Use a 1/2" socket with a short extension and remove this bolt. Failure to remove this bolt from inside the trim tab cavity may result in an expensive part being broken in an attempt to separate the lower unit from the exhaust housing. Remove the 5/8" countersunk bolt located just ahead of the trim tab position.

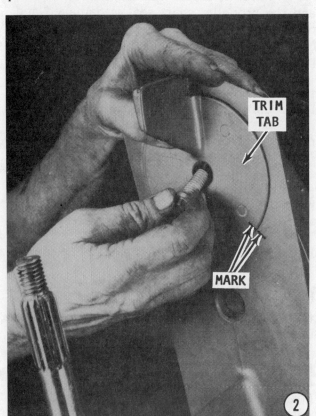

TRIM TAB

MARK

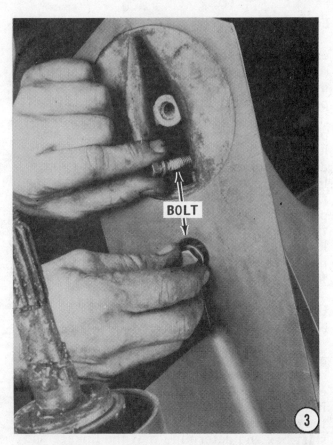

BOLT

4- Remove the four 9/16" bolts, two on each side, securing the lower unit to the exhaust housing. Work the lower unit free of the exhaust housing. If the unit is still mounted on a boat, tilt the engine forward to gain clearance between the lower unit and the deck (floor, ground, whatever). **EXERCISE CARE** to withdraw the lower unit straight away from the exhaust housing to prevent bending the driveshaft.

BOLT

WATER PUMP REMOVAL

5- Position the lower unit in the vertical position on the edge of the work bench resting on the cavitation plate. Secure the lower unit in this position with a C-clamp. The lower unit will then be held firmly in a favorable position during the service work. An alternate method is to cut a groove in a short piece of 2" x 6" wood to accommodate the lower unit with the cavitation plate resting on top of the wood. Clamp the wood in a vise and service work may then be performed with the lower unit erect (in its normal position), or inverted (upside down). In both positions, the cavitation plate is the supporting surface.

6- Remove the O-ring from the top of the driveshaft. Remove the bolts securing the water pump to the lower unit housing. Pull the water pump housing up and free of the driveshaft. Slide the water pump impeller up and free of the driveshaft.

7- Pop the impeller Woodruff key out of the driveshaft keyway. Slide the water pump base plate up and off of the driveshaft.

GOOD WORDS

If the only work to be performed is service of the water pump, proceed directly to Page 8-193, Water Pump Installation.

Bearing Carrier — Removal

8- Remove the four 5/16" bolts from inside the bearing carrier. Notice how each bolt has an O-ring seal. These O-rings should be replaced each time the bolts are removed. In most cases, the word **UP** is

embossed into the metal of the bearing carrier rim. This word must face **UP** in relation to the lower unit during installation.

9- Remove the bearing carrier using one of the methods described in the following paragraphs, under Special Words.

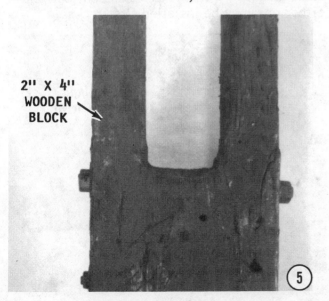

SPECIAL WORDS

Several models of bearing carriers are used on the lower units covered in this section.

The bearing carriers fit very tightly into the lower unit opening. Therefore, it is not uncommon to apply heat to the outside surface of the lower unit with a torch, at the same time the puller is being worked to remove the carrier. **TAKE CARE** not to overheat the lower unit.

One model carrier has two threaded holes on the end of the carrier. These threads permit the installation of two long bolts. These bolts will then allow the use of a flywheel puller to remove the bearing carrier, illustration **"9"**.

Another model does not have the threaded screw holes. To remove this type bearing carrier, a special puller with arms must be used. The arms are hooked onto the carrier web area, and then the carrier removed, reference illustration **"C"**.

WARNING

The next step involves a dangerous procedure and should be executed with care while wearing **SAFETY GLASSES**. The retaining rings are under tremendous tension in the groove and while they are being removed. If a ring should slip off the Truarc pliers, it will travel with incredible speed causing personal injury if it should strike a person. Therefore, continue to hold the ring and pliers firm after the ring is out of the groove and clear of the lower unit. Place

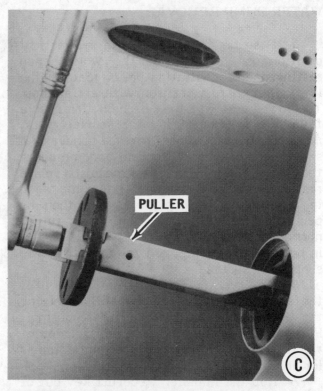

the ring on the floor and hold it securely with one foot before releasing the grip on the pliers. An alternate method is to hold the ring inside a trash barrel, or other suitable container, before releasing the pliers.

10- Obtain a pair of Truarc pliers. Insert the tips of the pliers into the holes of the first retaining ring. Now, **CAREFULLY** remove the retaining ring from the groove and gear case without allowing the pliers to slip. Release the grip on the pliers in the manner described in the above **WARNING**. Remove the second retaining ring in the same manner. The rings are identical and either one may be installed first.

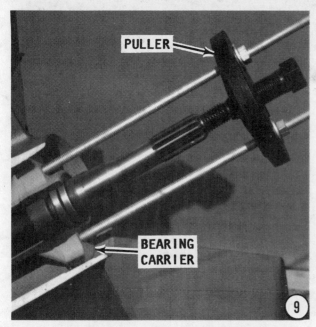

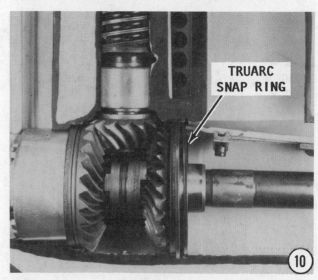

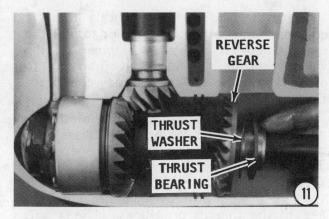

11- Remove the retainer plate, thrust washer, thrust bearing, and reverse gear from the propeller shaft.

12- Remove the four bolts from the driveshaft bearing housing. Pull up on the shift rod. This action will place shift dog into the forward gear position. Rotate the propeller shaft a bit as a check to be sure the unit is in forward gear.

Pinion Gear — Removal

Special tool, OMC No. 316612 is required to turn the driveshaft in order to remove the pinion gear nut.

13- Obtain the special tool and slip it over the end of the driveshaft with the splines of the tool indexed with the splines on the driveshaft. Hold the pinion gear nut with the proper size wrench, and at the same time rotate the driveshaft, with the special tool and wrench **COUNTERCLOCKWISE** until the nut is free. If the special tool is not available, clamp the driveshaft in a vise equipped with soft jaws, in an area below the splines but not in the water pump impeller area. Now, with the proper size wrench on the pinion gear nut, rotate the complete lower unit **COUNTERCLOCKWISE** until the nut is free. This procedure will require the lower unit to be rotated, then

the wrench released, the lower unit turned back, the wrench again attached to the nut, and the unit rotated again. Continue with this little maneuver until the nut is free. After the nut is free, proceed with the next step. The driveshaft will be withdrawn from the pinion gear.

Driveshaft — Removal

14- **CAREFULLY** pry the bearing housing upward away from the lower unit, then slide it free of the driveshaft. An alternate method is to again clamp the driveshaft in a vise equipped with soft jaws. Use a soft-headed mallet and tap on the top side of the bearing housing. This action will jar the housing loose from the lower unit. Continue tapping with the mallet and the bearing housing, O-rings, shims, thrust washer, thrust bearing, and the driveshaft will all breakaway from the lower unit and may be removed as an assembly. As the driveshaft is removed, reach into the lower unit, catch, and remove the pinion gear.

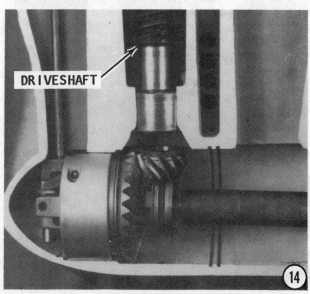

15- Push on the shift rod to move the unit into the reverse gear position. Remove the four bolts securing the shift rod cover. Rotate the shift rod **COUNTERCLOCKWISE** until it is free of the shifter detent in the lower unit. Remove the shift rod and cover as an assembly. As the plate is removed, notice which surface is facing into the housing, as an aid during installation.

Propeller Shaft — Removal

16- Grasp the propeller shaft firmly with one hand and remove the propeller shaft from the lower unit. The forward gear, and bearing housing will come out with the shaft as an assembly.

"FROZEN" PROPELLER SHAFT

On rare occasions, especially if water has been allowed to enter the lower unit, it may not be possible to withdraw the propeller shaft as described in Step 16. The shaft may be "frozen" in the hydraulic pump due to corrosion.

If efforts to remove the propeller shaft after the bearing carrier has been removed fail: Two methods are available to free the propeller shaft from the lower unit. The first and safest method is to obtain a block of wood 2"x 4" approx. one foot in length. Drill a hole in the center of the flat side, large enough for the propeller shaft to pass through. Place the block over the shaft. Slide some thick large washers over the shaft and thread the propeller nut onto the shaft.

With the skeg clamped securely in a vise equipped with soft jaws, attempt to pull the shaft free. If necessary hammer on the wood, rotating the block at intervals to prevent wedging the shaft in any one direction.

The second and less desirable method is as follows: Clamp the propeller shaft horizontally in a vise equipped with soft jaws. Two blocks of wood may be substituted for the soft jaws, but **NEVER** clamp the shaft in the vise without protection, to prevent damage to the threads or splines. Use a soft head mallet and strike the lower unit with quick sharp blows midway between the anti-cavitation plate and the propeller shaft. This action will drive the lower unit from the propeller shaft. **TAKE CARE** to hold the lower unit to keep it in line with the propeller shaft and prevent wedging the shaft in any one direction. Holding the lower unit will also prevent the housing from falling to the floor when the housing comes free of the shaft.

Driveshaft Disassembling
For Unit On Page 184 ONLY

17- The driveshaft upper bearing and cone are replaced as an assembly. If replacement is required, obtain special tool OMC No. 387131. Clamp the tool below the bearing, and then place the unit in an arbor press equipped with a deep throat pedestal. Press the shaft from the bearing. An alternate method is to clamp special tool OMC No. 387206 to the driveshaft just above the shoulder. Now, invert the driveshaft and

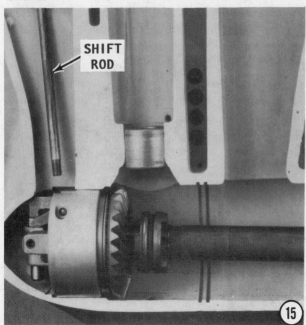

SHIFT ROD
15

16

place it in an arbor press with special tool OMC No. 387206 seated on the press. Place a piece of pipe over the driveshaft, and then press against tool OMC No. 387131 to remove the bearing.

Bearing Carrier -- Disassembling
INSPECTION FIRST

The bearings in the carrier need **NOT** be removed unless they are unfit for further service. Insert a finger and rotate the bearing. Check for "rough" spots or binding. Inspect the bearing for signs of corrosion or other types of damage. If the bearings must be replaced, proceed with the next step.

18- Use a seal remover to remove the two back-to-back seals or clamp the carrier in a vise and use a pry bar to pop each seal out.

19- Use a drift punch to drive the bearings free of the carrier. The bearings are being removed because they are unfit for service, therefore, additional damage is of no consequence.

Propeller Shaft -- Disassembling

20- Remove the coil spring from the outside groove of the clutch dog. **EXERCISE CARE** not to distort the spring as it is removed. Remove the clutch dog pin by

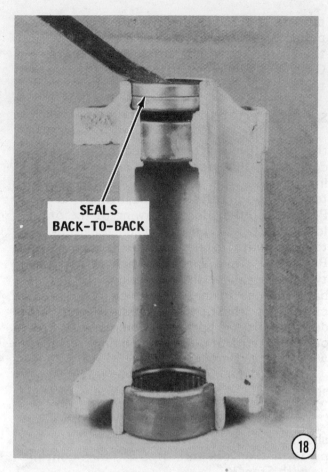

SEALS
BACK-TO-BACK

(18)

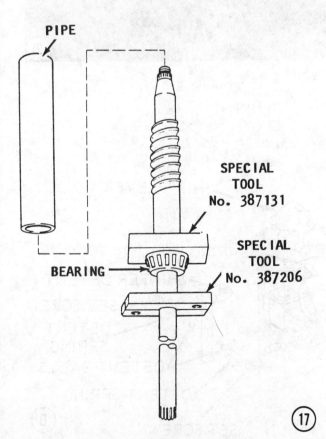

PIPE

SPECIAL
TOOL
No. 387131

SPECIAL
TOOL
No. 387206

BEARING

(17)

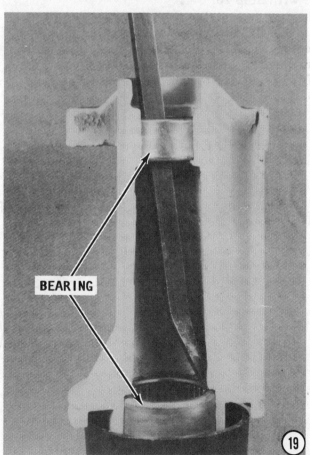

BEARING

(19)

pushing it through the clutch dog and propeller shaft. This pin is not a tight fit,
therefore, it is not difficult to remove.
Grasp the propeller shaft and pull it free of
the bearing housing.

SPECIAL WORDS

Two different shifter arrangements on
the propeller shaft are used on the units
covered in this section. An exploded drawing of one type is shown on Page 8-183
together with reference illustration **"D"** on
Page 180. An exploded drawing of the other
lower unit will be found on Page 8-184
together with reference illustration **"E"** on
Page 181. Take special notice of how one
type has a shifter shaft, bearing, one detent
ball, and one spring. The other model has a
clutch dog shaft, thrust washer, and two
detent balls and springs. Compare the propeller shaft from the unit being serviced
with the two drawings.

Now, if the unit being serviced is the one
shown on Page 8-183, perform Steps 21 and
22, then skip to Step 26.

If the unit being serviced is the one
shown on Page 8-184, skip to Step 23 and
perform the work thru Step 25, then continue with Step 26.

NOTE: The following two steps are to be
performed if servicing a unit as illustrated
on Page 8-183 and reference illustration
"D".

21- Remove the two Allen screws from
the back side of the bearing housing and at
the same time be prepared to catch the two
detent balls and springs.

22- Remove the shift lever pin. Remove
the shift lever and shifter detent, from the
bearing housing. Remove the shifter shaft.

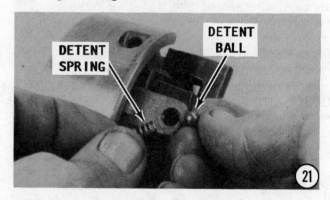

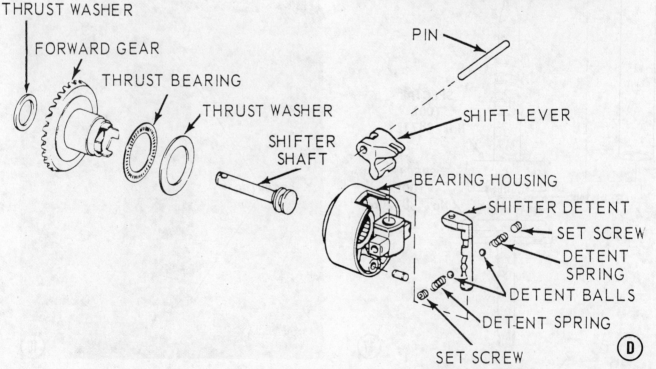

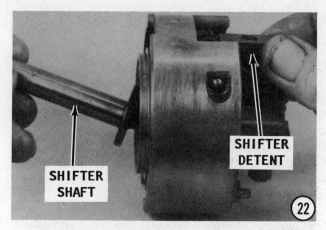

SHIFTER SHAFT

SHIFTER DETENT

22

SHIFT LEVER PIN

23

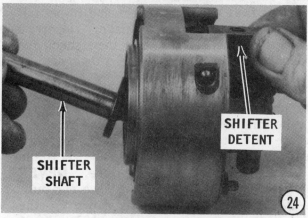

SHIFTER SHAFT

SHIFTER DETENT

24

NOTE: The following three steps are to be performed if servicing a unit as illustrated on Page 8-184 and reference illustration "E".

23- Remove the shift lever pin and disengage the shift lever from the cradle in the shift shaft.

24- Remove the shifter shaft and cradle. Remove the shift lever.

SAFETY WORD

The next step is dangerous. The detent balls are under tension from the springs. The balls are released with pressure. Therefore, **SAFETY GLASSES** should be worn as personal protection for the eyes.

25- Observe the short arm at the upper end of the shifter detent. Now, rotate the shifter detent until this arm is 180° from its original position (facing toward the opposite direction). Rotating the shifter detent 180° will depress the detent ball and spring. From this position, work the shifter detent up out of the housing.

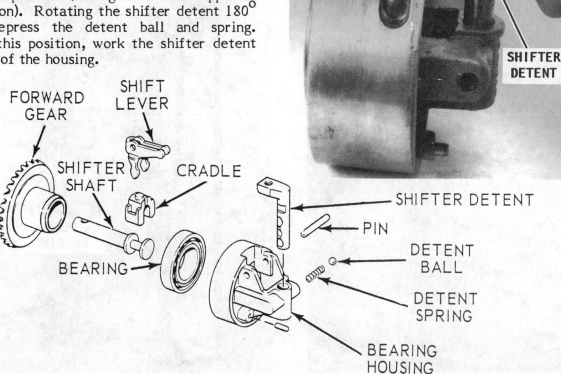

SHIFTER DETENT

25

FORWARD GEAR   SHIFT LEVER

SHIFTER SHAFT   CRADLE

BEARING

SHIFTER DETENT

PIN

DETENT BALL

DETENT SPRING

BEARING HOUSING

E

Lower Driveshaft Bearing — Removal
SPECIAL WORDS

Two different type bearings are used on the lower unit models covered in this section. Exploded drawings of both type bearings are given on Page 8-183 and Page 8-184. To determine which type bearing is used on the unit being serviced, check the shifter arrangement on the propeller shaft and compare it with the two illustrations.

A special tool **MUST** be used to remove either type bearing. Therefore, **DO NOT** attempt to remove this bearing unless it is unfit for further service. To check the bearing, first use a flashlight and inspect it for corrosion or other damage. Insert a finger into the bearing, and then check for "rough" spots or binding while rotating it.

26- Remove the Allen screw, from the water pickup slots in the starboard side of the lower unit housing. This screw secures the bearing in place and **MUST** be removed before an attempt is made to remove the bearing. The bearing must be actually **"PULLED"** upward to come free. **NEVER** make an attempt to "drive" it down and out.

27- If servicing a unit shown in the exploded illustration on Page 183, obtain special tool, OMC No. 385546. If servicing a unit shown in the exploded drawing on Page 8-184, obtain special tool, OMC No.

391257. Use the special tool and "pull" the bearing from the lower unit.

CLEANING AND INSPECTING

Wash all parts in solvent and dry them with compressed air. Discard all **O**-rings and seals that have been removed. A new seal kit for this lower unit is available from the local dealer. The kit will contain the necessary seals and **O**-rings to restore the lower unit to service.

Inspect all splines on shafts and in gears for wear, rounded edges, corrosion, and damage.

Carefully check the driveshaft and the propeller shaft to verify they are straight and true without any sign of damage. A complete check must be performed by turning the shaft in a lathe. This is only necessary if there is evidence to suspect the shaft is not true.

Check the water pump housing for corrosion on the inside and verify the impeller and base plate are in good condition. Actually, good shop practice dictates to rebuild or replace the water pump each time the lower unit is disassembled. The small cost is rewarded with "peace of mind" and satisfactory service.

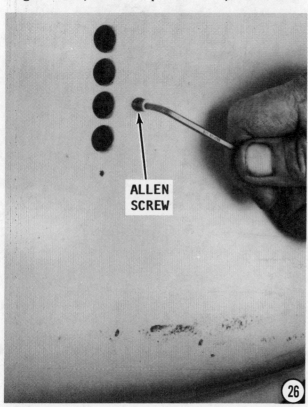

ALLEN
SCREW

26

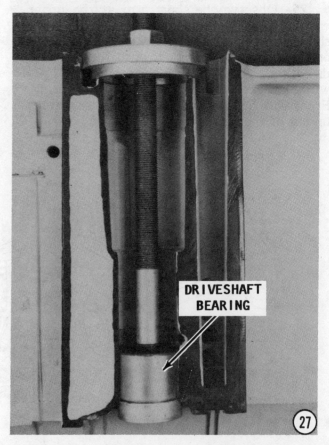

DRIVESHAFT
BEARING

27

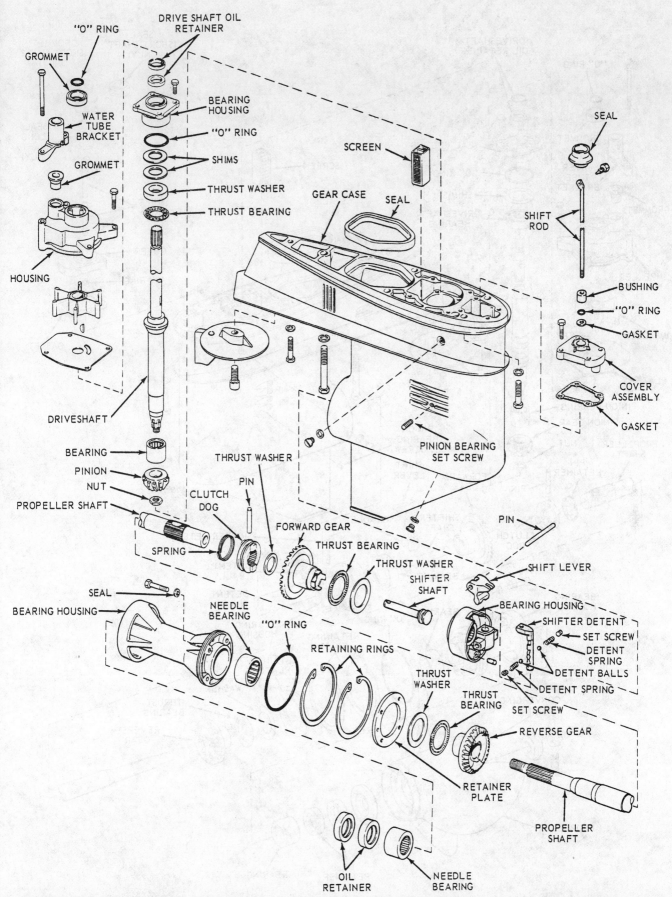

Exploded drawing of one of two lower units covered in this section. This unit has a thrust bearing and thrust washer on the driveshaft and two detent balls and springs for the shift mechanism. Enlarged details of the shift mechanism are shown in reference illustration "D" on Page 8-180. Numerous references in the text are made to these two illustrations.

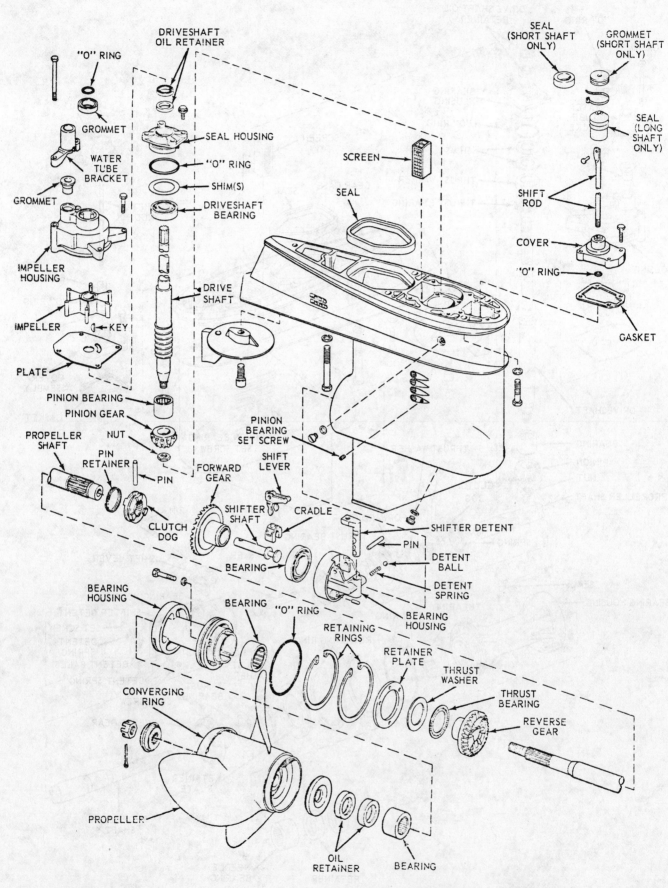

Exploded drawing of one of two lower units covered in this section. This unit has a single bearing on the driveshaft and one detent ball and spring for the shift mechanism. Enlarged details of the shift mechanism are shown in reference illustration "E" on Page 8-181. Numerous references in the text are made to these two illustrations.

Inspect the lower unit housing for nicks, dents, corrosion, or other signs of damage. Nicks may be removed with No. 120 and No. 180 emery cloth. Make a special effort to ensure all old gasket material has been removed and mating surfaces are clean and smooth.

Inspect the water passages in the lower unit to be sure they are clean. The screen may be removed and cleaned.

Check the gears and clutch dog to be sure the ears are not rounded. If doubt exists as to the part performing satisfactorily, it should be replaced.

Inspect the bearings for "rough" spots, binding, and signs of corrosion or damage.

ASSEMBLING

READ AND BELIEVE

The lower unit should **NOT** be assembled in a dry condition. Coat all internal parts with OMC HI-VIS lube oil as they are assembled. All seals should be coated with OMC Gasket Seal Compound. When two seals are installed back-to-back, use Triple Guard Grease between the seal surfaces.

Propeller Shaft Assembling
SPECIAL WORDS

Two different shifter arrangements on the propeller shaft are used on the units covered in this section. A different type of

lower driveshaft bearing is also used. An exploded drawing of one type is shown on Page 8-183, together with reference illustration **"F"**. An exploded drawing of the other lower unit will be found on Page 8-184 together with reference illustration **"G"**. Take special notice of how one type has a shifter shaft, bearing, one detent ball, and one spring. The other model has a shifter shaft, thrust washer, and two detent balls and springs. Compare the propeller shaft from the unit being serviced with the two drawings.

Now, if the unit being serviced is the one shown on Page 8-183, perform Steps 1 thru 4, then skip to Step 10.

If the unit being serviced is the one shown on Page 8-184, skip to Step 5 and perform the work thru Step 9, then continue with Step 10.

NOTE: The following four steps are to be performed if servicing a unit illustrated on Page 8-183 and reference illustration **"F"**.

1- Install the shift lever and shifter detent. Install the shifter shaft and then the pin.

2- Insert the two detent balls into the shifter detent, then the springs. Coat the threads of the set screws with Loctite, and then secure the springs and detent balls in place with the screws. The screws should be tightened until each head is flush with the housing.

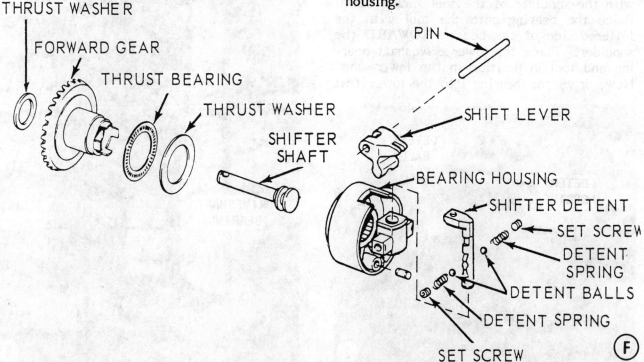

THRUST WASHER
FORWARD GEAR
THRUST BEARING
THRUST WASHER
SHIFTER SHAFT
PIN
SHIFT LEVER
BEARING HOUSING
SHIFTER DETENT
SET SCREW
DETENT SPRING
DETENT BALLS
DETENT SPRING
SET SCREW

(F)

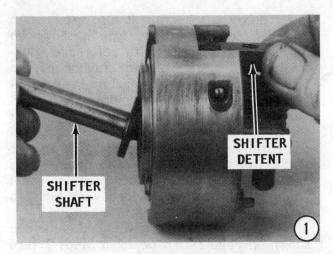

SHIFTER SHAFT

SHIFTER DETENT

3- Install the thrust bearing and thrust washer onto the shank of the forward gear. Now, slide the assembled forward gear into the bearing housing. Slide the small thrust washer onto the propeller shaft. Insert the propeller shaft into the forward gear with the slot in the shaft aligned with the hole in the shifter shaft. Insert the pin, and then install the coil spring around the outside groove of the clutch dog. Check to be sure one coil of the spring does not overlap another coil. Press the shifter detent down to move the clutch dog into the reverse gear position. Set the unit aside for later installation.

4- Obtain bearing installer, special tool, OMC No. 385546. Assemble the washer, plate, guide sleeve, and installer portion of the tool to the screw in the order given, and with the shoulder of the tool facing down. Place the bearing onto the tool with the lettered side of the bearing **TOWARD** the shoulder. Place the lower driveshaft bearing and tool in position in the lower unit. Now, drive the bearing into the lower unit

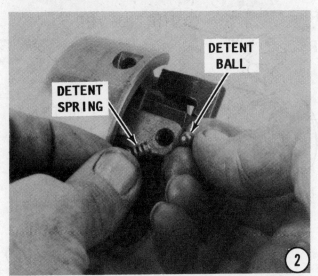

DETENT BALL

DETENT SPRING

PIN

until the plate makes contact with the lower unit surface. Coat the threads of the setscrew with Loctite and then secure the bearing in place with the setscrew.

Perform the following 5 steps if servicing a unit as shown in the exploded drawing on Page 8-184 together with reference illustration **"G"**.

5- Apply a thin coating of Needle Bearing Grease onto the detent ball and spring. Insert the detent ball and spring into the bearing housing. Observe the short arm on the shifter detent. Position the detent with

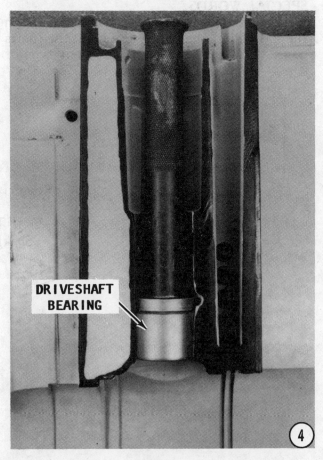

DRIVESHAFT BEARING

SHIFTER
DETENT

⑤

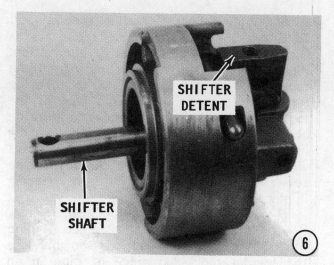

SHIFTER
DETENT

SHIFTER
SHAFT

⑥

this short arm facing forward. Now, use a punch, or similar tool, and depress the ball and spring, and **AT THE SAME TIME** press the shift detent down into the bearing housing. Rotate the shifter detent 180°.

6- Place the cradle onto the shifter shaft and in position in the bearing housing. Install shifter lever with the cradle engaged with the shifter detent. Work the pin through the shifter detent and cradle.

7- Install the clutch dog onto the propeller shaft with the **PROP END** identification on the dog facing the propeller end of the shaft. Align the holes in the dog with the slot in the shaft. Slide the forward gear bearing onto the forward gear shoulder. Install the thrust washer into the bearing

housing. Align the hole in the shifter shaft with the hole in the clutch dog. Now, slide the propeller shaft into the assembled forward gear and bearing housing. Insert the clutch dog retaining pin. Install a **NEW** clutch dog retaining spring. Check to be sure that one coil of the spring does not overlap another coil. Move the shifter detent down into the reverse gear position. Set the assembly aside for later installation.

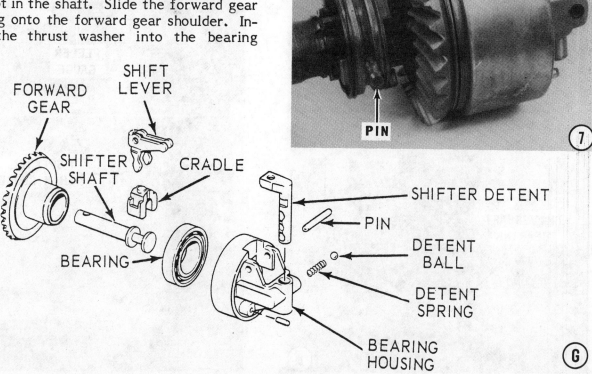

PIN

⑦

FORWARD
GEAR

SHIFT
LEVER

SHIFTER
SHAFT

CRADLE

BEARING

SHIFTER DETENT

PIN

DETENT
BALL

DETENT
SPRING

BEARING
HOUSING

Ⓖ

8- Obtain bearing installer, special tool, OMC No. 391257. Assemble the parts as shown in the accompanying illustration. Place the lower driveshaft bearing on the tool with the lettered side of the bearing facing **UP** towards the driving face of the tool. Use Needle Bearing Grease to hold the bearing on the tool. Now, drive the bearing into place until the washer makes contact with the spacer. Coat the threads of the setscrew with Loctite. Secure the bearing in place with the screw.

9- Obtain special tool, OMC No. 387131, OMC No. 387206, and a short piece of pipe to fit over the driveshaft, as shown. Use the special tools and piece of pipe in an arbor press to install the upper driveshaft bearing. Press with the tool against the inner race until the bearing is seated on the driveshaft shoulder.

10- Clamp the driveshaft in a vise equipped with soft jaws and in such a manner that the splines, water pump area, or other critical portions of the shaft cannot be damaged. Slide the pinion gear onto the driveshaft with the bevel of the gear teeth facing toward the lower end of the shaft. Install the pinion gear nut and tighten it to a torque value of 40 to 45 ft-lbs. No parts should be installed on the upper end of the driveshaft at this point. Slide the shims removed during disassembling onto the driveshaft and seat them against the driveshaft shoulder. If servicing a unit as shown

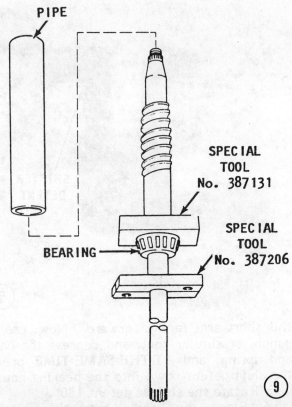

in exploded drawing on Page 8-183, obtain special shimming tool, OMC No. 315767. If servicing a unit as shown in the exploded drawing on Page 8-184, obtain special shimming tool, OMC No. 320739.

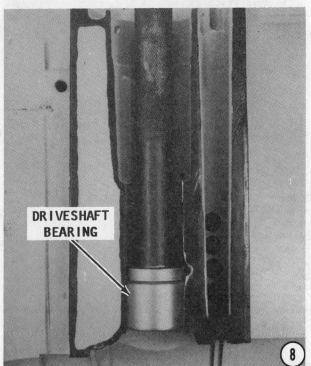

Slip the special tool down over the drive-shaft and onto the upper surface of the top shim. Measure the distance between the top of the pinion gear and the bottom of the tool. The tool should just barely make contact with the pinion gear surface for **ZERO** clearance. Add or remove shims from the upper end of the driveshaft to obtain the required **ZERO** clearance. If servicing the model as shown on Page 184, remove 0.007" of shim material **AFTER** the zero clearance has been obtained. Remove the tool and set the shims aside for installation later. Back off the pinion gear nut and remove the pinion gear. Remove the driveshaft from the vise.

11- Insert the complete assembled propeller shaft into the lower unit housing with pin on the back side of the forward gear housing indexed into hole in the lower unit.

12- Thread the shift rod into the shift detent assembly. Coat both sides of the shift rod gasket with sealer. Place the gasket in position, and then install the shift rod cover onto the housing. The shift rod is adjusted in the next step. Slide the shift rod boot down over the shift rod cover.

Shift Rod Adjustment

13- Measure the distance from the top of the lower unit housing to the center of the hole in the shift rod. This distance should be as follows for a standard short shaft unit. Add exactly 5" (12.7cm), for a unit with a long shaft.

40 & 50 hp
1972 -- 16-5/32" (41.0cm)
1973 -- 16-9/64" (40.9cm)
1975 -- 16-5/16" (41.4cm)
1980 and on -- 15-29/32" (40.4cm)

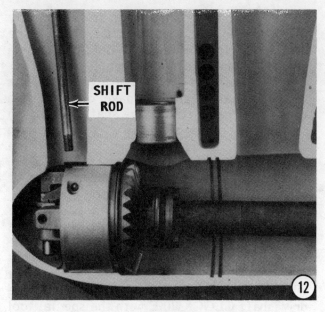

55 hp
1976-78 -- 15-7/8" (40.3cm)
1979-83 -- 15-29/32" (40.4cm)

60 hp
1980-85 -- 15-29/32" (40.4cm)

Rotate the shift rod clockwise or counterclockwise until the required dimension is obtained.

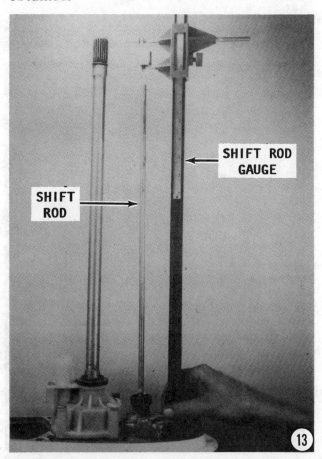

14- Insert the pinion gear into the cavity in the lower unit with the flat side of the bearing facing **UPWARD**. Hold the pinion gear in place and at the same time lower the driveshaft down into the lower unit. As the driveshaft begins to make contact with the pinion gear, rotate the shaft slightly to permit the splines on the shaft to index with the splines of the pinion gear. After the shaft has indexed with the pinion gear, thread the pinion gear nut onto the end of the shaft. Obtain special tool, OMC No. 316612. Slide the special tool over the upper end of the driveshaft with the splines of the tool indexed with the splines on the shaft. Attach a torque wrench to the special tool. Now, hold the pinion gear nut with the proper size wrench and rotate the driveshaft **CLOCKWISE** with the special tool until the pinion gear nut is tightened to a torque value of 40 to 45 ft-lbs. Remove the special tool.

15- If servicing a unit as shown on Page 183: Slide the thrust bearing, thrust washer, and the shims, set aside after the shim gauge procedure in Step 10, onto the driveshaft. If servicing a unit shown on Page 184: Slide the shims only onto the driveshaft. (This unit does not have the other items.)

16- Install the two seals back-to-back into the opening on top of the bearing

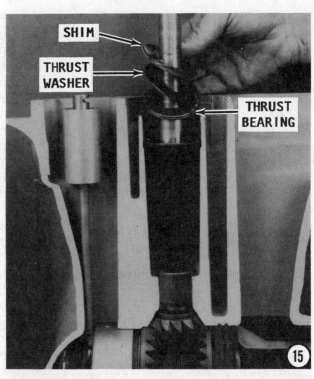

housing. Coat the outside surface of the **NEW** seals with OMC Lubricant. Press the first seal into the housing with the flat side of the seal facing **OUTWARD**. After the seal is in place, apply a coating of Triple Guard Grease to the flat side of the installed seal and the flat side of the second seal. Press the second seal into the bearing housing with the flat side of the seal facing **INWARD**. Insert a **NEW O**-ring into the bottom opening of the bearing housing.

17- Wrap friction tape around the splines of the driveshaft to protect the seals in the bearing housing as the housing is

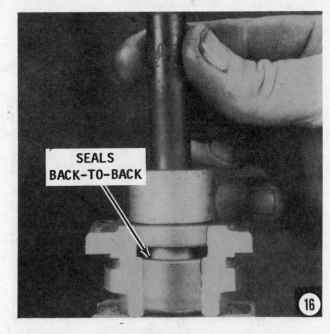

installed. Now, slide the assembled bearing housing down the driveshaft and seat it in the lower unit housing. Secure the housing in place with the four bolts. Tighten the bolts **ALTERNATELY** and **EVENLY.**

Reverse Gear — Installation

18- Apply a light coating of grease to the inside surface of the reverse gear to hold the thrust washer in place. Place the thrust washer onto the front side of the reverse gear. Install the thrust bearing and thrust washer onto the shank on the back side of the reverse gear. Slide the reverse gear down the propeller shaft and index the teeth of the reverse gear with the teeth of the pinion gear.

19- Insert the retainer plate into the lower unit against the reverse gear.

WARNING

This next step can be dangerous. Each snap ring is placed under tremendous tension with the Truarc pliers while it is being placed into the groove. Therefore, wear **SAFETY GLASSES** and exercise care to prevent the snap ring from slipping out of the pliers. If the snap ring should slip out, it would travel with incredible speed and cause personal injury if it struck a person.

20- Install the Truarc snap rings one at-a-time following the precautions given in

the **WARNING** and the **ADVICE** given in the following paragraph.

WORDS OF ADVICE

The two snap rings index into separate grooves in the lower unit housing. As the first ring is being installed, depth perception may play a trick on your eyes. It may

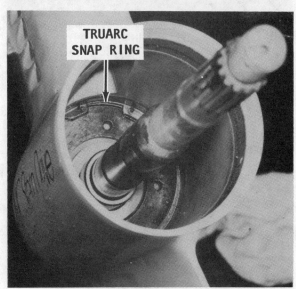

View into the lower unit showing the Truarc snap rings properly locked in the grooves.

appear that the first ring is properly indexed all the way around in the proper groove, when in reality, a portion may be in one groove and the remainder in the other groove. Should this happen, and the Truarc pliers be released from the ring, it is extremely difficult to get the pliers back into the ring to correct the condition. If necessary use a flashlight and carefully check to be sure the first ring is properly seated all the way around **BEFORE** releasing the grip on the pliers. Installation of the second ring is not so difficult because the one groove is filled with the first ring.

Bearing Carrier Bearings & Seals Installation

21- Install the reverse gear bearing into the bearing carrier by pressing against the **LETTERED** side of the bearing with the proper size socket. Press the forward gear bearing into the bearing carrier in the same manner. Press against the **LETTERED** side of the bearing.

22- Coat the outside surfaces of the seals with HI-VIS oil. Install the first seal with the flat side facing **OUT**. Coat the flat surface of both seals with Triple Guard Grease, and then install the second seal with the flat side going in **FIRST**. The seals are then back-to-back with the grease between the two surfaces. The outside seal prevents

A set of double seals showing the back side (left) and the front side (right). These seals are installed back-to-back (flat side-to-flat side) with Triple Guard Grease between the surfaces. This arrangement prevents fluid from passing in either direction.

water from entering the lower unit and the inside seal prevents the lubricant in the lower unit from escaping.

Bearing Carrier Installation

23- Obtain two long 1/4" rods with threads on one end. Thread the rods into the retainer plate opposite each other to act as guides for the bearing carrier. Check the bearing carrier to be sure a **NEW O**-ring has been installed. Position the carrier over the guide pins with the embossed word **UP** on the rim of the carrier facing **UP** in relation to the lower unit housing, reference illustration "H". Now, lower the bearing carrier down over the guide pins and into place in the lower unit housing.

BEARING

21

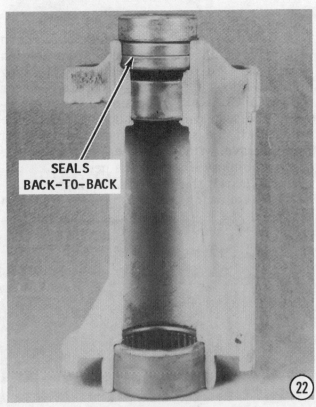

SEALS
BACK-TO-BACK

22

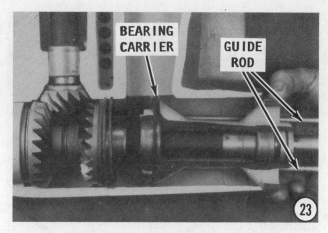

24- Slide **NEW** little O-rings onto each bolt, and apply some OMC Sealer onto the threads. Install the bolts through the carrier and into the retaining plate. After a couple bolts are in place, remove the guide pins and install the remaining bolts. Tighten the bolts **EVENLY** and **ALTERNATELY** to the torque value given in the Appendix.

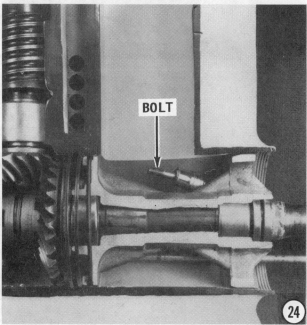

WATER PUMP INSTALLATION

FIRST, THESE WORDS

An improved water pump is available as a replacement. If the old water pump housing is unfit for further service, only the new pump housing can be purchased. It is strongly recommended to replace the water pump with the improved model while the lower unit is disassembled. The accompanying illustration shows the original equipment (left) compared with the improved pump (right), reference illustration **J**.

The new pump must be assembled before it is installed. Therefore, the following steps outline procedures for both pumps.

To assemble and install a replacement pump, perform steps 25 thru 33, then jump to Step 36.

To install an original equipment pump, proceed directly to Step 34.

Assembling an Improved Pump Housing

25- Remove the water pump parts from the container. Insert the plate into the housing, as shown. The tang on the bottom side of the plate **MUST** index into the short slot in the pump housing.

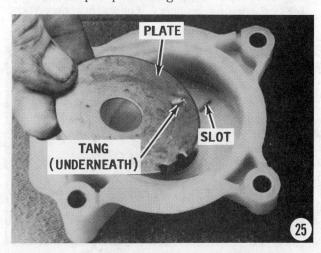

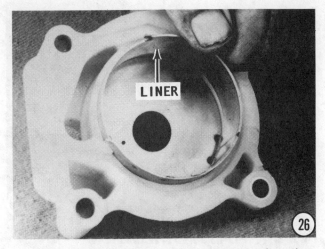

LINER

26- Slide the pump liner into the housing with the two small tabs on the bottom side indexed into the two cutouts in the plate.

27- Coat the inside diameter of the liner with light-weight oil. Work the impeller into the housing with all of the blades bent back to the right, as shown. In this position, the blades will rotate properly when the pump housing is installed. Remember, the pump and the blades will be rotating **CLOCKWISE** when the housing is turned over and installed in place on the lower unit.

28- Coat the mating surface of the lower unit with 1000 Sealer. Slide the water pump base plate down the driveshaft and into place on the lower unit. Insert the Woodruff key into the key slot in the driveshaft.

29- Lay down a very thin bead of 1000 Sealer into the irregular shaped groove in the housing. Insert the seal into the groove, and then coat the seal with the 1000 Sealer.

Water Pump Installation

30- Begin to slide the water pump down the driveshaft, and at the same time observe the position of the slot in the impeller. Continue to work the pump down the

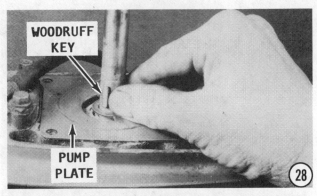

WOODRUFF KEY

PUMP PLATE

driveshaft, with the slot in the impeller indexed over the Woodruff key. The pump must be fairly well aligned before the key is covered because the slot in the impeller is not visible as the pump begins to come close to the base plate.

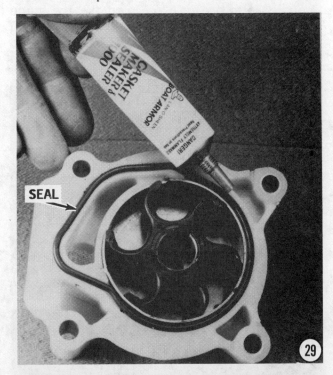

SEAL

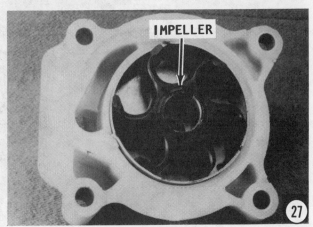

IMPELLER

WATER PUMP HOUSING

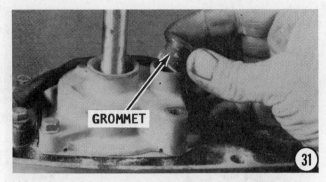

31- Install the short forward bolt through the pump and into the lower unit. **DO NOT** tighten this bolt at this time. Insert the grommet into the pump housing.

32- Install the grommet retainer and water tube guide onto the pump housing. Install the remaining pump attaching bolts. Tighten the bolts **ALTERNATELY** and **EVENLY,** and at the same time rotate the driveshaft **CLOCKWISE.** If the driveshaft is not rotated while the attaching bolts are being tightened, it is possible to pinch one of the impeller blades underneath the housing.

33- Slide the large grommet down the driveshaft and seat it over the pump collar. This grommet does not require sealer. Its function is to prevent exhaust gases from entering the water pump. Proceed directly to Lower Unit Installation, Step 36.

Water Pump Installation
Original Equipment

Perform the following two steps to install an original water pump.

34- Coat the water pump plate mating surface on the lower unit with 1000 Sealer. Slide the water pump plate down the driveshaft and **BEFORE** it makes contact with the sealer check to be sure the bolt holes in the plate will align with the holes in the housing. The plate will only fit one way. If the holes will not align, remove the plate, turn it over and again slide the plate down

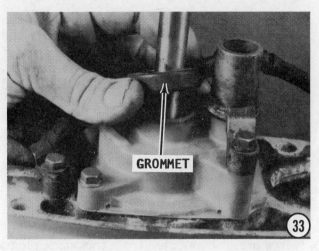

the driveshaft and into place on the housing. This checking will prevent accidently getting the sealer on both sides of the plate. If by chance sealer does get on the top surface it **MUST** be removed before the water pump impeller is installed.

Slide the water pump impeller down the driveshaft. Just before the impeller covers the cutout for the Woodruff key, install the key, and then work the impeller on down, with the slot in the impeller indexed over the Woodruff key. Continue working the impeller down until it is firmly in place on the surface of the pump plate.

35- Check to be sure **NEW** seals and O-rings have been installed in the water pump. Lubricate the inside surface of the water pump with light-weight oil. Lower the water pump housing down the driveshaft and over the impeller. **ALWAYS** rotate the driveshaft slowly **CLOCKWISE** as the housing is lowered over the impeller to allow the impeller blades to assume their natural and proper position inside the housing. Continue to rotate the driveshaft and work the water pump housing downward until it is seated on the plate. Coat the threads of the water pump attaching screws with sealer, and then secure the pump in place with the screws. Install the solenoid cable bracket with the

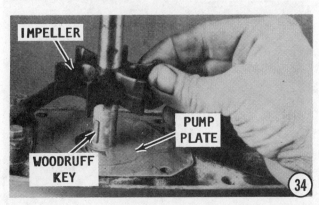

same bolt and in the same position from which it was removed. On some units the solenoid cable fits into a recess of the water pump and is held in place in that manner. Tighten the screws **ALTERNATELY** and **EVENLY**.

LOWER UNIT INSTALLATION

36- Check to be sure the water tubes are clean, smooth, and free of any corrosion. Coat the water pickup tubes and grommets with lubricant as an aid to installation. Check to be sure the spark plug wires are disconnected from the spark plugs. Bring the lower unit housing together with the exhaust housing, and at the same time, guide the water tube into the rubber grommet of the water pump. Continue to bring the two units together, at the same time guiding the water pickup tubes into the rubber grommet of the water pump, and, simultaneously rotating the flywheel slowly to permit the splines of the driveshaft to index with the splines of the crankshaft. This may sound like it is necessary to do four things at the same time, and so it is. Therefore, make an earnest attempt to secure the services of an assistant for this task.

37- After the surfaces of the lower unit and exhaust housing are close, dip the attaching bolts in OMC Sealer and then start them in place. Install the two bolts on each side of the two housings. Install the bolt in the recess of the trim tab. Install the bolt just forward of the trim tab that extends up through the cavitation plate. Tighten the bolts **ALTERNATELY** and **EVENLY** to the torque value given in the Appendix.

38- Install the trim tab with the mark made on the tab during disassembling aligned with the mark made on the lower unit housing.

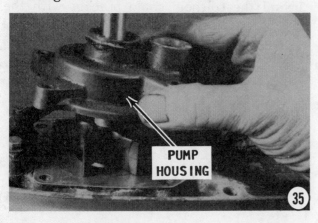

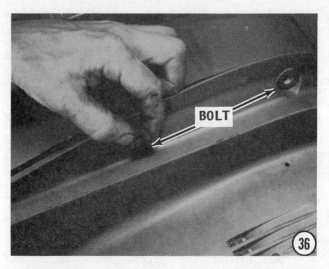

39- The shift rod is visible beneath and to the rear of the lower carburetor. Install the shift connector with a bolt or pin, depending on the model being serviced.

Filling the Lower Unit

Fill the lower unit with lubricant according to the procedures in Section 8-3.

Propeller Installation

Install the propeller, see Section 8-2.

FUNCTIONAL CHECK

Perform a functional check of the completed work by mounting the engine in a test tank, in a body of water, or with a flush attachment connected to the lower unit. If

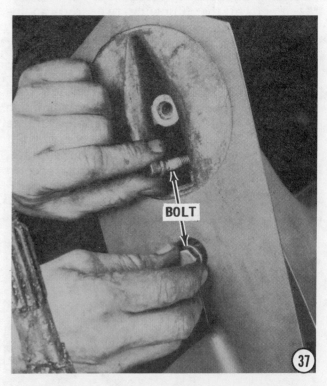

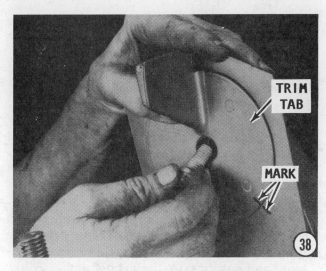

the flush attachment is used, **NEVER** operate the engine above an idle speed, because the no-load condition on the propeller would allow the engine to **RUNAWAY** resulting in serious damage or destruction of the engine.

CAUTION: Water must circulate through the lower unit to the engine any time the engine is run to prevent damage to the water pump in the lower unit. Just five seconds without water will damage the water pump.

Start the engine and observe the tattletale flow of water from idle relief in the exhaust housing. The water pump installation work is verified. If a "Flushette" is connected to the lower unit, **VERY LITTLE** water will be visible from the idle relief port. Shift the engine into the three gears and check for smoothness of operation and satisfactory performance. Remember, when the unit is in forward gear, it is at rest with no current flow to either solenoid.

8-14 "FROZEN" PROPELLER

DESCRIPTION

If an exhaust propeller is "frozen" to the propeller shaft and the usual methods of removal fail, using a mallet to beat on the propeller blades in an effort to dislodge the propeller will have little if any affect. This is due to the cushioning affect the rubber hub has on the blow being struck.

A "frozen" propeller is caused by the inner sleeve becoming corroded and stuck to the propeller shaft. Therefore, special procedures are required to free the propeller and remove it from the shaft.

The following detailed procedures are presented as the only practical method to remove a "frozen" propeller from the shaft without damaging other more expensive parts. In almost all cases it is successful.

REMOVAL

1- Remove the cotter pin, and then the castellated nut and splined washer from the propeller shaft.

2- Heat the inside diameter of the propeller with a torch. Do not apply the heat to the outside surface of the propeller. Concentrate the heat in the area of the hub and shaft where the nut was removed and as far into the hub as possible. Continue applying heat, and at the same time have an assistant use a piece of 2" x 4" wooden block

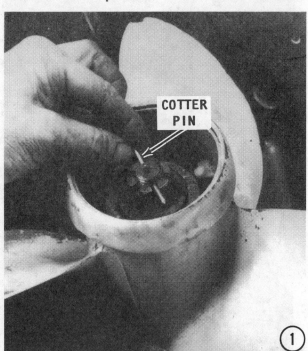

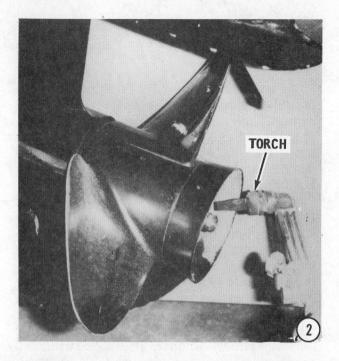

wedged between one of the blades and the lower unit housing. Use a prying force on the propeller while the heat is being applied. As the heat melts the inner rubber hub, the propeller will come free.

WARNING

As the force and heat are applied, the propeller may "pop" loose suddenly and without warning. Therefore, stand to one side while applying the heat as a precaution against personal injury.

3- After the propeller has been removed, the sleeve and what's left of the rubber hub will still be stuck to the propeller shaft. Attach a puller to the thrust washer. Apply more heat to the sleeve and rubber hub, and at the same time take up on the puller. When the sleeve reaches the proper temperature, it will be released and come free.

4- If a puller is not available, as described in Step 3, use a sharp knife and cut the rubber hub from the sleeve. An alternate method is to use canned heat, or the equivalent, and set fire to the rubber hub. Allow the hub to burn away from the sleeve. When the fire burns out, only a small amount may be left on the sleeve. Heat the sleeve again, and then while it is still hot, use a chisel, punch, or similar tool with a hammer, and drive the sleeve free of the shaft. Allow the propeller shaft to cool, and then clean the splines thoroughly. Take time to remove any corrosion. Install the propeller with a **NEW** hub and sleeve according to the instructions outlined in Section 8-2.

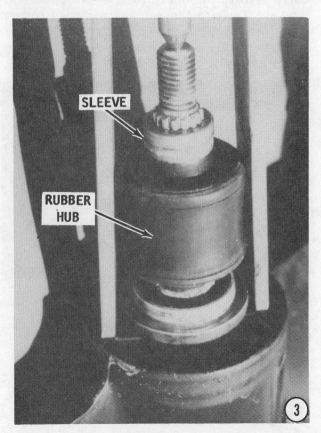

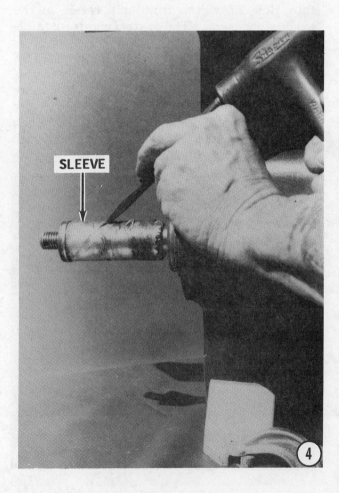

9
HAND STARTER

9-1 INTRODUCTION

The hand rewind starters installed on the Johnson/Evinrude outboards covered in this manual may be one of three basic designs:

One is a vertical spool type mounted on the side of the powerhead with a drive gear engaging the teeth of the flywheel.

The second design is a disc type which engages the teeth of the flywheel and may be mounted either horizontally or vertically.

The third design is a flat disc type mounted atop the flywheel. This design starter engages a ratchet plate on the flywheel.

UNFORTUNATELY, engineering changes have resulted in several models being used for each design. The only logical and practical method of designating different procedures for each model for each design was to asign a type number for each different starter. Therefore, the hand rewind starters in this chapter are designated Type I through Type VII with identification of the outboard models using each type.

Design 1

The first design is a cylinder with a pinion gear arrangement similar to an automotive starter motor. The unit is mounted vertically on the side of the powerhead.

Type I hand rewind starter installation on the 5hp, 6hp, 7.5hp, and 8hp powerheads. This starter is similar in operation to an automotive type starter.

Type II hand rewind starter installation on all 9.5 hp powerheads. Notice the difference with the unit in the left column, this page.

Type III swing arm hand rewind starter mounted on the port side on 4hp Delux and 4.5hp powerheads.

When the starter rope is pulled, a nylon drive gear slides upward and engages the flywheel ring gear. After the powerhead starts, the drive gear automatically disengages and retracts to the "rest" position.

Two models of this hand starter are installed on the outboards covered in this manual. Service procedures for one model, identified as Type I and used on the 5hp thru 8hp, are presented in Section 9-2. Instructions for the second model, identified as Type II, and used on the 9.5hp, are listed in Section 9-3.

Design 2

This second design is available in two types. One type is mounted horizontally to one side of the flywheel with the drive gear engaging the flywheel directly. The other type is mounted on the port side of the powerhead and the drive gear works on an axis. As the rope is pulled, a swing arm moves the drive gear upward to engage with the teeth of the flywheel ring gear. A coil spring winds and tightens as the rope unwinds. The spring then coils the rope around a pulley as the rope handle is returned to the control panel.

Two models of this hand starter are installed on the outboards covered in this manual. Service procedures for one model, identified as Type III, and used on the 4hp 1971-78, the Delux 4hp 1982 and on, and the 4.5hp 1980-84, are presented in Section 9-4. Instructions for the second model, identified as Type IV, and used on the 9.9hp and 15hp, are listed in Section 9-5.

Design 3

This design starter is usually mounted atop the flywheel with three mounting legs

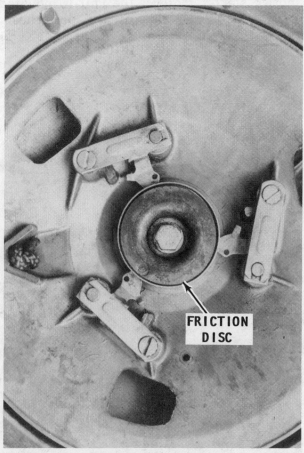

Type V hand rewind starter mounted atop the flywheel. This model starter has three pawls, with a friction disc and no return springs. This unit may be found on early 40hp powerheads.

Type IV hand rewind starter mounted flat (horizontally) next to the flywheel on all 9.9hp and 15hp powerheads. The drive gear moves upward and engages the ring gear of the flywheel when the starter rope is pulled.

attached to the powerhead. Three types of this design hand starter are installed on various Johnson/Evinrude powerheads.

Service procedures for one model, identified as Type V, and used on the model 40hp 1971-76 and 1981-85, are presented in Section 9-6.

Service procedures for the second model, identified as Type VI, and used on 2hp 1971-85, 2.5hp 1986 and on, 4hp 1980-85, 18hp 1971-73, 20hp 1971-73 and 1985 and on, 25hp 1971 and on, 35hp 1976-84 and the Colt, Ultra and Junior models for all years, are presented in Section 9-7.

Instructions for the third model, identified as Type VII, and used on the 40hp 1986 and on, are listed in Section 9-8.

EMERGENCY STARTING

The 50hp, 55hp, and 60hp powerheads are equipped with an electric starter motor only. However, cutouts are manufactured into the flywheel to permit the use of a rope for emergency start.

SAFETY WORD

If an emergency rope is used on **ANY** powerhead **NEVER** wrap the rope end around your hand for a better grip. If the powerhead should happen to backfire, the sudden jerk on the rope would severly **INJURE** your hand.

OPERATION

Normally, very few problems are encountered with the hand starter. It is strictly a mechanical device to crank the powerhead for starting. The spring will last an incredibly long time, if used properly. The greatest enemy of the spring is the operator.

Three causes contribute to starter failure. Two may be prevented, the third cannot.

The most common problem is the result of the operator pulling the starter rope too far outward. If the operator places one hand on the powerhead and pulls the rope with the other hand, it is physically impossible, in this positon, to pull the rope too far. Problems develop when the operator uses both hands to pull on the rope, with no control on how far the rope can be extended. The rope may be broken or the knot released from the starter disc. In either case, the spring rewinds with tremendous speed and in almost all cases travels past its normal rewind position bending the end of the spring in reverse. Therefore, more maintenance work is involved than merely replacing the rope.

Another bad habit, while using the hand starter, is to release the grip on the rope when it is in the extended position, allowing the rope to freely rewind. The operator should **NEVER** release his grip, but hold onto the rope, and thus control the rewind. The owner should always be alert to any wear on

Type VI hand rewind starter with single pawl. This starter was used on the Colt, Junior and Ultra model 2hp, 4hp, 18hp and early 20hp, 25hp, and 35hp powerheads. Since 1986, this unit has a second pawl opposite the one shown.

The starter rope broke on this Type VI hand rewind starter causing the spring to rewind with such incredible speed, the spring actually doubled back in a reverse direction, as shown.

the rope and replace it long before the possibility of breaking might occur. If the rope should break, the spring would rewind with incredible speed, the same as if the rope were released, causing damage to the spring and other starter parts.

The third cause of spring failure cannot be prevented -- age. As the outboard continues to perform year after year, the age of the spring steel will finally take its toll.

The rewind spring is made of spring steel. Depending on the model and the powerhead, from 6 feet to 12 feet of spring length is wound into about a 4 inch diameter. This places the spring under unbelievable tension, making it a highly **DANGEROUS** force. Therefore, any time the hand starter is serviced, especially during work on the spring, **SAFETY GLASSES** should be worn and the work performed with the utmost care.

Any time the rope is broken, the starter spring will rewind with incredible speed. Such action will cause the spring to rewind past its normal travel and the end of the spring will be bent back out of shape. Therefore, if the rope has been broken, the starter should be completely disassembled and the spring repaired or replaced.

9-2 TYPE I STARTER
CYLINDER WITH PINION GEAR
ALL 5HP, 6HP, 7.5HP, AND 8HP

This gear-drive starter is a new design employing the principle of an automotive type starter motor. When the starter rope is pulled, the starter rotates, and a nylon pinion gear slides upward and engages the flywheel ring gear. The gear automatically disengages when the powerhead starts. The ratio between the pinion gear and the ring on the flywheel has been selected to provide maximum cranking speed with minimum pulling effort to ensure fast, easy powerhead start.

STARTER ROPE REPLACEMENT

REMOVAL

1- Disconnect the high-tension leads from the spark plugs. Ground the high-tension leads. Pull the starter rope out until it is fully extended. Now, allow the rope to retract just a little, until the knot end on the spool is facing the port side of the

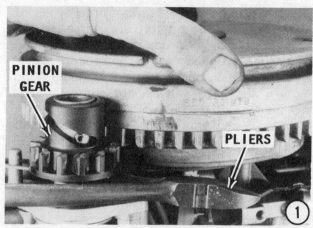

powerhead. Lift the pinion gear to engage the flywheel ring gear. Hold the pinion gear engaged with the ring gear, and at the same time, slide the handles of a pair of pliers under the pinion gear to lock the pinion gear with the ring gear, as shown.

2- Remove the handle from the end of the starter rope. **OBSERVE** how the rope is wound onto the spool and how the rope is secured by a loop formed in a slot in the spool. Remove the rope from the starter spool.

INSTALLATION

Rope Purchase Instructions

The length and diameter of the starter rope required will vary depending on the horsepower size of the model being serviced. Therefore, check the Hand Starter Rope Specifications in the Appendix, and then purchase a quality nylon piece of the

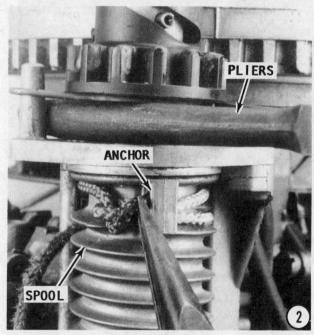

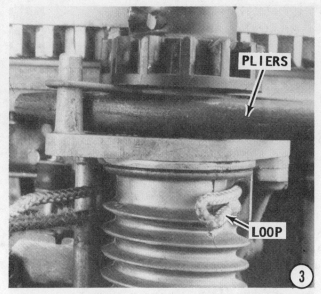

LOOP

PLIERS

③

PLIERS

ANCHOR

⑤

proper length and diameter size. Only with the proper rope, will you be assured of efficient operation following installation.

Each end of the nylon rope should be "fused" by burning them slightly with a very small flame (a match flame will do) to melt the fibers together. After the end fibers have been "fused" and while they are still hot, use a piece of cloth as protection and pull the end out flat to prevent a "glob" from forming.

3- Feed one end of the new rope through the spool anchor, make a loop, and then thread the end back through the hole in the anchor, but **DO NOT** pull it tight at this time, leave a loop.

4- Bring the short end of the rope through the loop just formed.

5- Work the short end back through the anchor, as shown.

6- Now, pull both ends of the rope tight. Feed the rope through the front engine

cowling, and then install the starter rope handle. Pull and hold tension on the rope, and at the same time remove the pliers from under the pinion gear. Allow the starter rope to rewind in a normal manner. After the rope is fully wound onto the starter spool, the rope handle should be up tight against the cowling. If the handle is not up tight, the rope was installed to long or the starter spring is weak and should be replaced.

STARTER REMOVAL

AUTHOR'S NOTE

For photographic clarity, the accompanying pictures were taken servicing a starter from a powerhead removed from the

PLIERS

ROPE END

④

PLIERS

ROPE END

⑥

exhaust housing. The hood need only be removed to work on the starter.

1- Pull approximately 3/4 of the starter rope out, and then form a knot in the rope to prevent it from recoiling. Allow the rope to recoil until the knot is tight against the cowling. Remove the rope handle.

2- The rope may be removed now, or later. To remove the rope now, first pull the rope all the way out. Slide the handles of a pair of pliers under the pinion gear to hold the gear engaged with the flywheel. Remove the rope from the starter spool.

3- Grasp the spool firmly. Remove the pliers and allow the spool to slip a little at a time until the spring is completely unwound.

4- Remove the two retaining bolts on top of the starter.

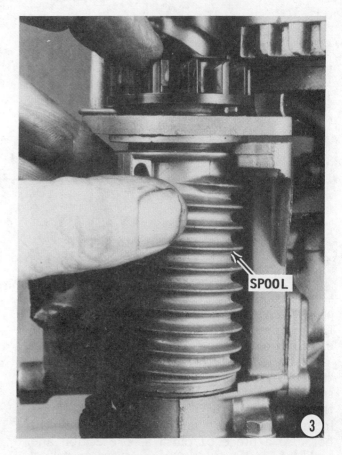

5- Loosen, but **DO NOT** remove the bolts on the bottom and on each side of the starter spool. When the bottom two bolts are loosened, the retainer will separate from the lower cap. Lift the starter from the powerhead. If the spring is still clipped into the lower retainer, release the spring by disengaging the spring tang from the retainer.

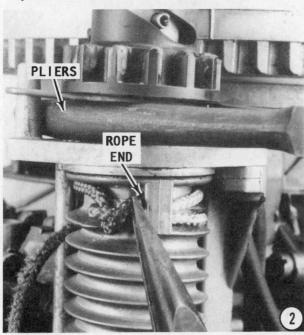

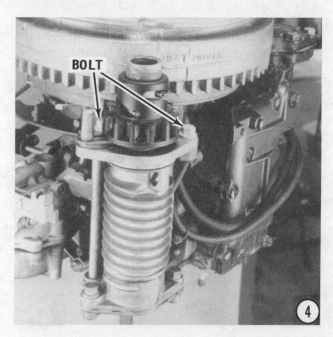

BOLT

⑤

DISASSEMBLING

6- Remove the pin in the pinion gear, and then remove the pinion gear from the collar. Slip the bearing head off the spool. Remove the retainer, installed under the pinion gear, from the starter.

7- Remove the spring retainer (the long tube) from the center of the spool.

8- Pull the starter spring from the spool.

PINION
GEAR

BEARING
HEAD

⑥

SPRING
RETAINER ⑦

CLEANING AND INSPECTING

Wash all parts in solvent, and then dry them with compressed air.

Inspect and replace the main spring if it is damaged or worn. Check the bottom end of the spring very carefully to be sure the two tangs (one on the inner and the other on the outer spring) are in good condition with no sign of distortion.

Inspect the bushing in the bottom collar of the starter housing. This bushing was not removed in the disassembling procedures. Feel with a finger for any roughness, burrs, or other evidence of excessive wear or damage. If the bushing is in good condition, it need not be removed.

Check the rope condition. If the rope is frayed or shows any sign of weakness, it should be replaced. There will never be an easier time to replace the rope than while the starter is disassembled.

Inspect the teeth of the pinion gear. The teeth will show some signs of normal wear. A broken tooth or excessive wear on one side of the teeth is justification for replacement.

Inspect the groove through the pinion gear. This is the groove to accommodate the roll pin. Check to be sure the upper part of the pinion gear is not cracked or distorted.

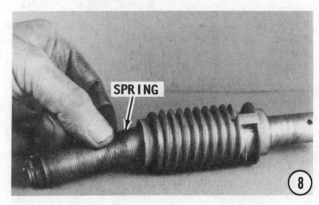

SPRING

⑧

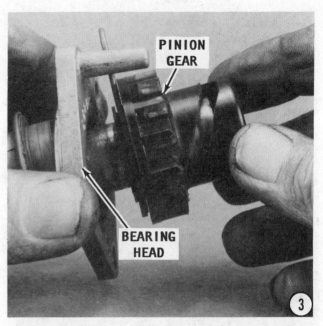

DO NOT lubricate the pinion gear. Oil applied to the pinion gear will attract dirt causing the gear to bind on the spool. Lubricate the upper and lower spool bearing surfaces with just a drop of outboard lubricant. Apply outboard oil to the spring on the pinion gear. **DO NOT** oil the pinion gear bearing or the surfaces of the spring.

ASSEMBLING

GOOD NEWS

Two methods of assembling and installing this starter are presented. The first is the factory suggested procedure and begins with the following steps on this page.

An alternate method is also outlined which many professional mechanics feel is much simplier, easier, and quicker. The alternate method is given on the following page, Steps 1A thru 5A.

After the starter is assembled and installed on the powerhead, continue the work with Step 7.

Factory Method

1- Install the spring retainer from the bottom side. Slide the spring onto the

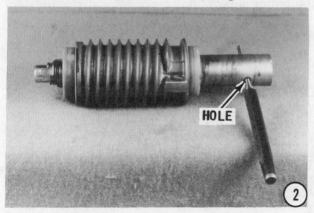

spring retainer. Work the spring upward until the inner spring tang engages the slot on the bottom of the retainer.

2- Align the hole in the retainer with the hole in the spool sleeve.

3- Slip the bearing head and pinion gear down over the spool shaft. Install the roll pin through the pinion gear and sleeve.

4- If the lower bushing was removed, install a **NEW** bushing into the bottom collar on the powerhead.

HELPFUL WORD

As an assist to installation, first soak the bushing in hot water for about ten minutes, and then lubricate it with just a drop of outboard oil.

5- The tang on the outer spring must hook into the slot of the lower spring retainer plate. Pull on the outer spring to elongate the spring, and at the same time, lower

SPRING TANG

5

the spring into the spring retainer plate and hook the tang into the slot in the plate. Rotate the spring **CLOCKWISE** to lock the tang in the plate. Hold upward and turn the spring **CLOCKWISE,** and at the same time tighten the two screws in the lower spring retainer plate.

6- Place the starter assembly in position on the powerhead and start the two upper screws through the upper bearing support. Check to be sure the guide is in place in the bottom and top retainers. Tighten the two bottom retainer screws.

Alternate Assembling Method

1A- If the lower bushing was removed, install a **NEW** bushing into the bottom collar on the powerhead.

BOLT

6

BOTTOM COLLAR

BUSHING

1A

HELPFUL WORD

As an assist to installation, first soak the bushing in hot water for about ten minutes, and then lubricate it with just a dorp of outboard oil.

2A- Install the spring retainer from the bottom side of the spool. Slip the bearing head and pinion gear down the spool shaft.

PINION GEAR

BEARING HEAD

2A

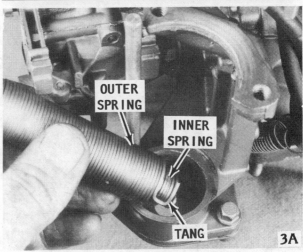

OUTER SPRING

INNER SPRING

TANG

3A

4A

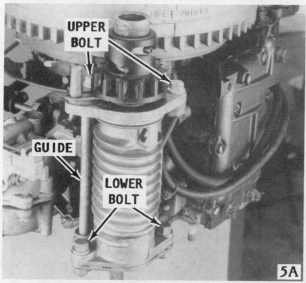

5A

7

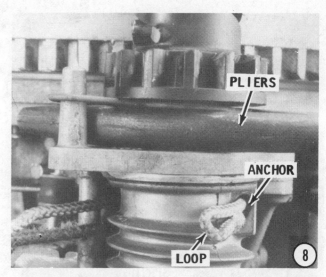

8

Align the holes and install the roll pin through the pinion gear and spool.

3A- Take the spring and lower it into the bottom retainer. Hook the outer spring into the retainer.

4A- Lower the spool assembly down over the spring and engage the spring retainer into the tang of the inner spring.

5A- Start the two upper bolts through the upper bearing suport. Check to be sure the guide is in place in the bottom and top retainers. Tighten the two bottom retainer bolts.

And now the work continues from Step 6

7- Insert a large size screwdriver into the top of the spool, and then rotate the spool, by count, exactly 16-1/2 complete turns.

8- Lift the pinion gear to engage the flywheel ring gear. Hold the pinion gear engaged with the ring gear, and at the same time, slide the handles of a pair of pliers

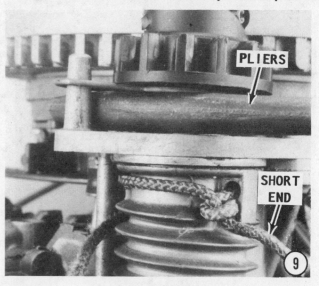

9

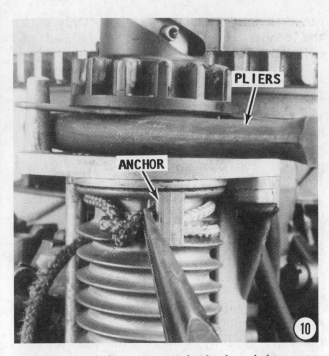

under the pinion gear to lock the pinion gear with the ring gear, as shown.

Feed one end of the new rope through the spool anchor, make a loop, and then thread the end back through the hole in the anchor, but **DO NOT** pull it tight at this time, leave a loop.

9- Bring the short end of the rope through the loop just formed.

10- Work the short end back through the anchor, as shown.

11- Now, pull both ends of the rope tight.

12- Feed the rope through the front engine cowling, and then install the starter rope handle. Hold tension on the rope with the handle and at the same time, remove the pliers from underneath the pinion gear. Allow the starter rope to wind onto the spool. After the rope has been wound onto the spool, the starter handle should be up tight against the cowling. If the handle is not up tight against the cowling the rope was installed too long and needs to be shortened.

9-3 TYPE II STARTER CYLINDER WITH PINION GEAR ALL 9.5 HP

This gear-drive starter is a new design employing the principle of an automotive type starter motor. When the starter rope is pulled, the starter rotates, and a nylon pinion gear slides upward and engages the flywheel ring gear. The gear automatically disengages when the powerhead starts. The ratio between the pinion gear and the ring on the flywheel has been selected to provide maximum cranking speed with minimum pulling effort to ensure fast, easy start.

ROPE REMOVAL

1- Disconnect the high-tension leads from the spark plugs. Ground the high

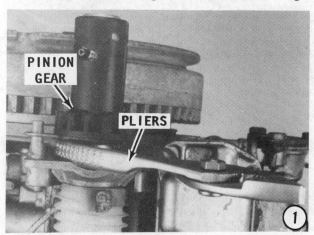

tension leads. Pull the starter rope out until it is fully extended. Now, allow the rope to retract just a little, until the knot end on the spool is facing the port side of the powerhead. Lift the pinion gear to engage the flywheel ring gear. Hold the pinion gear engaged with the ring gear, and at the same time, slide the handles of a pair of pliers under the pinion gear to lock the pinion gear with the ring gear, as shown.

2- Remove the handle from the end of the starter rope. **OBSERVE** how the rope is wound onto the spool and how the rope is secured by a knot. Pull the knot and the rope from the spool.

ROPE INSTALLATION

Rope Purchase Instructions

The length and diameter of the starter rope required will vary depending on the horsepower size of the model being serviced. Therefore, check the Hand Starter Rope Specifications in the Appendix, and then purchase a quality nylon piece of the

proper length and diameter size. Only with the proper rope, will you be assured of efficient operation following installation.

Each end of the nylon rope should be "fused" by burning them slightly with a very small flame (a match flame will do) to melt the fibers together. After the end fibers have been "fused" and while they are still hot, use a piece of cloth as protection and pull the end out flat to prevent a "glob" from forming.

3- Tie a figure **8** knot in one end of the new rope. Feed the rope through the spool anchor around the back of the spool and out the hole in the cowling. Install the handle on the the end of the rope.

4- Pull and hold tension on the rope, and at the same time remove the pliers from under the pinion gear. Allow the starter rope to rewind in a normal manner. After the rope is fully wound onto the starter spool, the rope handle should be up tight against the cowling. If the handle is not up tight, the rope was installed to long or the starter spring is weak and should be replaced.

AUTHOR'S WORD

The exhaust shroud has been removed only for photographic clarity in the accompanying illustrations.

STARTER REMOVAL

1- Pull the starter rope out until it is fully extended. Now, allow the rope to

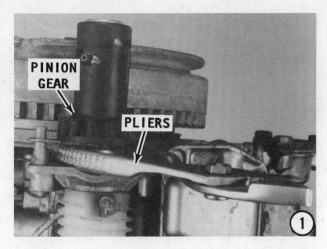

retract just a little, until the knot end on
the spool is facing the port side of the
powerhead. Lift the pinion gear to engage
the flywheel ring gear. Hold the pinion gear
engaged with the ring gear, and at the same
time, slide the handles of a pair of pliers
under the pinion gear to lock the pinion gear
with the ring gear, as shown.

2- Remove the handle from the end of
the starter rope. **OBSERVE** how the rope is
wound onto the spool and how the rope is
secured by a knot. Pull the knot and the
rope from the spool.

3- Grasp the spool firmly and remove
the pliers from under the pinion gear. Now,
allow the spool to slip a little at a time until
the spring is completely unwound.

4- Remove the two retaining bolts on
top of the starter. Lift the starter assembly
from the powerhead.

OBSERVE

As the starter is removed, take special
note of how the upper bearing retainer extends over a hole in the powerhead. This

hole is a water passage. A gasket is installed under the retainer to form a seal for the
water passage. This gasket may remain on
the block or come with the starter retainer
as the starter is removed. To ensure a good
seal, the gasket should be discarded and
replaced with a new one at time of installation.

5- Remove the roll pin extending
through the pinion gear.

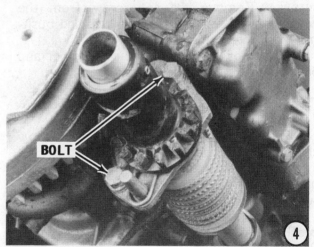

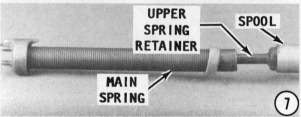

6- Remove the pinion gear, spring and bearing head from the spool.

7- Pull the main spring and upper spring retainer from the spool.

8- Notice the lower spring retainer on the bottom of the spool secured with a set screw. Remove the set screw from the retainer and then pull the retainer and bushing from the spring.

9- Remove the upper spring retainer and outer bearing from the spring.

CLEANING AND INSPECTING

Wash all parts in solvent, and then blow them dry with compressed air.

Inspect the main spring. Check the tab on the bottom end of the inner spring to be sure it is not bent or cracked.

Inspect the teeth of the pinion gear. The teeth will show some signs of normal wear. A broken tooth or excessive wear on one

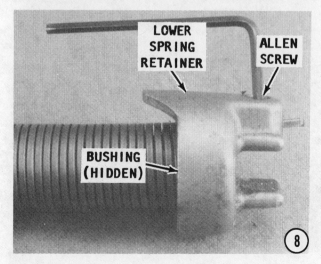

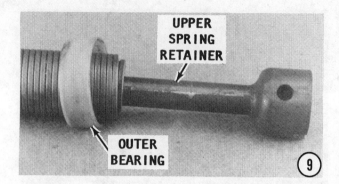

side of the teeth is justification for replacement.

Inspect the groove through the pinion gear. This is the groove to accommodate the roll pin. Check to be sure the upper part of the pinion gear is not cracked or distorted.

DO NOT lubricate the pinion gear, spring, or spool. Oil applied to these parts will attract dirt causing the gear to bind on the spool. Lubricate the upper and lower spool bearing surfaces with just a drop of outboard lubricant. Apply outboard oil to the spring on the pinion gear. **DO NOT** oil the pinion gear bearing or the surfaces of the spring.

Check the rope condition. If the rope is frayed or shows any sign of weakness, it should be replaced. There will never be an

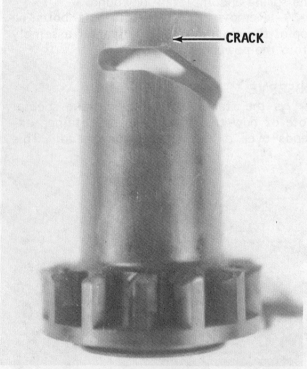

Pinion gear from a hand starter. Notice the crack on the top side, indicated by the callout. This unit must be replaced.

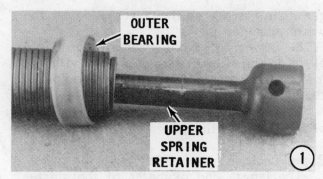

easier time to replace the rope than while the starter is disassembled.

ASSEMBLING

1- Place the outer bearing and the upper spring retainer onto the main spring.

2- Slide the bushing over the main spring. Guide the tang on the main spring through the hole in the main spring retainer.

3- Install the set screw securing the inner spring to the lower retainer. Set the completed spring assembly aside.

4- Slide the outer bearing down over the outer spring, as shown. Insert the spring retainer through the center of the inner spring from the **TOP**. Rotate the inner spring until the "hook" on the end of the spring seats in the groove of the spring retainer.

Insert the assembled springs and retainer from Step 3, into the bottom of the spool.

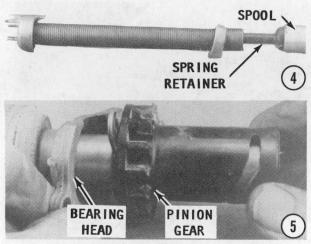

5- Install the bearing head, spring, and pinion gear down over the spool shaft.

6- Align the hole in the upper spring retainer with the hole in the starter spool and with the slot in the pinion gear.

INSTALLATION

AUTHOR'S WORD

For the following illustrations, the engine exhaust shroud was removed, only for photographic clarity.

7- Cover both sides of a **NEW** gasket with Perfect Seal No. 4, and then place the

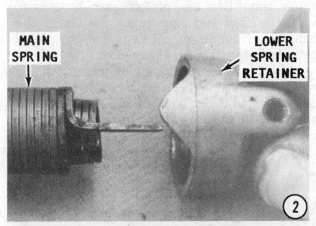

LOWER SPRING RETAINER

PIN

8

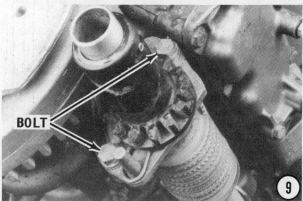

BOLT

9

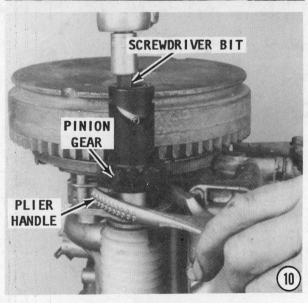

SCREWDRIVER BIT

PINION GEAR

PLIER HANDLE

10

gasket in position on the powerhead over the water passage.

8- Lower the spool assembly down into place on the side of the powerhead with the lower spring retainer indexed over the pin in the bottom portion of the housing.

9- Install the two bolts securing the bearing head to the powerhead.

Rope Purchase Instructions

The length and diameter of the starter rope required will vary depending on the horsepower size of the model being serviced. Therefore, check the Hand Starter Rope Specifications in the Appendix, and then purchase a quality nylon piece of the proper length and diameter size. Only with the proper rope, will you be assured of efficient operation following installation.

Each end of the nylon rope should be "fused" by burning them slightly with a very small flame (a match flame will do) to melt the fibers together. After the end fibers have been "fused" and while they are still hot, use a piece of cloth as protection and pull the end out flat to prevent a "glob" from forming.

10- Insert a large screwdriver blade into the top of the spool, and then rotate the spool **COUNTERCLOCKWISE**, by count, 20-1/2 complete turns. After the required numbers of turns have been made, hold pressure on the screwdriver, and at the same time lift the pinion gear to engage with the flywheel ring gear, and insert the handles of a pair of pliers under the pinion gear to hold the gear engaged.

11- Tie a figure **8** knot in one end of the new rope. Feed the rope through the spool anchor around the back of the spool and out the hole in the cowling. Install the handle on the the end of the rope.

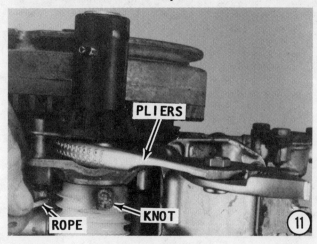

PLIERS

ROPE

KNOT

11

12- Pull and hold tension on the rope, and at the same time remove the pliers from under the pinion gear. Allow the starter rope to rewind in a normal manner. After the rope is fully wound onto the starter spool, the rope handle should be up tight against the cowling. If the handle is not up tight, the rope was installed too long or the starter spring is weak and should be replaced.

9-4 TYPE III STARTER
COIL SPRING WITH SWING ARM
VERTICAL MOUNT
4HP 1971-78
DELUX 4HP 1982 AND ON
4.5HP 1980-84

This type hand starter is a flat type mounted on the port side of the powerhead. As the rope is pulled, a swing arm moves the drive gear upward to engage with the teeth of the flywheel ring gear. A coil spring winds and tightens as the rope unwinds. The spring then coils the rope around a pulley as the rope handle is returned to the front of the cowling.

The coil spring consists of a lengthy piece of spring steel (approximately 12-feet, tightly wound inside a housing (the cup and stop assembly). Movement of the drive gear to the retracted position is accomplished through a second spring.

The starter must be disassembled to replace the rope.

WARNING
The rewind spring is under tremendous tension and is a potential hazard. Therefore, **SAFETY GLASSES** should be worn and extreme **CARE** exercised to follow the procedures carefully during disassembling and asembling work with the starter.

SAFETY WORDS
Work on the starter can be very dangerous. Because approximately 12-feet of spring steel is tightly wound into about a 4-inch housing, the spring is placed under tremendous tension -- a real tiger in a cage. If the spring should accidently be released, severe personal injury could result from being struck by the spring with force. Therefore, the service instructions **MUST**, and we say again **MUST**, be followed closely to prevent release of the spring at the wrong time. Such action would be a **BAD SCENE**, a very **BAD SCENE**, because serious personal injury could result.

The starter rope should **NEVER** be released from the extended position. Such action would allow the spring to wind with incredible speed resulting in serious damage to the starter mechanism.

REMOVAL

1- Remove the spark plugs and ground the high tension leads. Pull the starter rope out, and then tie a knot in the rope behind the handle. Allow the rope to rewind until the knot is against the cowling. Untie the knot in the end of the rope, and then remove the handle and the rubber bumper.

2- Remove the knot tied in the rope in Step 1. Allow the rope to **SLOWLY** wind into the starter. Before the rope end passes

ROPE HANDLE

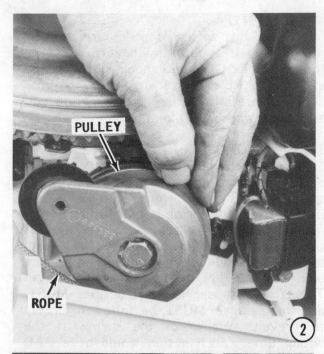

PULLEY

ROPE

②

SPRING

③

BOLT

④

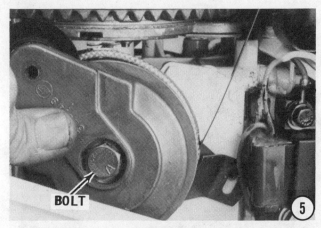

BOLT

⑤

the cowling, firmly grasp the starter pulley, and then allow the starter to unwind.

3- Observe the back side of the starter. Notice the hook of the starter spring protruding out of a hole in the starter. Grasp the spring hook with a pair of needle-nose pliers, and then pull the spring out as far as possible, to relieve tension on the spring.

4- Remove the 3/8" bolt from the bracket between the starter and the exhaust housing.

5- Hold the starter together with one hand, and at the same time **LOOSEN** the large bolt from the center of the starter. **DO NOT** remove this bolt at this time. Remove the starter from the powerhead.

6- If the starter is only removed in order to accomplish other work, install a 3/8" x 16 nut onto the far side of the thru-bolt to hold the starter together and prevent the spring from escaping.

DISASSEMBLING

7- Remove the center bolt, idler gear arm, and the idler gear arm spring.

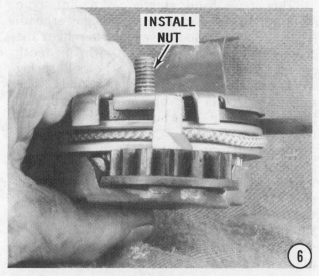

INSTALL NUT

⑥

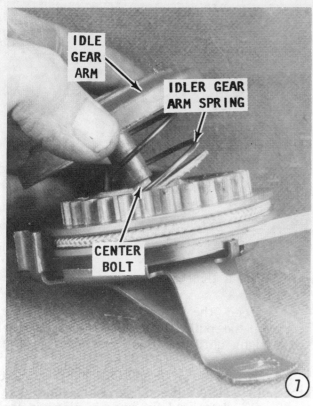

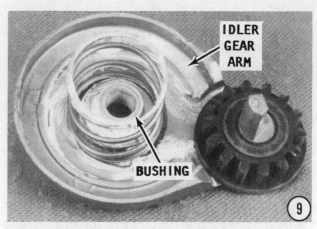

be released from the cup. After the pulley has been removed, notice the position of the spring loop. Remove the rope from the pulley. Notice how the rope unwinds from the pulley **COUNTERCLOCKWISE**.

9- Remove the bushings from the idler gear arm and the bushings installed one on each side of the pulley.

10- Two different methods are suggested to remove the spring from the starter cup. One method involves pulling continuously on the end of the spring that contains the loop. The second method is to simply toss the cup a safe distance onto carpeting or a lawn, allowing the spring to be released instantly from the cup.

If this second method is used, be sure the spring will not cause a threat to any individual in the area when it is released.

CLEANING AND INSPECTING

Wash all parts except the rope in solvent and then blow them dry with compressed air.

SAFETY WORDS

The next step could be dangerous. Removing the pulley from the cup **MUST** be done with care to prevent personal injury.

8- Lift the pulley **SLIGHTLY** and then use a screwdriver and work the spring free of the pulley. **DO NOT** allow the spring to

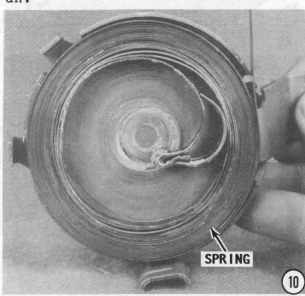

Remove any trace of corrosion and wipe all metal parts with an oil dampened cloth.

Inspect the starter spring end loops. Replace the spring if it is weak, corroded or cracked.

Inspect the rope. Replace the rope if it appears to be weak or frayed. If the rope is frayed, check the hole through which the rope passes for rough edges or burrs. Remove the rough edges or burrs with a file, and polish the surface until it is smooth.

Inspect the dog ears of the pulley gears to be sure they are not worn and are free of burrs. Check the idler gear for cracks and missing teeth.

ASSEMBLING

Rope Purchase Instructions

The length and diameter of the starter rope required will vary depending on the horsepower size of the model being serviced. Therefore, check the Hand Starter Rope Specifications in the Appendix, and then purchase a quality nylon piece of the proper length and diameter size. Only with the proper rope, will you be assured of efficient operation following installation.

Each end of the nylon rope should be "fused" by burning them slightly with a very small flame (a match flame will do) to melt the fibers together. After the end fibers have been "fused" and while they are still hot, use a piece of cloth as protection and pull the end out flat to prevent a "glob" from forming.

1- Tie a figure **8** knot in one end of the rope.

2- Feed the other end of the rope through the pulley hole, and then pull the rope tight until the knot is seated in the pulley.

3- Wind the rope **CLOCKWISE** around the pulley. Use a piece of masking tape or a rubber band to hold the rope in place in the pulley.

4- Coat the bushings with a light film of OMC Type A lubricant. Insert one bushing into the pulley, another bushing into the idler gear arm, and two more bushings, one on each side of the pulley.

5- If the spring was **NOT** removed from the cup, lower the pulley down over the spring and insert the end of the spring into the spring anchor post of the pulley. If the spring **WAS** removed from the cup, hook the end of the spring into the pulley and then

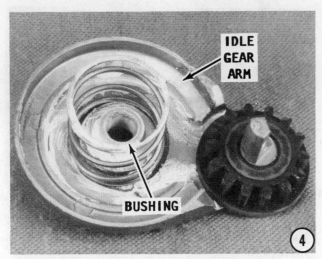

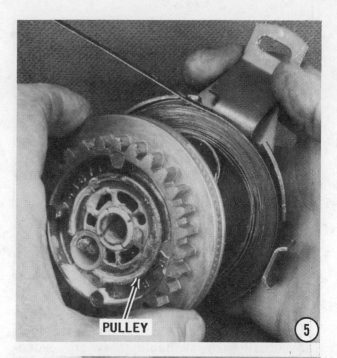

allow the spring to come out the slot in the cup. Turn the pulley and wind the spring into the cup until about 1/2 of the spring length has been wound. Hold the gear and allow it to back off slowly. **DO NOT** allow the spring to rewind quickly. The remainder of the spring will be installed later.

6- Assemble the idler gear with the shoulder against the idler gear arm. Install the idler gear arm and spring to the pulley and cup with the stop on the underside of the idler gear shaft located between the upper stop and the lower stop on the cup and stop assembly.

7- Hold the assembly together and install the assembly onto the powerhead. Thread the shoulder bolt into place first. This bolt will hold the starter assembly together.

8- Install the bolt through the idler arm and into the exhaust manifold. **DO NOT** tighten this bolt at this time. Apply a light coating of OMC Type A lubricant to the portion of the spring extending out of the starter.

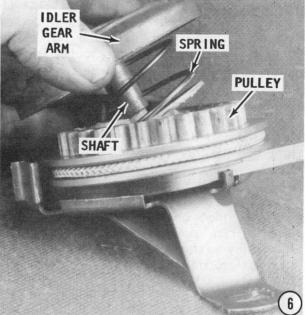

Housing (right) with the spring properly installed and the spring end bent toward the center. The pin in the pulley (left) must index into the loop on the spring end during installation.

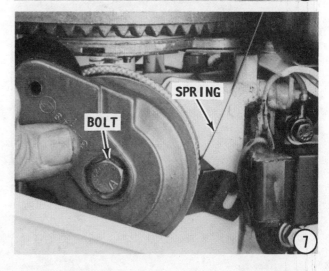

BOLT

⑧

SPECIAL WORDS

Special tools are available to lock-in the starter to the flywheel. However, the tools are usually not available; they are expensive; and professional mechanics have developed an alternate method. The procedure will take time and patience, but it is the only way without the special tools. To work without the special tools proceed as follows:

9- Remove the rubber band or the masking tape from the coiled rope. Working from the rear of the powerhead, pull on the rope and the idler gear will engage with the flywheel ring gear. Continue to pull the rope, and at the same time, work the spring down into the cup. If the rope becomes fully extended, before the spring is installed into the cup, allow the rope to rewind onto the pulley as far as possible and then wind the rope around the pulley again. Now, pull

SPRING

⑨

ROPE

⑩

on the rope again from the back side of the rear of the powerhead, and continue to work the spring into the cup until it is completely installed.

10- Ease back on the rope until there is no spring tension on the starter. Thread the rope into the pulley **CLOCKWISE** around the starter. Two, or possibly more, loops may be required to accomplish the task. Use all of the rope in the pulley with the starter in the relaxed position. After all of the rope has been fed into the pulley, grab the end of the rope in front of the starter and pull it out, then feed it through the cowling at the front of the powerhead. Continue to pull the rope until about two feet is extending out through the cowling. Tie a slip knot in the rope.

11- Install the rubber bumper and handle onto the rope. Tie a figure **8** knot in the end of the rope, and then pull the knot into the handle.

⑪

ROPE HANDLE

12

12- Untie the slip knot and ease the rope back into the starter. The starter handle must be up tight against the cowling when the rope is completely rewound on the starter pulley. If the rope is not tight against the cowling, remove the knot and handle from the rope, and then wind the rope around the pulley one complete turn. Tie another knot in the end of the rope as described earlier in this step and then check to be sure the handle is tight against the cowling when the rope is wound onto the pulley.

Starter Adjustment

13- Hold the idler gear arm stop against the cup stop. Fully engage the idler gear teeth with the teeth in the flywheel ring gear. Tighten the cup and stop assembly screw. Tighten the shoulder screw securely.

BOLT

13

9-5 TYPE IV STARTER COIL SPRING WITH SLIDING GEAR HORIZONTAL MOUNT ALL 9.9 HP ALL 15 HP

This pinion gear starter is a design employing the principles of an automotive-type starter motor. A nylon pinion gear slides upward and engages the flywheel ring gear as the starter rope is pulled. The pinion gear automatically disengages when the engine starts. The ratio between the pinion gear and the ring gear was selected to provide maximum cranking speed with minimum pulling effort to ensure fast and easy powerhead start.

A lockout pawl linked to the cam follower prevents manual starter engagement if the throttle is advanced beyond the start position.

REMOVAL

SPECIAL NOTE

If the only work to be performed on the starter is replacement of the rope, perform Steps 1 thru 17, then jump to Page 9-31, Starter Installation. If the starter is to rebuilt, perform Steps 1 thru 4, then jump to Step 18.

1- Disconnect the high-tension leads from the spark plugs. Ground the high-tension leads. Pull the rope out far enough, and then tie a knot in the rope. Allow the rope to rewind until the knot is against the front of the cowling.

2- Remove the handle rope anchor and then the rope from the handle. Remove the rubber bumper.

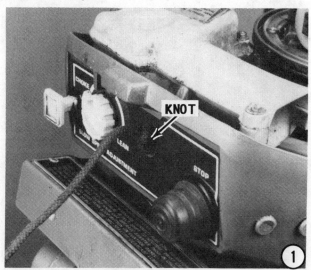

KNOT

1

3— Untie the knot in the rope and allow the rope to **SLOWLY** wind onto the pulley and the spring to unwind. When the rope is about to pass through the opening in the cowling, hold the pulley, and then allow your grip to slip to permit the rope to continue winding onto the pulley, and the spring to unwind slowly. **DO NOT** release the grip on the pulley completely. If the spring rewinds too rapidly the spring may be damaged along with other parts.

WARNING

The rewind spring is under tremendous tension and is a potential hazard. Therefore, **SAFETY GLASSES** should be worn and extreme **CARE** exercised to follow the procedures carefully during removal, disassembling, and assembling work with the starter.

4— HOLD the pulley and cup together, and at the same time, loosen the center mounting bolt and back it out of the intake manifold. **DO NOT** remove the center bolt from the starter. Continue to hold the starter **FIRMLY** together and carefully re-

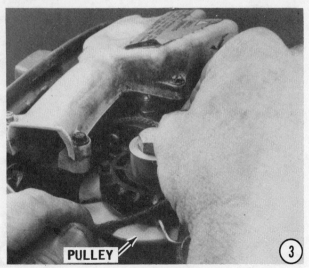

move it from the powerhead. Take care not to damage the starter pawl.

5— If the starter is only removed in order to accomplish other work, install a 3/8" x 16 nut onto the far side of the thru-bolt to hold the starter together and prevent the spring from escaping.

STARTER ROPE REPLACEMENT

REMOVAL

6— Clamp the starter in a vise, as shown. Tighten the vise just **SLIGHTLY** onto the cup bushing. Remove the center bolt.

7— Remove the pinion spring and gear from the pulley.

8— **EXERCISE CARE** during this next procedure. Slide a putty knife or other similar flat tool in between the bottom side of the pulley and top side of the spring. Using the tool, work the pulley off the

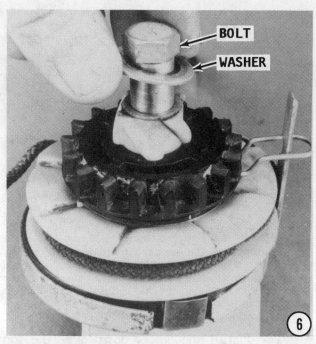

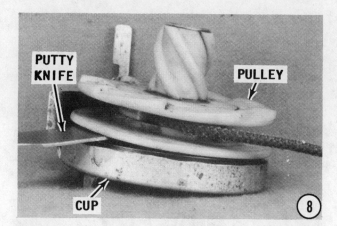

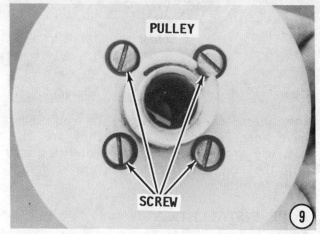

spring **WITHOUT** allowing the spring to escape from the cup.

9- Remove the four screws from the back side of the pulley, and then separate the pulley. Remove the rope.

INSTALLATION

Rope Purchase Instructions

The length and diameter of the starter rope required will vary depending on the horsepower size. of the model being serviced. Therefore, check the Hand Starter

Rope Specifications in the Appendix, and then purchase a quality nylon piece of the proper length and diameter size. Only with the proper rope, will you be assured of efficient operation following installation.

Each end of the nylon rope should be "fused" by burning them slightly with a very small flame (a match flame will do) to melt

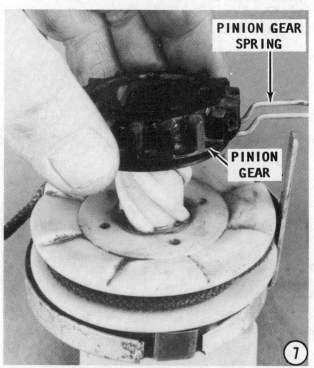

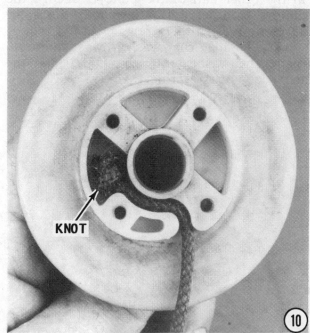

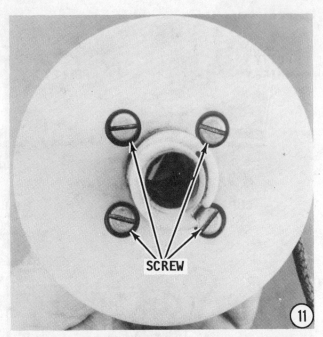

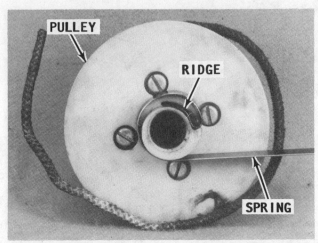

View showing the spring indexed into the pulley. Installation is not performed in this manner. This photo merely illustrates how the pulley ridge indexes into the loop in the spring when the pulley is installed, as described in the text.

the fibers together. After the end fibers have been "fused" and while they are still hot, use a piece of cloth as protection and pull the end out flat to prevent a "glob" from forming.

ROPE INSTALLATION

10- Tie a figure **8** knot as close to the end of the rope as possible, as shown. After the knot has been tied, feed the knot into the recess of the pulley.

11- Position the other half of the pulley over the rope. Rotate the cap slightly to align the four holes in the cap with the holes in the other half of the pulley. Secure the pulley together with the four retaining screws. Tighten the screws securely.

12- If the cup washer was removed, place the washer into the cup under the inner loop of the spring.

13- Install the pulley down over the top of the spring. Check to be sure the end of the spring is engaged in the pulley slot where the rope is installed. **DO NOT** lubricate the pulley or the pinion gear.

14- Install the pinion gear and the pinion spring onto the pulley.

15- Lubricate the mounting screw and washer with OMC outboard oil, and then install the screw through the pulley and cup.

16- Thread a 3/8" x 16 nut onto the mounting bolt to hold the pulley and spring in the cup until the starter is installed on the powerhead.

17- Wind the rope onto the pulley **COUNTERCLOCKWISE.**

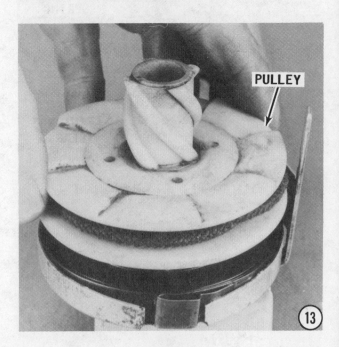

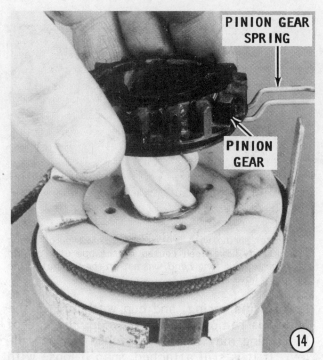

SPECIAL NOTE

If the only work to be performed is replacement of the starter rope, proceed directly to Step 8, under Starter Installation, to install the starter onto the powerhead.

DISASSEMBLING

18– After the starter has been removed as outlined in Steps 1 thru 4 of this section,

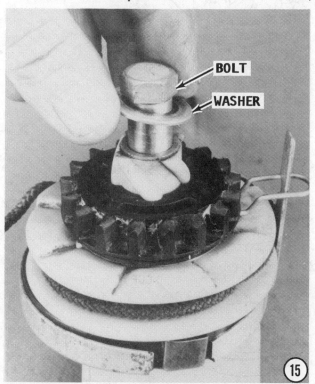

thread a 3/8" x 16 nut onto the center thru-bolt. Clamp the starter in a vise with the mounting screw secured between the vise jaws.

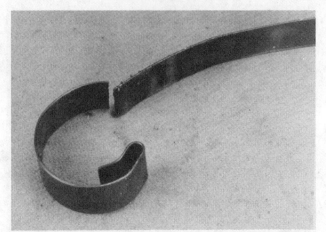

Starter spring with broken end. This type of spring damage may have been caused by the operator pulling the rope too far (one hand was not on the engine) or the rope may have broken.

19- Use a screwdriver to engage the spring loop, and then pull the spring out of the cup until it is completely unwound.

20- Release the starter from the vise. Remove the nut, mounting bolt, and washer. Remove the pinion gear, pinion spring, pulley, rewind spring, and cup washer. As the pulley is removed from the cup, the end of the spring may remain attached to the pulley. If it is still attached, snap it loose with a screwdriver.

ROPE REPLACEMENT

If the rope is to be replaced during the disassembling work, perform Steps 4 thru 16, at the begining of this section.

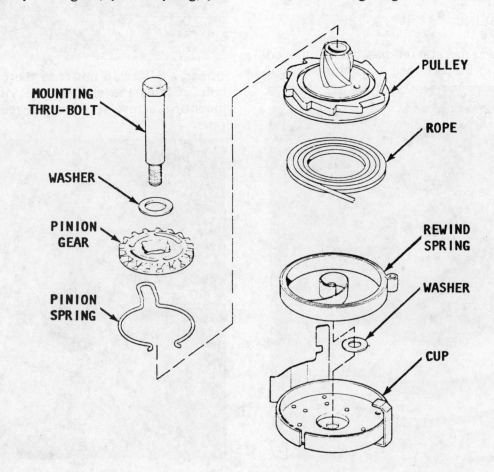

Broken outside end of a spring. This failure was most likely the result of metal fatigue -- age, and cannot be prevented.

CLEANING AND INSPECTING

Wash all parts except the rope in solvent and then blow them dry with compressed air.

Remove any trace of corrosion and wipe all metal parts with an oil dampened cloth.

Inspect the starter spring end loops. Replace the spring if is is weak, corroded or cracked.

Inspect the rope. Replace the rope if it appears to be weak or frayed. If the rope is frayed, check the hole through which the rope passes for rough edges or burrs. Remove the rough edges or burrs with a file, and polish the surface until it is smooth.

Inspect the pinion gear and pulley for wear, chipped or broken teeth. Inspect the

A distorted starter cup no longer fit for service. This type damage was probably caused because the rope broke and the spring rewound with such incredible speed the cup was damaged. The cup material is not capable of withstanding such force from within.

cup for corrosion or damage, such as being warped out of shape.

SPECIAL GOOD WORDS

If the rope is to be replaced during the disassembling work, perform Steps 4 thru 16, at the beginning of this section.

STARTER ASSEMBLING

The following procedures pickup the work after a new rope has been installed according to Steps 4 thru 16 of this section.

1– Coat the inside surface of the cup with OMC Type A Lubricant. Position the rewind spring into the cup, as shown. Place the cup washer into the cup.

2– Install the pulley with the spring loop engaging the pulley, as shown. Check to be sure the spring feeds out of the cup slot.

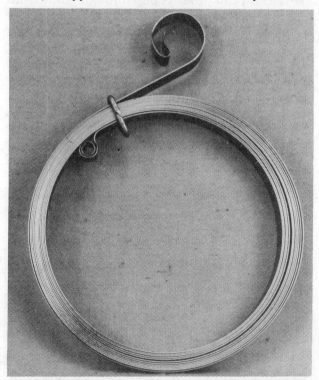

A new spring as it appears when purchased from the marine dealer. Notice the loop on the outside. Prior to installation, the spring must be rewound with the large loop on the inside.

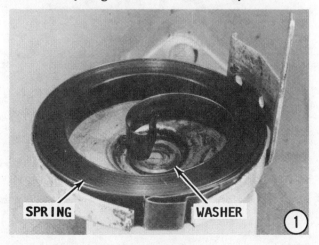

SPRING   WASHER   ①

SLOT

2

3- Install the pinion gear onto the pulley.

4- Coat the threads of the mounting screws with OMC Outboard Oil. Insert the center thru-bolt through the pulley and cup.

5- Thread a 3/8" x 16 nut onto the bolt to hold the parts together.

6- Hold the cup and at the same time wind the spring into the cup by rotating the pulley **COUNTERCLOCKWISE** as viewed from the top of the pulley. As soon as excessive resistance is felt during the winding, feed the spring into the cup through the slot to relieve spring tension. Continue to wind and feed the spring into the cup until the loop on the end of the spring is drawn up tight against the side of the cup.

7- Wind the rope **COUNTERCLOCKWISE** around the pulley. Install the pinion gear spring. Hold the spring in place with a rubberband or piece of string.

STARTER INSTALLATION

8- Hold the starter pulley and cup together and position the assembly in place on the powerhead, as shown. Thread the mounting screw into the manifold and tighten the screw securely.

9- Thread the rope through the front cowling and pull it all the way out. With the rope fully extended, the starter spring end must be free to extend a minimum of 1/2" from the cup. If the spring is not free to extend 1/2" from the cup, allow the starter rope to fully rewind onto the pulley. After the rope has rewound onto the pulley, release one full turn of the rope from the pulley and make the test again. Repeat the procedure until the spring is free to extend

PINION GEAR

PINION GEAR SPRING

3

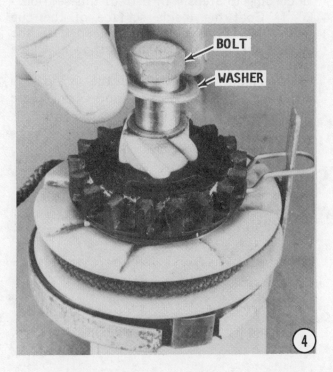

BOLT

WASHER

4

NUT

BOLT

5

a minimum of 1/2" from the cup when the rope is fully extended. With the rope still through the front of the cowling tie a knot in the rope. Allow the rope to rewind until the knot is tight against the cowling.

10- Install the rubber bumper, handle and anchor onto the rope. Secure the rope by pressing the anchor into the handle. Remove the knot from the rope and allow the rope to rewind onto the pulley. When the rope is fully rewound, the handle should be up tight against the cowling.

STARTER ADJUSTMENT

11- Place the shift lever in the **NEU-TRAL** position, and the throttle in the **START** position. Check to see if the lockout pawl clears the highest point on the starter pulley by 0.050" to 0.110" (1.27 to 2.79 mm). This clearance is required to allow power-head start with the shift mechanism in gear and the throttle in the start position. To adjust, loosen the nut and rotate the adjust-

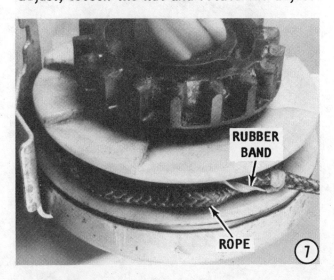

ing screw inward or outward until the proper clearance is obtained. Tighten the nut securely to hold the adjustment. To check, the hand starter should crank the powerhead with the shift mechanism in either **FOR-WARD** or **REVERSE** gear and with the throt-

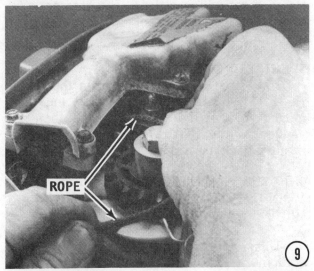

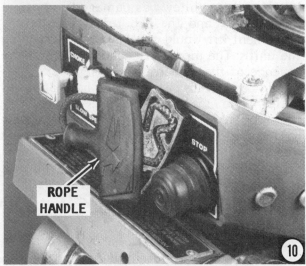

ADJUSTMENT

FEELER GAUGE

11

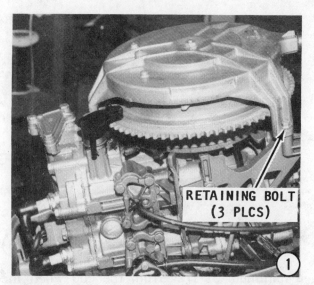

RETAINING BOLT (3 PLCS)

1

tle in the **START** position. Move the throttle to the **FAST** position. The starter lockout pawl should prevent the starter from cranking the powerhead.

SPECIAL WORDS

To restart an powerhead from the throttle **FAST** position, the throttle must be moved to the **SLOW** position and then advanced to the **START** position in order to remove backlash from the lockout linkage.

9-6 TYPE V MOUNTED ATOP FLYWHEEL MODEL WITH NO RETURN SPRINGS 40 HP 1971-76 AND 1981-85

This type starter is installed as original equipment by the manufacturer. However, if the starter pulley was damaged sometime in the past, and the hub replaced, the replacement kit would contain parts modifying the unit. The most noticeable change is the absence of the pawl return springs.

However, it is possible, and has happened quite often that the previous owner may have replaced the starter unit with a unit having the pawl return springs. Possibly this was the only unit available from the local dealer. Procedures to service this type unit is covered in Seloc's Johnson/Evinrude Volume I manual.

WARNING

As with other types of hand starters, the rewind spring is a potential hazard. The

spring is under tremendous tension when it is wound -- a real tiger in a cage. If the spring should accidentally be released, severe personal injury could result from being struck by the spring with force. Therefore, the service instructions **MUST**, and we say again **MUST**, be followed closely to prevent release of the spring at the wrong time. Such action would be a **BAD SCENE**, a very **BAD SCENE**, because serious personal injury could result.

The starter rope should **NEVER** be released from the extended position. Such action would allow the spring to wind with incredible speed, resulting in serious damage to the starter mechanism.

Any time the rope is broken, the starter spring will rewind with incredible speed. Such action will cause the spring to rewind past its normal travel and the end of the

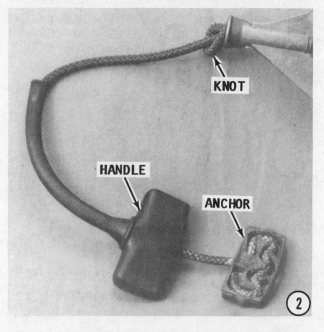

KNOT

HANDLE

ANCHOR

2

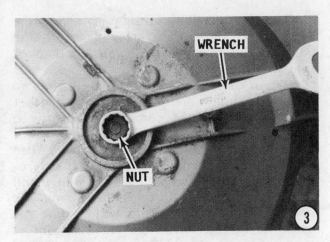

spring will be bent back out of shape. Therefore, if the rope has been broken, the starter must be completely disassembled and the spring repaired or replaced.

STARTER REMOVAL

1- Disconnect any linkage between the starter and the carburetor. Move the linkage out of the way. Remove the starter leg retaining bolts, and then lift the complete starter from the powerhead.

2- Pull the rope out far enough, and then tie a knot in the rope. Allow the rope to rewind to the knot. Work the rope anchor out of the rubber covered handle, then remove the rope from the anchor. Remove the handle from the rope. Untie the knot in the rope, and then hold the disc pulley, but permit it to turn and thus allow the rope to wind back onto the pulley **SLOWLY**. Continue to allow the spring in the pulley to unwind **SLOWLY** until all tension has been released.

3- Remove the center nut from the top side of the starter. Some models do not

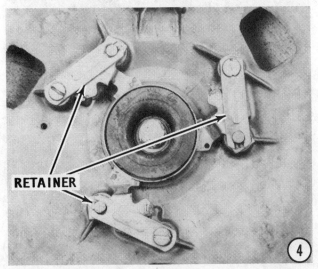

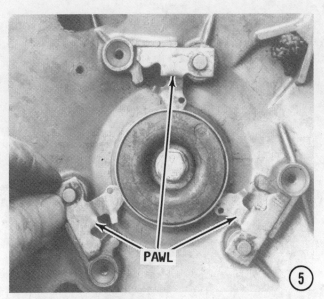

have a center nut. On other models, the nut may have vibrated loose, but if the center bolt has threads showing, a nut **MUST** be installed during assembling.

4- Lay the starter on its back on a work surface. Remove the three screws securing the pawl retainers to the pulley.

5- Remove the three pawls from their retainers.

6- Remove the center bolt from the hub.

7- Hold the pulley, and at the same time, remove the spindle, the wavy washer, friction ring, and the nylon bushing from the center of the pulley.

WARNING

The rewind spring is a potential hazard. The spring is under tremendous tension when

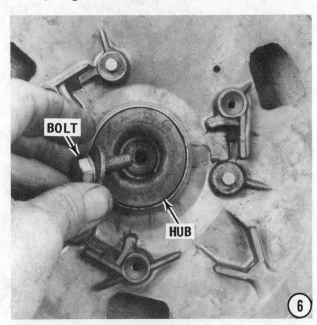

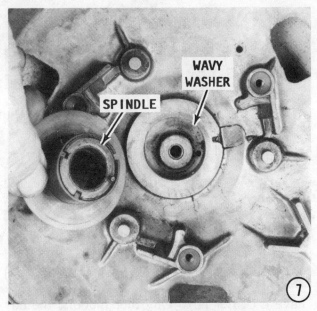

it is wound -- a real tiger in a cage. If the spring should accidentally be released, severe personal injury could result from being struck by the spring with force. Therefore, the following step **MUST** be performed with care to prevent personal injury to self and others in the area. If the spring should be accidently released at the wrong time, such action would be a **BAD SCENE**, a very **BAD SCENE**, because serious personal injury could result.

8- Lift the pulley straight up and at the same time work the spring free of the pulley. The spring has a small loop hooked into the pulley.

9- An alternate and safe method is to hold the pulley and the housing together tightly and turn the complete assembly so the legs are facing downward. Now, lower the complete assembly to the floor. When the legs make contact with the floor, release the grip on the pulley. The pulley will fall and the spring will be released from the housing, but the three legs will contain the spring and prevent it from traveling across the room. If the spring was not released from the housing, the only safe method is to jar the three legs on the floor to release the spring. Unwind the rope out of the pulley groove, and then pull it free.

SAFETY WORD

If the spring was not released from the housing, the only safe method is to jar the three legs on the floor a second or third time to release the spring. Check to be sure **ALL** of the spring has been released. If some of the spring is hung up in the housing, tap on the top of the housing with a mallet while the legs are still on the floor.

CLEANING AND INSPECTING

If the rope was broken and the spring is bent backward, as shown in the accompanying illustration, it is a simple matter to bend the spring end back to its normal position. The next illustration clearly shows a spring end properly positioned in the housing.

Wash all parts except the rope in solvent and then blow them dry with compressed air.

Remove any trace of corrosion and wipe all metal parts with an oil dampened cloth.

Inspect the rope. Replace the rope if it appears to be weak or frayed. If the rope is frayed, check the hole through which the rope passes for rough edges or burrs. Remove the rough edges or burrs with a file, and polish the surface until it is smooth.

Insepct the starter spring end loops. Replace the spring if it is weak, corroded or cracked. Check the spring pin located at the back side of the pulley to be sure it is straight and solid.

Check the inside surface of the housing and remove any burrs.

Check the condition of the pawl springs to be sure they are not stretched out of shape. The end of each spring should be bent back toward the coil of the spring. Inspect the pawls for wear and that the edges are not rounded.

Inspect the hub center locating pin to be sure it is straight and tight.

STARTER ASSEMBLING

GOOD WORDS

The accompanying illustration shows a new starter spring as it is purchased. Note how the spring is held wound with "hog rings". The spring **MUST** be released to its full extended position before it can be installed. Therefore, use care and remove the "hog rings" and allow the spring to unwind

The rope on this unit broke, causing the spring to rewind with incredible speed. The end of the spring was bent back in the wrong direction.

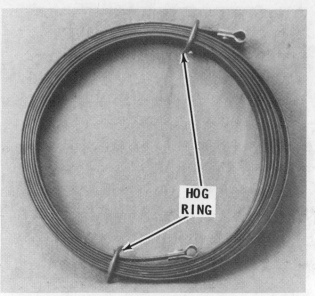

A new spring as it appears direct from the marine store. The hog rings must be CAREFULLY removed, as described in the text.

until it is a straight piece of spring steel.

SAFETY WORD

Wear a good pair of gloves while unwinding and installing the spring. The spring will develop tension and the edges of the spring steel are sharp. The gloves will prevent cuts on your hands and fingers.

A safe method is for one person to remove the hog rings while an assistant holds the spring. After the rings have been removed, both persons work to unwrap the spring, one coil at-a-time.

1- Slide the spring onto the outer pin and then start the spring from the outside

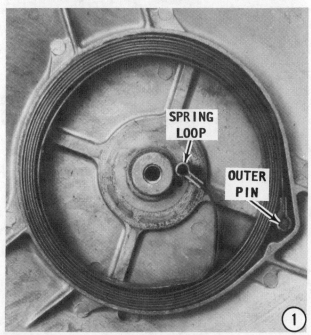

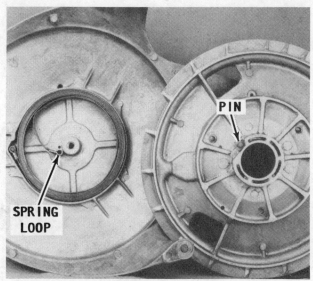

Housing (left) with the spring properly installed and the spring end bent toward the center. The pin in the pulley (right) must index into the loop on the spring end during installation.

edge of the housing and insert it into the housing **COUNTERCLOCKWISE**, as shown in the accompanying illustration. Notice the small hump in the housing. This hump prevents the spring from being wound in the wrong direction. Work the first turn into the housing, and then hold the spring down with one hand and continue to wind the spring into the housing. Patience and time are required to work the spring completely into the housing. After the last portion is in

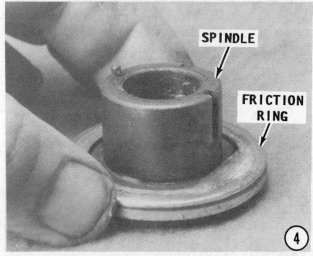

place, bend the end of the spring towards the center of the housing. This position will allow the pulley pin to align with the loop in the end of the spring, when the pulley is installed.

2- Lower the pulley down over the top of the spring with the pulley pin indexing into the loop in the end of the spring. In the accompanying illustration, notice the callout for the boss on the backside of the pulley. The pin is located directly under the boss. The boss can, therefore, be a guide during pulley installation.

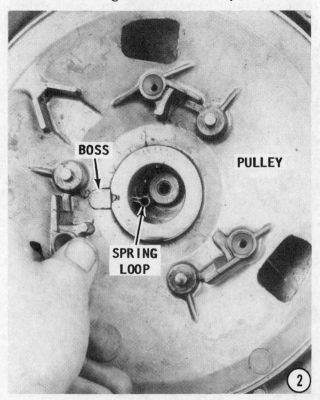

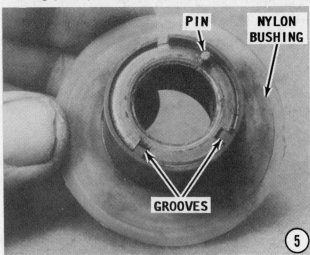

3- Coat the spindle with a thin film of OMC Type A lubricant. Place the wavy washer onto the spindle.

4- Slide the friction ring onto the spindle with the flats in the washer indexed with the flats on the spindle.

5- Install the nylon bushing onto the spindle with the protrusions on the bushing indexed into the slots in the spindle. Notice the pin protruding from the bottom of the spindle. This pin **MUST** drop into the hole in the starter housing. Observe into the housing and visually locate this hole.

6- Lower the spindle assembly down through pulley and index the pin into the hole in the housing.

7- Place the washer inside the spindle housing and then install the bolt through the washer into the housing. Tighten the bolt securely.

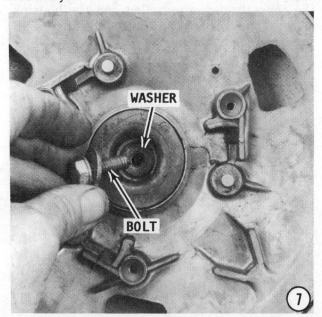

8- Install the three pawls into their retainers with the tip of each pawl laying over the top of the nylon bushing.

9- Install the retainers securing the pawls to the pulley.

10- Turn the starter over and install the retaining nut onto the thru-bolt (if a nut is used). Tighten the nut securely. Check to be sure the pulley will rotate smoothly and does not bind on the spindle. Rotate the pulley slightly **COUNTERCLOCKWISE**, then release it to be sure the spring is properly engaged with the pulley and that the pulley has good spring tension.

ROPE INSTALLATION

Rope Purchase Instructions

The length and diameter of the starter rope required will vary depending on the

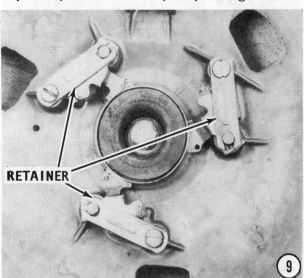

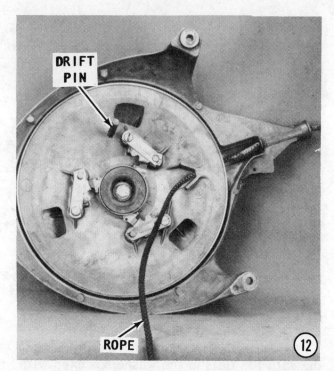

horsepower size of the model being serviced. Therefore, check the Hand Starter Rope Specifications in the Appendix, and then purchase a quality nylon piece of the proper length and diameter size. Only with the proper rope, will you be assured of efficient operation following installation.

Each end of the nylon rope should be "fused" by burning them slightly with a very small flame (a match flame will do) to melt the fibers together. After the end fibers have been "fused" and while they are still

hot, use a piece of cloth as protection and pull the end out flat to prevent a "glob" from forming.

11- Tie a figure **8** knot in one end of the rope. Set the rope aside, but handy, to be picked up with one hand.

12- Hold the housing with one hand and rotate the pulley three complete turns **COUNTERCLOCKWISE** with the other hand. After three complete turns have been made, align the rope outlet in the pulley with the outlet in the housing. Insert a drift pin or other suitable tool through the hole in the pulley and the hole in the housing to hold the pulley in the desired position. Feed the the rope through the pulley and the housing and out the other side of the housing. Pull the rope tight until the knot is seated against the pulley.

13- A special tool is manufactured by OMC to install the handle onto the rope. This tool has three prongs and is inserted through the handle and then attached to the

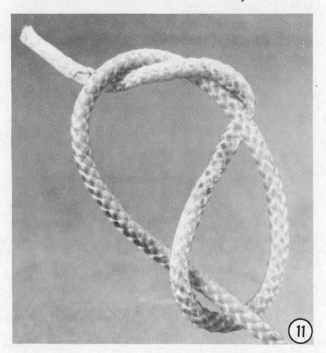

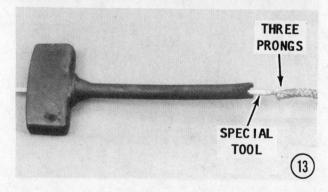

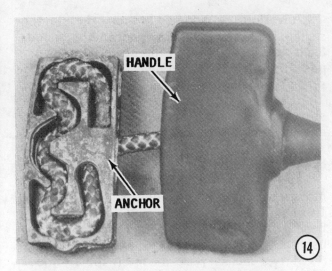

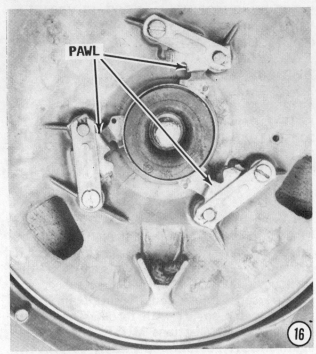

rope and pulled back through the handle. If the special tool is not available take a stiff piece of wire; insert it through the handle; thread it through the rope; apply just a little oil to the rope; then pull the wire and rope through the handle.

14- Work the end of the rope into the handle anchor. Secure the rope in place by pushing the anchor into the rubber handle.

15- Lightly pull on the rope to relieve tension on the pin installed through the pulley and housing in Step 12. Maintain some tension on the rope, remove the pin, and allow the spring to **SLOWLY** wind the rope onto the pulley. Check the bolt through the spindle to be sure it is tight.

16- Lay the starter on its back and pull the rope with quick movements, and at the same time check the pawls to be sure they move towards the center of the pulley. Release the rope slowly and check to be

sure the pawls return to their original position under the retainers.

STARTER INSTALLATION

17- Position the starter over the flywheel with the three legs aligned over the holes in the powerhead for the retaining bolts. Install the retaining bolts and tighten them securely. Connect the linkage from the carburetor.

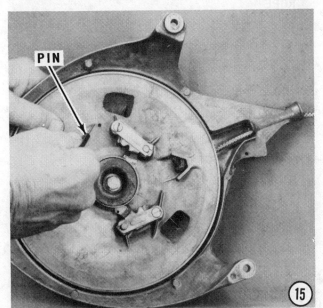

9-7 TYPE VI MOUNTED ATOP FLYWHEEL WITH ONE OR TWO NYLON PAWLS
COLT, JUNIOR, ULTRA, & EXCEL
  ALL YEARS
2 HP 1971-85
2.5 HP ALL YEARS
4 HP 1980-85
18 HP 1971-73
20 HP 1971-73 AND 1985 AND ON
25 HP 1971 AND ON
35 HP 1976-84

WARNING

As with other types of hand starters, the rewind spring is a potential hazard.   The spring is under tremendous tension when it is wound -- a real tiger in a cage. If the spring should accidentally be released, severe personal injury could result from being struck by the spring with force. Therefore, the service instructions **MUST**, and we say again **MUST**, be followed closely to prevent release of the spring at the wrong time. Such action would be a **BAD SCENE**, a very **BAD SCENE**, because serious personal injury could result.

The starter rope should **NEVER** be released from the extended position.   Such action would allow the spring to wind with incredible speed, resulting in serious damage to the starter mechanism.

Any time the rope is broken, the starter spring will rewind with incredible speed. Such action will cause the spring to rewind past its normal travel and the end of the spring will be bent back out of shape. Therefore, if the rope has been broken, the starter must be completely disassembled and the spring repaired or replaced.

Hand Starter Timing

Surprising as it may sound, this starter, mounted on top of the powerhead over the flywheel, can actually timed to the powerhead. This timing can best be described by using an example.

If two marks were made on the flywheel 180° apart, and matching marks made on the powerhead, then each time the powerhead was shut down, one set of marks on the flywheel would align very closely with one of the marks on the powerhead. What is actually happening, is the powerhead is stopping with either the top piston at TDC (top dead center) or the bottom piston at the TDC position.

Now, assume the powerhead has been operating at idle speed and then suddenly stops for any number of reasons. The problem is corrected and the powerhead is once again ready to be started. Two notches are manufactured into the inside diameter of the flywheel. A single dog on the pulley engages with one of these dogs when the rope is pulled. Now, if it is necessary to pull an excessive amount of

Closeup view of the pulley and the housing with the arrow on the housing aligned with the marks on the pulley. This alignment is necessary to "time" the starter with the engine.

The starter rope on this engine has been pulled much too far before the flywheel begins to rotate. This starter is, therefore, not timed properly with the engine, as described in the text.

rope before the dog is able to engage the flywheel, full starting rotation power would not be available in the rope.

Therefore, the starter is timed to engage the starter with the flywheel after the rope has been pulled exactly the same amount each time. This distance is very short to allow as much rope pull as possible to rotate the crankshaft for fast start. This "timing" will assure an adequate amount available for starting **AND** that one of the pistons will return to TDC when the pull is completed, if the powerhead fails to start. If an excessive amount of pull is necessary before the flywheel begins to rotate, the starter was not assembled properly -- the arrow on the pulley was not aligned with the two marks on the starter housing when the spring is relaxed and the rope handle is retracted.

2.5 HP, COLT, JUNIOR, AND EXCEL PRELIMINARY TASKS

WORDS OF CAUTION
DO NOT ATTEMPT to remove the large slotted head screw from atop the starter housing. Loosening this screw will cause the starter spring to disengage and forcably unwind within the confines of the starter housing.

ALL MODELS EQUIPPED WITH REMOTE CONTROL OR TILLER HANDLE
Remove the screw retaining the throttle cable trunnion to the powerhead. Pull the throttle cable down to release it from the armature plate. Remove the one nut and two bolts securing the starter housing to the powerhead.

BOLT (3 PLCS)

①

To remove the throttle cable from the starter housing, remove the two screws on the small retaining bracket across the throttle cable and then twist the cable **COUNTERCLOCKWISE** to release the cable from the throttle knob.

STARTER REMOVAL

1- Remove the attaching bolts securing the three legs of the starter housing to the powerhead. On some smaller horsepower powerheads, the starter housing is attached to the fuel tank with screws. Remove the hand starter and lay it on the bench with the pulley facing toward you.

DISASSEMBLING

2- Pull the rope out enough to tie a knot in the rope. Tie a knot, and then allow the rope to rewind to the knot. Work the rope anchor out of the rubber covered handle, then remove the rope from the anchor. Remove the handle from the rope. Untie the knot in the rope, and then hold the disc pulley, but permit it to turn and thus allow the rope to wind back onto the pulley **SLOWLY**. Continue to allow the spring in the pulley to unwind **SLOWLY** until all tension has been released.

SPECIAL WORDS
This series of photographs depicts a hand rewind starter with a single pawl. If servicing a starter with two pawls, merely repeat Steps 3 and 4 to remove the other

ANCHOR   HANDLE

②

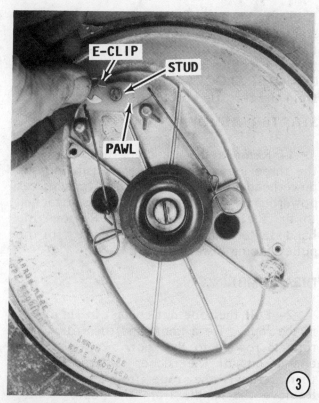

E-CLIP
STUD
PAWL

(3)

pawl. All other procedural steps are valid for both starters.

3- Remove the E-clip from the nylon pawl. Lift the pawl from the stud.

4- Remove the friction spring and friction link from the pawl.

5- Remove the bolt, lockwasher, and washer from the center of the pulley spin-

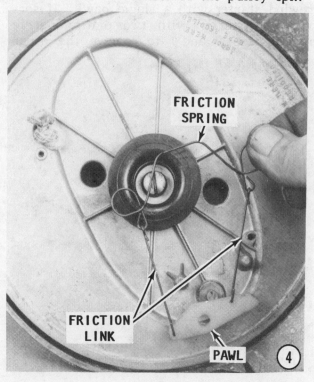

FRICTION
SPRING

FRICTION
LINK

PAWL

(4)

SPINDLE

(5)

dle. Lift the spindle out of the pulley, and at the same time hold the pulley firmly together with the housing.

SPECIAL WORDS
FOR TWO PAWL STARTERS

A starter spring shield is installed between the pulley and the spring on these units.

6- Lift the pulley straight up and at the same time work the spring free of the pulley. The spring has a small loop hooked into the pulley. An alternate and safe method is to hold the pulley and the housing together tightly and turn the complete assembly with the legs extending downward in

PULLEY
SPRING

(6)

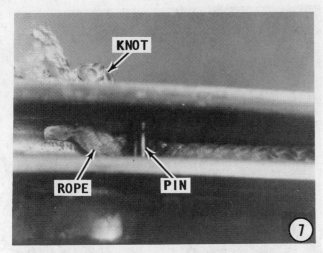

Damaged pawl unfit for further service.

the normal manner. Now, lower the complete assembly to the floor. When the legs make contact with the floor, release your grip. The pulley will fall and the spring will be released from the housing, but the three legs will contain the spring and prevent it from traveling across the room. If the spring was not released from the housing, the only safe method is to again make contact with the three legs on the floor and jar the spring free.

7- Unwind the rope out of the pulley groove. Notice the pin next to the knot in the rope and how the rope feeds **BEHIND** the pin. Pull the knot and the rope out far enough to untie the knot, and then pull the rope free of the pulley.

CLEANING AND INSPECTING

If the rope was broken and the spring is bent backward, as shown in the accompany-

ing illustration, it is a simple matter to bend the spring end back to its normal position. The next illustration clearly shows a spring end properly positioned in the housing.

Wash all parts except the rope in solvent and then blow them dry with compressed air.

Remove any trace of corrosion and wipe all metal parts with an oil dampened cloth.

Inspect the rope. Replace the rope if it appears to be weak or frayed. If the rope is frayed, check the hole through which the rope passes for rough edges or burrs. Remove the rough edges or burrs with a file, and polish the surface until it is smooth.

The rope on this unit broke, causing the spring to rewind with incredible speed. The end of the spring was bent back in the wrong direction.

Knot tied in the end of the starter rope to prevent the rope from being pulled from the starter.

Inspect the starter spring end loops. Replace the spring if it is weak, corroded or cracked. Check the spring pin located at the back side of the pulley to be sure it is straight and solid.

Check the inside surface of the housing and remove any burrs.

Check the condition of the pawl spring to be sure it is not stretched out of shape. The end of the spring should be bent back toward the coil of the spring. Inspect the pawl for wear and that the edges are not rounded.

Check the friction spring and link to be sure they are not distorted.

Inspect the spindle. The spindle must be straight and tight.

SPECIAL WORDS

This series of photographs depicts a hand starter with a single pawl. If servicing a unit with two pawls, merely repeat Steps 5 and 6 to install the other pawl. All other procedural steps are valid for both starters.

STARTER ASSEMBLING

ROPE INSTALLATION

Rope Purchase Instructions

The length and diameter of the starter rope required will vary depending on the horsepower size of the model being serviced. Therefore, check the Hand Starter Rope Specifications in the Appendix, and then purchase a quality nylon piece of the proper length and diameter size. Only with the proper rope, will you be assured of efficient operation following installation.

Each end of the nylon rope should be "fused" by burning them slightly with a very small flame (a match flame will do) to melt

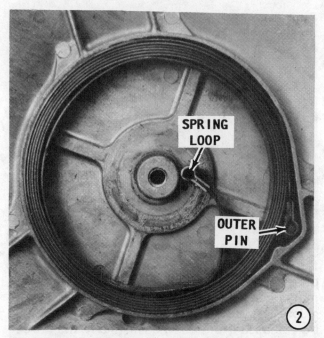

the fibers together. After the end fibers have been "fused" and while they are still hot, use a piece of cloth as protection and pull the end out flat to prevent a "glob" from forming.

1- Tie a figure 8 knot in the end of the rope. Insert one end of the new rope through the hole in the pulley and housing and on the back side of the pin, as shown. Continue to wrap the remainder of the rope COUNTERCLOCKWISE around the pulley.

SAFETY WORD

Wear a good pair of gloves while installing the spring. The spring will develop tension and the edges of the spring steel are

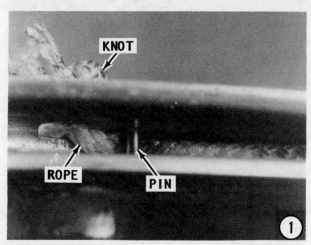

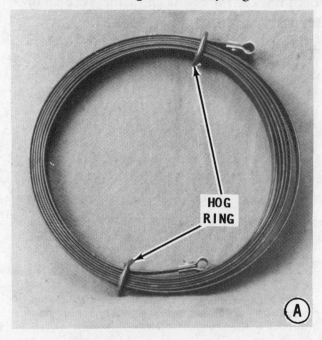

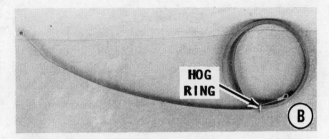

HOG RING

sharp. The gloves will prevent cuts on your hands and fingers.

2- Slide the spring onto the outer pin and then start the spring from the outside edge of the housing and insert it into the housing **COUNTERCLOCKWISE**, as shown in the accompanying illustration. Notice the small hump in the housing. This hump prevents the spring from being wound in the wrong direction. Work the first turn into the housing, and then hold the spring down with one hand and continue to wind the spring into the housing. Patience and time are required to work the spring completely into the housing. After the last portion is in place, bend the end of the spring towards the center of the housing. This position will allow the pulley pin to align with the loop in the end of the spring, when the pulley is installed.

Alternate Method

An alternate method of installing a **NEW** spring into the housing with less risk of personal injury is presented with accompanying illustrations.

HOG RING

A- Remove **ONLY** the hog ring next to the end of the outside wrap of the spring.

B- Pull on the outside end of the spring. As the spring is pulled, the inside diameter will get smaller and smaller, as shown. When the diameter is a bit smaller than the inside diameter of the starter housing, wrap the entire free end of the spring around the coiled portion.

C- CAREFULLY lower the coiled spring into the starter housing with the loop on the free end of the spring indexed over the peg in the housing and the spring feeding **COUNTERCLOCKWISE**, as shown. Remove the second hog ring **WITHOUT** allowing the spring to escape from the housing. An easy and safe method is to cut the hog ring with a pair of "dikes". Bend the inside end of the spring toward the center of the housing to permit the pin in the pulley to index into the loop.

SPECIAL WORDS
FOR TWO PAWL STARTERS

A starter spring shield is installed between the pulley and the spring on these units. Place the shield against the underside of the pulley aligning the holes.

3- Lower the pulley down over the top of the spring with the pulley pin indexing into the loop in the end of the spring. In the accompanying illustration, notice the callout for the boss on the backside of the pulley. The pin is located directly under the boss. The boss can, therefore, be a guide during pulley installation.

4- If servicing a model 20 or 30hp 1986 and later: Install the bushing and shim material onto the pulley followed by the square friction plate. Then install the

SPRING LOOP
BOSS

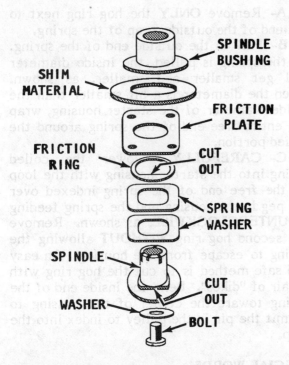

SPINDLE
BUSHING

SHIM
MATERIAL

FRICTION
PLATE

FRICTION
RING

CUT
OUT

SPRING
WASHER

SPINDLE

CUT
OUT

WASHER

BOLT

Arrangement of parts comprising the spindle for Models 20hp and 30hp since 1986.

spindle, friction ring and two wavy washers, indexing the friction ring with the spindle, flat-spot-to-flat-spot, as shown in the accompanying illustration. Install the washer and bolt and tighten the bolt to a torque value of 10 ft lb (13.6Nm).

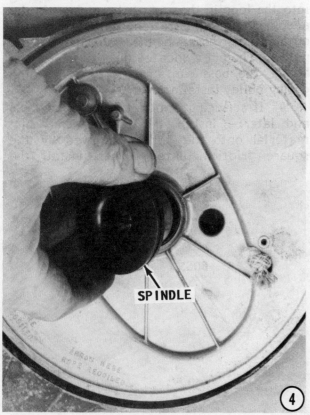

SPINDLE

FRICTION
SPRING

STUD

FRICTION
LINK

PAWL

All other models: Lower the spindle assembly down through pulley. Place the washer and lockwasher inside the spindle housing and then install the bolt through the washer into the housing. Tighten the bolt securely. Check to be sure the pulley will rotate smoothly and does not bind on the

PAWL

E-CLIP

spindle. Rotate the pulley slightly **COUNTERCLOCKWISE** and then release it to be sure there is proper engagement with the spring and the pulley has good spring tension.

5- Install the friction spring and link and the nylon pawl onto the starter hub. The friction spring fits into a groove in the spindle. Pull just a little on the pawl and set it over the stud on the flywheel pulley.

6- Snap the **E-clip** over the top of the pawl to secure it in place.

7- Rotate the pulley three complete revolutions, and then work the rope out through the hole in the pulley and the housing. Pull on the rope until a couple of feet are exposed. Tie a knot in the rope and allow the rope to rewind until the knot is tight against the housing. Work the end of the rope into the handle anchor. Secure the rope in place by pushing the anchor into the rubber handle.

Pull on the rope enough to untie the knot, and then allow the rope to slowly recoil into the pulley.

8- Lay the starter on its back with the pulley facing toward you. Notice the imprint on the pulley: **ARROW HERE ROPE-RECOILED.** Also notice the arrow on the housing. On Johnson powerheads, when the rope is fully coiled (the starter pulley completely wound) the arrow must fall between the marks on the pulley. Further up on the pulley you will notice the letter **E** (for Evinrude). On the Evinrude powerheads, the arrow must fall between the two marks on

the pulley. If the arrow is not properly aligned the starter rope is not the proper length. It is either too long or too short. The arrow must align properly for the starter to be timed with the powerhead. See the beginning of this section.

9- With the starter still on its back, pull the rope with quick movements, and at the same time check the pawl to be sure it moves toward the center of the pulley. Release the rope slowly and check to be sure the pawl returns to its original position.

STARTER INSTALLATION

Position the starter over the flywheel with the three legs aligned over the holes in the powerhead for the retaining bolts. Install the retaining bolts and tighten them to a torque value of 6.8 ft lb (8Nm).

Throttle Cable Installation
Models Equipped with Remote Control
or Tiller Handle

Insert the throttle cable through the cable guide in the housing. Rotate the cable **CLOCKWISE** to engage the cable with the throttle knob. Place the small retaining bracket across the cable and secure it to the housing with two screws.

To adjust the throttle cable, back off the toothed idle stop wheel away from the throttle knob untill the throttle cable can be snapped back into place on the armature plate. Push the throttle knob toward the starter housing while pushing the armature plate **CLOCKWISE** as far as possible. Adjust the position of the throttle cable trunnion to allow the cable to slide into the side retaining bracket and then tighten the bracket retaining screw.

9-8 TYPE VII MOUNTED ATOP FLYWHEEL WITH LARGE PAWL PLATE
40 HP 1986 AND ON

This design hand rewind starter is usually mounted atop the flywheel with three mounting legs attached to the powerhead. The unit has a large rectangular pawl plate visible on the underside surface.

This type hand rewind starter is used on the 40hp powerhead -- 1986 and on.

STARTER REMOVAL

1- Remove the screw retaining the lockout cable clamp to the housing, and then remove the lockout slide. Remove the attaching bolts securing the three starter housing legs to the powerhead. Remove the two screws securing the starter handle bracket to the powerhead. Remove the hand starter and lay it on the bench upside down with the pulley and legs facing up.

DISASSEMBLING

2- Pull the rope out enough to tie a knot in the rope. Tie a knot, and then allow the rope to rewind to the knot. Work the rope anchor out of the rubber covered handle, and then remove the rope from the anchor. Remove the handle from the rope. Untie the knot in the rope, but hold and prevent the disc pulley from rotating. Now, slack the hold on the pulley, and permit it to turn thus winding the rope back onto the pulley **SLOWLY**. Continue to allow the spring in

the pulley to unwind **SLOWLY** until all tension has been released.

3- Remove the bolt retaining the lockout lever, the lockout lever spring, and the washer to the housing.

Place one hand under the housing to prevent all the large pawl plate and other small parts from flying loose. Keep one finger over the center bolt.

Remove the large center nut and washer from atop the starter housing. Carefully turn the starter housing over without disturbing any parts.

WARNING

THE REWIND SPRING IS A POTENTIAL HAZARD. The spring is under tremendous tension when it is wound — a real **"tiger"** in a cage! If the spring should accidentally be released, severe personal injury could result from being struck by the spring with force. Therefore, the following steps **MUST** be performed with care to prevent personal injury to self and others in the area.

4- Lift off the center bolt and washer, and then lift off the large pawl plate and the plate return spring from the side of the starter housing.

Remove the center spring. Use a screwdriver to pry out the pulley lockring from under the spring. Lift off the friction plate and the spring washer beneath the plate.

5- Remove the starter pawl and spring washer from the pawl pivot post.

6- Lift the pulley straight up and at the same time work the spring free of the pulley. The spring has a small loop hooked into the pulley. An alternate and safe method is to hold the pulley and the housing together tightly and turn the complete assembly over with the legs extending downward in the normal manner. Now, lower the complete assembly to the floor. When the legs make contact with the floor, release your grip. The pulley will fall and the spring will be released from the housing almost instantly and with considerable **FORCE**. However, the three legs will contain the spring and prevent it from lashing out causing possible injury to self or others in the area. If the spring was not released from the housing as just described, the only safe method is to again jar the three legs on the floor and dislodge the spring.

7- Unwind the rope out of the pulley groove.

CLEANING AND INSPECTING

If the rope was broken bending the spring backward, it is a simple matter to bend the spring end back to its normal position.

Wash all parts except the rope in solvent and then blow them dry with compressed air. Remove any trace of corrosion and wipe all metal parts with an oil dampened cloth.

Inspect the rope. Replace the rope if it appears to be weak or frayed. If the rope is frayed, check the hole through which the rope passes for rough edges or burrs. Remove the rough edges or burrs with a file, and polish the surface until it is smooth.

Inspect the starter spring end loops. Replace the spring if it is weak, corroded or cracked.

Check the inside surface of the housing and remove any burrs.

Check the condition of the pawl spring to be sure it is not stretched out of shape.

Inspect the pawl for wear and that the edges are not rounded.

Check the friction spring to be sure they are not distorted.

Rope Purchase Instructions

Purchase a quality piece of nylon rope 96 1/2" (245cm) long. Only with the proper rope, can efficient operation be ensured following installation.

Each end of the nylon rope should be "fused" by burning the last half inch slightly with a very small flame (a match flame will do) to melt the fibers together. After the end fibers have been "fused" and while they are still hot, use a piece of cloth as protection and pull the end out flat to prevent a "glob" from forming.

Tie a figure **8** knot in the end of the rope and keep the rope close at hand for installation in Step 4.

STARTER ASSEMBLING

SAFETY WORDS

Wear a good pair of gloves while installing the spring. The spring will develop tension and the edges of the spring steel are sharp. The gloves will prevent cuts on hands and fingers.

It is **STRONGLY** recommended a pair of safety goggles or a face shield be worn while the spring is being installed. As the work progresses a "tiger" is being forced

into a cage. If the spring is accidentally released, it will lash out with tremendous ferocity and very likely could cause personal injury to the installer or other persons nearby.

1- Slide the spring onto the outer pin and then start the spring from the outside edge of the housing and insert it into the housing **COUNTERCLOCKWISE**. Work the first turn into the housing, and then hold the spring down with one hand and continue to wind the spring into the housing. Patience and time are required to work the spring completely into the housing. After the last portion is in place, bend the end of the spring towards the center of the housing. This position will allow the long slot in the center hub to align with the loop in the end of the spring, when the pulley is installed.

2- Lower the pulley down over the top of the spring with the long slot in the center hub indexing into the loop in the end of the spring.

Install the friction plate spring washer and friction plate over the pulley hub. Position the lockring over the plate and tap around the circumference of the ring until it is well seated into the groove of the hub. Place the starter housing spring over the lockring.

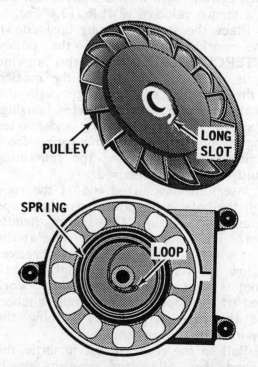

*The long slot at the pulley center **MUST** engage the inner hook of the rewind spring for the starter to function correctly.*

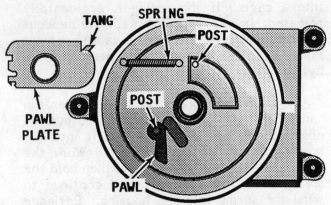

The pawl plate return spring is positioned and attached between a post on the pulley and a tang on the pawl plate.

Apply a coat of OMC Triple-Guard grease onto the pawl pivot post before installing the spring washer and pawl onto the post.

Hook one end of the pawl plate return spring over the pulley post. Hook the other end of the spring onto the pawl plate tang. Slide the pawl plate into position to align with the pulley hub. Place the washer over the pawl plate and install the center bolt.

Place one hand over the entire pawl plate and keep one finger over the center bolt to keep all the parts together while the assembly is inverted.

3- Place the washer over the end of the installed center bolt. Install and tighten the nut to a torque value of 12 ft lb (13.6Nm).

4- Place the starter housing upsidedown on the workbench. Rotate the pulley **COUNTERCLOCKWISE** until the rewind spring is under tension. Stop the motion when the pulley rope notch aligns with the rope guide mounted on the starter housing. Keep a tight hold on the pulley and starter housing to prevent the rewind spring from unwinding while performing the remaining work until the rope is secured.

5- Insert the unknotted end of the rope through the hole in the pulley, past the rope guide and out through the starter handle bracket. Tie a slip knot in the rope at the point at which it emerges from the bracket and allow the rope to rewind slightly until the knot is tight against the housing. Work the end of the rope into the handle anchor. Secure the rope in place by pushing the anchor into the rubber handle.

6- Pull on the rope enough to untie the knot, and then allow the rope to slowly recoil into the pulley.

7- With the starter still on its back, pull

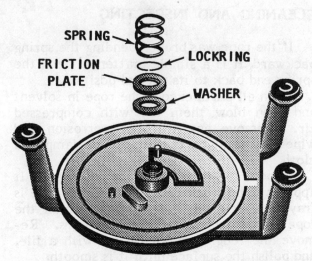

The friction plate spring washer and friction plate are secured with the pulley lockring. The starter housing spring rests on top of the lockring.

the rope with quick movements, and at the same time check the pawl to be sure it extends outward when the rope is pulled out.

Release the rope slowly and check to be sure the pawl retracts and returns to its original position.

8- Install the starter lockout lever, spring, and washer onto the housing. Tighten the bolt securely.

STARTER INSTALLATION

9- Position the starter over the flywheel with the three legs aligned over the holes in the powerhead for the retaining bolts. Install the retaining bolts and tighten them to a torque value of 6.8 ft lb (8Nm). Install and tighten the two starter handle bracket screws to the same torque value.

Position the lockout slide in the channel of the housing. Lightly secure the lockout cable to the starter housing using the cable clamp and bolt. Check to be sure the lower unit is in **NEUTRAL** gear and adjust the lockout cable to align the slide with the lockout lever. Tighten the cable clamp securely to hold this adjustment.

Shift the lower unit into **FORWARD** and then **REVERSE** gear and make a check of the starter lockout system. The starter **MUST** be **LOCKED**, and unable to rotate when the lower unit is in any gear other than **NEUTRAL**.

If the starter fails this test, loosen the cable clamp and adjust the position of the slide until no motion of the starter is possible when the lower unit is in **FORWARD** or **REVERSE** gear.

10
TRIM/TILT

10-1 SYSTEM DESCRIPTION

The power trim and tilt unit is a hydraulic/mechanical unit mounted port and starboard of the stern brackets. The unit consists of a single trim cylinder (port) and a single tilt cylinder (starboard), a combination pump, fluid reservoir and bi-directional electric motor in a single unit is installed on the starboard side of the transom brackets, and the necessary tubing, valves, check valves and relief valves, to make it all function properly.

Two versions of this trim/tilt system were used. One manufactured by Prestolite and the other by Calco, as indicated in the accompanying illustrations. The exploded drawing of the trim cylinder and the exploded drawing of the tilt cylinder are valid for both manufacturers.

The upper ends of both the trim and tilt cylinders are attached to a removable bracket mounted to the motor swivel bracket. When the cylinders are extended, they push up on this bracket. When the cylinders retract, they pull down on this bracket.

When the boat operator activates an electric switch on the control panel to the **UP** position, power is supplied to the electric motor which drives the hydraulic pump. The pump forces hydraulic fluid into the trim cylinder and into the tilt cylinder below the pistons. The trim cylinder moves the outboard motor the first 15° of movement, considered the "tilt range". At this position, the tilt cylinder takes over and moves the outboard through the final 50° of motion, the "tilt range".

The outboard may be operated up to approximately 1500 rpm while it is elevated past the 15° position for boat movement in very shallow water. However, if powerhead speed is increased above 1500 rpm, a relief valve will automatically open causing the outboard to lower to the fully trimmed out position. The outboard cannot be tilted while operating above 1500 rpm.

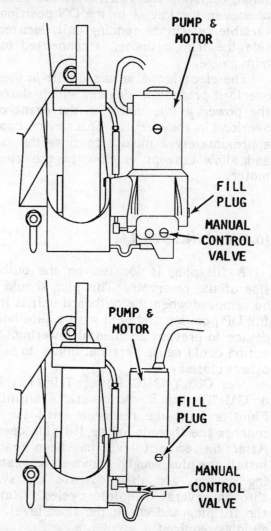

PUMP & MOTOR

FILL PLUG

MANUAL CONTROL VALVE

PUMP & MOTOR

FILL PLUG

MANUAL CONTROL VALVE

Line drawing to depict the exterior appearance of the Prestolite pump and motor (top), and the Calco pump and motor (bottom).

When the boat operator activates an electric switch on the control panel to the **DOWN** position, power is supplied to the electric motor which drives the hydraulic pump in the opposite direction. The pump forces hydraulic fluid into the trim cylinder and into the tilt cylinder above the pistons. The outboard is moved in the down direction, movement of the cylinders are all in the opposite direction to the up movement. The tilt cylinder moves first until the swivel bracket is resting on the trim rods. At that point the trim cylinder continue movement in the down direction.

A manual control valve/screw is located at the lower end of the pump. This screw may be rotated **COUNTERCLOCKWISE** to raise or lower the outboard unit in the event a malfunction in the system prevents movement using the trim/tilt system.

A trim gauge, mounted on the control panel, registers the position of the outboard whenever the key is in the **ON** position. A variable resistance sending unit, located inside the trim cylinder, is connected to the trim gauge.

The electric motor has a built-in thermal overload protection device, which shuts off the power to the motor in the event of an overload in the circuit. This device cools in approximately a minute to close the switch and allow current to flow to the electric motor.

10-2 FILLING SYSTEM

A fill plug is located on the outboard side of the reservoir. This plug should only be removed when the outboard unit is in the full **UP** position and held with a safe holding device to prevent accidental lowering. Such action could cause personal injury to self or others close by.

Use **ONLY** OMC Power Trim/Tilt Fluid or GM "Dexron II" Automatic Transmission Fluid and fill the reservoir until the fluid reaches the threads in the fill plug opening. After the correct level has been reached, install the plug snugly, remove the restraining device, and then operate the system through several complete cycles. Remove the fill plug and check the fluid level. Add fluid as required.

The total capacity of the unit is 25 fl. oz. (740 mL).

10-3 TROUBLESHOOTING

Troubleshooting **MUST** be done **BEFORE** the system is opened in order to isolate the problem to one area. Always attempt to proceed with troubleshooting in a definite, orderly, and directed manner. The "shot in the dark" approach will only result in wasted time, incorrect diagnosis, replacement of unnecessary parts, and frustration.

The following procedures are presented in a logical sequence to check the mechanical components, with the most prevalent, easiest, and less costly items to be checked listed first.

PRELIMINARY CHECKS AND INSPECTION

The following items are logical areas causing problems with the trim/tilt system. All may be checked and corrected without the use of special tools or equipment, with the exception of a torque wrench.

1- Check to be sure the battery is up to a full charge and the terminal connections are clean and tight. Make a visual inspection of all exposed wiring for an open circuit or other damage that might cause a problem.

2- Check the hydraulic fluid level at the reservoir fill plug **AFTER** all air has been removed from the unit and when the tilt and trim cylinders are fully extended. The outboard unit must be in the vertical position. The fluid level should be even with the bottom of the fill hole with the motor full tilted up. Top off the reservoir to the plug level. Operate the motor and then again check the oil level. Attempt to cycle the unit several times and again check the level when the cylinders are fully extended. The system should be cycled through at least five complete movements to ensure all air has been purged from the system.

3- Add **ONLY** OMC Power Trim/Tilt Fluid or GM "Dexron II" Automatic Transmission Fluid as required. Total capacity of the is 25 fl. oz. (740 mL).

4- Make an external (outside) inspection of the system for damage or signs of a fluid leak.

5- Seat the manual release valve by tightening the screw to a torque value of 45 to 55 in lbs (5.1 to 6.2 Nm).

6- Inspect the stern brackets for signs of binding with the swivel brackets in the

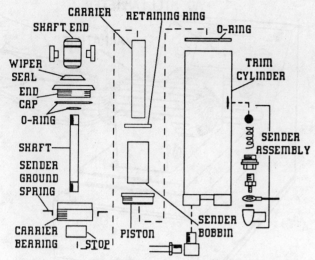

Exploded line drawing of the trim cylinder covered in this section, with major parts identified.

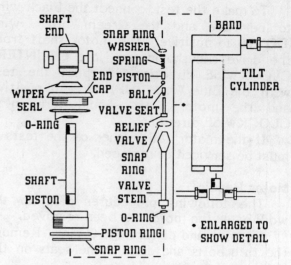

Exploded line drawing of the tilt cylinder covered in this section, with major parts identified.

thrust rod area. Check the tilt tube nut for a torque value of 24 to 26 ft. lbs. (32.5 to 35.2 N m), then back off (loosen) the nut 1/8 to 1/4 turn. Inspect trim and tilt cylinders for bent rods.

7- Trailering the boat with the motor in the full tilt position and unsupported can cause a hydraulic "lock-up". To relieve such a "lock-up", loosen one trim cylinder end cap 1/4 turn with a socket type spanner wrench. Operate the unit down then up slightly. Tighten the end cap.

10-4 SERVICING

Raise the outboard unit to the full **UP** position. If the trim/tilt system is not operative, rotate the manual release valve/-screw **COUNTERCLOCKWISE**, and then lift the unit manually. Secure the outboard in a safe manner using a restraint or support.

Obtain a suitable container to receive the hydraulic fluid from the reservoir.

Always use flare wrenches when disconnecting or connecting hydraulic lines at the fittings to prevent "rounding" the corners, which is likely if a standard wrench is used.

Begin by disconnecting the two hydraulic lines at the trim/tilt housing base. Remove the fill plug at the reservoir and allow the hydraulic fluid to drain into the container.

After the fluid has drained, temporarily install the fill plug to prevent contaminates from entering the system.

Disconnect the trim motor electrical harness at the connector plug.

Remove the hardware securing the trim and tilt shaft ends to the outboard. Remove

the bolts securing the system to the transom, and then **CAREFULLY** remove the complete unit.

The oil pump cannot be serviced. If defective, it must be replaced. The brushes in the electric motor should be replaced if they are worn to 1/4" (6mm) or less.

Follow the procedures in the following section to service the electric motor.

NOW, THESE WORDS

In the majority of cases, service of the hydraulic items removed thus far will solve any rare problems encountered with the trim/tilt system.

The reservoir can be removed through the attaching hardware and cleaned if the system is consider to be contaminated with foreign material. The trim/tilt components comprise what is considered a "closed" system. The only route for entry of foreign material is through the fill opening.

Further disassembly and service to the system would best be left to a shop properly equipped with the proper test equipment and trained personnel with the expertise to work with high pressure hydraulic systems.

10-5 ELECTRIC MOTOR TESTING AND REPAIR

Motor Testing

The condition of the motor can be bench tested with a current draw test on a no load test.

On a no load test, the motor should have a maximum current draw of 18 amps at a minimum of 8500 rpm at 12-volts.

To make the test, connect the black wire to negative and the Green/White wire **(DOWN)** to positive. The motor shaft from the drive end should turn in a **COUNTER-CLOCKWISE** direction. Repeat the test with the Blue/White **(UP)** wire to positive and the motor shaft should turn in a **CLOCKWISE** direction.

If the motor fails either of the tests it must be serviced or replaced.

Motor Repair

The following procedures pick up the work after the motor has been removed.

Remove and discard the O-ring. Remove the thru bolts and discard the seals on the bolts.

Refer to illustration **A** for the following instructions. Remove the drive end cap from the motor. Discard the gasket. Removing the seal exercising **CARE** not to scratch the casting surfaces to ensure the new seal will seat properly. Discard the old seal.

Remove the armature from the motor housing. **TAKE CARE** not to lose the fiber washer on each end of the shaft. Tip the end cap free of the motor housing. Discard the springs and the end cap gasket.

SPECIAL WORD

The end cap is serviced as a complete assembly.

CLEANING AND INSPECTION

Clean all parts with a dry cloth. **DO NOT** clean either head in solvent, because the solvent will remove the lubricating oils in the armature shaft bushings. **DO NOT** clean the armature in solvent, because the solvent will leave traces of oil residue on the commutator segments. Oil will cause arcing between the commutator and the brushes.

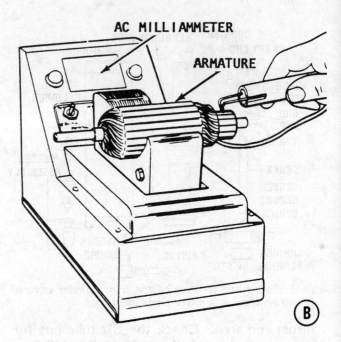

(B)

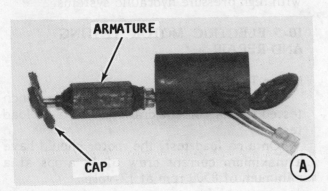

(A)

Brush Replacement

A new brush head may be purchased from the Local OMC dealer as a complete assembly with new brushes installed.

Armature Shorted

The armature **CANNOT** be checked in the usual manner on a growler, because the internal connections and the low resistance of the windings. If a growler is used, all the coils will check out shorted.

HOWEVER, the armature can be tested using an AC milliammeter, five milliamperes with 100 scale divisions, and making tests between the commutator segments. Refer to illustration **B** for the following instructions. Move from one segment to the next and watch closely for changes in the meter readings. The segments should all check out with almost the same reading. If a test between two segments indicates a significant lower reading, the winding is shorted.

Armature Grounded

Refer to illustration **C** for the following instructions. Connect one lead of a continuity tester to a good ground, and then move the other lead around the entire surface of the commutator. Any indication of continuity means the armature is grounded and **MUST** be replaced. If the commutator segments are dirty or show signs of wear (roughness), clean between the bars, and then true it in a lathe. **NEVER** undercut the mica because the brushes are harder than the insulation.

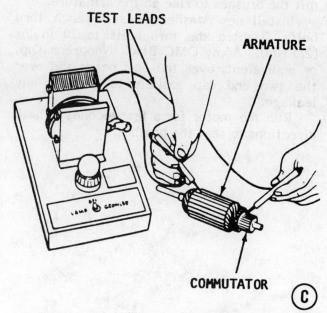

TEST LEADS

ARMATURE

COMMUTATOR

(C)

After turning the armature, the insulation between the segments MUST be under cut to a depth of 1/32" (0.79 mm). The undercut MUST be flat at the bottom and should extend the full width of each insulated groove and beyond the brush contact in both directions. This will prevent the segment insulation from being smeared over the commutator as the segments wear.

After undercutting, the commutator should be sanded to remove the ridges left during the undercutting. Now, clean the commutator thoroughly to remove any metal chips or sanding grit.

Again perform the shorting and grounding tests on the armature.

End Head Bushings

Side play of each end head on the armature should be carefully checked. Any side play indicates bearing wear and the end head MUST be replaced, because the bearings are not serviced separately. If the heads with worn bearings are returned to service, the armature will rub against the pole shoes, or the armature shaft may actually bind.

To replace the commutator end head, first cut the lead connecting it to the field coil as close to the end head as possible. Next, solder the new end lead to the brush holder.

Field Coils

The field coils are series-wound and are NOT grounded to the frame. To test the field coils for a short, make contact with one probe of a test light to a good ground on the frame, as shown in illustration D. Make contact with the other probe to the Blue/-White or Green/White tilt motor lead. If the test light comes on, the field coil is grounded and MUST be replaced. The field coils are only available as a complete field coils and frame assembly.

ASSEMBLING

Press a NEW seal into the end cap with the lip and seal spring facing UP, toward the tool. Use a flat ended bar. Continue to press the seal into place until the seal is below the chamber in the end cap.

Install a new gasket onto the end cap. Split the end of NEW brush lead CAREFULLY to 1/8" (3.18 mm) to fit over the piece of brush lead left on the motor head. Fit the end of new split brush leads over the old brush lead ends on the motor head, and then twist them slightly.

Hold the new leads with a pair of pliers and solder them using rosin core solder. The pliers will act as a heat barrier, so the solder will not spread and the rest of the brush leads will stay flexible.

Install a NEW end cap gasket. Slide the new gasket over the wires and the end head into position.

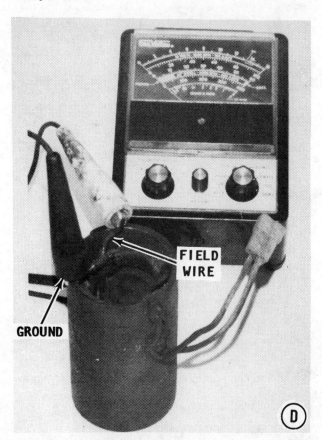

FIELD WIRE

GROUND

(D)

Use two paper clips bent to make brush and spring holders. Install the new brush springs and insert the brushes into position using the paper clip as a tool to hold the brushes retracted in place.

Install the armature into the motor housing and into the motor head, with a fiber washer on the shaft at each end of the armature. Remove the paper clips to per-

mit the brushes to ride on the armature.

Install new washer seals on each thru bolt. Tighten the thru bolts to 20 in lbs (2.5 Nm). Apply OMC Black Neoprene Dip, or equivalent, over the bolt heads and over the two end cap gaskets to prevent any leakage.

Run the motor for a few seconds in both directions to seat the brushes.

II
MAINTENANCE

11-1 INTRODUCTION

The authors estimate 75% of engine repair work can be directly or indirectly attributed to lack of proper care for the engine. This is especially true of care during the off-season period. There is no way on this green earth for a mechanical engine, particularily an outboard motor, to be left sitting idle for an extended period of time, say for six months, and then be ready for instant satisfactory service.

Imagine, if you will, leaving your automobile for six months, and then expecting to turn the key, have it roar to life, and be able to drive off in the same manner as a daily occurrence.

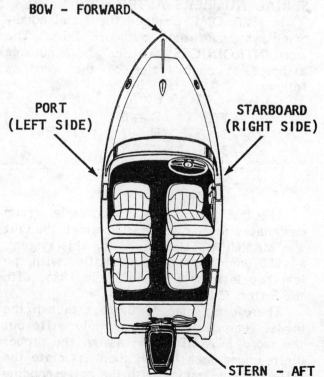

Common terminology used throughout the world for reference designation on boats. These are the terms used in this book.

Labels in figure: BOW - FORWARD; PORT (LEFT SIDE); STARBOARD (RIGHT SIDE); STERN - AFT

It is critical for an outboard engine to be run at least once a month, preferably, in the water. At the same time, the shift mechanism should be operated through the full range several times and the steering operated from hard-over to hard-over.

Only through a regular maintenance program can the owner expect to receive long life and satisfactory performance at minimum cost.

Many times, if an outboard is not performing properly, the owner will "nurse" it through the season with good intentions of working on the unit once it is no longer being used. As with many New Year's resolutions, the good intentions are not completed and the outboard may lie for many months before the work is begun or the unit is taken to the marine shop for repair.

Imagine, if you will, the cause of the problem being a blown head gasket. And let us assume water has found its way into a cylinder. This water, allowed to remain over a long period of time, will do considerably more damage than it would have if the unit had been disassembled and the repair work performed immediately. **THEREFORE**, if an outboard is not functioning properly, **DO NOT** stow it away with promises to get at it when you get time, because the work and expense will only get worse, the longer corrective action is postponed. In the example of the blown head gasket, a relatively simple and inexpensive repair job could very well develop into major overhaul and rebuild work.

Outboards On Sail Boats

Owners of sail boats pride themselves in their ability to use the wind to clear a harbor or for movement from Port A to Port B, or maybe just for a day sail on a lake. The outboard is carried only as a last resort -- in case the wind fails completely, or in an emergency situation.

As a result, the outboard is stowed below, usually in a very poorly ventilated area, and subjected to moisture, stale air --in short, an excellent enviroment for "sweating" and corrosion.

If the owner could just take the time about once every month or two, to pull out his outboard, clean it up, and give it a short run, not only would he have "peace of mind" knowing it **WILL** start in an emergency, but also his maintenance cost will be drastically reduced.

Chapter Coverage

The material presented in this chapter is divided into five general areas.

1- General information every boat owner should know.

2- Maintenance tasks that should be performed periodically to keep the boat operating at minimum cost.

3- Care necessary to maintain the appearance of the boat and to give the owner that "Pride of Ownership" look.

4- Winter storage practices to minimize damage during the off-season when the boat is not in use.

5- Preseason preparation work that should be performed to ensure satisfactory performance the first time it is put in service.

In nautical terms, the front of the boat is the **bow** and the direction is **forward**; the rear is the **stern** and the direction is **aft**; the right side, when facing forward, is the **starboard** side; and the left side is the **port** side. All directional references in this manual use this terminology. Therefore, the

direction from which an item is viewed is of no consequence, because **starboard** and **port NEVER** change no matter where the individual is located or in which direction he may be looking.

11-2 ENGINE SERIAL NUMBERS

The engine serial numbers are the manufacturer's key to engine changes. These numbers identify the year of manufacture, the qualified horsepower rating, and the parts book identification. If any correspondence or parts are required, the engine model number **MUST** be used or proper identification is not possible. The accompanying illustrations will be very helpful in locating the engine identification tag for the various models.

The model number establishes the year in which the engine was produced and not neccessarily the year of first installation.

On some model engines, the serial number and model number were stamped on a plate mounted between the two swivel brackets underneath the hood.

On other model engines, the plate is mounted on the port side of the engine on the front or side of the swivel bracket. The hp and rpm range will also be found on the plate.

SERIAL NUMBERS AFTER 1979

In 1980, OMC changed the serial numbering system of their outboard units. The word **INTRODUCES** was used and a number assigned to each letter of the word as follows:

I -- 1	D -- 6
N -- 2	U -- 7
T -- 3	C -- 8
R -- 4	E -- 9
O -- 5	S -- 0

The last two letters of the serial group designates the model year in which the unit was **MANUFACTURED**. Using this system, a 1980 engine would be identified with the last two letters "C" and "S"; in 1985, with the letters C and O, etc.

Therefore, since 1980, to establish the model year of an engine, simply write out the word INTRODUCES; assign the proper digits under each letter; then associate the letters on the engine with the corresponding letters of "INTRODUCES" and the model year is established.

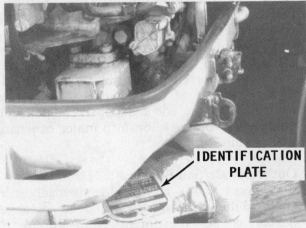

IDENTIFICATION PLATE

Manufacturer's identification plate installed between the transom brackets.

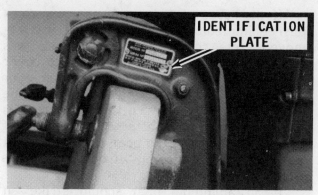

Manufacturer's identification plate installed on the port side of the transom bracket.

A new fiberglass boat and trailer outfit ready for an owner and a power package.

11-3 FIBERGLASS HULLS

Fiberglass reinforced plastic hulls are tough, durable, and highly resistant to impact. However, like any other material they can be damaged. One of the advantages of this type of construction is the relative ease with which it may be repaired. Because of its break characteristics, and the simple techniques used in restoration, these hulls have gained popularity throughout the world. From the most congested urban marina, to isolated lakes in wilderness areas, to the severe cold of far off northern seas, and in sunny tropic remote rivers of primitive islands or continents, fiberglass boats can be found performing their daily task with a minimum of maintenance.

A fiberglass hull has almost no internal stresses. Therefore, when the hull is broken or stove-in, it retains its true form. It will not dent to take an out-of-shape set. When the hull sustains a severe blow, the impact will be either absorbed by deflection of the laminated panel or the blow will result in a definite, localized break. In addition to hull damage, bulkheads, stringers, and other stiffening structures attached to the hull, may also be affected and therefore, should be checked. Repairs are usually confined to the general area of the rupture.

11-4 ALUMINUM HULLS

Aluminum boats have become popular in recent years because they are so lightweight and may be carried with ease atop an automobile or other vehicle. These aluminum

An aluminum boat ready for an engine. The owner of this type boat will probably carry it atop his vehicle and be saved the expense and trouble of trailering to the water.

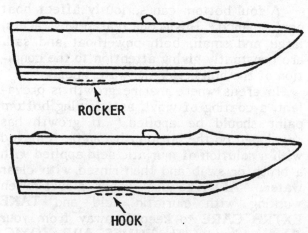

Simple drawing to illustrate two types of possible damage to the hull. Such injury to the boat will affect the boat's performance and subtract from the owner's enjoyment.

A boat and outboard used in salt water. Notice the marine growth on the lower unit and the anti-fouling bottom paint on the hull which prevented the marine growth.

craft are available in sizes ranging from small 8-foot prams to twin-hulled pontoon houseboats or swimming "rafts" in excess of 30 feet. Naturally, the large units cannot be carried atop a vehicle.

One of the advantages of an aluminum hull is the easy maintenance program required, and the ability of the material to resist corrosion.

As an added protection against marine growth, the below the waterline area may be painted with an anti-fouling paint. Bottom paint sold for use on a wooden or fiberglass hull is **NOT** suitable. At the time of purchase, check to be sure the paint contains the chemical properties required for an aluminum surface. The label should clearly indicate the intended use is specifically for aluminum.

If the aluminum hull does not have antifouling paint but requires cleaning to remove marine growth, one method is to rub the hull with a gunny sack just as soon as the boat is removed from the water and while it is still wet. The roughness of the sack is fairly effective in cleaning the surface of marine growth, including crustaceans (barnacles for instance) that have attached themselves to the hull. As soon as the rubdown has been completed the hull should be washed with high-pressure fresh water.

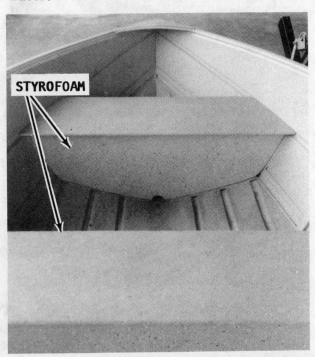

Aluminum boat with the wooden seat removed exposing the Styrofoam blocks for flotation. The seat should be removed at least once each season and the blocks thoroughly dried.

If the rubdown and wash was not accomplished immediately after the boat was removed from the water and the hull was allowed to dry, it will be necessary to cover the hull with wet blankets, gunny sacks or other suitable material and to continue soaking the covering until the growth is loosened. An easy alternate method, of course, is to return the boat to the water, if possible, and then too pull it out after it has been allowed to soak.

If an aluminum boat should strike an underwater object resulting in damage to the hull and a leak develops, the only emergency action possible is to make an attempt to reduce the amount of water being taken on by stuffing any type of available material into the opening until the boat is returned to shore. The aluminum cannot be repaired while it is wet. Repair of a damaged hull must be performed by a shop equipped for heliarc welding and other aluminum work.

Styrofoam blocks are installed under the seats of all aluminum boats. The foam blocks are designed for flotation to prevent the boat from sinking even if it should fill with water. Once each season, the wooden seat should be removed and the foam allowed to dry. Some manufacturers enclose the foam blocks in plastic bags prior to intallation to protect them from moisture and loss of their flotation ability. New blocks may be purchased in a wide range of sizes. If new blocks are obtained, make an attempt to enclose the block in some form of plastic covering, then seal the package before installing it under the seat.

11-5 BELOW WATERLINE SERVICE

A foul bottom can seriously affect boat performance. This is one reason why racers, large and small, both powerboat and sail, are constantly giving attention to the condition of the hull below the waterline.

In areas where marine growth is prevalent, a coating of vinyl, anti-fouling bottom paint should be applied. If growth has developed on the bottom, it can be removed with a solution of muriatic acid applied with a brush or swab and then rinsed with clear water. **ALWAYS** use rubber gloves when working with muriatic acid and **TAKE EXTRA CARE** to keep it away from your face and hands. The **FUMES ARE TOXIC.** Therefore, work in a well-ventilated area, or if outside, keep your face on the windward side of the work.

Barnacles have a nasty habit of making their home on the bottom of boats which have not been treated with anti-fouling paint. Actually they will not harm the fiberglass hull, but can develop into a major nuisance.

If barnacles or other crustaceans have attached themselves to the hull, extra work will be required to bring the bottom back to a satisfactory condition. First, if practical, put the boat into a body of fresh water and allow it to remain for a few days. A large percentage of the growth can be removed in this manner. If this remedy is not possible, wash the bottom thoroughly with a high-pressure fresh water source and use a scraper. Small particles of hard shell may still hold fast. These can be removed with sandpaper.

11-6 SUBMERGED ENGINE SERVICE

A submerged engine is always the result of an unforeseen accident. Once the engine is recovered, special care and service procedures **MUST** be closely followed in order to return the unit to satisfactory performance.

NEVER, again we say **NEVER** allow an engine that has been submerged to stand more than a couple hours before following the procedures outlined in this section and making every effort to get it running. Such delay will result in serious internal damage. If all efforts fail and the engine cannot be started after the following procedures have been performed, the engine should be disassembled, cleaned, assembled, using new gaskets, seals, and O-rings, and then started as soon as possible.

Submerged engine treatment is divided into three unique problem areas: submersion in salt water; submerged engine while running; and a submerged engine in fresh water, including special instructions.

Crankshaft from a submerged two-cylinder engine recovered from salt water. In a very short time the crank was severely damaged by corrosion.

Rod bearing and cages badly damaged by salt water corrosion.

The most critical of these three circumstances is the engine submerged in salt water, with submersion while running a close second.

Salt Water Submersion

NEVER attempt to start the engine after it has been recovered. This action will only result in additional parts being damaged and the cost of restoring the engine increased considerably. If the engine was submerged in salt water the complete unit **MUST** be disassembled, cleaned, and assembled with new gaskets, O-rings, and seals. The corrosive effect of salt water can only be eliminated by the complete job being properly performed.

Cleaner to restore an engine recovered from fresh water after it has been submerged.

Submerged While Running
Special Instructions

If the engine was running when it was submerged, the chances of internal engine damage is greatly increased. After the engine has been recovered, remove the spark plugs and attempt to rotate the flywheel with the rewind starter. On larger horsepower engines without a rewind starter, use a socket wrench on the flywheel nut. If the attempt to rotate the flywheel fails, the chances of serious internal damage, such as, bent connecting rod, bent crankshaft, or damaged cylinder, is greatly increased. If all attempts to rotate the flywheel fail, the powerhead must be completely disassembled.

Submerged Engine — Fresh Water
SPECIAL WORD

As an aid to performing the restoration work, the following steps are numbered and should be followed in sequence. However, illustrations are not included with the procedural steps because the work involved is general in nature.

1- Recover the engine as quickly as possible.

2- Remove the cowling and the spark plugs.

3- Remove the carburetor. To rebuild the carburetor, see Chapter 4.

4- Flush the outside of the engine with fresh water to remove silt, mud, sand, weeds, and other debris. **DO NOT** attempt to start the engine if sand has entered the powerhead. Such action will only result in serious damage to powerhead components. Sand in the powerhead means the unit must be disassembled.

5- Remove as much water as possible from the powerhead. Most of the water can be eliminated by first holding the engine in a horizontal position with the spark plug holes **DOWN,** and then cranking the engine with the rewind starter or with a socket wrench on the flywheel nut.

6- Alcohol will absorb water. Therefore, pour alcohol into the carburetor throat and again crank the engine.

7- Lay the engine in a horizontal position, and then roll it over until the spark plug openings are facing **UPWARD.** Pour alcohol into the spark plug openings and again crank the engine.

8- Roll the engine in the horizontal position until the spark plug openings are again facing **DOWN.** Pour engine oil into the

carburetor throat and, at the same time, crank the engine to distribute oil throughout the crankcase.

9- With the engine still in a horizontal position, roll it over until the spark plug holes are again facing **UPWARD.** Pour approximately one teaspoon of engine oil into each spark plug opening. Crank the engine to distribute the oil in the cylinders.

10- Install the spark plugs and tighten them to the torque value given in the Appendix. Connect the high-tension leads to the spark plugs.

11- Install the carburetor onto the engine with a **NEW** gasket on the intake manifold.

12- Mount the engine in a test tank or body of water.

CAUTION: Water must circulate through the lower unit to the engine any time the engine is run to prevent damage to the water pump in the lower unit. Just five seconds without water will damage the water pump.

Obtain **FRESH** fuel and attempt to start the engine. If the engine will start, allow it to run for approximately an hour to eliminate any water remaining in the engine.

Rust preventative to be sprayed inside the engine in preparation for storage, as explained in the text.

13- If the engine fails to start, determine the cause, electrical or fuel, correct the problem, and again attempt to get it running. **NEVER** allow an engine to remain unstarted for more than a couple hours without following the procedures in this section and attempting to start it. If attempts to start the engine fail, the unit should be disassembled, cleaned, assembled, using new gaskets, seals, and O-rings, just as soon as possible.

11-7 WINTER STORAGE

Taking extra time to store the boat properly at the end of each season, will increase the chances of satisfactory service for the next season. **REMEMBER**, idleness is the greatest enemy of an outboard motor. The unit should be run on a monthly basis. The boat steering and shifting mechanism should also be worked through complete cycles several times each month. The owner who spends a small amount of time involved in such maintenance will be rewarded by satisfactory performance, and greatly reduced maintenance expense for parts and labor.

ALWAYS remove the drain plug and position the boat with the bow higher than the stern. This will allow any rain water

and melted snow to drain from the boat and prevent "trailer sinking". This term is used to describe a boat that has filled with rain water and ruined the interior, because the plug was not removed or the bow was not high enough to allow the water to drain properly.

Proper storage for the engine involves adequate protection of the unit from physical damage, rust, corrosion, and dirt.

The following steps provide an adequate maintenance program for storing the unit at the end of a season.

1- Remove the cowling. Start the powerhead and allow it to warm to operating temperature.

CAUTION: Water must circulate through the lower unit to the engine any time the engine is run to prevent damage to the water pump in the lower unit. Just five seconds without water will damage the water pump.

Disconnect the fuel line from the engine and allow the unit to run at **LOW** rpm and, at the same time, inject about 4 ounces of rust preventative spray through each carburetor throat. Allow the engine to run until it shuts down from lack of fuel.

2- Drain the fuel tank and the fuel lines. Pour approximately one qt. of benzol (benzene) into the fuel tank, and then rinse the tank and pickup filter with the benzol. Drain the tank. Store the fuel tank in a cool dry area with the vent **OPEN** to allow air to circulate through the tank. **DO NOT** store the fuel tank on bare concrete. Place the tank to allow air to circulate around it. If

Engine mounted in a test tank. The engine can be safely operated at idle speeds in preparation for winter storage, as explained in the text.

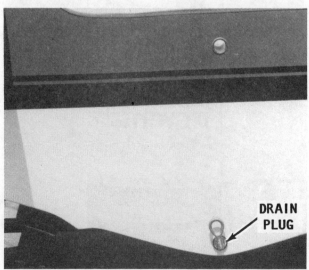

DRAIN PLUG

Drain plug removed from the transom to allow rain and melted snow to drain from the boat. Failure to remove this plug during long periods of storage can cause "beaucoup" problems.

Standard OMC fuel tank. During periods of storage the tank should be empty and the cap "cracked" open to allow the tank to "breathe".

the fuel tank containing fuel is to be stored for more than a month, a commercial additive such as Sta-Bil should be added to the fuel. This type of additive will maintain the fuel in a "fresh" condition for up to a full year.

3- Clean the carburetor fuel filter/s with benzol, see Chapter 4, Carburetor Repair Section.

OMC fuel conditioner added to the fuel will keep it fresh for up to one full year.

Rust preventative to be used when preparing the engine for long periods of non-use and storage.

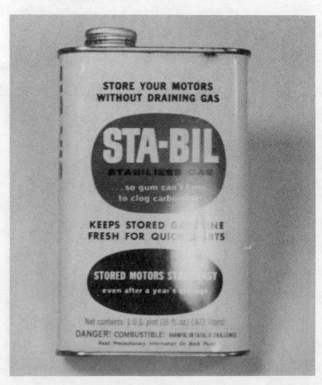

Chemical additives, such as Sta-Bil and the OMC fuel conditioner at the top of the page will prevent fuel from "souring" for up to twelve months.

4- Drain, and then fill the lower unit with OMC Lower Unit Gear Lubricant, as outlined in Section 11-8.

5- Lubricate the throttle and shift linkage. Lubricate the swivel pin and the tilt tube with Multi-purpose Lubricant, or equivalent.

Clean the engine thoroughly. Coat the powerhead with Corrosion and Rust Preventative spray. Install the cowling and then apply a thin film of fresh engine oil to all painted surfaces.

Remove the propeller. Apply anti-sieze or a waterproof sealer to the propeller shaft, and then install the propeller back in position.

FINAL WORDS: Be sure all drain holes in the gear housing are open and free of obstruction. Check to be sure the **FLUSH** plug has been removed to allow all water to drain. Trapped water could freeze, expand, and cause expensive castings to crack.

ALWAYS store the engine off the boat with the lower unit below the powerhead to prevent any water from being trapped inside. The ideal storage position for an outboard is to hang it with the lower unit down. If hanging is not practical, lay the outboard on its back. This will place the lower unit below the powerhead.

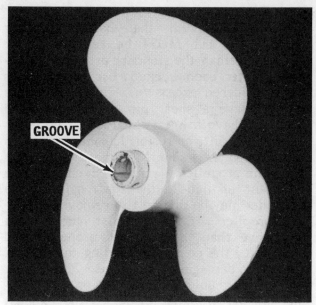

Propeller with grooves worn into the hub.

11-8 LOWER UNIT SERVICE

PROPELLERS

With Shear Pin

The propeller should be checked regularly to be sure all the blades are in good condition. If any of the blades become bent or nicked, such damage will set up vibrations in the motor. Remove and inspect the propeller. Use a file to trim nicks and burrs. **TAKE CARE** not to remove any more material than is absolutely necessary. For a complete check, take the propeller to your marine dealer where the proper equipment and knowledgeable mechanics are available to perform a proper job at modest cost.

Inspect the propeller shaft to be sure it is still true and not bent. If the shaft is not perfectly true, it should be replaced.

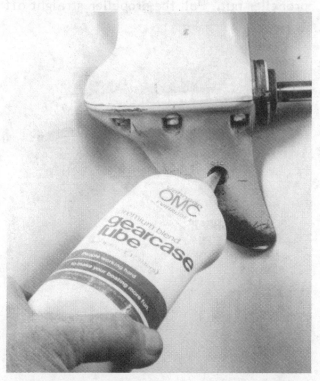

Adding lubricant to the lower unit on a small horsepower engine. The lubricant must always be added through the drain plug after the upper vent plug has been removed.

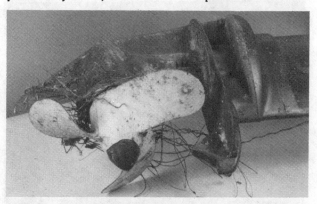

Sea weed entangled in the propeller. The propeller should be removed frequently and any foreign material, especially fish line removed before it can cause damage.

With Exhaust

Propellers with the exhaust passing through the hub **MUST** be removed more frequently than the standard propeller. Removal after each weekend use or outing is not considered excessive. These propellers do not have a shear pin. The shaft and propeller have splines which **MUST** be coated with an anti-corrosion lubricant prior to installation as a aid to removal the next time the propeller is pulled. Even with the lubricant applied to the shaft splines, the propeller may be difficult to remove.

The propller with the exhaust hub is more expensive than the standard propeller and therefore, the cost of rebuilding the unit, if the hub is damaged, is justified.

A replaceable diffuser ring on the backside of the propeller disperses the exhaust away from the propeller blades. If the ring becomes broken or damaged "ventillation" would be created pulling the exhaust gases back into the negative pressure area behind the propeller. This condition would create considerable air bubbles and reduce the effectiveness of the the propeller.

Standard Propeller Removal

On some model engines, the shear pin is installed behind the propeller. First, pull the cotter key, and then remove the propeller nut, drive pin, and washer. Because the drive pin is not a tight fit, the propeller is able to move on the pin and cause burrs on the hole. These burrs may make removing the propeller difficult. To overcome this problem, the propeller hub has two grooves running the full length of the hub. Hold the shaft from turning, and then rotate the propeller 1/4 turn to position the grooves over the drive pin holes. The propeller can then be pulled straight off the shaft. After the propeller has been removed, file the drive pin holes on both sides of the shaft to remove the burrs.

Standard Propeller Installation

On some model engines, the shear pin is installed behind the propeller. On such units the propeller shaft should be coated with Perfect Seal, and then the propeller installed. After the propeller is on the shaft, install the washer, shear pin, propeller nut, and finally a **NEW** cotter pin.

Propeller With Exhaust Removal

First, disconnect the high tension leads to the spark plugs to prevent accidental engine start. Next, pull the cotter pin from the propeller nut. Wedge a piece of wood between on of the propeller blades and the cavitation plate to prevent the propeller from rotating. Back off the castellated propeller nut. Pull the propeller straight off

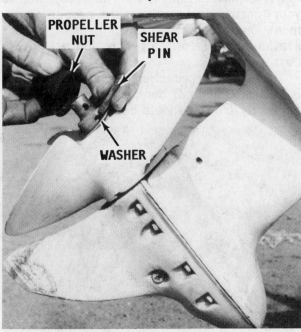

Propeller installation with a washer and shear pin. The cotter pin is installed through the propeller nut. This type of installation is the most popular.

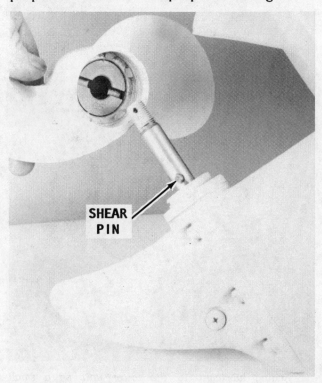

Propeller with a slot for the shear pin. The pin is inserted first, then the propeller is installed onto the propeller shaft and over the shear pin.

the shaft. It may be necessary to carefully tap on the front side of the propeller with a soft headed mallet to jar it loose. If the propeller appears to be "frozen" to the shaft, see Chapter 8 for special removal instructions. The thrust washer does not have to be removed unless it appears damaged.

Propeller Exhaust Installation

First, check to be sure the high tension leads have been disconnected from the spark plugs to prevent accidental engine start. Install the thrust washer onto the propeller shaft, if it was removed. Coat the splines of the driveshaft with anti-corrosion lubricant. The lubricant **MUST** be applied to the shaft **EACH** time the propeller is installed to prevent it from "freezing" to the shaft. The propeller may "freeze" to the shaft in a short time in fresh water and much sooner in salt water.

Slide the propeller onto the shaft with the splines on the shaft indexed with the splines in the propller hub. Force the propeller onto the shaft until it is tight against

Applying gasket sealer to the shaft splines prior to installing a "prop exhaust" propeller.

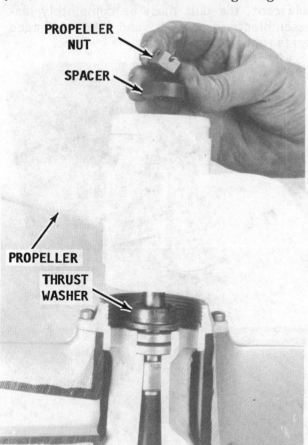

Installing a "prop exhaust" propeller. The thrust washer is installed first, then the propeller, spacer, nut and finally the cotter pin through the propeller nut.

OMC anti-corrosion lubricant that should be applied to the propeller shaft before the propeller is installed. Such lubricant on the shaft will not only fight corrosion, but assist in propeller removal.

the thrust washer. If the propeller cannot be moved tight against the thrust washer, the splines in the hub or on the shaft are dirty must be cleaned.

After the propeller is in place, wedge a piece of wood between one of the blades and the cavitation plate to prevent the propeller from rotating. Thread the castellated nut onto the shaft and bring it up tight against the propeller. Insert the cotter pin through the nut and propeller shaft. If the hole in the shaft does not align with one of the holes in the nut, **TIGHTEN** the nut until one of the holes is aligned. **NEVER** loosen the nut to align the holes. Install the cotter pin, remove the piece of wood, and connect the high tension leads to the spark plugs.

Draining Lower Unit

Remove the **VENT** plug just above the anti-cavitation plate **FIRST,** and then the **FILL** plug from the gear housing. **NEVER** remove the vent or filler plug when the drive unit is hot. Expanded lubricant would be released through the plug hole.

CRITICAL WORD

The Phillips screw securing the shift fork in place is located very close to the drain plug. If the wrong screw is removed,

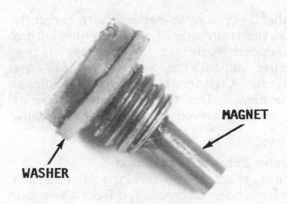

MAGNET

WASHER

Late model lower unit drain plugs have a magnet to attract metal particles in the lubricant before they cause damage. If an unusual amount of particles are discovered, disassembly and search should be made to discover the source.

BAD NEWS, VERY BAD NEWS. The lower unit will have to be disassembled in order to return the shift fork to its proper location.

Allow the gear lubricant to drain into the container. As the lubricant drains, catch some with your fingers, from time-to-time, and rub it between your thumb and finger to determine if any metal particles are present. If metal is detected in the lubricant, the unit must be completely disassembled, inspected, and the damaged parts replaced.

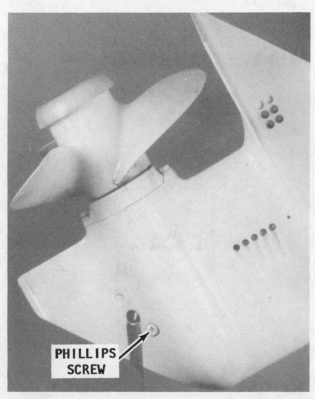

PHILLIPS SCREW

*Draining lubricant from a lower unit. **TAKE CARE** not to remove the Phillips screw by mistake. This screw secures the shift fork in position.*

Lubricant being drained from a lower unit without shift capabilities. Therefore, the Phillips screw mentioned in the illustration in the left column is not present.

If the lubricant appears milky brown, or if large amounts of lubricant must be added to bring the lubricant up to the full mark, a thorough check should be made to determine the cause of the loss.

Filling Lower Unit

Add only OMC lower unit lubricant. Lubricant, for engines covered in this manual, is as follows: Use **ONLY** Type C, now known as Premium Blend Gearcase Lube, in all electric shift models. Use either Premium Blend Lube or the OMC Hi-Vis Gearcase Lube for all other engines. **NEVER** use regular automotive-type grease in the lower unit because it expands and foams too much. Lower units do not have provisions to accommodate such expansion.

The gearcase lubricant should be changed twice each year or season. If the lubricant is purchased in a large container, say the one-gallon size, a considerable savings can be realized. What is not used this season will be used in the next or the one after. A small inexpensive pump can be be

Filling the lower unit on a small horsepower engine through the lower hole after the vent plug has been removed.

purchased to move the lubricant from the large container to the lower unit.

Position the drive unit approximately vertical and without a list to either port or starboard. Insert the lubricant tube into the **FILL/DRAIN** hole at the bottom plug hole, and inject lubricant until the excess begins to come out the **VENT** hole. Install the **VENT** and **FILL** plugs with **NEW** gaskets.

The two types of lubricant used for Johnson/Evinrude engines. The text clearly identifies the lubricant to be used on the different model engines covered in this manual.

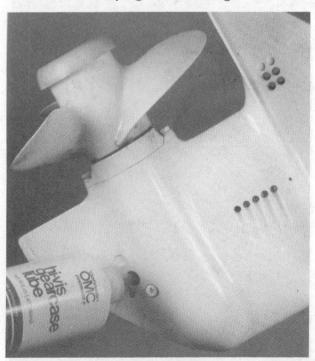

Filling a lower unit with a "prop exhaust" propeller. The lower unit should be "topped off" as outlined in the text.

Filling a non-shifting lower unit with lubricant.

After the lower plug has been installed, remove the vent plug again and using a squirt-type oil can, add lubricant through this vent hole. A squirt-type oil can must be used to allow the trapped air in the lower unit to escape at the same time the final lubricant is added. Once the unit is completely full, install and tighten the vent plug.

Using a squirt can to "top off" the lubricant in the lower unit, as explained in the text.

Check to be sure the vent and drain plug gaskets are properly positioned to prevent water from entering the housing.

See the Appendix for lower unit capacities.

11-9 BATTERY STORAGE

Remove the batteries from the boat and keep them charged during the storage period. Clean the batteries thoroughly of any dirt or corrosion, and then charge them to full specific gravity reading. After they are fully charged, store them in a clean cool dry place where they will not be damaged or knocked over.

NEVER store the battery with anything on top of it or cover the battery in such a manner as to prevent air from circulating around the filler caps. All batteries, both new and old, will discharge during periods of storage, more so if they are hot than if they remain cool. Therefore, the electrolyte

A check of the electrolyte in the battery should be on the maintenance schedule for any boat. A hydrometer reading of 1.300 or in the green band, indicates the battery is in satisfactory condition. If the reading is 1.150 or in the red band, the battery needs to be charged.

level and the specific gravity should be checked at regular intervals. A drop in the specific gravity reading is cause to charge them back to a full reading.

In cold climates, **EXERCISE CARE** in selecting the battery storage area. A fully-charged battery will freeze at about 60 degrees below zero. A discharged battery, almost dead, will have ice forming at about 19 degrees above zero.

11-10 PRESEASON PREPARATION

Satisfactory performance and maximum enjoyment can be realized if a little time is spent in preparing the engine for service at the beginning of the season. Assuming the unit has been properly stored, as outlined in Section 11-7, a minimum amount of work is required to prepare the engine for use.

The following steps outline an adequate and logical sequence of tasks to be performed before using the engine the first time in a new season.

1- Lubricate the engine according to the manufacturer's recommendations. Remove, clean, inspect, adjust, and install the spark plugs with new gaskets if they require gaskets. Make a thorough check of the ignition system. This check should include: the points, coil, condenser, condition of the wiring, and the battery electrolyte level and charge.

2- If a built-in fuel tank is installed, take time to check the tank and all of the

Today, numerous type spark plugs are available for service. **ALWAYS** *check with the local OMC dealer to be sure you are purchasing the proper plugs for the engine being serviced.*

fuel lines, fittings, couplings, valves, and the flexible tank fill and vent. Turn on the fuel supply valve at the tank. If the fuel was not drained at the end of the previous season, make a careful inspection for gum formation. If a six-gallon fuel tank is used, take the same action. When gasoline is allowed to stand for long periods of time, particularly in the presence of copper, gummy deposits form. This gum can clog the filters, lines, and passageways in the carburetor. See Chapter 4, Fuel System Service.

3- Check the oil level in the lower unit by first removing the vent screw on the port side just above the anti-cavitation plate. Insert a short piece of wire into the hole and check the level. Fill the lower unit according to procedures outlined in Section 11-8.

4- Close all water drains. Check and replace any defective water hoses. Check to be sure the connections do not leak.

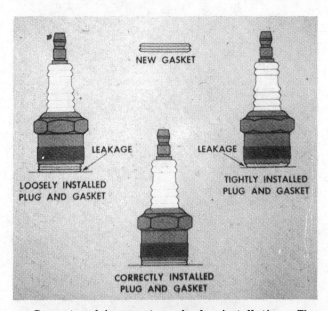

NEW GASKET

LEAKAGE   LEAKAGE

LOOSELY INSTALLED   TIGHTLY INSTALLED
PLUG AND GASKET   PLUG AND GASKET

CORRECTLY INSTALLED
PLUG AND GASKET

Correct and incorrect spark plug installation. The plugs **MUST** *be installed properly and tightened to the proper torque value for satisfactory performance.*

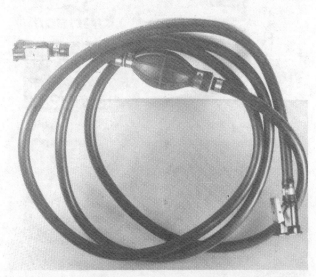

Typical fuel hose with squeeze bulb. The hose and bulb must remain flexible. The O-rings **MUST** *prevent fuel leakage.*

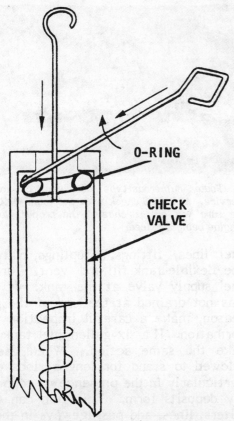

Method of removing an O-ring from a connector. The connector is also replaceable.

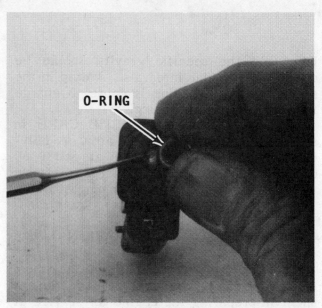

Using a punch to depress the check ball while installing an oiled O-ring.

Replace any spring-type hose clamps, if they have lost their tension, or if they have distorted the water hose, with band-type clamps.

5- The engine can be run with the lower unit in water to flush it. If this is not practical, a flush attachment may be used. This unit is attached to the water pick-up in the lower unit. Attach a garden hose, turn on the water, allow the water to flow into the engine for awhile, and then run the engine.

CAUTION: Water must circulate through the lower unit to the engine any time the engine is run to prevent damage to the water pump in the lower unit. Just five seconds without water will damage the water pump.

Adding OMC oil to the fuel. Only a high grade oil should be added to the fuel to ensure proper lubrication.

OMC lubricants for Johnson/Evinrude outboards.

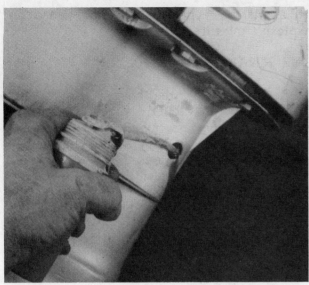

Using a squirt can to "top off" the lubricant in a lower unit, as described in the text.

Check the idle exhaust port for water discharge. At idle speed, only a fine water mist will be visible. Check for leaks. Check operation of the thermostat. After the engine has reached operating temperature, tighten the cylinder head bolts to the torque value given in the Specifications in the Appendix.

The battery should be located near the engine and well secured to prevent even the slightest amount of movement. The battery, including the terminals MUST be kept clean for maximum performance.

Testing a 25 hp unit, mounted on an aluminum boat, in a test tank.

An OMC degreaser widely used for cleaning the boat and engine.

6- Check the electrolyte level in the batteries and the voltage for a full charge. Clean and inspect the battery terminals and cable connections. **TAKE TIME** to check the polarity, if a new battery is being installed. Cover the cable connections with grease or special protective compound as a prevention to corrosion formation. Check all electrical wiring and grounding circuits.

7- Check all electrical parts on the engine and electrical fixture or connections in the lower portions of the hull inside the boat to be sure they are not of a type that could cause ignition of an explosive atmosphere. Rubber caps help keep spark insulators clean and reduce the possibility of arcing. Starters, generators, distributors, alternators, electric fuel pumps, voltage regulators, and high-tension wiring harnesses should be of a marine type that cannot cause an explosive mixture to ignite.

ONE FINAL WORD

Before putting the boat in the water, **TAKE TIME** to **VERIFY** the drain plugs are installed. Countless number of boating excursions have had a very sad beginning because the boat was eased into the water only to have water begin filling the inside.

Keep the gas tank full, the trim tab trimmed, the fuel pump pumping, the spark plugs sparking, and the pistons, well, keep them working too.

Joan & Clarence

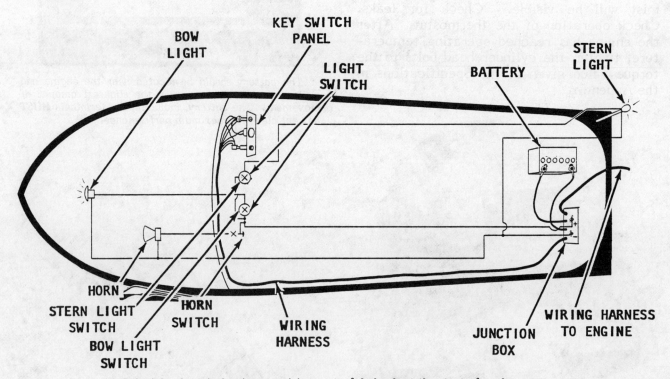

Principle electrical points requiring a careful check at the start of each season.

APPENDIX

METRIC CONVERSION CHART

LINEAR
inches	X 25.4	= millimetres (mm)
feet	X 0.3048	= metres (m)
yards	X 0.9144	= metres (m)
miles	X 1.6093	= kilometres (km)
inches	X 2.54	= centimetres (cm)

AREA
inches^2	X 645.16	= millimetres^2 (mm^2)
inches^2	X 6.452	= centimetres^2 (cm^2)
feet^2	X 0.0929	= metres^2 (m^2)
yards^2	X 0.8361	= metres^2 (m^2)
acres	X 0.4047	= hectares (10^4 m^2) (ha)
miles^2	X 2.590	= kilometres^2 (km^2)

VOLUME
inches^3	X 16387	= millimetres^3 (mm^3)
inches^3	X 16.387	= centimetres^3 (cm^3)
inches^3	X 0.01639	= litres (l)
quarts	X 0.94635	= litres (l)
gallons	X 3.7854	= litres (l)
feet^3	X 28.317	= litres (l)
feet^3	X 0.02832	= metres^3 (m^3)
fluid oz	X 29.60	= millilitres (ml)
yards^3	X 0.7646	= metres^3 (m^3)

MASS
ounces (av)	X 28.35	= grams (g)
pounds (av)	X 0.4536	= kilograms (kg)
tons (2000 lb)	X 907.18	= kilograms (kg)
tons (2000 lb)	X 0.90718	= metric tons (t)

FORCE
ounces - f (av)	X 0.278	= newtons (N)
pounds - f (av)	X 4.448	= newtons (N)
kilograms - f	X 9.807	= newtons (N)

ACCELERATION
feet/sec^2	X 0.3048	= metres/sec^2 (m/S^2)
inches/sec^2	X 0.0254	= metres/sec^2 (m/s^2)

ENERGY OR WORK (watt-second - joule - newton-metre)
foot-pounds	X 1.3558	= joules (j)
calories	X 4.187	= joules (j)
Btu	X 1055	= joules (j)
watt-hours	X 3500	= joules (j)
kilowatt - hrs	X 3.600	= megajoules (MJ)

FUEL ECONOMY AND FUEL CONSUMPTION
miles/gal	X 0.42514	= kilometres/litre (km/l)

Note:
235.2/(mi/gal) = litres/100km
235.2/(litres/100 km) = mi/gal

LIGHT
footcandles	X 10.76	= lumens/metre^2 (lm/m^2)

PRESSURE OR STRESS (newton/sq metre - pascal)
inches HG (60 F)	X 3.377	= kilopascals (kPa)
pounds/sq in	X 6.895	= kilopascals (kPa)
inches H2O (60° F)	X 0.2488	= kilopascals (kPa)
bars	X 100	= kilopascals (kPa)
pounds/sq ft	X 47.88	= pascals (Pa)

POWER
horsepower	X 0.746	= kilowatts (kW)
ft-lbf/min	X 0.0226	= watts (W)

TORQUE
pound-inches	X 0.11299	= newton-metres (N·m)
pound-feet	X 1.3558	= newton-metres (N·m)

VELOCITY
miles/hour	X 1.6093	= kilometres/hour (km/h)
feet/sec	X 0.3048	= metres/sec (m/s)
kilometres/hr	X 0.27778	= metres/sec (m/s)
miles/hour	X 0.4470	= metres/sec (m/s)

TEMPERATURE

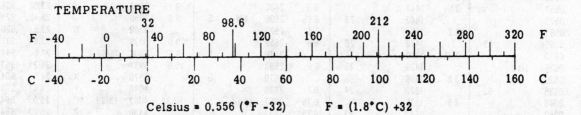

Celsius = 0.556 (°F -32)   F = (1.8°C) +32

DRILL SIZE CONVERSION CHART

SHOWING MILLIMETER SIZES, FRACTIONAL AND DECIMAL INCH SIZES AND NUMBER DRILL SIZES

Milli-Meter	Dec. Equiv.	Frac-tional	Num-ber
.1	.0039		
.15	.0059		
.2	.0079		
.25	.0098		
.3	.0118		
....	.0135		80
.35	.0138		
....	.0145		79
.39	.0156	1/64	
.4	.0157		
....	.0160		78
.45	.0177		
....	.0180		77
.5	.0197		
....	.0200		76
....	.0210		75
.55	.0217		
....	.0225		74
.6	.0236		
....	.0240		73
....	.0250		72
.65	.0256		
....	.0260		71
....	.0280		70
.7	.0276		
....	.0292		69
.75	.0295		
....	.0310		68
.79	.0312	1/32	
.8	.0315		
....	.0320		67
....	.0330		66
.85	.0335		
....	.0350		65
.9	.0354		
....	.0360		64
....	.0370		63
.95	.0374		
....	.0380		62
....	.0390		61
1.0	.0394		
....	.0400		60
....	.0410		59
1.05	.0413		
....	.0420		58
....	.0430		57
1.1	.0433		
1.15	.0452		
....	.0465		56
1.19	.0469	3/64	
1.2	.0472		
1.25	.0492		
1.3	.0512		
....	.0520		55
1.35	.0513		
....	.0550		54
1.4	.0551		
1.45	.0570		
1.5	.0591		
....	.0595		53
1.55	.0610		
1.59	.0625	1/16	
1.6	.0629		
....	.0635		52
1.65	.0649		
1.7	.0669		
....	.0670		51

Milli-Meter	Dec. Equiv.	Frac-tional	Num-ber
1.75	.0689		
....	.0700		50
1.8	.0709		
1.85	.0728		
....	.0730		49
1.9	.0748		
....	.0760		48
1.95	.0767		
1.98	.0781	5/64	
....	.0785		47
2.0	.0787		
2.05	.0807		
....	.0810		46
....	.0820		45
2.1	.0827		
2.15	.0846		
....	.0860		44
2.2	.0866		
2.25	.0885		
....	.0890		43
2.3	.0905		
2.35	.0925		
....	.0935		42
2.38	.0937	3/32	
2.4	.0945		
....	.0960		41
2.45	.0964		
....	.0980		40
2.5	.0984		
....	.0995		39
....	.1015		38
2.6	.1024		
....	.1040		37
2.7	.1063		
....	.1065		36
2.75	.1082		
2.78	.1094	7/64	
....	.1100		35
2.8	.1102		
....	.1110		34
....	.1130		33
2.9	.1141		
....	.1160		32
3.0	.1181		
....	.1200		31
3.1	.1220		
3.18	.1250	1/8	
3.2	.1260		
3.25	.1279		
....	.1285		30
3.3	.1299		
3.4	.1338		
....	.1360		29
3.5	.1378		
....	.1405		28
3.57	.1406	9/64	
3.6	.1417		
....	.1440		27
3.7	.1457		
....	.1470		26
3.75	.1476		
....	.1495		25
3.8	.1496		
....	.1520		24
3.9	.1535		
....	.1540		23
3.97	.1562	5/32	

Milli-Meter	Dec. Equiv.	Frac-tional	Num-ber
....	.1570		22
4.0	.1575		
....	.1590		21
....	.1610		20
4.1	.1614		
4.2	.1654		
....	.1660		19
4.25	.1673		
4.3	.1693		
....	.1695		18
4.37	.1719	11/64	
....	.1730		17
4.4	.1732		
....	.1770		16
4.5	.1771		
....	.1800		15
4.6	.1811		
....	.1820		14
4.7	.1850		13
4.75	.1870		
4.76	.1875	3/16	
4.8	.1890		12
....	.1910		11
4.9	.1929		
....	.1935		10
....	.1960		9
5.0	.1968		
....	.1990		8
5.1	.2008		
....	.2010		7
5.16	.2031	13/64	
....	.2040		6
5.2	.2047		
....	.2055		5
5.25	.2067		
5.3	.2086		
....	.2090		4
5.4	.2126		
....	.2130		3
5.5	.2165		
5.56	.2187	7/32	
5.6	.2205		
....	.2210		2
5.7	.2244		
5.75	.2263		
....	.2280		1
5.8	.2283		
5.9	.2323		
....	.2340		A
5.95	.2344	15/64	
6.0	.2362		
....	.2380		B
6.1	.2401		
....	.2420		C
6.2	.2441		
6.25	.2460		D
6.3	.2480		
6.35	.2500	1/4	E
6.4	.2520		
6.5	.2559		
....	.2570		F
6.6	.2598		
....	.2610		G
6.7	.2638		
6.75	.2657	16/64	
6.75	.2657		
....	.2660		H

Milli-Meter	Dec. Equiv.	Frac-tional	Num-ber
6.8	.2677		
6.9	.2716		
....	.2720		I
7.0	.2756		
....	.2770		J
7.1	.2795		
....	.2811		K
7.14	.2812	9/32	
7.2	.2835		
7.25	.2854		
7.3	.2874		
....	.2900		L
7.4	.2913		
....	.2950		M
7.5	.2953		
7.54	.2968	19/64	
7.6	.2992		
....	.3020		N
7.7	.3031		
7.75	.3051		
7.8	.3071		
7.9	.3110		
7.94	.3125	5/16	
8.0	.3150		
8.1	.3189		
8.2	.3228		
....	.3230		P
8.25	.3248		
8.3	.3268		
8.33	.3281	21/64	
8.4	.3307		
....	.3320		Q
8.5	.3346		
8.6	.3386		
....	.3390		R
8.7	.3425		
8.73	.3437	11/32	
8.75	.3445		
8.8	.3465		
....	.3480		S
8.9	.3504		
9.0	.3543		
9.1	.3583		
9.13	.3594	23/64	
9.2	.3622		
9.25	.3641		
9.3	.3661		
....	.3680		U
9.4	.3701		
9.5	.3740		
9.53	.3750	3/8	
....	.3770		V
9.6	.3780		
9.7	.3819		
9.75	.3838		
9.8	.3858		
....	.3860		W
9.9	.3898		
9.92	.3906	25/64	
10.0	.3937		
....	.3970		X
....	.4040		Y
10.32	.4062	13/32	
....	.4130		Z
10.5	.4134		

Milli-Meter	Dec. Equiv.	Frac-tional
10.72	.4219	27/64
11.0	.4330	
11.11	.4375	7/16
11.5	.4528	
11.51	.4531	29/64
11.91	.4687	15/32
12.0	.4724	
12.30	.4843	31/64
12.5	.4921	
12.7	.5000	1/2
13.0	.5118	
13.10	.5156	33/64
13.49	.5312	17/32
13.5	.5315	
13.89	.5469	35/64
14.0	.5512	
14.29	.5624	9/16
14.5	.5709	
14.68	.5781	37/64
15.0	.5906	
15.08	.5937	19/32
15.48	.6094	39/64
15.5	.6102	
15.88	.6250	5/8
16.0	.6299	
16.27	.6406	41/64
16.5	.6496	
16.67	.6562	21/32
17.0	.6693	
17.06	.6719	43/64
17.46	.6875	11/16
17.5	.6890	
17.86	.7031	45/64
18.0	.7087	
18.26	.7187	23/32
18.5	.7283	
18.65	.7344	47/64
19.0	.7480	
19.05	.7500	3/4
19.45	.7656	49/64
19.5	.7677	
19.84	.7812	25/32
20.0	.7874	
20.24	.7969	51/64
20.5	.8071	
20.64	.8125	13/16
21.0	.8268	
21.04	.8281	53/64
21.43	.8437	27/32
21.5	.8465	
21.83	.8594	55/64
22.0	.8661	
22.23	.8750	7/8
22.5	.8858	
22.62	.8906	57/64
23.0	.9055	
23.02	.9062	29/32
23.42	.9219	59/64
23.5	.9252	
23.81	.9375	15/16
24.0	.9449	
24.21	.9531	61/64
24.5	.9646	
24.61	.9687	31/32
25.0	.9843	
25.03	.9844	63/64
25.4	1.0000	1

TORQUE SPECIFICATIONS

All torque values given in in lb. Divide by **12** for ft lb.

HP	YEAR	FLYWHEEL NUT	CONNECTING ROD SCREW	CYLINDER HEAD SCREWS	CRANKCASE TO CYL SCREWS UPPER LOWER	CRANKCASE TO CYL SCREWS CENTER
1.25	1987 & on	240-300	60-66	60-80	60-80	60-80
2	1971-85	240-300	60-66	60-80	60-80	60-80
2.5	1987 & on	240-300	60-66	60-80	60-80	60-80
4	1971 & on	360-480	60-66	60-80	60-80	60-80
4.5	1980-85	360-480	60-66	60-80	60-80	60-80
5	1984-85	480-600	60-66	144-168	144-168	144-168
6	1971-79	480-540	60-66	60-80	60-80	60-80
6	1982 & on	480-600	60-66	144-168	144-168	144-168
7.5	1980-83	480-600	60-66	144-168	144-168	144-168
8	1984 & on	480-600	60-66	144-168	144-168	144-168
9.5	1971-73	480-540	90-100	96-120	120-132	120-144
9.9	1974-81	540-600	48-60	144-168	144-168	144-168
9.9	1982 & on	540-600	48-60	216-240	144-168	144-168
15	1974-85	540-600	48-60	216-240	144-168	144-168
18	1971-73	480-540	180-186	96-120	110-130	120-130
20	1971-73	480-540	180-186	96-120	110-130	120-130
20	1981 & on	1200-1260	348-372	216-240	168-192	168-192
25	1971-76	480-540	180-186	96-120	120-144	120-144
25	1977 & on	1200-1260	348-272	216-240	168-192	168-192
30	1985 & on	1200-1260	348-372	216-240	168-192	168-192
35	1976-81	1200-1260	348-372	158-192	150-170	162-168
35	1982-84	1200-1260	348-372	216-240	168-192	168-192
40	1971-76	1200-1260	348-372	168-192	150-170	162-168
40	1981-82	1200-1260	348-372	168-192	150-170	162-168
40	1983 & on	1200-1260	348-372	216-240	216-240	216-240
50	1971-76	1200-1260	348-372	168-192	216-240	216-240
50	1977-81	1200-1260	348-372	216-240	216-240	216-240
50	1982-84	960-1020	348-372	216-240	216-240	216-240
50	1985 & on	1200-1260	348-372	216-240	216-240	216-240
55	1976-83	960-1020	348-372	216-240	216-240	216-240
60	1980-81	960-1020	348-372	216-240	216-240	216-240
60	1982-85	1200-1260	348-372	216-240	216-240	216-240

All spark plugs -- 1971-81: 210-246 in lbs, 1982 & on: 216-252 in lbs

STANDARD BOLTS AND NUTS
Torque value for special bolts may vary.

Bolt Size	In./Lbs	Ft/Lbs	Newton Meters
No. 6	7-10	--	0.8-1.2
No. 8	15-22	--	1.6-2.4
No. 10	25-35	2-3	2.8-4.0
No. 12	35-40	3-4	4.0-4.6
1/4"	60-80	5-7	7-9
5/16"	120-140	10-12	14-16
3/8"	220-240	18-20	24-27
7/16"	340-360	28-30	38-40

TUNE-UP SPECIFICATIONS

SEE GENERAL AND SPECIAL NOTES APPENDIX PAGE A-17

JOHNSON MODEL	EVINRUDE MODEL	NO. CYL	HP	CU IN DISPL	WOT RPM	SPARK PLUG	TIMING AT WOT	PRIMARY PICKUP LOCATION	PRIMARY ADJUSTMENT NOTE
1971	**1971**							**1971**	**1971**
2R-71	2102	1	2	2.64	4500	J6J	—	A	1
4R, 4W-71	4106-4136	2	4	5.28	4500	J6J	—	B	1
6R, 6RL-71	6102-03	2	6	8.84	4500	J6J	—	B	1
9R, 9RL-71	9122-23	2	9.5	15.2	4500	J4J	—	C	2
—	18102-03	2	18	22.0	4500	J4J	—	D	3
22R, 22RL-71	25102-03	2	20	22.0	4500	J4J	—	D	3
25R, 25RL-71		2	25	22.0	5500	J4J	—	D	3
40RL, 40E, 40EL-71	40102-03, 40152-53	2	40	43.9	4500	J4J	—	E	5
50ES, 50ESL-71	50172-73	2	50	41.5	5500	L77J4	19°	C	4
1972	**1972**							**1972**	**1972**
2R-72	2202	1	2	2.64	4500	J6J	—	B	1
4R, 4W-72	4206, 4236	2	4	5.28	4500	J6J	—	B	1
6R, 6RL-72	6202-03	2	6	8.84	4500	J6J	—	B	1
9R, 9RL-72	9222-23	2	9.5	15.2	4500	J4J	—	C	2
—	18202-03	2	18	22.0	4500	J4J	—	C	3
20R, 20RL-72	—	2	20	22.0	4500	J4J	—	C	3
25R, 25RL, 25E, 25EL-72	25202-03, 25252-53	2	25	22.0	5500	J4J	—	C	3
40R, 40RL, 40E, 40EL-72	40202-02, 40252-53	2	40	43.9	4500	J4J	—	E	3
50R, 50RL, 50ES, 50ESL-72	50202-03, 50272-73	2	50	41.5	5500	L77J4	19°	C	4
1973	**1973**							**1973**	**1973**
2R-73	2302	1	2	2.64	4500	J6J	—	B	1
4R, 4W-73	4306, 4336	2	4	5.28	4500	J6J	—	B	1
6R, 6RL-73	6302-03	2	6	8.84	4500	J6J	—	B	1
9R, 9RL-73	9322, 9323	2	9.5	15.2	4500	J4J	—	C	2
—	18304-05	2	18	22.0	4500	J4J	—	C	6
20R, 20RL-73	—	2	20	22.0	4500	J4J	—	C	6
25R, 25RL, 25E, 25EL-73	25302-03, 25352-53	2	25	22.0	5500	J4J	—	C	6
40R, 40RL, 40E, 40EL-73	40304-05, 40354-55	2	40	43.9	4500	J4J	—	C	3
50ES, 50ESL	50302-03, 50372-73	2	50	41.5	5500	L77J4	19°	C	4

NOTE: See Appendix Page A-17 for primary pickup location and primary pickup adjustment note called out in this table.

TUNE-UP SPECIFICATIONS

SEE GENERAL AND SPECIAL NOTES APPENDIX PAGE A-17

JOHNSON MODEL	EVINRUDE MODEL	NO. CYL	HP	CU IN DISPL	WOT RPM	SPARK PLUG	TIMING AT WOT	PRIMARY PICKUP LOCATION	PRIMARY ADJUSTMENT NOTE
1974	**1974**							**1974**	
2R-74	2402	1	2	2.64	4500	J6J	—	B	1
4W, 4R-74	4406, 4436	2	4	5.28	4500	J6J	—	B	1
6R, 6RL-74	6402-03	2	6	8.84	4500	J6J	—	B	1
10R, 10RL, 10E, 10EL-74	10424-25, 10454-55	2	9.9	13.2	5000	UL81J	—	C	1
15R, 15RL, 15E, 15EL-74	15404-05, 15454-55	2	15	13.2	6000	UL81J	—	C	1
25R, 25RL, 25E, 25EL-74	25402-03, 25452-53	2	25	22	5500	J4J	—	C	6
40R, 40RL, 40E, 40EL-74	40404-05, 40454-55	2	40	43.9	4500	UL81J	—	C	3
50ES, 50SL-74	50442-43, 50472-73	2	50	41.5	5500	L77J4	19°	C	4
1975	**1975**							**1975**	
2R-75	2502	1	2	2.64	4500	J6J	—	B	1
4W, 4R-75	4506, 4536	2	4	5.28	4500	J6J	—	B	1
6R, 6RL-75	6504-05	2	6	8.84	4500	J6J	—	B	1
10R, 10RL, 10E, 10EL-75	10524-25, 10554-55	2	9.9	13.2	5000	UL81J	—	C	1
15R, 15RL, 15E, 15EL-75	15504-05, 15554-55	2	15	13.2	6000	UL81J	—	C	1
25R, 25RL, 25E, 25EL-75	25502-03, 25552-53	2	25	22	5500	UL81J	—	C	6
40R, 40RL, 40E, 40EL-75	40504-05, 40554-55	2	40	43.9	4500	UL81J	—	C	3
50ES, 50ESL-75	50572-73, 50542-43	2	50	41.5	5500	L77J4	19°	C	4
1976	**1976**							**1976**	
2R-76	2602	1	2	2.64	4500	J6J	—	A	1
4W, 4R-76	4606, 4636	2	4	5.28	4500	J6J	—	B	1
6R, 6RL-76	6604-05	2	6	8.84	4500	J6J	—	D	1
10R, 10RL, 10E, 10EL-76	10624-25, 10654-55	2	9.9	13.2	5000	UL81J	—	C	1
15R, 15RI, 15E, 15EL-76	15604-05, 15654-55	2	15	13.2	6000	UL81J	—	C	1
25R, 25RL, 25E, 25EL-76	25602-03, 25652-53	2	25	22	5500	J4J	—	D	6
35R, 35RL, 35E, 35EL-76	35602-03, 35652-53	2	35	31.8	5500	UL81J	34°	C	1
40R, 40RL, 40E, 40EL-76	40604-05, 40654-55	2	40	43.9	4500	UL81J	—	E	3
55E, 55EL-76	55672-73, 55642-43	2	55	44.99	5500	L77J4	19°	C	4

NOTE: See Appendix Page A-17 for primary pickup location and primary pickup adjustment note called out in this table.

TUNE-UP SPECIFICATIONS

SEE GENERAL AND SPECIAL NOTES APPENDIX PAGE A-17

JOHNSON MODEL	EVINRUDE MODEL	NO. CYL	HP	CU IN DISPL	W O T RPM	SPARK PLUG	TIMING AT WOT	PRIMARY PICKUP LOCATION	PRIMARY ADJUSTMENT NOTE
1977	**1977**							**1977**	
2R-77	2702	1	2	2.64	4500	J6J	—	A	1
4W, 4R-77	4706, 4736	2	4	5.28	4500	L77J4	—	D	1
6R, 6RL-77	6704-05	2	6	8.84	4500	L77J4	—	D	1
10R, 10RL, 10E, 10EL-77	10724-25, 10754-55	2	9.9	13.2	5000	L77J4	—	C	1
15R, 15RL, 15E, 15EL-77	15704-05, 15754-55	2	15	13.2	6000	L77J4	—	C	1
25R, 25RL, 25E, 25EL-77	25702-03, 25752-53	2	25	31.8	5000	L77J4	34°	C	4
35R, 35RL, 35E, 35EL-77	35720-03, 35752-53	2	35	31.8	5500	L77J4	30°	C	4
55E, 55EL-77	55772-73	2	55	44.99	5500	L77J4	19°	C	4
1978	**1978**							**1978**	
2R-78	2802	1	2	2.64	4500	J6J	—	A	1
4W, 4R, 4RL-78	4806, 4836, 4837	2	4	5.28	4500	L77J4	—	C	1
6R, 6RL-78	6804-05	2	6	8.84	4500	L78V	—	D	7
10R, 10RL, 10EL, 10SL-78	10824-25, 10835-55	2	9.9	13.2	5000	L78V	—	C	1
15R, 15RL, 15E, 15EL-78	15804-05, 15854-55	2	15	13.2	6000	L78V	—	C	1
25R, 25RL, 25E, 25EL-78	25802-03, 25852-53	2	25	31.8	5000	L78V	34°	C	4
35R, 35RL, 35E, 35EL-78	35802-03, 35852-53	2	35	31.8	5500	L78V	30°	C	4
50MR, 50MRL-78	50802-03	2	50	44.99	5000	L77J4	19°	C	4
55E, 55EL-78	55874-75	2	55	44.99	5500	L77J4	19°	C	4
1979	**1979**							**1979**	
2R-79	2902	1	2	2.64	4500	J6J	—	D	7
4W, 4R, 4RL-79	4904, 4932, 4933	2	4	5.28	4500	L77J4	—	C	7
6R, 6RL-79	6904-05	2	6	8.4	4500	L77J4	—	D	7
10R, 10RL, 10EL, 10SEL-79	10924-25, 10935-55	2	9.9	13.20	5000	L77J4	—	D	7
15R, 15RL, 15E, 15EL-79	15904-05, 15954-55	2	15	13.20	6000	L77J4	—	D	7
25R, 25RL, 25E, 25EL-79	25904-05, 25952-53	2	25	31.80	5000	L77J4	34°	C	4
35R, 35RL, 35E, 35EL-79	35902-03, 35952-53	2	35	31.80	5500	L77J4	30°	C	4
50R, 50RL-79	50902-03	2	50	44.99	5000	L77J4	19°	C	4
55E, 55EL-79	55974-75	2	55	44.99	5500	L77J4	19°	C	4

NOTE: See Appendix Page A-17 for primary pickup location and primary pickup adjustment note called out in this table.

TUNE-UP SPECIFICATIONS

SEE GENERAL AND SPECIAL NOTES APPENDIX PAGE A-17

JOHNSON MODEL 1980	EVINRUDE MODEL 1980	NO. CYL	HP	CU IN DISPL	W O T RPM	SPARK PLUG	TIMING AT WOT	PRIMARY PICKUP LOCATION 1980	PRIMARY ADJUSTMENT NOTE
2RCS	E2RCS	1	2	2.64	4500	J6J	—	D	7
4WCS, 4RLCS	E4WCS, E4RLCS	2	4	5.28	4500	L77J4	—	C	7
5RCS, 5RLCS, 5RHCS, 5RHLCS	E5RCS, E5RLCS	2	4.5	5.28	5000	L77J4	—	D	7
8RCS, 8RLCS	E8RCS, E8RLCS	2	7.5	10.0	5000	L77J4	—	D	7
10RCS, 10RLCS, 10ELCS, 10SELCS	E10RCS, E10RLCS, E10SELCS, E10ELCS	2	9.9	13.20	5000	L77J4	—	D	7
15RCS, 15RLCS, 15ECS, 15ELCS	E15RCS, E15RLCS, E15ECS, E15ELCS	2	15	13.20	6000	L77J4	—	D	7
25RCS, 25TECS, 25TELCS	E25RCS, E25RLCS, E25TCS, E25TELCS	2	25	31.8	5000	L77J4	34°	B	4
35RCS, 35RLCS, 35ECS, 35ELCS	E35RCS, E35RLCS, E35ECS, E35ELCS	2	35	31.8	5500	L77J4	30°	D	4
50ECS, 50ELCS	—	2	50	44.99	5000	L77J4	19°	C	4
55RCS, 55RLCS	E55RCS, E55RLCS	2	55	44.99	5000	L77J4	19°	C	4
60ECS, 60ELCS	E60ECS, E60ELCS	2	60	44.99	5500	L77J4	21°	C	4

NOTE: See Appendix Page A-17 for primary pickup location and primary pickup adjustment note called out in this table.

TUNE-UP SPECIFICATIONS

SEE GENERAL AND SPECIAL NOTES APPENDIX PAGE A-17

JOHNSON MODEL	EVINRUDE MODEL	NO. CYL	HP	CU IN DISPL	W O T RPM	SPARK PLUG	TIMING AT WOT	PRIMARY PICKUP LOCATION	PRIMARY ADJUSTMENT NOTE
1981	**1981**								
J2RCIB, J2RSV	E2RCIB, E2RSV	1	2	2.64	4500	J6J	—	D	7
J4BRCIC, BRLCIC, WCIC	E4BRCIC, BRLCIC, WCIC	2	4	5.28	4500	J6J	—	C	7
J5RCIC, RLCIC, RHCIC, RHLCIC	E5RCIC, RLCIC, RHCIC, RHLCIC	2	4.5	5.28	5000	L77J4	—	D	7
J8RCIC, RLCIC, BACIC, BALCIC	E8RCIC, RLCIC, BACIC, BALCIC	2	7.5	10.0	5000	L77J4	—	D	7
J10RCID, RLCID, ELCID, BACID, BALCID, SELCID	E10RCID, RLCID, ELCID, BACID, BALCID, SELCID	2	9.9	13.2	5000	L77J4	—	D	7
J15RCIS, RLCIS, ECIS, BACIS, BALCIS	E15RCIS, RLCIS, ECIS, BACIS, BALCIS	2	15	13.2	6250	L77J4	—	D	7
J20CRCIM, CRLCIM, BFCIM, BFLCIM	E20CRCIM, CRLCIM, BFCIM, BFLCIM	2	20	31.8	5000	L77J4	—	D	4
J25ECIM, ELCIM, RCIM, RLCIM, BACIM, BALCIM, TECIM, TELCIM, BFCIM, BFLCIM, BECIM, BELCIM, RSA, RSLA, RWCIM, RWLCIM	E25ECIM, ELCIM, RCIM, RLCIM, BACIM, BALCIM, TECIM, TELCIM, BFCIM, BFLCIM, BECIM, BELCIM, RSA, RSLA, RWCIM, RWLCIM	2	25	31.8	5000	L77J4	34°	B	4
J35RCIG, RLCIG, ECIG, ELCIG, BACIG, BALCIG	E35RCIG, RLCIG, ECIG, ELCIG, BACIG, BALCIG	2	35	31.8	5500	L77J4	30°	D	4
J40RSD, RSLD, RSR, RSLR, RWCIS, RWLCIS	E40RSD, RSLD, RSR, RSLR, RWCIS, RWLCIS	2	40	31.8	5500	L77J4	—	D	4
J50BECIC, BELCIC	E50BECIC, BELCIC	2	50	45.0	5000	L77J4	19°	C	4
J55RCIM, RLCIM	E55RCIM, RLCIM	2	55	45.0	5000	L77J4	19°	C	4
J60ECIH, ECIA, ELCIH, ELCIA, TLCIA	E60ECIH, ECIA, ELCIH, ELCIA, TLCIA	2	60	45.0	5500	L77J4	21°	C	4

NOTE: See Appendix Page A-17 for primary pickup location and primary pickup adjustment note called out in this table.

TUNE-UP SPECIFICATIONS

SEE GENERAL AND SPECIAL NOTES APPENDIX PAGE A-17

JOHNSON MODEL	EVINRUDE MODEL	NO. CYL	HP	CU IN DISPL	WOT RPM	SPARK PLUG	TIMING AT WOT	PRIMARY PICKUP LOCATION	PRIMARY ADJUSTMENT NOTE
1982	**1982**								
J2RCN	E2RCN	1	2	2.64	4200	J6J	—	D	7
J4WCN, BRCN, RLCN, RHCN, RHLCN & RSY, RSLY	E4WCN, BRCN, RLCN, RHCN, RHLCN & RSY, RSLY	2	4	5.28	4000	L86	—	C	7
J5RCN, RLCN, RHCN, RHLCN	E5RCN, RLCN, RHCN, RHLCN	2	4.5	5.28	4500	L77J4	—	D	7
J6BFCN, BFLCN, RCN, RLCN & RSM, RSLM	E6BFCN, BFLCN, RCN, RLCN & RSM, RSLM	2	6	10.0	4500	L77J4	—	D	7
J8RCN, RLCN, BACN, BALCN, SRLCN	E8RCN, RLCN, BACN, BALCN, SRLCN	2	7.5	10.0	4500	L77J4	—	D	7
J10RCN, RLCN, BACN, BALCN, ELCN & RSW, RSLW	E10RCN, RLCN, BACN, BALCN, ELCN & RSW, RSLW	2	9.9	13.2	4500	L77J4	—	D	7
J15RCN, RLCN, ECN, BACN, ALCN, RSP, RSLP	E15RCN, RLCN, ECN, BACN, ALCN, RSP, RSLP	2	15	13.2	5500	L77J4	—	D	7
J20CRCN, CRLCN, BFCN, BFLCN	E20CRCN, CRLCN, BFCN, BFLCN	2	20	31.8	4500	L77J4	—	D	4
J25RCN, RLCN, ECN, ELCN, BACN, BALCN, TECN, TELCN, BFCN, BFLCN, BECN, BELCN & RSK, RSLK, RWCN, RWLCN	E25RCN, RLCN, ECN, ELCN, BACN, BALCN, TECN, TELCN, BFCN, BFLCN, BECN, BELCN & RSK, RSLK, RWCN, RWLCN	2	25	31.8	4500	L77J4	30°	B	4
J35RCN, RLCN, ECN, ELCN, BACN, BALCN	E35RCN, RLCN, ECN, ELCN, BACN, BALCN	2	35	31.8	4500	L77J4	30°	D	4
J40RSR, RSLR, RWCN, RWLCN	E40RSR, RSLR, RWCN, RWLCN	2	40	43.9	4000	L77J4	—	D	4
J50BECN, BELCN	E50BECN, BELCN	2	50	45	4500	L77J4	19°	C	4
J55RCN, RLCN, RSJ, RSLJ	E55RCN, RLCN, RSJ, RSLJ	2	55	45	4500	L77J4	19°	C	4
J60ECN, ELCN, TLCN	E60ECN, ELCN, TLCN	2	60	45	5000	L77J4	21°	C	4

NOTE: See Appendix Page A-17 for primary pickup location and primary pickup adjustment note called out in this table.

TUNE-UP SPECIFICATIONS

SEE GENERAL AND SPECIAL NOTES APPENDIX PAGE A-17

1983

JOHNSON MODEL	EVINRUDE MODEL	NO. CYL	HP	CU IN DISPL	WOT RPM	SPARK PLUG	TIMING AT WOT	PRIMARY PICKUP LOCATION	PRIMARY ADJUSTMENT NOTE
J2RCT	E2RCT	1	2	2.64	4200	J6J	—	D	7
J4BRCT, RLCT, RHCT, RHLCT	E4BRCT, RLCT, RHCT, RHLCT	2	4	4.28	4000	L86	—	C	7
J5RCT, RHCT, RLCT, RHLCT	E5RCT, RHCT, RLCT, RHLCT	2	4.5	5.28	4500	L77J4	—	D	7
J6BFCT, BACT, BFLCT BALCT, RCT, RLCT	E6BFCT, BACT, BFLCT BALCT, RCT, RLCT	2	6	10.00	4500	L77J4	—	D	7
J8BACT, RCT, BALCT, RLCT, SRLCT	E8BACT, RCT, BALCT, RLCT, SRLCT	2	7.5	10.00	4500	L77J4	—	D	7
J10RCT, RLCT, BACT, BALCT, SELCT	E10RCT, RLCT, BACT, BALCT, SELCT	2	9.9	13.20	4500	L77J4	—	D	7
J15RCT, RLCT, ECT, ELCT, BACT, BALCT	E15RCT, RLCT, ECT, ELCT, BACT, BALCT	2	15	13.20	4500	L77J4	—	D	7
J20CRCT, CRLCT, BECT, BELCT, BTCT, BFCT, BFLCT, BICT, BILCT	E20CRCT, CRLCT, BECT, BELCT, BTCT, BFCT, BFLCT, BICT, BILCT	2	20	31.80	4500	L77J4	30°	D	4
J25RCT, RLCT, ECT, ELCT, BACT, BALCT, TECT, TELCT & RWCTE, RWLCTE	E25RCT, RLCT, ECT, ELCT, BACT, BALCT, TECT, TELCT & RWCTE, RWLCTE	2	25	31.80	4500	L77J4	30°	B	4
J35RCT, RLCT, ECT, ELCT, TELCT, BACT, BALCT	E35RCT, RLCT, ECT, ELCT, TELCT, BACT, BALCT	2	35	31.80	4500	L77J4	30°	D	4
J40RWCTR, RWLCTR, RCV, RCLV	E40RWCTR, RWLCTR, RCV, RCLV	2	40	43.90	4000	UL81J	19°	D	4
J50BECT, BELCT, TELCT	E50BECT, BELCT, TELCT	2	50	45	4500	L77J4	19°	C	4
J55RCTE, RLCTE	E55RCTE, RLCTE	2	55	45	4500	L77J4	19°	C	4
J60ECT, ELCT, TLCT	E60ECT, ELCT, TLCT	2	60	45	5000	L77J4	21°	C	4

NOTE: See Appendix Page A-17 for primary pickup location and primary pickup adjustment note called out in this table.

TUNE-UP SPECIFICATIONS

SEE GENERAL AND SPECIAL NOTES APPENDIX PAGE A-17

1984

JOHNSON MODEL	EVINRUDE MODEL	NO. CYL	HP	CU IN DISPL	WOT RPM	SPARK PLUG	TIMING AT WOT	PRIMARY PICKUP LOCATION	PRIMARY ADJUSTMENT NOTE
J2RCR	E2RCR	1	2	2.64	4500	J6J	--	D	7
J4BRCR, RLCR, RHCR, RHLCR	E4BRCR, RLCR, RFHCR, RHLCR	2	4	5.28	4500	L86	--	C	7
J4RDCR, RDLCR, RDHCR, RDHLCR	E4RDCR, RDLSR, RDHCR, RDHLCR	2	4.5	5.28	4500	L77J4	—	D	7
J5RHCR, BFCR, BFLCR	E5RHCR, BFCR, BFLCR	2	5	10.0	4500	L77J4	--	D	7
J6BACR, BALCR, BFCR, FLCR, FLCR, RCR, RLCR, SLCR	E6BACR, BALCR, BFCR,	2	6	10.0	4500	L77J4	--	D	7
J8RCR, RLCR, BACR, BALCR, SRLCR	E8RCR, RLCR, BACR, BALCR, SRLCR	2	8	10.0	4500	L77J4	—	D	7
J10RCR, RLCR, BACR, BALCR, SELCR, ELCR	E10RCR, RLCR, BACR, BALCR, SELCR, ELCR	2	9.9	13.2	5000	L77J4	—	D	7
J15RCR, RLCR, ECR, ELCR, BACR, BALCR	E15RCR, RLCR, ECR, ELCR, BACR, BALCR	2	15	13.2	5000	L77J4	—	D	7
J20CRCR, CRLCR, BFCR, BFLCR, BICR, BILCR & J25BECR, J25BELCR, J25BFCR, J25BFLCR	E20CRCR, CRLCR, BFCR, BFLCR, BICR, BILCR & E25BECR, E25BELCR, E25BFCR, E25BFLCR	2	20	31.8	4500	L77J4	30°	D	4
J25RCR, RLCR, BACR, BACR, ECR, TECR, TELCR, ELCR	E25RCR, RLCR, BACR, ECR, TECR, TELCR, ELCR	2	25	31.8	4500	L77J4	30°	B	4
J30ECR, ELCR	E30ECR, ELCR	2	30	31.8	5200	L77J4	—	D	4
J35RCR, RLCR, ECR, ELCR, BACR, BALCR, TELCR	E35RCR, RLCR, ECR, ELCR, BACR, BALCR, TELCR	2	35	31.8	5200	L77J4	30°	D	4
J40TACR, TALCR, ECR, ELCR, TELCR	E40TACR, TALCR, ECR, ELCR, TELCR	2	40	45.0	4500	L77J4	19°	D	4
J50BECR, BELCR, TELCR, TLCR	E50BECR, BELCR, TELCR, TLCR	2	50	45.0	4500	L77J4	19°	C	4
J60ECR, ELCR, TLCR	E60ECR, ELCR, TLCR	2	60	45.0	5000	L77J4	21°	C	4

NOTE: See Appendix Page A-17 for primary-pickup-location and primary-pickup adjustment note called out in this table.

TUNE-UP SPECIFICATIONS

SEE GENERAL AND SPECIAL NOTES APPENDIX PAGE A-17

JOHNSON MODEL	EVINRUDE MODEL	NO. CYL	HP	CU IN DISPL	WOT RPM	SPARK PLUG	TIMING AT WOT	PRIMARY PICKUP LOCATION	PRIMARY ADJUSTMENT NOTE
1985	**1985**								
J2RCO	E2RCO	1	2	2.64	4000	J6J	—	D	7
J4BRCO, BRLCO, BRHCO, BRHLCO	E4BRCO, BRLCO, BRHCO, BRHLCO	2	4	5.28	4500	L86	—	C	7
J4RDCO, RDLCO, RDHCO, RDHLCO	E4RDCO, RDLCO, RDHCO, RDHLCO	2	4.5	5.28	4500	L77J4	—	D	7
J5RHCO, BFCO, BFLCO	E5RHCO, BFCO, BFLCO	2	5	10.0	4500	L77J4	—	D	7
J6BACO, BALCO, BFCO,RCO BFLCO, RLCO, SLCO	E6BACO, BALCO, BFCO,RCO BFLCO, RLCO, SLCO	2	6	10.0	4500	L77J4	—	D	7
J8RCO, RLCO, BACO, BALCO, SRLCO	E8RCO, RLCO, BACO, BALCO, SRLCO	2	8	10.0	4500	L77J4	—	D	7
J10RCO, RLCO, BACO, ECO, BALCO, SELCO, ELCO	E10RCO, RLCO, BACO, ECO, BALCO, SELCO, ELCO	2	9.9	13.2	5000	L77J4	—	D	7
J15RCO, RLCO, ECO, ELCO, BACO, BALCO	E15RCO, RLCO, ECO, ELCO, BACO, BALCO	2	15	13.2	5500	L77J4	—	D	7
J20CRCO, CRLCO, ELCO, ECO, BECO, BELCO, BFCO, BFLCO, BICO, BILCO & J25BFCO, 25BFLCO	E20CRCO, CRLCO, ELCO, ECO, BECO, BELCO, BFCO, BFLCO, BICO, BILCO & E25BFCO, E25BFLCO	2	20	31.8	4500	L77J4	30°	D	4
J25RCO, RLCO, BACO, ECO, BALCO, TECO, TELCO, ELCO	E25RCO, RLCO, BACO, ECO, BALCO, TECO, TELCO, ELCO	2	25	31.8	4500	L77J4	30°	B	4
J30RCO, RLCO, ECO, ELCO, BACO, BALCO, TECO, TELCO	E30RCO, RLCO, ECO, ELCO, BACO, BALCO, TECO, TELCO	2	30	31.8	4500	L77J4	30°	B	4
J40RCO, RLCO, BACO, ECO, BALCO, ELCO, TECO, TELCO	E40RCO, RLCO, BACO, ECO, BALCO, ELCO, TECO, TELCO	2	40	45.0	4500	L77J4	19°	D	4
J50BECO, BELCO, TELCO, TLCO	E50BECO, BELCO, TELCO, TLCO	2	50	45.0	4500	L77J4	19°	C	4
J60ECO, ELCO, TLCO	E60ECO, ELCO, TLCO	2	60	45.0	5000	L77J4	21°	C	4

NOTE: See Appendix Page A-17 for primary pickup location and primary pickup adjustment note called out in this table.

TUNE-UP SPECIFICATIONS

SEE GENERAL AND SPECIAL NOTES APPENDIX PAGE A-17

JOHNSON MODEL	EVINRUDE MODEL	NO. CYL	HP	CU IN DISPL	WOT RPM	SPARK PLUG	TIMING AT WOT	PRIMARY PICKUP LOCATION	PRIMARY ADJUSTMENT NOTE
1986	**1986**								
J2RCD	E2RCD	1	2	2.64	4500	RJ6C	—	D	7
J4BRHLCD, BRHLCD, RDHCD, RDHLCD	E4BRHCD, BRHLCD, RDHCD, RDHLCD	2	4	5.28	5000	RL86	—	D	7
J6RCD, RLCD, SLCD	E6RCD, RLCD, SLCD	2	6	10.0	5000	L77JC4	—	D	7
J8RCD, RLCD, SRLCD,	E8RCD, RLCD, SRLCD,	2	8	10.0	5500	L77JC4	—	D	7
J10RCD, RLCD, ECD, SELCD, ELCD	E10RCD, RLCD, ECD, SELCD, ELCD	2	9.9	13.2	5500	L77JC4	—	C	7
J15RCD, RLCD, ECD, ELCD	E15RCD, RLCD, ECD, ELCD	2	15	13.2	6250	L77JC4	—	C	7
J20CRCD, CRLCD, ELCD, ECD	E20CRCD, CRLCD, ELCD, ECD	2	20	31.8	5000	L77JC4	30°	D	4
J25RCD, RLCD, ECD, ELCD, TECD, TELCD	E25RCD, RLCD, ECD, ELCD, TECD, TELCD	2	25	31.8	5000	L77JC4	30°	D	4
J30RCD, RLCD, ECD, ELCD, TECD, TELCD	E30RCD, RLCD, ECD, ELCD, TECD, TELCD	2	30	31.8	5500	L77JC4	30°	D	4
J40RCD, RLCD, ECD, ELCD, TECD, TELCD	E40RCD, RLCD, ECD, ELCD, TECD, TELCD	2	40	45.0	5000	L77JC4	19°	D	4
J50BECD, BELCD, TELCD, TLCD	E50BECD, BELCD, TELCD, TLCD	2	50	45.0	5000	L77JC4	19°	D	4

NOTE: See Appendix Page A-17 for primary pickup location and primary pickup adjustment note called out in this table.

TUNE-UP SPECIFICATIONS

SEE GENERAL AND SPECIAL NOTES APPENDIX PAGE A-17

JOHNSON MODEL	EVINRUDE MODEL	NO. CYL	HP	CU IN DISPL	WOT RPM	SPARK PLUG	TIMING AT WOT	PRIMARY PICKUP LOCATION	PRIMARY ADJUSTMENT NOTE
1987	**1987**								
JCO-CU, COLT	EJRCU	1	1.25	2.64	4500	RJ36C	—	D	7
J3RCU	E3RCU	2	2.5	5.28	4500	RL86	—	D	7
J4RCU, RLCU, RDHCU, RDHLCU	E4BRHCU, RLCU, RDHCU, RDHLCU	2	4	5.28	5000	RL86	—	D	7
J6RCU, RLCU, SLCU	E6RCU, RLCU, SLCU	2	6	10.0	5000	L77JC4	—	D	7
J8RCU, RLCU, SRLCU	E8RCU, RLCU, SRLCU	2	8	10.0	5500	L77JC4	—	D	7
J10RCU, RLCU, ECU, SELCU, ELCU	E10RCU, RLCU, ECU, SELCU, ELCU	2	9.9	13.2	5500	L77JC4	—	C	7
J15RCU, RLCU, ECU, ELCU	E15RCU, RLCU, ECU, ELCU	2	15	13.2	6250	L77JC4	--	C	7
J20CRCU, CRLCU, ELCU, ECU	E20CRCU, CRLCU, ELCU, ECU	2	20	31.8	5000	L77JC4	30°	D	4
J25RCU, RLCU, ECU, ELCU, TECU, TELCU	E25RCU, RLCU, ECU, ELCU, TECU, TELCU	2	25	31.8	5000	L77JC4	30°	D	4
J30RCU, RLCU, ECU, ELCU, TECU, TELCU	E30RCU, RLCU, ECU, ELCU, TECU, TELCU	2	30	31.8	5500	L77JC4	30°	D	4
J40RCU, RLCU, ECU, ELCU, TECU, TELCU, TLCU PT/T	E40RCU, RLCU, ECU, ELCU, TECU, TELCU, TLCU PT/T	2	40	45.0	5000	L77JC4	19°	D	4
J50BECU, BELCU, TELCU, TLCU, TLCU PT/T	E50BECU, BELCU, TELCU, TLCU PT/T	2	50	45.0	5000	L77JC4	19°	D	4

NOTE: See Appendix Page A-17 for primary pickup location and primary pickup adjustment note called out in this table.

TUNE-UP SPECIFICATIONS

SEE GENERAL AND SPECIAL NOTES APPENDIX PAGE A-17

JOHNSON MODEL	EVINRUDE MODEL	NO. CYL	HP	CU IN DISPL	WOT RPM	SPARK PLUG	TIMING AT WOT	PRIMARY PICKUP LOCATION	PRIMARY ADJUSTMENT NOTE
1988	**1988**								
JCOCC, COLT	EJRCC, JUNIOR	1	1.25	2.64	4500	RJ6C	--	D	7
J3RCC	E3RCC, EXCEL	2	2.5	5.28	4500	RL86	--	D	7
J4RCC, RLCC, RDHCC, RDHLCC	E4BRHCC, RCC, RDHCC, RDHLCC	2	4	5.28	5000	RL87	--	D	7
J6RCC, RLCC, SLCC	E6RCC, RLCC, SLCC	2	6	10.0	5000	L77JC4	--	D	7
J8RCC, RLCC, SRLCC	E8RCC, RLCC, SRLCC	2	8	10.0	5500	L77JC4	--	D	7
J10RCC, RLCC, ECC, SELCC, ELCC	E10RCC, RLCC, ECC, SELCC, ELCC	2	9.9	13.2	5500	L77JC4	--	C	7
J15RCC, RLCC, ECC, ELCC	E15RCC, RLCC, ECC, ELCC	2	15	13.2	6250	L77JC4	--	C	7
J20CRCC, CRLCC, ECC, ELCC	E20CRCC, CRLCC, ECC, ELCC	2	20	31.8	5000	L77JC4	30°	D	4
J25RCC, RLCC, ECC, ELCC, TECC, TELCC	E25RCC, RLCC, ECC, ELCC, TECC, TELCC	2	25	31.8	5000	L77JC4	30°	D	4
J30RCC, RLCC, ECC, ELCC, TECC, TELCC	E30RCC, RLCC, ECC, ELCC, TECC, TELCC	2	30	31.8	5500	L77JC4	30°	D	4
J40RCC, RLCC, ECC, ELCC, TECC, TELCC, TLCC PT/T	E40RCC, RLCC, ECC, ELCC, TECC, TELCC, TLCC PT/T	2	40	45.0	5000	L77JC4	19°	D	4
J45RWLE, RWYE, RSE, RSLE, RSYE, RCR, RCLR	E45RWLE, RWYE, RSE, RSLE, RSYE, RCR, RCLR	2	45	45.0	5000	QL78V	19°	D	4
J48ESLCC	E48ESLCC	2	48	45.0	5000	QL77JC4	19°	D	4
J50BECC, BELCC, TELCC, TLCC, TLCC PT/T	E50BECC, BELCC, TELCC, TLCC, TLCC PT/T	2	50	45.0	5000	L77JC4	19°	D	4

NOTE: See Appendix Page A-17 for primary pickup location and primary pickup adjustment note called out in this table.

TUNE-UP SPECIFICATIONS

JOHNSON/EVINRUDE
1989 and 1990 Models

NO. CYL	HP	CU IN DISPL	WOT RPM	SPARK PLUG	TIMING AT WOT	PRIMARY PICKUP LOCATION	PRIMARY ADJUSTMENT NOTE
1	1.25	2.64	4500	RJ6C	—	D	7
2	2.5	5.28	4500	RL86	—	D	7
2	4	5.28	5000	RL86	—	D	7
2	6	10.0	5000	L77JC4	—	D	7
2	8	10.0	5500	L77JC4	—	D	7
2	9.9	13.2	5500	L77JC4	—	C	7
2	15	13.2	6250	L77JC4	—	C	7
2	20	31.8	5000	L77JC4	30°	D	4
2	25	31.8	5000	L77JC4	30°	D	4
2	30	31.8	5500	L77JC4	30°	D	4
2	40	45.0	5000	L77JC4	19°	D	4
2	50	45.0	5000	L77JC4	19°	D	4

NOTE: See Appendix Page A-17 for primary pickup location and primary pickup adjustment note called out in this table.

TUNE-UP SPECIFICATIONS

GENERAL NOTES:

Top cylinder -- No. 1
Bottom cylinder -- No. 2
All powerheads equipped with point set, adjust at 0.020".
Set all spark plugs -- 1971-81 and for all 2 hp & 4 hp at 0.030".
Set all spark plug plugs 4.5 hp thru 60 hp -- 1982 and on, at 0.040".
Timing **NOT** adjustable, except on models 20 hp and larger.

SPECIAL NOTES

Primary Pickup Adjustment

1- Loosen two screws under armature plate and move plate in or out to make contact with roller.
2- Loosen center screw on throttle lever and move lever inward or outward to align with mark on the cam.
3- Loosen clamps on throttle shaft and move roller to make contact with armature cam.
4- On the **STARBOARD SIDE,** loosen the throttle arm screw and move roller inward or outward to align with the mark on the cam.
5- On the **PORT SIDE,** loosen eccentric lock screw on the throttle shaft and turn the eccentric to move the roller to make contact with the armature cam.
6- On the **STARBOARD SIDE,** loosen throttle linkage screw and move roller to make contact with the cam.
7- Advance throttle control with adjusting screw at base of follower.

Primary Pickup Location

A- **PORT** side of mark.
B- **STARBOARD** side of mark.
C- **CENTER** of mark.
D- **BETWEEN** marks.
E- Pointer **BETWEEN** marks.

POWERHEAD SPECIFICATIONS

H.P.	YEAR	CU. IN. DISPL.	STROKE	BORE STD.	BORE OVERSIZE	PISTON TO CYL CLEARANCE MAX.	MIN.	PISTON RING GROOVE MAX.	MIN.	WIDTH OF RING MAX.	MIN.	PISTON RING GAP MAX.	MIN.
1.25	1987 & on	2.64	1.374"	1.567"	0.030"	0.0055"	0.0043"	0.0040"	0.0020"	0.0625"	0.0615"	0.025"	0.015"
2	1971-84	2.64	1.374"	1.5625"	0.030"	0.0055"	0.0043"	0.0040"	0.0020"	0.0625"	0.0615"	0.025"	0.015"
2	1985 & on	2.64	1.374"	1.567"	0.030"	0.0055"	0.0043"	0.0040"	0.0020"	0.0625"	0.0615"	0.025"	0.015"
2.5	1987 & on	5.28	1.374"	1.567"	0.030"	0.0030"	0.0018"	0.0040"	0.0020"	0.0625"	0.0615"	0.015"	0.005"
4	1971-84	5.28	1.374"	1.5625"	0.030"	0.0030"	0.0018"	0.0040"	0.0020"	0.0625"	0.0615"	0.015"	0.005"
4	1985 & on	5.28	1.374"	1.567"	0.030"	0.0030"	0.0018"	0.0040"	0.0020"	0.0625"	0.0615"	0.015"	0.005"
4.5	1980-85	5.28	1.374"	1.5625"	0.030"	0.0030"	0.0018"	0.0040"	0.0020"	0.0625"	0.0615"	0.015"	0.005"
5	1984-85	10.00	1.700"	1.9375"	0.030"	0.0030"	0.0018"	0.0035"	0.0010"	0.0625"	0.0615"	0.015"	0.005"
6	1971-79	8.84	1.500"	1.9375"	0.030"	0.0030"	0.0018"	0.0035"	0.0020"	0.0935"	0.0925"	0.015"	0.005"
7.5	1982 & on	10.00	1.700"	1.9375"	0.030"	0.0030"	0.0018"	0.0035"	0.0020"	0.0625"	0.0615"	0.015"	0.005"
8	1980-83	10.00	1.700"	1.9375"	0.030"	0.0030"	0.0018"	0.0035"	0.0010"	0.0625"	0.0615"	0.015"	0.005"
9.5	1971-73	15.20	1.810"	2.310"	0.030"	0.0050"	0.0035"	0.0035"	0.0020"	0.0935"	0.0925"	0.017"	0.007"
9.9	1974 & on	13.20	1.760"	2.188"	0.030"	0.0038"	0.0025"	0.0035"	0.0020"	NOTE 2	NOTE 2	0.015"	0.005"
15	1974 & on	13.20	1.760"	2.188"	0.030"	0.0038"	0.0025"	0.0035"	0.0010"	NOTE 2	NOTE 2	0.015"	0.005"
18	1971-73	22.00	2.250"	2.500"	0.030"	0.0048"	0.0033"	0.0040"	0.0025"	NOTE 3	NOTE 3	0.017"	0.007"
20	1971-73	22.00	2.250"	2.500"	0.030"	0.0048"	0.0033"	0.0040"	0.0025"	NOTE 3	NOTE 3	0.017"	0.007"
20	1971-73	31.80	2.250"	3.000"	0.030"	0.0065"	0.0035"	0.0040"	0.0020"	NOTE 3	NOTE 3	0.017"	0.007"
20	1981-83	31.80	2.250"	3.000"	0.030"	0.0044"	0.0024"	*0.0040"	*0.0020"	NOTE 3	NOTE 3	0.017"	0.007"
25	1971-76	22.00	2.250"	2.500"	0.030"	0.0048"	0.0033"	0.0040"	0.0015"	NOTE 3	NOTE 3	0.017"	0.007"
25	1977-83	31.80	2.250"	3.000"	0.030"	0.0065"	0.0035"	*0.0040"	*0.0015"	NOTE 3	NOTE 3	0.017"	0.007"
25	1984 & on	31.80	2.250"	3.000"	0.030"	0.0044"	0.0024"	*0.0040"	*0.0015"	NOTE 3	NOTE 3	0.017"	0.007"
25	1985 & on	31.80	2.250"	3.000"	0.030"	0.0044"	0.0024"	0.0040"	0.0015"	NOTE 3	NOTE 3	0.017"	0.007"
30	1976-84	31.80	2.250"	3.000"	0.030"	0.0065"	0.0035"	0.0040"	0.0015"	NOTE 3	NOTE 3	0.017"	0.007"
35	1976-84	31.80	2.250"	3.000"	0.030"	0.0050"	0.0030"	0.0040"	0.0015"	NOTE 3	NOTE 3	0.017"	0.007"
40	1971-76	45	2.820"	3.188"	0.030"	0.0050"	0.0030"	0.0040"	0.0015"	NOTE 3	NOTE 3	0.017"	0.007"
40	1981-83	43.90	2.750"	3.188"	0.030"	0.0050"	0.0030"	*0.0040"	*0.0015"	NOTE 3	NOTE 3	0.017"	0.007"
40	1984 & on	45	2.820"	3.188"	0.030"	0.0018"	0.0018"	*0.0040"	*0.0015"	NOTE 3	NOTE 3	0.017"	0.007"
50	1971-75	45	2.820"	3.188"	0.030"	0.0065"	0.0045"	0.0040"	0.0015"	NOTE 3	NOTE 3	0.017"	0.007"
50	1978-80	45.1	2.820"	3.190"	0.030"	0.0065"	0.0045"	0.0040"	0.0015"	NOTE 3	NOTE 3	0.017"	0.007"
50	1981 & on	45	2.820"	3.188"	0.030"	0.0018"	0.0018"	0.0040"	0.0015"	NOTE 3	NOTE 3	0.017"	0.007"
55	1976-83	45	2.820"	3.188"	0.030"	0.0049"	0.0018"	*0.0040"	*0.0015"	NOTE 3	NOTE 3	0.017"	0.007"
60	1980-85	45	2.820"	3.188"	0.030"	0.0049"	0.0018"	*0.0040"	*0.0015"	NOTE 3	NOTE 3	0.017"	0.007"

NOTES:

1- Dimensions given are for top and bottom ring, unless otherwise noted.

2- Top ring: Max. 0.0700", Min. 0.0695". Bottom ring: Max. 0.0625", Min. 0.0615".

3- Top ring: Max. 0.0900", Min. 0.0895". Bottom ring: Max. 0.0625", Min. 0.615".

4- Asterisk (*): OMC states -- "Piston ring groove, Lower" 1983 and on,

OIL/FUEL MIXTURE

OIL/FUEL MIXTURE

1- For all 1971-1984 engines in this manual -- use 50:1 oil mixture.

2- Since 1985 -- 2 hp thru 35 hp: first 5 hours of break-in -- use 50:1 oil mixture. Recommendation is 1/6 pint of oil per gallon of gasoline (50:1). After "break-in", use 100:1 oil mixture.

3- Since 1985 -- 40 hp thru 60 hp without VRO system: during break-in and after -- use 50:1 oil mixture.

4- **ALWAYS** use OMC or BIA certified TC-W oil lubricant.

5- **NEVER** use: automotive oils, premixed fuel of unknown oil quantity, or premixed fuel richer than 50:1.

GEAR OIL CAPACITIES

MODEL SIZE	YEAR	CAPACITY OUNCES	MODEL SIZE	YEAR	CAPACITY OUNCES
1.25	1987 & on	1.3	20	1971-85	8.3
2	1971-85	1.3	20	1986 & on	10.4
2.5	1987 & on	2.7	25	1971-85	8.3
4R	1971-81	1.3	25	1986 & on	10.4
4R	1982 & on	2.7	30	1984-85	8.3
4 Deluxe	1982-85	10.4	30	1986 & on	10.4
4W	1971-81	3.4	35	1976-82	9.8
4.5	1980-82	14.7	35	1983-84	12.2
4.5	1983-84	10.4	40	1971-83	13.9
5	1984-85	10.4	40	1984 & on	21.3
6	1971-81	8.5	50	1971-72	25.3
6	1982	14.7	50	1973-75	21.3
6	1983	10.4	50	1978-80	20.5
6	1984-85	14.7	50 Std	1981-82	21.3
6	1986 & on	10.4	50 Long	1982	25.4
7.5	1980-82	14.7	50 Std	1983 & on	21.3
7.5	1983	10.4	55 Std	1976-80	20.3
8	1984-85	14.7	55 Std	1981-82	21.3
8	1986 & on	10.4	55 Long	1982	25.4
9.5	1971-73	9.7	60	1980	20.3
9.9	1974 & on	8.8	60	1981-85	21.3
15	1974 & on	8.8	60 Long	1982	25.4
18	1971-73	8.3			

STARTER MOTOR SPECIFICATIONS

Model Number	Brush Spring Tension (Ounces)	Armature End Play (Inches)	Volts	Max. Amperes	Min. RPM	Volts	Max. Amperes	Min. Lbs. Ft.
MDO-0	42-66	0.005 min.	10.0	38	10,000	4.0	170	1.5
MDO-1	42-66	.010-.035	10.0	38	10,000	4.0	170	1.5
MDW-0	42-66	.010-.035	10.0	26	8,500	4.0	160	2.1
MDW-1	42-66	.005 min.	10.0	26	8,500	4.0	160	2.1

Pinion position, 1-25/32 $\pm$ 1/16 inches from face of mounting flange to edge of pinion.

MDO-4002M	Use Test	MDO-0	CCW Rotation
MDO-4003M	Use Test	MDO-1	CW Rotation
MDW-4001M	Use Test	MDW-0	CW Rotation
MDW-4002M	Use Test	MDW-1	CCW Rotation

REGULATOR SPECIFICATIONS

Part No.	VRU-6101A
System Voltage	12
Ground Polarity	Negative
Armature Air Gap	
Circuit Breaker	.031-.034 in.
Voltage Reg.	.048-.052 in.
Current Regulator	.048-.052 in.
Current Regulator	
Setting Amps	9.0-11.0

CB Shunt Winding 107 to 121 ohms
VR Winding 43.7 to 49.3 ohms

Circuit Breaker	
Close Volts	12.6-13.6
Open Amps	
Discharge	3.0-5.0

Regulator (Hot) Operating Voltages
  Tolerance $\pm$ .4 Volt

50°F	15.2 Volts
80°F	15.0 Volts
110°F	14.8 Volts
140°F	14.6 Volts

These figures are for a unit in normal operation while charging at 1/2 rated output or with 1/4 ohm fixed resistor in series with the battery.

GENERATOR SPECIFICATIONS

Generator	GJG-4001M
	GJG-4002M
Rot. D.E.	C
Ground Polarity	Negative
Brush Spring	
Tension	12-24 oz.
Field Coil Draw	
Volts	10.0
Amps	1.7-1.9
Monitoring Draw	
Volts	10.0
Amps	5.0-6.0
Generator Output	
Volts	15.0
Max. Amps	10.0
Max. RPM	7,000

CONDENSER SPECIFICATIONS

ENGINE SIZE	MODEL	CONDENSER OMC PART NO.	MFD CAPACITY
1.25 hp	All	580321	18-22
2 hp	All	580321	18-22
4 hp	1971-76	580321	18-22
6 hp	1971-76	580321	18-22
9.5 hp	1971-73	580321	18-22
18 hp	1971-73	580419	25-29
20 hp	1971-72	580419	25-29
25 hp	1971-72	580419	25-29
40 hp	1971-72	580419	25-29

STARTER ROPE SPECIFICATIONS

ENGINE SIZE	MODEL	DIAMETER INCHES	LENGTH INCHES
1.25 hp	1987 & on	1/8	62-1/4
2 hp	1971-85	1/8	64
2.5 hp	1987 & on	1/8	62-1/4
4 hp	1971-75	1/8	64
4 hp	1976	9/64	64
4 hp	1977-78	9/64	61
4 hp	1979-83	1/8	64
4 hp	1984 & on	1/8	59-1/2
4.5 hp	1980-85	9/64	61
6 hp	1971-73	1/8	56
6 hp	1974-83	1/8	51-3/4
6 hp	1984 & on	1/8	59-1/2
7.5 hp	1980-83	1/8	59-1/2
8 hp	1984 & on	1/8	59-1/2
9.5 hp	1971-73	5/32	65-1/4
9.9 hp	1974-75	1/8	64
9.9 hp	1976-78	9/64	71-1/2
9.9 hp	1979 & on	9/64	68-1/2
15 hp	1974-75	1/8	64
15 hp	1976-78	9/64	71-1/2
15 hp	1979 & on	9/64	68-1/2
18 hp	1971-73	7/32	72-1/4
20 hp	1971 & on	7/32	72-1/4
25 hp	1971 & on	7/32	72-1/4
30 hp	1985 & on	7/32	72-1/4
35 hp	1977-84	7/32	72-1/4
40 hp	1971-72	7/32	73-3/4
40 hp	1973-76	7/32	70-3/16
40 hp	1981-83	7/32	68-3/4
40 hp	1984 & on	7/32	96-1/2

NOTES

1- Purchase a good grade of nylon rope.

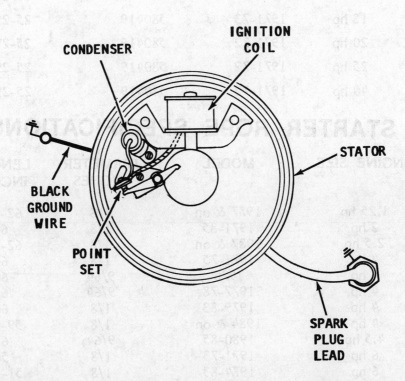

Typical wiring diagram for a one-cylinder powerhead.

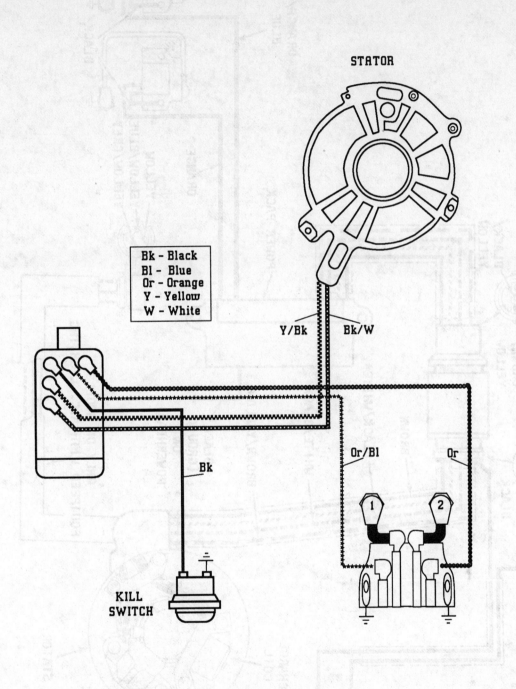

STATOR

Bk - Black
Bl - Blue
Or - Orange
Y - Yellow
W - White

Y/Bk   Bk/W

Or/Bl   Or

Bk

1   2

KILL
SWITCH

Wire Identification — All 2.5 hp, Excel, Ultra, and 4 hp -- 1984 and on.

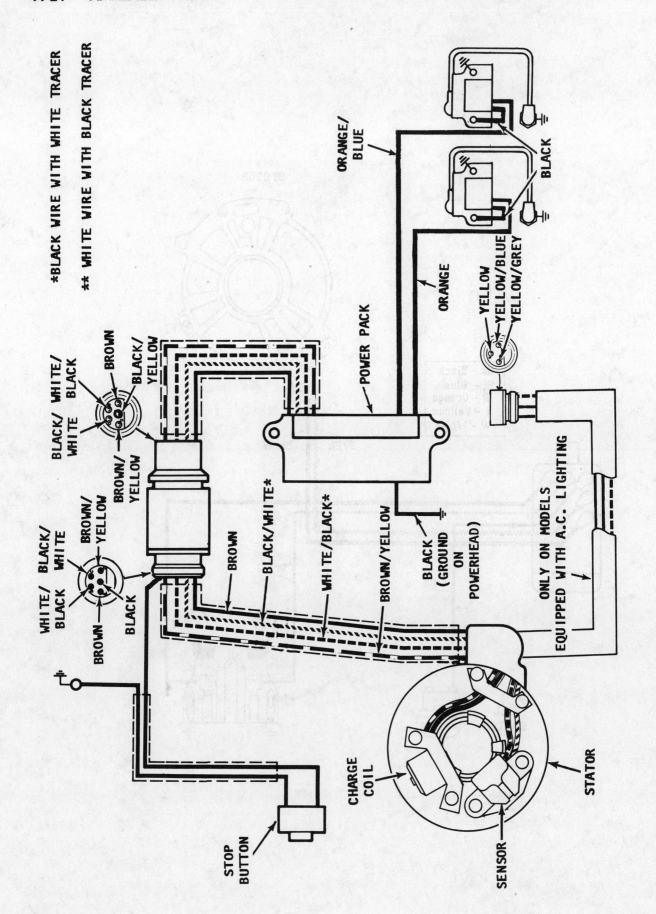

*BLACK WIRE WITH WHITE TRACER

** WHITE WIRE WITH BLACK TRACER

ORANGE/BLUE

BLACK

ORANGE

POWER PACK

YELLOW
YELLOW/BLUE
YELLOW/GREY

BROWN
BLACK/WHITE
BROWN/YELLOW
BLACK/WHITE*
WHITE/BLACK*
BROWN/YELLOW
BLACK (GROUND ON POWERHEAD)
ONLY ON MODELS EQUIPPED WITH A.C. LIGHTING

BLACK/WHITE/BLACK
BROWN
BLACK/YELLOW
BROWN/YELLOW

WHITE/BLACK/WHITE
BROWN/YELLOW
BROWN
BLACK

CHARGE COIL
SENSOR
STATOR

STOP BUTTON

Wire Identification — 4 Delux thru 55hp with CD II ignition, manual start; also models equipped with A.C. lighting.

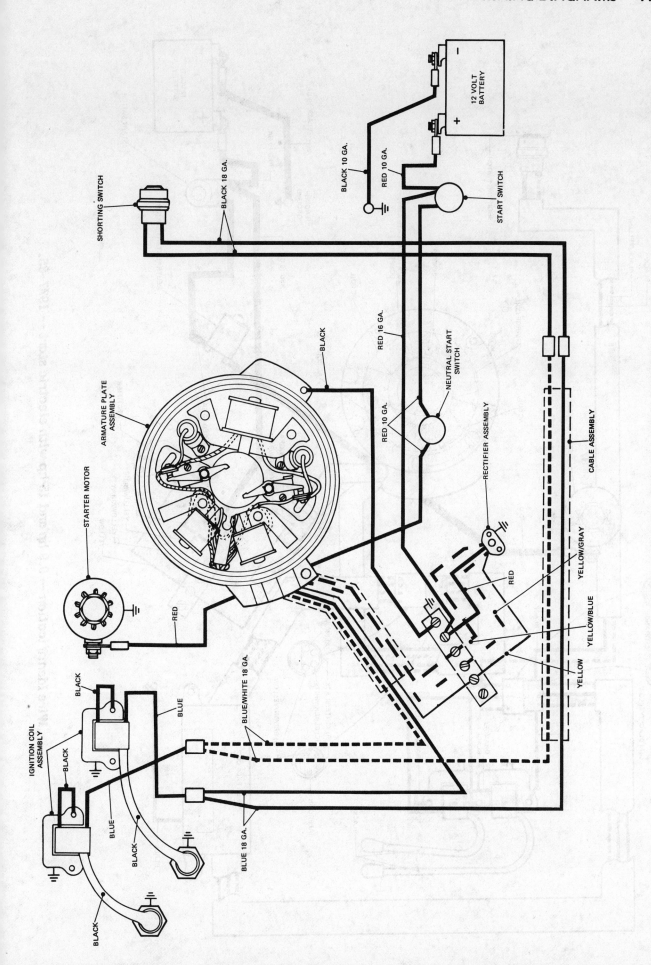

Wire Identification — 9.9 hp and 15 hp with electric start — 1974-76.

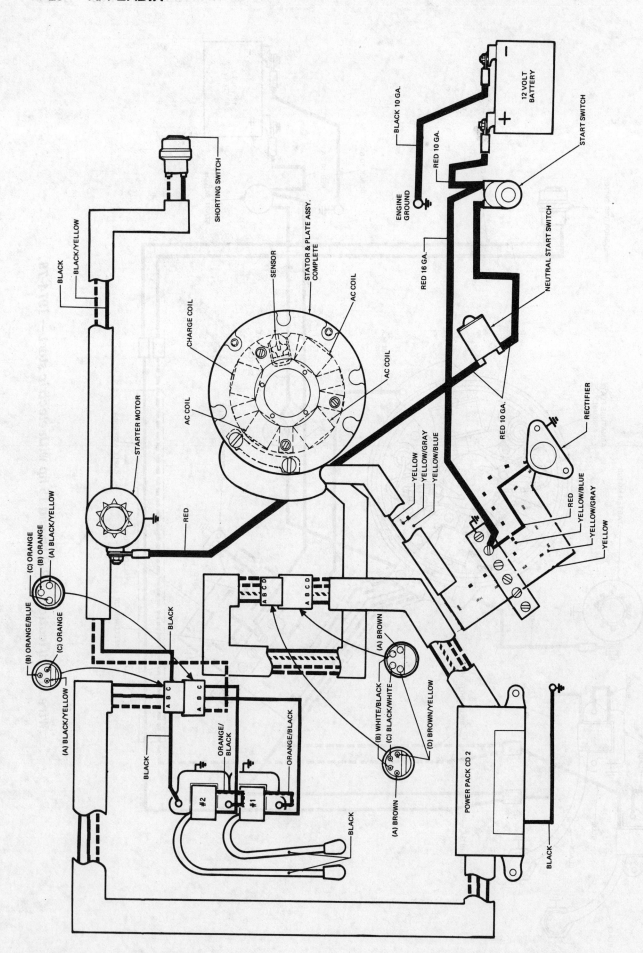

Wire Identification — 9.9 hp and 15 hp with electric start — 1977–85.

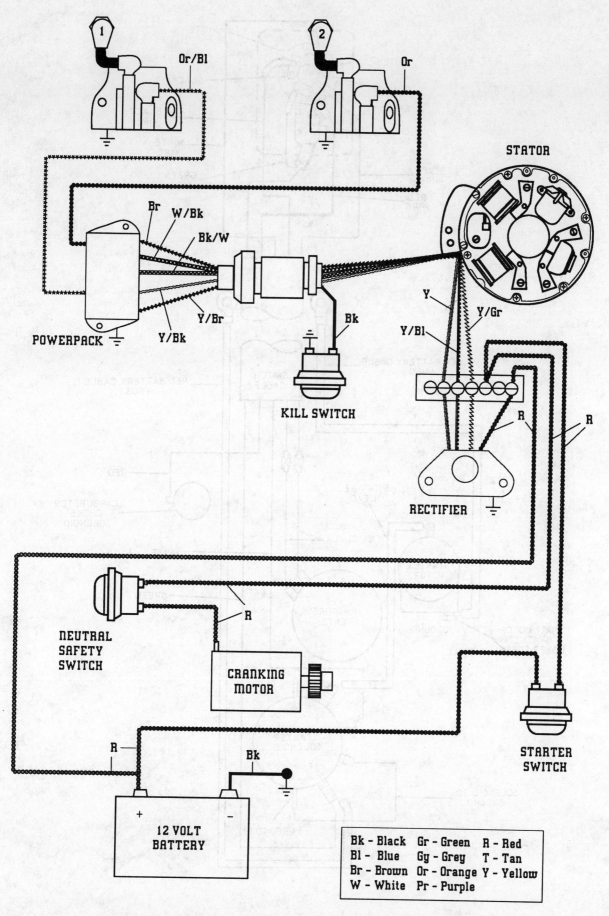

STATOR

1

Or/Bl

2

Or

Br
W/Bk
Bk/W

POWERPACK

Y/Br
Y/Bk

Bk

KILL SWITCH

Y
Y/Gr
Y/Bl

RECTIFIER

R

R

NEUTRAL
SAFETY
SWITCH

R

CRANKING
MOTOR

R

Bk

STARTER
SWITCH

12 VOLT
BATTERY

+   –

Bk – Black	Gr – Green	R – Red
Bl – Blue	Gy – Grey	T – Tan
Br – Brown	Or – Orange	Y – Yellow
W – White	Pr – Purple	

Wire Identification — 9.9 hp and 15 hp with electric start — 1986 and on.

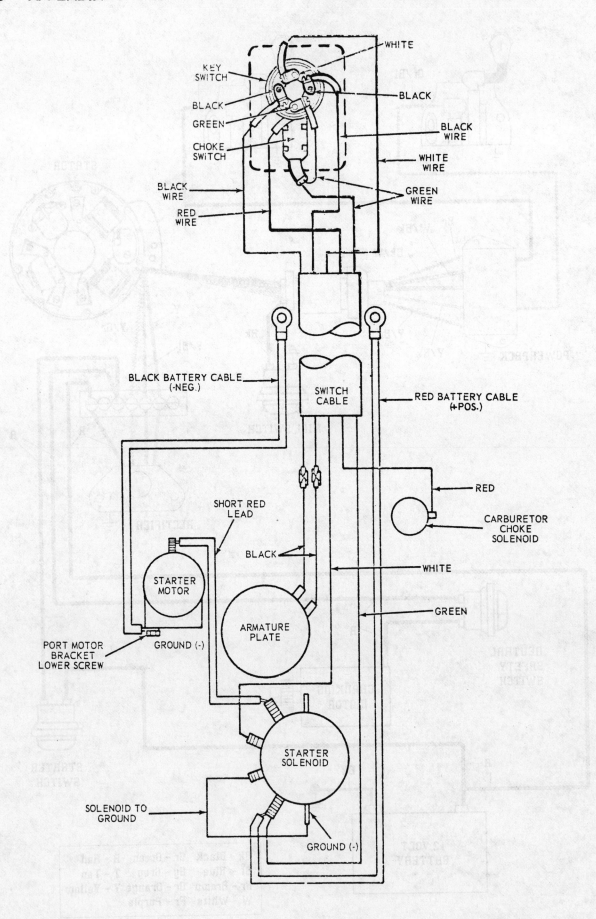

Wire Identification — 18 hp, 20 hp, and 25 hp with electric start — 1971 and 1972.

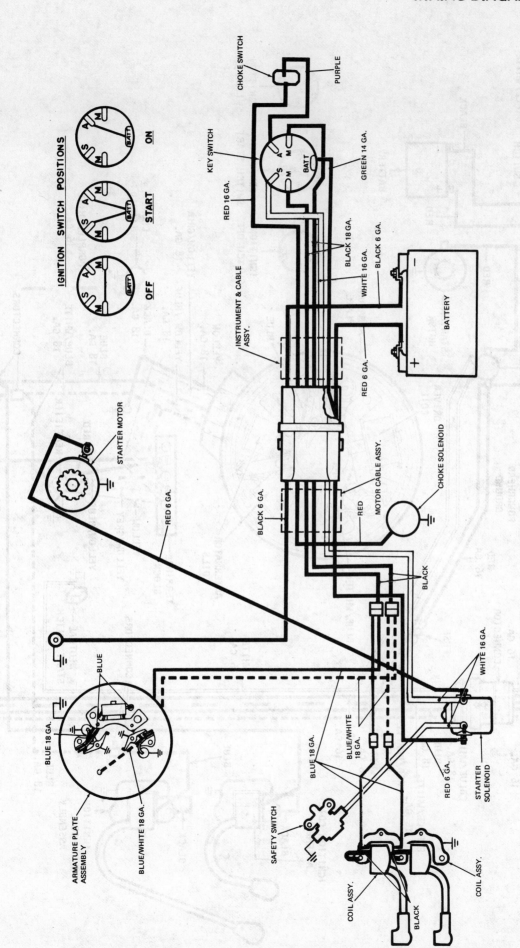

Wire Identification — 18 hp, 20 hp, and 25 hp with electric start — 1973-76.

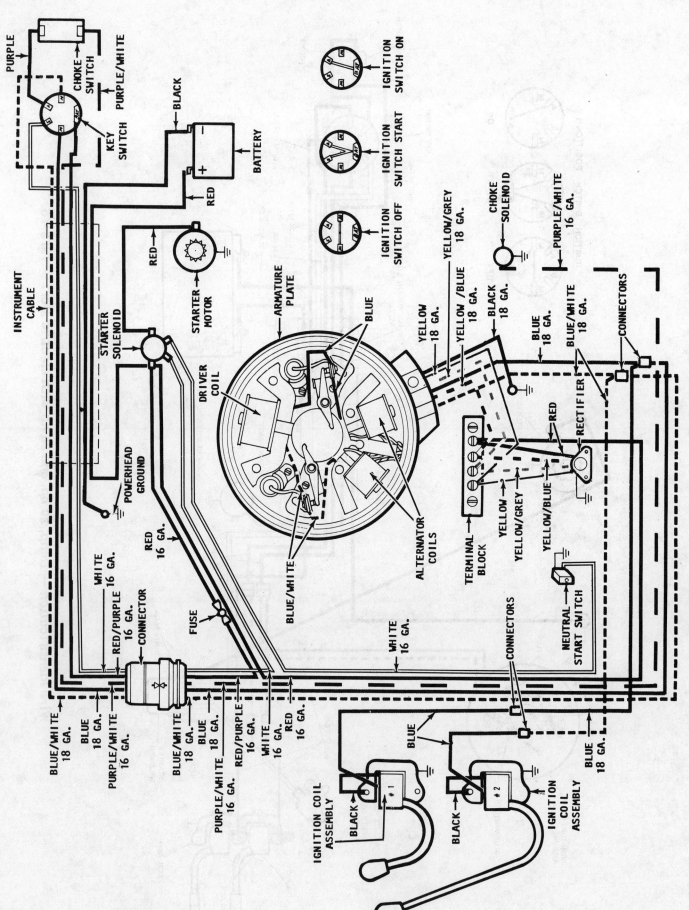

Wire Identification — 35 hp with electric start — 1976.

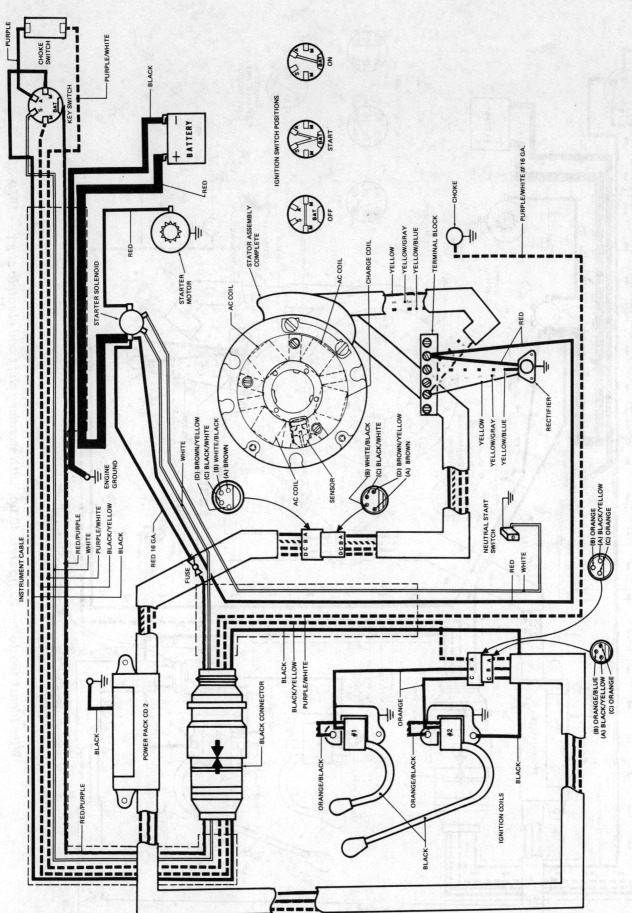

Wire Identification — 25 hp and 35 hp with electric start — 1977-78.

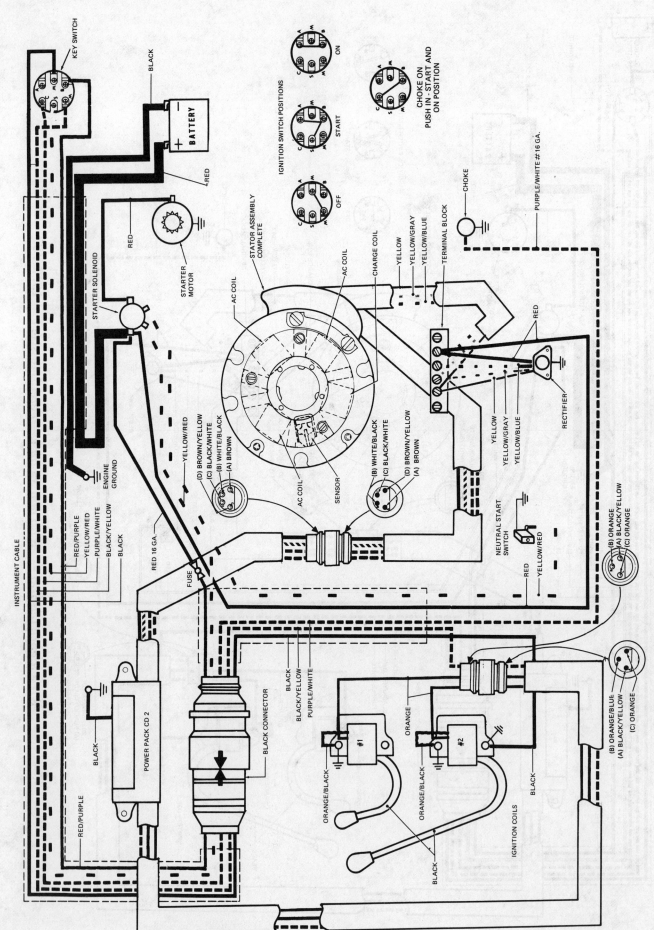

Wire Identification — 20 hp thru 40 hp, with standard shift and electric start — 1979-85.

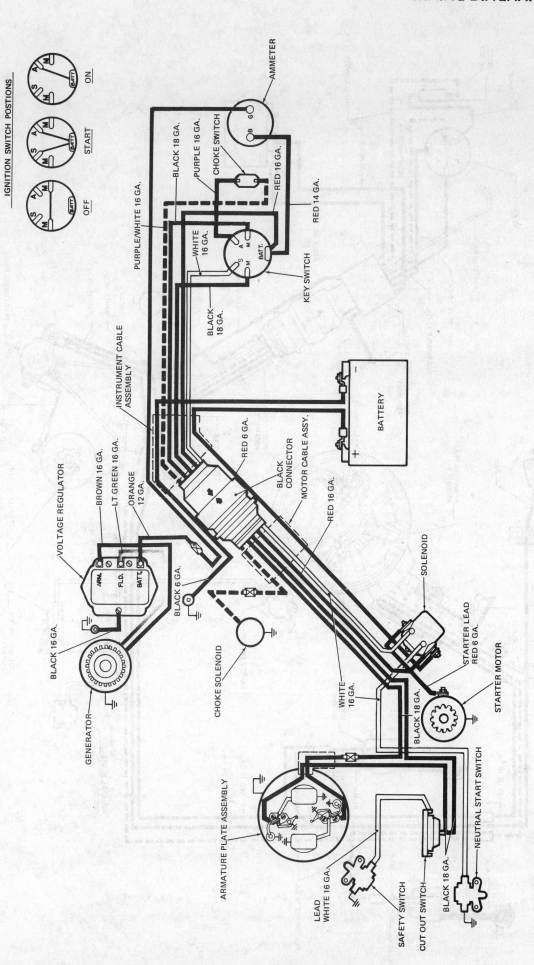

Wire Identification — 40 hp with generator, standard shift and electric start — 1971-73.

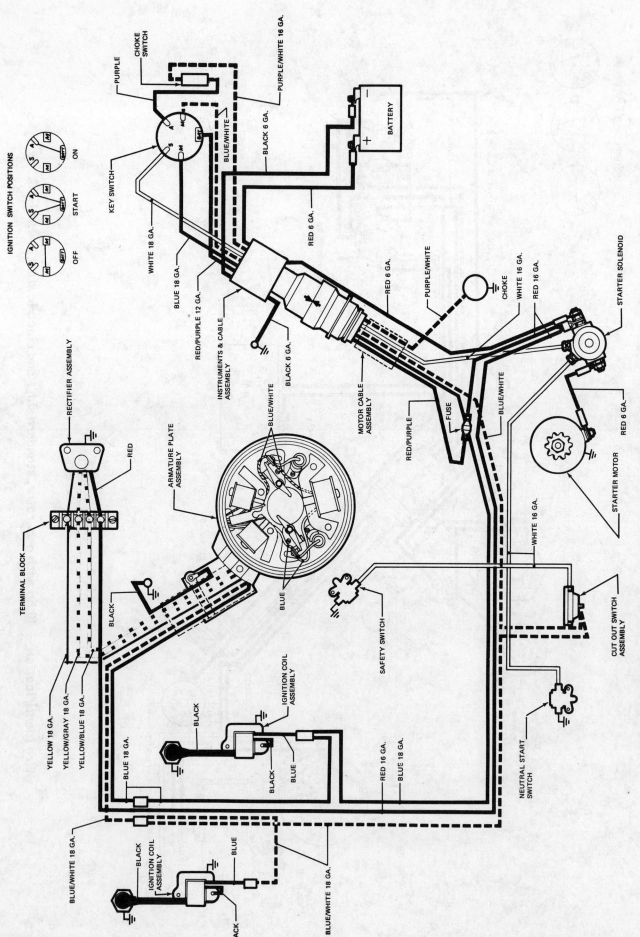

IGNITION SWITCH POSITIONS

ON

START

OFF

PURPLE

CHOKE SWITCH

PURPLE/WHITE 16 GA.

BATTERY

BLUE/WHITE

BLACK 6 GA.

KEY SWITCH

RED 6 GA.

WHITE 18 GA.

BLUE 18 GA.

RED/PURPLE 12 GA.

RED 6 GA.

PURPLE/WHITE

WHITE 16 GA.

RED 16 GA.

CHOKE

STARTER SOLENOID

INSTRUMENTS & CABLE ASSEMBLY

BLACK 6 GA.

MOTOR CABLE ASSEMBLY

RED/PURPLE

FUSE

BLUE/WHITE

RED 6 GA.

STARTER MOTOR

RECTIFIER ASSEMBLY

RED

ARMATURE PLATE ASSEMBLY

BLUE/WHITE

TERMINAL BLOCK

BLACK

BLUE

SAFETY SWITCH

WHITE 16 GA.

CUT OUT SWITCH ASSEMBLY

YELLOW 18 GA.

YELLOW/GRAY 18 GA.

YELLOW/BLUE 18 GA.

BLUE 18 GA.

BLACK

IGNITION COIL ASSEMBLY

BLACK

BLUE

RED 16 GA.

BLUE 18 GA.

NEUTRAL START SWITCH

BLUE/WHITE 18 GA.

BLACK

IGNITION COIL ASSEMBLY

BLUE

BLACK

BLUE/WHITE 18 GA.

Wire Identification — 40 hp with alternator, standard shift and electric start — 1974-76.

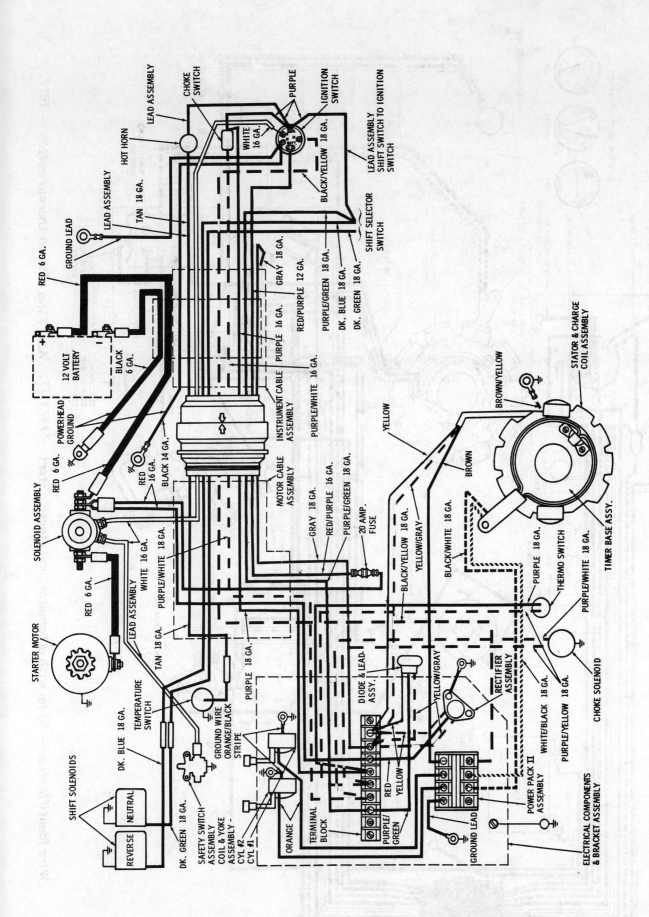

Wire Identification — 50 hp with alternator, and electric start — 1971-72.

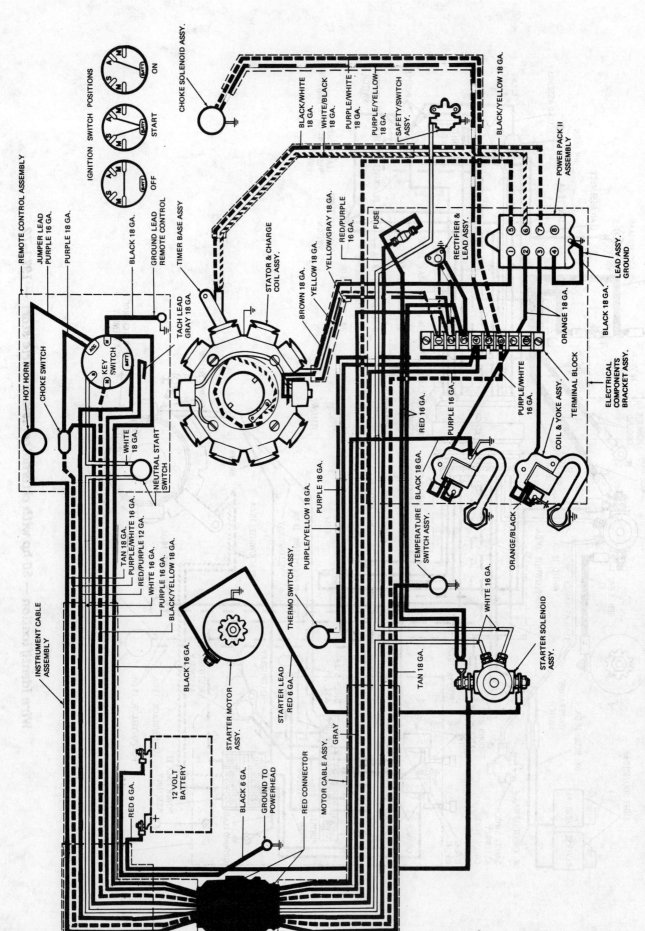

Wire Identification — 50 hp with alternator, and electric start -- 1973-77. Also 55hp with no thermo switch — 1977.

Wire Identification — 50 hp with manual start — 1980-81.

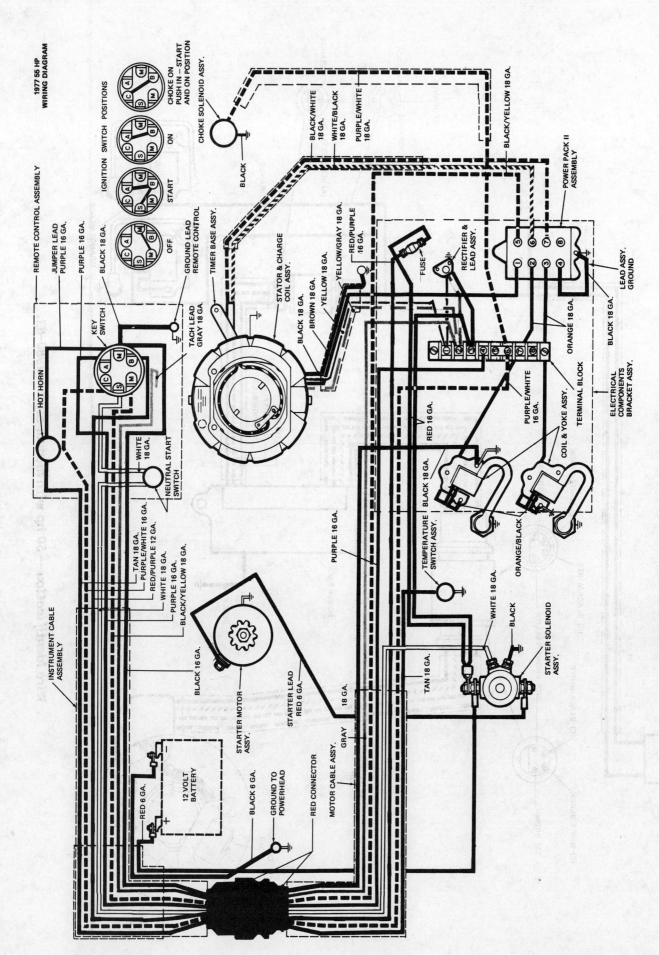

Wire Identification — 55 hp with electric start — 1977.

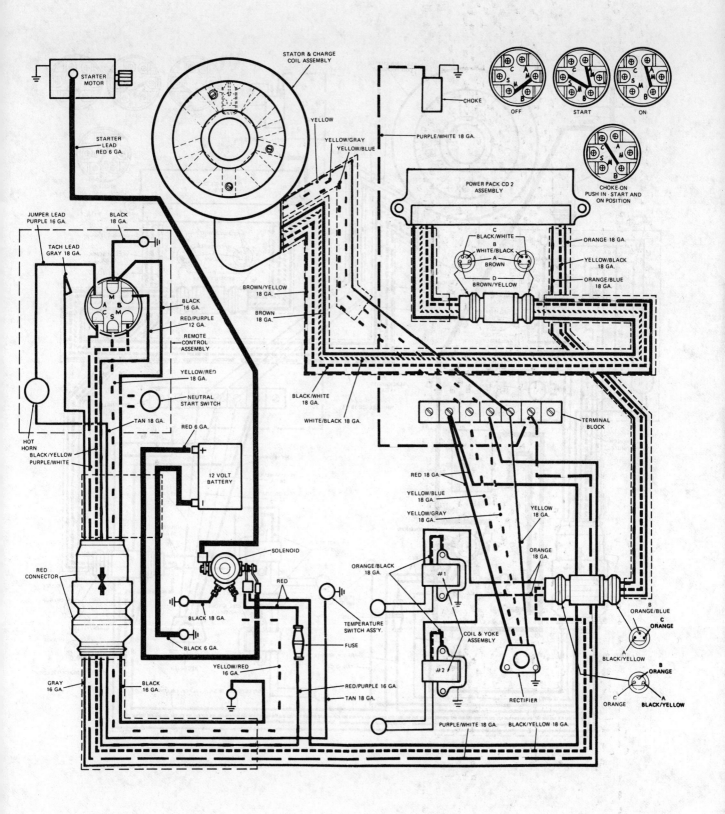

Wire Identification — 50 hp and 55 hp with electric start — 1978.

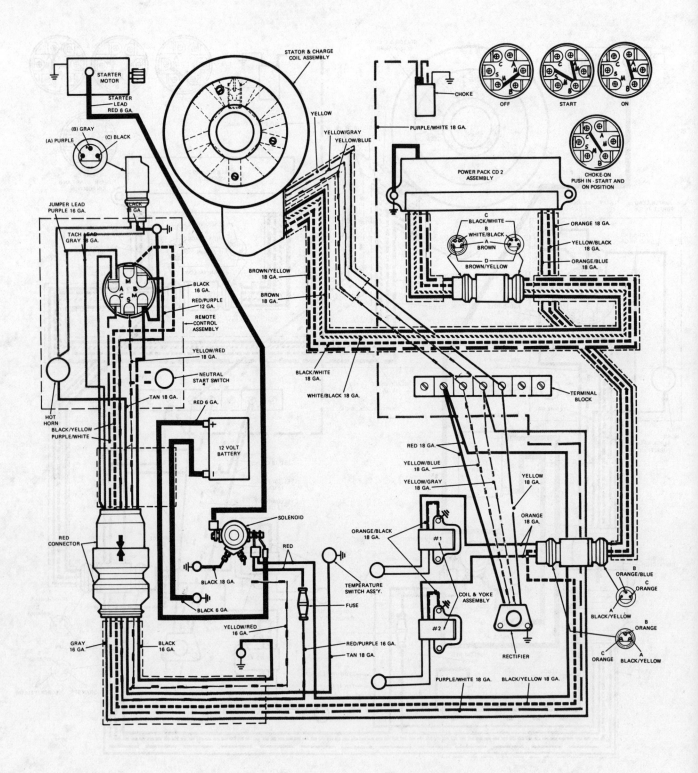

Wire Identification — 50 hp — 1979; 55 hp — 1979-83; and 60 hp 1980-81, all with electric start.

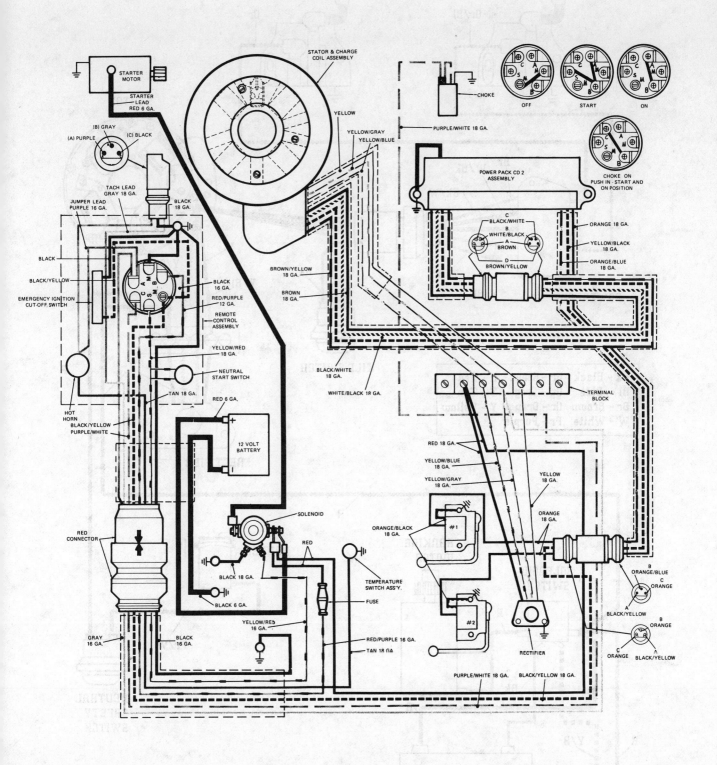

Wire Identification — 20 hp, 25 hp, 30 hp, 35 hp, 40 hp, 50 hp, and 60 hp with remote electric start — 1982-85.

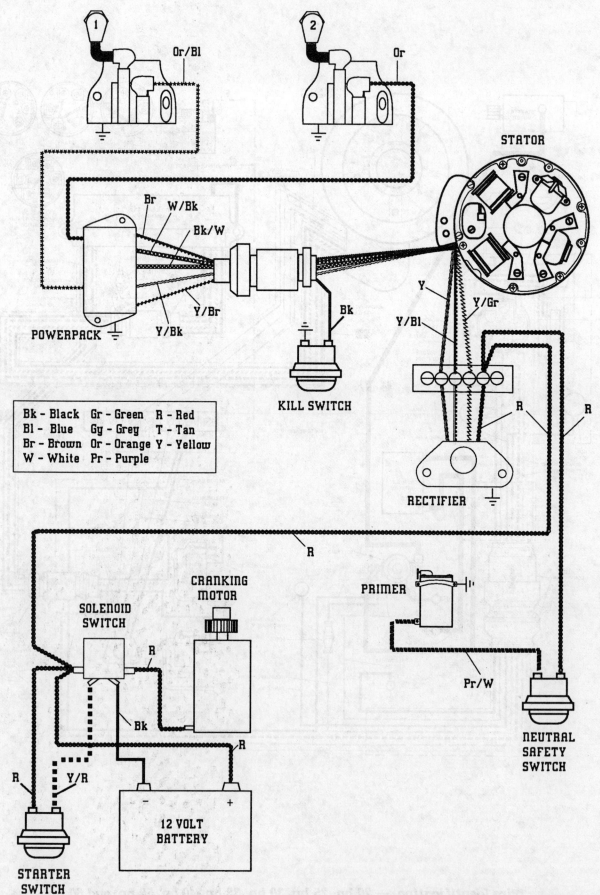

Bk – Black	Gr – Green	R – Red
Bl – Blue	Gy – Grey	T – Tan
Br – Brown	Or – Orange	Y – Yellow
W – White	Pr – Purple	

Wire Identification — 40 hp and 50 hp with tiller electric start — 1986 and on.

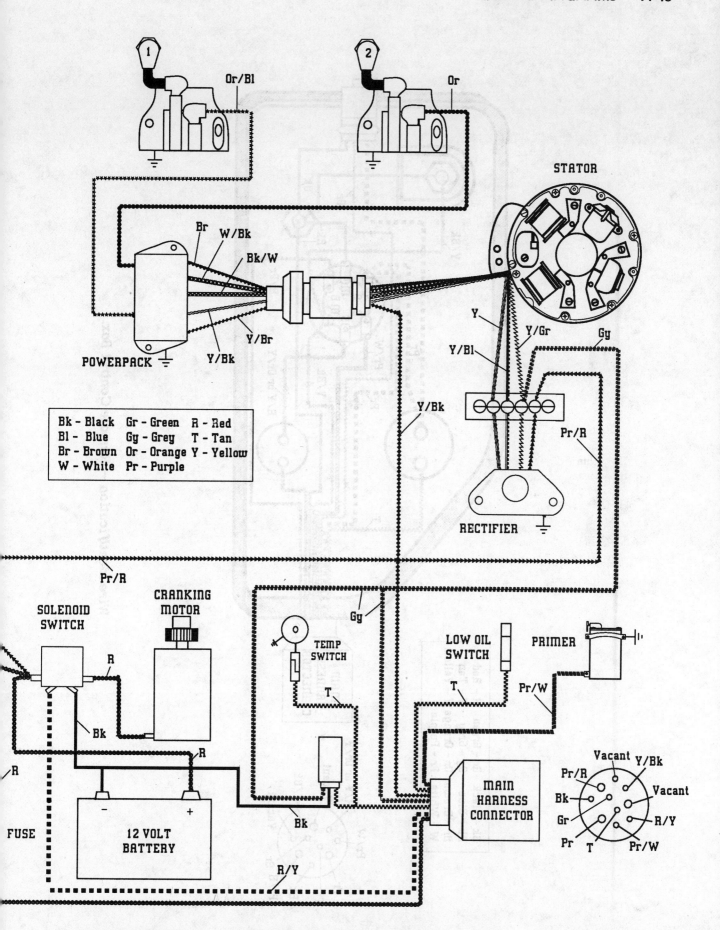

Wire Identification -- 40 hp and 50 hp with remote electric start -- 1986 and on.

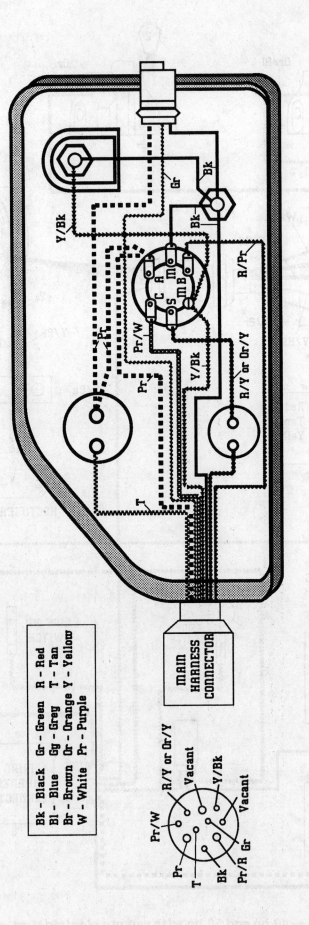

Wire Identification — Remote Control Box.

Bk - Black   Gr - Green   R - Red
Bl - Blue   Gg - Grey   T - Tan
Br - Brown   Or - Orange   Y - Yellow
W - White   Pr - Purple

MAIN
HARNESS
CONNECTOR

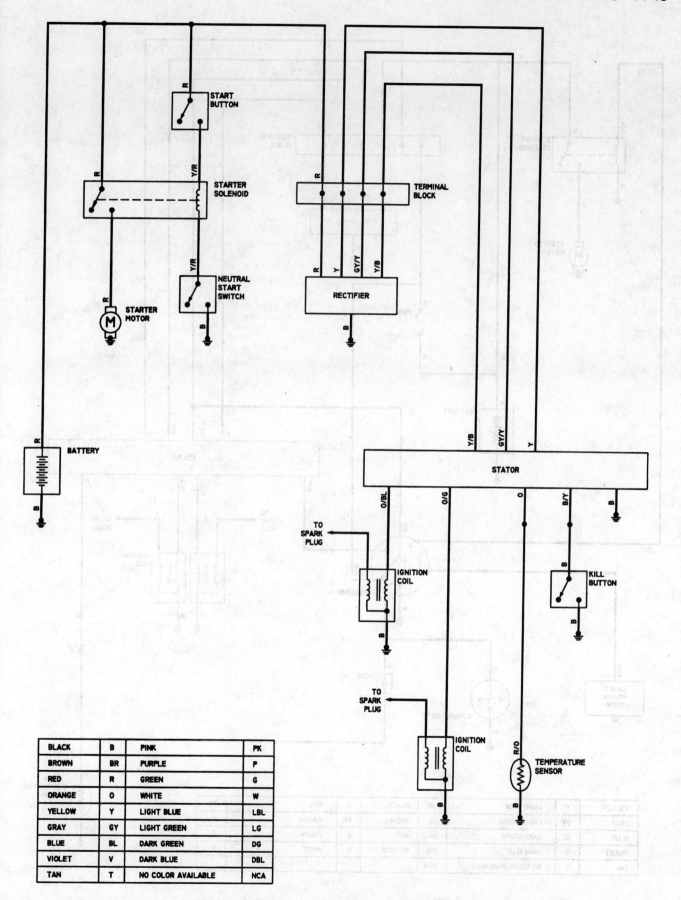

BLACK	B	PINK	PK
BROWN	BR	PURPLE	P
RED	R	GREEN	G
ORANGE	O	WHITE	W
YELLOW	Y	LIGHT BLUE	LBL
GRAY	GY	LIGHT GREEN	LG
BLUE	BL	DARK GREEN	DG
VIOLET	V	DARK BLUE	DBL
TAN	T	NO COLOR AVAILABLE	NCA

1989 Johnson/Evinrude 20-30 hp tiller electric.

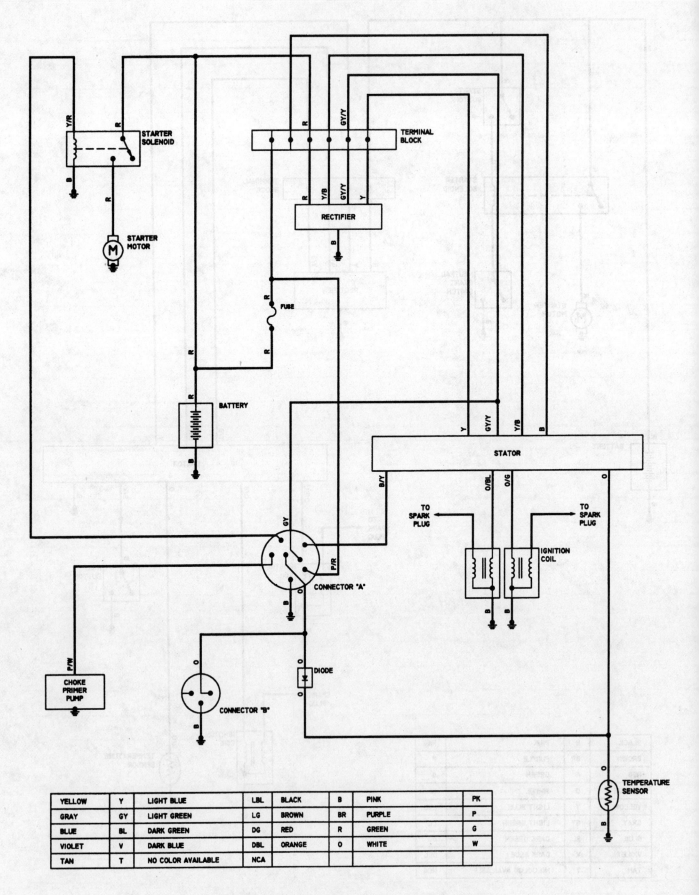

YELLOW	Y	LIGHT BLUE	LBL	BLACK	B	PINK	PK
GRAY	GY	LIGHT GREEN	LG	BROWN	BR	PURPLE	P
BLUE	BL	DARK GREEN	DG	RED	R	GREEN	G
VIOLET	V	DARK BLUE	DBL	ORANGE	O	WHITE	W
TAN	T	NO COLOR AVAILABLE	NCA				

1989 Johnson/Evinrude 20-30 hp remote start.

NOTES & NUMBERS

Other Seloc Marine Manuals

New titles are constantly being produced and the updating work on existing manuals never ceases.
All manuals contain complete detailed instructions, specifications, and wiring diagrams.